Proceedings
of the Steklov Institute of Mathematics

edited by
I. G. Petrovskiĭ and S. M. Nikol′skiĭ

Number 117 (1972)

Theory and applications of differentiable functions of several variables. IV

edited by

S. M. Nikol′skiĭ

AMERICAN MATHEMATICAL SOCIETY
Providence, Rhode Island
1974

Академия Наук
Союза Советских Социалистических Республик

ТРУДЫ
ордена Ленина
МАТЕМАТИЧЕСКОГО ИНСТИТУТА
имени В. А. СТЕКЛОВА
CXVII

Ответственный редактор академик И. Г. ПЕТРОВСКИЙ

Зам. ответственного редактора
член-корреспондент АН СССР С. М. НИКОЛЬСКИЙ

ИССЛЕДОВАНИЯ ПО ТЕОРИИ ДИФФЕРЕНЦИРУЕМЫХ ФУНКЦИЙ МНОГИХ ПЕРЕМЕННЫХ И ЕЕ ПРИЛОЖЕНИЯМ. IV

Сборник работ
под редакцией С. М. НИКОЛЬСКОГО

Издательство "Наука"
Москва 1972

AMS (MOS) subject classifications (1970). Primary 26A16, 26A33, 26A54, 35B30, 35B45, 35C15, 35D05, 35D99, 35J05, 35J15, 35J60, 35J70, 35P20, 41A30, 41A63, 42A08, 46E15, 46E35, 47F05, 65N15; Secondary 26A69, 35B45, 35C15, 35E05, 35J05, 35J20, 42A92, 46E15, 46E30, 46E35.

Library of Congress Cataloging in Publication Data (Revised) CIP

Akademii͡a nauk SSSR. Matematicheskiĭ institut.
Theory and applications of differentiable functions of several variables.
(Proceedings of the Steklov Institute of Mathematics, no. 77, 89, 105, 117)
"Work done in the function theory section of the Mathematical Institute of the Academy of Sciences of the USSR."
Translation of Issledovanii͡a po teorii different͡si-ruemykh funkt͡siĭ mnogikh peremennykh i ee prilozhenii͡am.
Includes bibliographies.
1. Functions of several real variables—Addresses, essays, lectures. 2. Differential equations, Partial—Addresses, essays, lectures. I. Nikol'skii, Sergeĭ Mikhailovich, ed. II. Title. III. Series: Akademii͡a nauk SSSR. Matematicheskii Institut. Proceedings no. 77 [etc.].
QA1.A413 no. 77, etc. 515'.84 68-1677
ISBN 0-8218-3017-1 (v. 117)

PROCEEDINGS OF THE STEKLOV INSTITUTE OF MATHEMATICS
IN THE ACADEMY OF SCIENCES OF THE USSR

(Труды математического института им. В. А. Стеклова, т. CXVII, 1972)

TABLE OF CONTENTS

Page

Trudy Mat. Inst. Steklov.
117 (1972)

Proc. Steklov Inst. Math.
117 (1972)

THE BEHAVIOR OF DIFFERENTIABLE FUNCTIONS ON A NONSMOOTH SURFACE

O. V. BESOV

UDC 517.518.22

Abstract. This paper considers the Sobolev space $W_p^l(\Omega)$ on an n-dimensional region Ω with boundary $\partial\Omega$ (of various dimensions) locally satisfying a Lipschitz condition. There is found a reversible characterization of the traces of these functions on the boundary (embedding and extension theorems).

Bibliography: 4 items.

1. We will consider differentiable functions defined on a region $\Omega \subset E^n$ and belonging to the Sobolev space $W_p^l(\Omega)$ [1], i.e. having finite norm

$$\|f\|_\Omega + \sum_{|\alpha|=l} \|D^\alpha f\|_\Omega < \infty,$$

where

$$\|f\|_\Omega = \left\{ \int_\Omega |f(x)|^p\, dx \right\}^{\frac{1}{p}}, \qquad 1 < p < \infty.$$

We will obtain here an exact (reversible) characterization for traces of functions in $W_p^l(\Omega)$ $(l = 1, 2, \cdots)$ on sections of the region (or on its boundary) of various dimensions locally satisfying (in a suitable rectangular coordinate system) a Lipschitz condition.

Such a reversible characterization for the case $l = 1, m = n - 1, 1 < p < \infty$ was obtained by Gagliardo [2]. In this case one can rectify a section by means of a transformation satisfying a Lipschitz condition and preserving the class W_p^1. This same device of rectification has been applied by Slobodeckiĭ, Solonnikov and Uspenskiĭ for classes $W_p^l(\Omega)$ and sections of smoothness l. Note that a reversible characterization of traces of functions in $W_p^l(E^n)$ on planes of dimension $m < n - 1$ was indicated by us in [3]. As V. P. Il'in has observed, necessary conditions on the trace of functions (embedding theorems), using a representation which he proposed, can also be obtained for sections of smoothness $l - (n - m)/p + \epsilon$.

Our results are obtained on the basis of a new way of norming traces, not related to rectifying the boundary, which permits one to consider, for any $l = 1, 2, \cdots$, traces

---

AMS (MOS) subject classifications (1970). Primary 46E35; Secondary 46E15.

on nonsmooth sections. For $l = 1$ our result reduces to that of Gagliardo [2].

First of all we consider the main case when the equation of the m-dimensional surface Γ has the form

$$x'' = \varphi(x'), \quad x' \in G' \subset E^m, \quad x'' \in E^{n-m}, \quad x = (x', x'') \in E^n,$$
$$|\varphi(x') - \varphi(y')| \leqslant M_0 |x' - y'|, \quad x' \in G', \quad y' \in G', \tag{1}$$

where the region $G' \subset E^m$ and $|\cdot|$ is Euclidean distance in the corresponding space or subspace.

For $x' \in G' \subset E^m$ let

$$\pi(x) = |x'' - \varphi(x')| \qquad \text{if} \quad m < n-1,$$
$$\pi(x) = x'' - \varphi(x') = x_n - \varphi(x_1, \ldots, x_{n-1}) \qquad \text{if} \quad m = n-1.$$

It is clear that $|\pi(x)|$ is equivalent to $\rho(x) = \rho(x, \Gamma)$, the distance from x to Γ.

For $M_1 > M_0$ let V denote the body

$$V = \{x : |x''| > M_1 |x'|\}, \quad m < n-1,$$
$$V = \{x : x_n > M_1 |x'|\}, \quad m = n-1.$$

It follows from (1), for example, that if $y \in \Gamma$ the intersection $\Gamma \cap (y + V)$ is empty (the symbols $\pm$ will denote the arithmetic sum or difference of sets).

Further let

$$\tilde{\Omega} = \{x : 0 < \pi(x) < H\},$$
$$\Omega = \tilde{\Omega} \cap \{x : (x - V) \cap \Gamma = (x - V) \cap \bar{\Gamma}\}, \tag{1'}$$
$$\Gamma(x) = \Gamma \cap (x - V). \tag{1''}$$

For a system of functions $\{f_\alpha\}$, $|\alpha| \leqslant s$, by

$$P_s(w; y) = P_s(w; \{f_\alpha(y)\}) = \sum_{|\alpha| \leqslant s} \frac{f_\alpha(y)}{\alpha!} w^\alpha \tag{2}$$

we will denote the polynomial formally corresponding to the Taylor polynomial of $f(x)$. The latter will be denoted by

$$P_s(w; y) = P_s(w; f(y)) = \sum_{|\alpha| \leqslant s} \frac{f^{(\alpha)}(y)}{\alpha!} w^\alpha.$$

By $C, C_1, C_2, \cdots$ we will denote certain positive constants not depending on functions. Throughout below s is a nonnegative integer,

$$0 \leqslant s < l - \frac{n-m}{p} < s+1, \quad 1 < p < \infty. \tag{3}$$

For $f \in W_p^l(\Omega)$, due to known theorems [1, 2], all the generalized derivatives are defined, $f^{(\beta)}(y) \in L_p^{\mathrm{loc}}(\Gamma)$, $|\beta| \leqslant s$, and thus also the Taylor polynomial $P_s(w; y)$, $y \in \Gamma$.

2. Let us prove an embedding theorem for the case $(n-m)/p < 1$. For $x \in \Omega$, $y \in \Gamma$, $y - x \in V$ and $|\beta| \leqslant s$ we have

$$f^{(\beta)}(x)-P_s^{(\beta)}(y-x;\ y)=\int_0^1 \frac{(1-t)^{s-|\beta|}}{(s-|\beta|)!}\frac{\partial^{s-|\beta|+1}}{\partial t^{s-|\beta|+1}} f^{(\beta)}[y+t(x-y)]\,dt$$

$$=\int_0^1 \frac{(1-t)^{s-|\beta|}}{(s-|\beta|)!}\sum_{|\alpha|=s+1} c_\alpha f^{(\alpha)}[y+t(x-y)](x-y)^{\alpha-\beta}\,dt. \qquad (4)$$

Using this equality, for $s=l-1$ we carry out the next estimate by adding and subtracting $f^{(\beta)}(x)$ within the absolute value appearing in the integrand:

$$I_s=\left\{\int_\Omega\int_{\Gamma(x)}\int_{\Gamma(x)}\left|\frac{P_s^{(\beta)}(x-y;\ y)-P_s^{(\beta)}(x-z;\ z)}{\rho(x)^{l+2\frac{m}{p}-|\beta|}}\right| dy'dz'dx\right\}^{\frac{1}{p}}$$

$$\leqslant C_1\sum_{|\alpha|=l}\left\{\int_\Omega\int_{\Gamma(x)}\rho(x)^{-m}\left[\int_0^1|f^{(\alpha)}[y+t(x-y)]|\,dt\right]^p dy'dx\right\}^{\frac{1}{p}}$$

$$\leqslant C_2\sum_{|\alpha|=l}\int_0^1\left\{\int_\Gamma\int_V |w|^{-m}|g_\alpha[y+tw]|^p\,dw\,dy'\right\}^{\frac{1}{p}} dt,$$

where $g_\alpha(x)=f^{(\beta)}(x)$ for $x\in\Omega$; $g_\alpha(x)=0$ otherwise.

Now we substitute $tw=v$; then $v'=u'|v''|$ (so that $|u'|<1/M_1$), and finally $x'=y'+u'|v''|$, $x''=\varphi(y')+v''$ with Jacobian $J=\partial(x',x'')/\partial(y',v'')$ existing at almost every point of the region and satisfying the estimate $1+M_0/M_1\geqslant J\geqslant 1-M_0/M_1>0$. This latter fact can be established, for example, as follows. One can suppose that $u'=(u_1,\cdots,u_m)=(u_1,0,\cdots,0)$, which can already be achieved by the orthogonal transformation $y'=(y_1,\cdots,y_m)$ defined by the corresponding rotation of the axes in $E^m\ni y'$. Writing out the Jacobian J explicitly in the form of a determinant of an $n\times n$ matrix and computing it by the Laplace theorem, we find that

$$J=1+\sum_{k=m+1}^{n}\pm\frac{\partial}{\partial y_1}\varphi_k(y')\,u_1\frac{v_k}{|v''|},$$

and

$$\sum_{k=m+1}\left|\frac{\partial}{\partial y'}\varphi_k(y')\,u_1\frac{v_k}{|v''|}\right|\leqslant\sqrt{\sum_{m+1}^{n}\left(\frac{\partial}{\partial y_1}\varphi_k(y')\right)^2}\,|u_1|\leqslant\frac{M_0}{M_1},$$

where $(\varphi_{m+1}(y'),\cdots,\varphi_n(y'))=\varphi(y')$.

As a result of the indicated substitutions we obtain that

$$I_{l-1}\leqslant C_2\sum_{|\alpha|=l}\int_0^1 t^{-\frac{n-m}{p}}\left\{\int_\Gamma\int_V|v|^{-m}|g_\alpha[y+v]|^p\,dv\,dy'\right\}^{\frac{1}{p}}dt$$

$$\leqslant C_3\sum_{|\alpha|=l}\left\{\int_{|u'|<\frac{1}{M_1}}\int_{G'}\int_{E^{n-m}}|g_\alpha[(y'+u'|v''|,\ \varphi(y')+v'')]|^p\,dv''dy'du'\right\}^{\frac{1}{p}}$$

and finally,

$$I_{l-1} \leqslant C_4 \sum_{|\alpha|=l} \|f^{(\alpha)}\|_\Omega. \tag{5}$$

3. Now let us prove the embedding theorem for the case $(n-m)/p > 1$. We need a Sobolev-type integral representation [1] of a function of one variable in terms of the derivative of given order. For completeness let us give the derivation of such a representation (taken from [4]).

Let

$$K(t) \in C_0^\infty(E^1), \ \operatorname{supp} K \subset \left(\frac{q}{3}, \frac{2q}{3}\right) \subset (0, 1),$$

$$\int_0^1 K(t)\,dt = 1, \quad \Phi(\xi) = \frac{d^l}{d\xi^l}\left[\frac{\xi^{l-1}}{(l-1)!}\int_0^\xi K(t)\,dt\right],$$

whence it follows that

$$\Phi(\xi) \in C_0^\infty(E^1), \ \operatorname{supp}\Phi \subset (0, 1), \ \int_0^1 \Phi(\xi)\,d\xi = 1.$$

For $g = g(t) \in L^{\mathrm{loc}}(t > 0)$ and $\eta > 0$, consider the averaging

$$g_\eta(t) = \int g(t+\xi)\frac{1}{\eta}\Phi\left(\frac{\xi}{\eta}\right)d\xi.$$

By the Leibniz-Newton formula, for almost all t (more precisely, at Lebesgue points of $g(t)$) we have the equality

$$g(t) = g_0(t) = g_h(t) - \int_0^h \frac{\partial}{\partial\eta} g_\eta(t)\,d\eta \qquad (h > 0).$$

Note that

$$\frac{\partial}{\partial\eta}\left[\frac{1}{\eta}\Phi\left(\frac{\xi}{\eta}\right)\right] = \frac{\partial^l}{\partial\xi^l}\frac{\partial}{\partial\eta}\left[\frac{\xi^{l-1}}{(l-1)!}\int_0^{\frac{\xi}{\eta}} K(t)\,dt\right]$$

$$= \frac{\partial^l}{\partial\xi^l}\left[\frac{\xi^l}{(l-1)!}\left(-\frac{1}{\eta^2}\right)K\left(\frac{\xi}{\eta}\right)\right] = -\frac{1}{\eta^2}M\left(\frac{\xi}{\eta}\right),$$

where

$$M(\xi) = \frac{\partial^l}{\partial\xi^l}L(\xi) = \frac{\partial^l}{\partial\xi^l}\left(\frac{\xi^l}{(l-1)!}K(\xi)\right).$$

Now if $l > s+1 \geqslant 1$ (see (3)), we have

$$\begin{aligned} g(t) &= g_h(t) + \int_0^h \int g(t+\xi)\frac{1}{\eta^2}L^{(l)}\left(\frac{\xi}{\eta}\right)d\xi d\eta \\ &= g_h(t) + \int_0^h \int g^{(l-s-1)}(\xi)\,\eta^{l-s-3}L^{(s+1)}\left(\frac{\xi-t}{\eta}\right)d\xi d\eta. \end{aligned} \tag{6}$$

Note that if $g(t)$ is defined on $(0, T)$ and equal to zero off $(0, T/3)$, and if $\operatorname{supp} K \subset (q/3, 2q/3) \subset (0, 1)$, then $g_h(t)$ (where $h = T/q$) is defined and is zero on $(0, T/3)$.

For $|\beta| \leqslant s < l - (n - m)/p < s + 1$ (see (3)) we substitute in the right side of (4), in place of $f^{(\beta)}[y + t(x - y)] = g(t)$, the representation (6), assuming that $x - y \in V$ and $f[y + t(x - y)] = 0$ if $t|x - y| \geqslant H/3$, and $h = H/q|x - y|$. Then

$$| f^{(\beta)}(x) - P_s^{(\beta)}(x - y;\ y)|$$
$$\leqslant C_5 \sum_{|\alpha|=l} \int_0^\infty \int_0^1 \int_0^\infty g_\alpha[y + \xi(x - y)]|x - y|^{l-|\beta|}\eta^{l-s-3}\left| L^{(s+1)}\left(\frac{\xi - t}{\eta}\right)\right| d\eta dt d\xi,$$

where $g_\alpha(x) = |f^{(\alpha)}(x)|$ if $x \in \widetilde{\Omega}$; $g_\alpha(x) = 0$ otherwise.

Let us perform on the right side the following estimate of the integral in η and in t. For $0 < t < \xi$,

$$\int_0^\infty \left| L^{(s+1)}\left(\frac{\xi - t}{\eta}\right)\right| \eta^{l-s-3} d\eta \leqslant (\xi - t)^{l-s-2} \int_0^\infty \left| L^{(s+1)}\left(\frac{1}{\eta}\right)\right| \eta^{l-s-3} d\eta = c_1 (\xi - t)^{l-s-2}.$$

Introducing the notation $[a]_1 = \min\{a, 1\}$, we have next

$$\int_0^{[\xi]_1} c_1 (\xi - t)^{l-s-2} dt \leqslant c_1 \xi^{l-s-2} [\xi]_1.$$

Combining the estimates, we obtain

$$| f^{(\beta)}(x) - P_s^{(\beta)}(x - y;\ y)|$$
$$\leqslant C_6 \sum_{|\alpha|=l} |x - y|^{l-|\beta|} \int_0^\infty | g_\alpha[y + \xi(x - y)]| \xi^{l-s-2}[\xi]_1 d\xi. \quad (7)$$

Now let us estimate for $|\beta| \leqslant s$ the value of I_s by adding and subtracting $f^{(\beta)}(x)$ within the absolute value appearing in the integrand and assuming that $f(x) = 0$ for $x \notin \Gamma + V(H/3)$. Let us replace I_s by the sum of two terms to each of which we apply estimate (7). We obtain

$$I_s \leqslant C_7 \sum_{|\alpha|=l} \left\{ \int_{\widetilde{\Omega}} \int_{\Gamma(x)} \rho(x)^{-m} \left[\int_0^\infty g_\alpha[y + \xi(x - y)]\, \xi^{l-s-2}[\xi]_1 d\xi \right]^p dy' dx \right\}^{\frac{1}{p}}.$$

We apply the generalized Minkowski inequality in ξ, change the order of integration in y' and x and set $x - y = v/\xi$. Then

$$I_s \leqslant C_8 \sum_{|\alpha|=l} \int_0^\infty \xi^{l-s-\frac{n-m}{p}-2} [\xi]_1 d\xi \left\{ \int_\Gamma \int_V |v|^{-m} [g_\alpha(y + v)]^p dv dy' \right\}^{\frac{1}{p}}.$$

The integral in ξ is finite by (3), and $\{\cdots\}$ has already been estimated in obtaining (5). Thus in this case for $f(x)=0$ off $\Gamma + V(H/3)$ we have

$$I_s \leqslant C_9 \sum_{|\alpha|=l} \| f^{(\alpha)} \|_{\widetilde{\Omega}}.$$

Now if $f(x), x \in \widetilde{\Omega}$, is not zero off $\Gamma + V(H/3)$, then introducing the cut-off function $\eta(x), x \in \widetilde{\Omega}$, having continuous and bounded derivatives up to order l inclusive, equal to zero off $\Gamma + V(H/3)$ and to one on $\Gamma + V\epsilon H$, from the previous estimate we obtain that

$$I_s \leqslant C_9 \sum_{|\alpha|=l} \| (\eta f)^{(\alpha)} \|_{\Omega} \leqslant C_{10} \| f \|_{\Omega} + C_{10} \sum_{|\alpha|=l} \| f^{(\alpha)} \|_{\Omega}.$$

Thus from this and (5) in the general case (3) we get

$$\sum_{|\beta|\leqslant s} \left\{ \int_{\Omega} \int_{\Gamma(x)} \int_{\Gamma(x)} \left| \frac{P_s^{(\beta)}(x-y;\ y) - P_s^{(\beta)}(x-z;\ z)}{\rho(x)^{l+2\frac{m}{p}-|\beta|}} \right|^p dy'dz'dx \right\}^{\frac{1}{p}} \tag{8}$$

$$\leqslant C \| f \|_{\widetilde{\Omega}} + C \sum_{|\alpha|=l} \| f^{(\alpha)} \|_{\widetilde{\Omega}}.$$

4. Consider the reverse problem of extending a system of functions $\{f_\alpha(y)\}$, $|\alpha| \leqslant s$, defined on the boundary Γ to the region Ω. First of all we construct on Ω a function $r(x) \sim \pi(x) \sim \rho(x)$ such that

$$| D^\alpha r(x) | \leqslant c_\alpha \rho(x)^{1-|\alpha|}, \qquad |\alpha| \leqslant l. \tag{9}$$

This construction can be done, for example, as follows. We take a function $\mu(x)$, not identically zero, with the properties

$$\mu(x) = \mu^*\left(\frac{|x'|}{|x|}\right), \quad \mu(x) \geqslant 0,$$

$$\mu(x) \in C^\infty(|x| > 0), \quad \mu(x) = 0 \quad \text{if} \quad |x'| > \varepsilon |x|,$$

where $\epsilon > 0$ is sufficiently small.

Now we set

$$r(x) = \int_\Gamma \mu\left(\frac{x-y}{|x-y|}\right) \frac{dy'}{|x-y|^{m-1}}.$$

The equivalence $r(x) \sim \rho(x)$ follows from the obvious equality

$$\int_{y''=0} \mu\left(\frac{x-y}{|x-y|}\right) \frac{dy'}{|x-y|^{m-1}} = q\,|x''|, \qquad q > 0,$$

and conditions (9) are verified by differentiation under the integral.

Now for the system $\{f_\alpha(y)\}, y \in \Gamma, |\alpha| \leqslant s$, we take the polynomial (2) $P_s(w;y) = P_s(w;\{f_\alpha(y)\})$ and construct on Ω the function

$$f(x) = \int_\Gamma P_s(x-y;\ y) \frac{1}{r(x)^m} \omega\left(\frac{y'-x'}{r(x)}\right) dy'. \tag{10}$$

We will assume here that $\omega(x') \in C_0^\infty(E^n)$ and

$$\int_{E^m} \omega(y')\,dy' = 1 \tag{11}$$

and that $\omega(x')$ is concentrated on a small enough neighborhood of zero in E^m so that

$$\omega\left(\frac{y'-x'}{r(x)}\right) = 0 \quad \text{for} \quad x \in \Omega,\ y \in \Gamma \setminus \Gamma(x).$$

Let us estimate the derivatives $f^{(\alpha)}(x)$ on Ω. By (10) and (11) for $z \in \Gamma$ and $|\alpha| = l$ we have

$$f(x) = P_s(x-z;\ z) + \int_\Gamma [P_s(x-y;\ y) - P_s(x-z;\ z)] \frac{1}{r(x)^m}\,\omega\left(\frac{y'-x'}{r(x)}\right) dy',$$

$$\begin{aligned} f^{(\alpha)}(x) &= \sum_{\beta\leqslant\alpha} c_\beta \int D_x^\beta [P_s(x-y;\ y) - P_s(x-z;\ z)]\, D_x^{\alpha-\beta}\left[r(x)^{-m}\,\omega\left(\frac{y'-x'}{r(x)}\right)\right] dy' \\ &= \sum_{\beta\leqslant\alpha} c_\beta \int_\Gamma\int_\Gamma [P_s^{(\beta)}(x-y;\ y) - P_s^{(\beta)}(x-z;\ z)]\, r(x)^{-m}\,\omega\left(\frac{z'-x'}{r(x)}\right) \\ &\quad \times D_x^{\alpha-\beta}\left[r(x)^{-m}\,\omega\left(\frac{y'-x'}{r(x)}\right)\right] dy'\,dz' = \sum_{\beta\leqslant\alpha} c_\beta F_\beta(x). \end{aligned} \tag{12}$$

By means of (9) and Hölder's inequality we get

$$|f^{(\alpha)}(x)| \leqslant C \sum_{\beta\leqslant\alpha} \left\{ \int_{\Gamma(x)}\int_{\Gamma(x)} \left| \frac{P_s^{(\beta)}(x-y;\ y) - P_s^{(\beta)}(x-z;\ z)}{\rho(x)^{|\alpha|+2\frac{m}{p}-2|\beta|}} \right|^p dy'\,dz' \right\}^{\frac{1}{p}},$$

whence (for $|\alpha| = l$)

$$\|f^{(\alpha)}\|_\Omega \leqslant C \sum_{\beta\leqslant\alpha} \left\{ \int_\Omega \int_{\Gamma(x)}\int_{\Gamma(x)} \left| \frac{P_s^{(\beta)}(x-y;\ y) - P_s^{(\beta)}(x-z;\ z)}{\rho(x)^{|\alpha|+2\frac{m}{p}-|\beta|}} \right|^p dy'\,dz'\,dx \right\}^{\frac{1}{p}}. \tag{13}$$

Immediately from (10) we also have

$$\|f\|_\Omega \leqslant C_1 \sum_{|\beta|\leqslant s} \left\{ \int_\Gamma |f_\beta(y)|^p\,dy' \right\}^{\frac{1}{p}}. \tag{14}$$

Now let us prove that under the condition of finiteness of the right side of (13) for all α, $|\alpha| = l$, the derivatives $f^{(\gamma)}(x)$, $x \in \Omega$, converge in a definite sense as $x \to y \in \Gamma$ to corresponding functions $f_\gamma(y)$ of the system $\{f_\gamma(y)\}$, $|\gamma| \leqslant s$, $y \in \Gamma$, with respect to which the extension (10) was constructed.

For $w \in V$, $w \to 0$ and $\Gamma_w = \{y \in \Gamma\colon y + w \in \Omega\}$ we consider

$$A_\gamma(w) = \left\{ \int_{\Gamma_w} | f^{(\gamma)}(z+w) - P_s^{(\gamma)}(w;\ z) |^p \, dz' \right\}^{\frac{1}{p}}.$$

From (12) we find that for $|\gamma| \leqslant s$

$$| f^{(\gamma)}(x) - P_s^{(\gamma)}(x-z;\ z) |$$

$$= \left| \sum_{\beta \leqslant \gamma} c_\beta \int_\Gamma [P_s^{(\beta)}(x-y;\ y) - P_s^{(\beta)}(x-z;\ z)] \, D_x^{\gamma-\beta} \left[\frac{1}{r(x)^m} \omega\left(\frac{y'-x'}{r(x)} \right) \right] dy' \right|$$

$$\leqslant C_2 \sum_{\beta \leqslant \gamma} \rho(x)^{-\frac{m}{p} - |\gamma-\beta|} \left\{ \int_{\Gamma(x)} | P_s^{(\beta)}(x-y;\ y) - P_s^{(\beta)}(x-z;\ z) |^p \, dy' \right\}^{\frac{1}{p}}.$$

Hence for $V(\epsilon) = V \cap \{x\colon |x| < \epsilon\}$ we have

$$\left\{ \int_{V(\varepsilon)} | A_\gamma(w) |^p \frac{dw}{| w |^{m+p(l-|\gamma|)}} \right\}^{\frac{1}{p}}$$

$$= \left\{ \int_\Gamma \int_{\Omega \cap (z+V(\varepsilon))} | x-z |^{-m-p(l-|\gamma|)} | f^{(\gamma)}(x) - P_s^{(\gamma)}(x-z;\ z) |^p \, dx dz' \right\}^{\frac{1}{p}}$$

$$\leqslant C_3 \left\{ \int_{\Omega \cap (\Gamma+V(\varepsilon))} \int_{\Gamma(x)} \rho(x)^{-m-p(l-|\gamma|)} | f^{(\gamma)}(x) - P_s^{(\gamma)}(x-z;\ z) |^p \, dz' dx \right\}^{\frac{1}{p}}$$

$$\leqslant C_4 \sum_{\beta \leqslant \gamma} \left\{ \int_{\Omega \cap (\Gamma+V(\varepsilon))} \int_{\Gamma(x)} \int_{\Gamma(x)} \left| \frac{P_s^{(\gamma)}(x-y;\ y) - P_s^{(\gamma)}(x-z;\ z)}{\rho(x)^{l+2\frac{m}{p}-|\beta|}} \right|^p dy' dz' dx \right\}^{\frac{1}{p}}.$$

Hence for $|\gamma| \leqslant s$ and $\delta = l - |\gamma| - (n-m)/p > 0$ (see (3)) we have

$$\left\{ \int_{V(\varepsilon)} | A_\gamma(w) |^p \frac{dw}{| w |^{n+p\delta}} \right\}^{\frac{1}{p}} \to 0 \quad \text{as} \quad \varepsilon \to 0, \tag{15}$$

and indeed this proves that in a definite sense (and in fact in what sense) $f^{(\gamma)}(x) \to f_\gamma(y)$ as $x \to y$ ($x \in \Omega$, $y \in \Gamma$).

5. Let us transfer our results to the case of a region $\Omega \subset E^n$ of a certain class. Let the open set $\Omega_0 \subset \Omega$, $\partial\Omega_0 \subset \Omega$ (for $k \geqslant 1$), of the region $\Omega_k \subset \Omega$, $\Gamma_k = \partial\Omega \cap \partial\Omega_k$ have dimension m_k:

$$\Omega_k^{(\delta)} = \{x : x \in \Omega_k,\ \rho(x,\ \partial\Omega_k \setminus \partial\Omega) > \delta\},$$

where for a suitable rotation of cartesian coordinates in E^n, Γ_k has the form (1), Ω_k the form (1′), and $\Gamma_k(x)$ the form (1″). We require that for $\Omega \subset E^n$ there exist a $\delta > 0$ and a natural number K such that

$$\Omega = \bigcup_{k=0}^{K} \Omega_k = \bigcup_{k=0}^{K} \Omega_k^{(\delta)}. \tag{16}$$

THEOREM. *Let the region Ω with boundary $\partial\Omega = \Gamma$ satisfy condition* (16) *for $s = s_k$ and $m = m_k$, subject to condition* (3). *Then*

$$\sum_{k=1}^{K} \sum_{|\beta| \leqslant s_k} \left\{ \int_{\Omega_k} \int_{\Gamma_k(\bar{x})} \int_{\Gamma_k(x)} \left| \frac{P_{s_k}^{(\beta)}(x-y;\, y) - P_{s_k}^{(\beta)}(x-z;\, z)}{\rho(x)^{l+2\frac{m_k}{p}-|\beta|}} \right|^p d_{(y)}\Gamma d_{(z)}\Gamma dx \right\}^{\frac{1}{p}} \tag{17}$$
$$+ \sum_{k=1}^{K} \sum_{|\beta| \leqslant s_k} \|f^{(\beta)}\|_{\Gamma_k} \leqslant C \sum_{|\alpha| \leqslant l} \|f^{(\alpha)}\|_{\Omega},$$

where the $P_{s_k}(w; y) = P_{s_k}(w; f(y))$ are the Taylor polynomials for f at the point y of degree w.

Conversely, on $\partial\Omega = \bigcup_1^K \Gamma_k$ let there be given the system $\{f_\alpha(y)\}$, $|\alpha| \leqslant s_k$ for $y \in \Gamma_k$. Then there exists a function on Ω such that

$$\sum_{|\alpha| \leqslant l} \|f^{(\alpha)}\|_{\Omega} \leqslant C \sum_{k=1}^{K} \sum_{|\beta| \leqslant s_k} \|f_{(\beta)}\|_{\Gamma_k}$$
$$+ C \sum_{k=1}^{K} \sum_{|\beta| \leqslant s_k} \left\{ \int_{\Omega_k} \int_{\Gamma_k(x)} \int_{\Gamma_k(x)} \left| \frac{P_{s_k}^{(\beta)}(x-y;\, y) - P_{s_k}^{(\beta)}(x-z;\, z)}{\rho(x)^{l+2\frac{m_k}{p}-|\beta|}} \right|^p d_{(y)}\Gamma d_{(z)}\Gamma dx \right\}^{\frac{1}{p}}, \tag{18}$$

where the $P_{s_k}(w; y) = P_{s_k}(w; \{f_\alpha(y)\})$ are constructed in correspondence with (2). *Here $f^{(\alpha)}(y) = f_\alpha(y)$, $|\alpha| \leqslant s_k$, $y \in \partial\Omega$ in the sense of* (15).

The proof follows from the previous estimates and use of a partition of unity.

BIBLIOGRAPHY

1. S. L. Sobolev, *Applications of functional analysis in mathematical physics,* Izdat. Sibirsk. Otd. Akad. Nauk SSSR, Novosibirsk, 1962; English transl., Transl. Math. Monographs, vol. 7, Amer. Math. Soc., Providence, R. I., 1963. MR 29 #2624.

2. E. E. Gagliardo, *Caratterizzazioni delle tracce sulla frontiera relative ad alcune classi di funzioni in n variabili,* Rend. Sem. Mat. Univ. Padova 27 (1957), 284–305. MR 21 #1525.

3. O. V. Besov, *Investigation of a family of function spaces in connection with theorems of imbedding and extension,* Trudy Mat. Inst. Steklov. 60 (1961), 42–81; English transl., Amer. Math. Soc. Transl. (2) 40 (1964), 85–126. MR 24 #A3501.

4. V. P. Il'in, *Integral representations of differentiable functions and their applications to questions of extension of functions of the $W_p^{(l)}(G)$-classes,* Sibirsk. Mat. Ž. 8 (1967), 573–586 = Siberian Math. J. 8 (1967), 421–432. MR 36 #5683.

Translated by J. ELLIS

Trudy Mat. Inst. Steklov.
117 (1972)

Proc. Steklov Inst. Math.
117 (1972)

ON TRACES ON A NONSMOOTH SURFACE OF CLASSES OF DIFFERENTIABLE FUNCTIONS

UDC 517.518.22

O. V. BESOV

Abstract. On a region Ω of n-dimensional Euclidean space with boundary (of various dimensions) locally satisfying a Lipschitz condition there are considered functions with a definite behavior of the L_p modulus of smoothness of their derivatives. There is established a reversible characterization of traces of these functions on the boundary (embedding and extension theorems).

Bibliography: 5 items.

We will study differentiable functions $f(x), x \in \Omega \subset E^n$, defined on an open set Ω of n-dimensional Euclidean space, whose differential-difference properties are characterized by the L_p-norm of the difference

$$\|\Delta^M(\Omega, z) f^{(\alpha)}(\cdot)\|_\Omega,$$

where $\alpha = (\alpha_1, \cdots, \alpha_n)$ is the multi-index of the derivative, $|\alpha| = \alpha_1 + \cdots + \alpha_n$, and

$$\|f\|_\Omega = \left\{ \int_\Omega |f(x)|^p \, dx \right\}^{\frac{1}{p}}, \quad 1 \leqslant p \leqslant \infty,$$

and, for $z \in E^n$, $\Delta^M(\Omega, z)f(x) = \Delta^M(z)f(x)$ is the difference of order M for z if $[x, x + Mz] \in \Omega$, while $\Delta^M(\Omega, z)f(x) = 0$ in the contrary case.

There will be considered here classes of functions satisfying the Hölder condition in L_p

$$\sup_{z \in E^n} \frac{\|\Delta^M(\Omega, z) f^{(\alpha)}\|_{p, \Omega}}{|z|^{l-|\alpha|}} \leqslant \text{const} \qquad (l > |\alpha|),$$

and their generalizations. We will be interested in the behavior of such functions on sections of Ω (or on its boundary) of various dimensions satisfying locally (in a suitable system of rectangular coordinates) a Lipschitz condition.

Note that the results derived herein in the case of plane sections or plane portions of the boundary are not new (see [1], for example), with the exception of the way of extending the system of functions and the way of norming the traces. The case of sections or portions of the boundary of sufficiently great smoothness ($\geqslant l$) is reduced to the plane case by a tranformation rectifying the section or boundary and preserving

---

AMS (MOS) subject classifications (1970). Primary 46E35; Secondary 46E15.

the class of functions. Note here the works of S. M. Nikol′skiĭ, V. A. Solonnikov and S. V. Uspenskiĭ. V. P. Il′in [2] proved that by means of representation proposed by him one can obtain an embedding theorem (for a trace characterization) on the boundary of a region of almost the same smoothness as that of the trace, i.e. $l - (n - m)/p + \epsilon$, $\epsilon > 0$.

Here we propose a way of characterizing the trace of functions of any smoothness on a nonsmooth surface (satisfying a Lipschitz condition).

1. We will consider first of all the main case when the equation of the m-dimensional surface Γ has the form

$$\begin{gathered} x'' = \varphi(x'), \quad x' \in G' \subset E^m, \quad x'' \in E^{n-m}, \quad x = (x', x'') \in E^n, \\ |\varphi(x') - \varphi(y')| \leqslant M_0 |x' - y'|, \quad x' \in G', \quad y' \in G', \end{gathered} \tag{1}$$

where the region $G' \subset E^m$, and $|\cdot|$ is Euclidean distance in the corresponding space or subspace. Let $\rho(x) = \rho(x, \Gamma)$ denote the distance from $x \in E^n$ to Γ, let V be some n-dimensional open convex cone with vertex at the origin contained in $\{x: |x''| \geqslant M_1 |x'|\}$ for some $M_1 > M_0$. From (1) it follows, in particular, that the cone $y + V$, $y \in \Gamma$, does not meet Γ (by $G_1 \pm G_2$ we will denote the arithmetic sum or difference of the sets G_1 and G_2).

Further let

$$V(H) = V \cap \{x : |x| < H\},$$

$$\tilde{\Omega}(H) = \Gamma + V(H), \tag{1'}$$

$$\Omega(H) = \tilde{\Omega}(H) \cap \{x : (x - V) \cap \partial\Gamma \text{ is void}\}, \tag{1''}$$

$$\Gamma(x) = \Gamma \cap (x - V), \tag{1'''}$$

$$\Gamma_H = \{y : y \in \Gamma,\ y + V(H) \subset \Omega(H)\}.$$

By $b = b(M_0, V) \geqslant 1$ we will denote a number such that

$$(z + V(bH)) \cap \tilde{\Omega}(H) = (z + V) \cap \tilde{\Omega}(H)$$

for any $z \in \Gamma$.

For a system of functions $\{f_\alpha\}$, $|\alpha| \leqslant s$, by

$$P_s(w; y) = P_s(w; \{f_\alpha(y)\}) = \sum_{|\alpha| \leqslant s} \frac{f_\alpha(y)}{\alpha!} w^\alpha \tag{2}$$

we will denote the polynomial formally corresponding to the Taylor polynomial of $f(x)$. The latter will be denoted by

$$P_s(w; y) = P_s(w; f(y)) = \sum_{|\alpha| \leqslant s} \frac{f^{(\alpha)}(y)}{\alpha!} w^\alpha.$$

The traces on Γ of the functions of the classes considered will be characterized by the behavior of the function

$$\omega_{s+1}(\delta) = \omega_{s+1}(\delta; f) = \sum_{|\beta|\leqslant s} \delta^{-\frac{m+n}{p}+|\beta|} \left\{ \int\limits_{\Omega(\delta)} \int\limits_{\Gamma(x)} \int\limits_{\Gamma(x)} | P_s^{(\beta)}(x+y; f(y)) - P_s^{(\beta)}(x-z; f(z))|^p \, dy' dz' dx \right\}^{\frac{1}{p}}, \tag{3}$$

which has the character of the modulus of smoothness in L_p of order $s+1$ in the case when the boundary Γ is planar. In terms of $\omega_{s+1}(\delta)$ we will give necessary and sufficient conditions characterizing the trace of a function of the classes considered. Thus, for example, for the class of functions with Hölder condition introduced above they have the form

$$\sup_{\delta>0} \frac{\omega_{s+1}(\delta)}{\delta^{l-\frac{n-m}{p}}} \leqslant \text{const.}$$

To formulate the result we need a

DEFINITION. By an *N-functional* we mean a nonnegative functional

$$N[\psi(T)] = N_T[\psi(T)],$$

defined on nonnegative measurable functions $\psi(T)$, $0 < T < H < \infty$, and satisfying the following conditions for any $a \geqslant 0$, $b \geqslant 0$ and $\delta > 0$:

1° $N[a\psi_1(T) + b\psi_2(T)] \leqslant CaN[\psi_1(T)] + CbN[\psi_2(T)]$.

2° $N[\psi_1(T)] \leqslant CN[\psi_2(T)]$ for $0 \leqslant \psi_1(T) \leqslant \psi_2(T)$.

3° $N_T\left[\int\limits_0^1 \psi(Tt)\, t^{-1+\delta}\, dt\right] + N_T\left[\int\limits_1^{\frac{H}{T}} \psi(Tt)\, t^{-1-\delta} dt\right] \leqslant C(\delta)\, N_T[\psi(T)]$.

4° $N[T^\delta] < \infty$.

5° $\int\limits_0^H \psi(t)\, t^{-1+\delta} dt \leqslant C(\delta)\, N[\psi(T)]$.

Note that embedding and extension theorems for classes of functions characterized by means of N-functionals were found by S. M. Nikol'skiĭ using

$$N[\psi(T)] = \sup_{0<T<H} \psi(T),$$

and then by the author for

$$N[\psi(T)]=I_\theta[\psi(T)]=\left\{\int_0^H \psi^\theta(T)\frac{dT}{T}\right\}^{\frac{1}{\theta}}, \qquad 1\leqslant\theta<\infty$$

(see [1] in this connection). Further, in such questions K. K. Golovkin [3] used a more general functional, introduced by him, of a maximization type. The notion proposed here of N-functional is broader than that of functional of maximization type, as is seen from the example of the N-functional

$$\left(\int_0^H |\psi(T)|^\theta \ln^r\left(\frac{2H}{T}\right)\frac{dT}{T}\right)^{\frac{1}{\theta}}, \qquad 1\leqslant\theta<\infty,\ -\infty<r<\infty.$$

By $C, C_1, C_2, \cdots$ we will denote within each chain of inequalities certain positive constants (not depending on functions). In distinct chains of inequalities we may denote in identical fashion unequal constants. Other constants $a, b, c, q, M_0, M_1, \cdots$ are determined only by the geometry of the region.

We will assume throughout that s is a nonnegative integer:

$$0\leqslant s<l-\frac{n-m}{p}<s+1,\ l-1\leqslant|\alpha|<l. \tag{4}$$

We will introduce a certain integral representation of a locally summable function in terms of a difference of a type of known representations of Il′in.

Let

$$\eta(x)\in C_0^\infty(E^n),\quad \int\eta(x)\,dx=1,\quad \int x_i\eta(x)\,dx=0\quad (i=1,\ldots,n),$$
$$\operatorname{supp}\eta\subset\left\{x:0<\varepsilon<|x|<\frac{1}{M}\right\}\cap V,$$

where the natural number $M\geqslant 1$, V is some convex cone with vertex at the origin, and

$$\Psi(x)=A\sum_{k=1}^{M}\frac{(-1)^{M-k}}{k}C_M^k\eta\left(\frac{x}{k}\right),$$

$$\frac{1}{A}=\sum_{k=1}^{M}\frac{(-1)^{M-k}}{k}C_M^k=\int_0^1\frac{(t-1)^M-(-1)^M}{t}\,dt\neq 0.$$

For $h>0$ we set

$$f_h(x)=h^{-n}\int\Psi\left(\frac{y}{h}\right)f(x+y)\,dy,$$

$$f_{hh}(x)=(f_h)_h(x)=h^{-2n}\iint\Psi\left(\frac{y}{h}\right)\Psi\left(\frac{z}{h}\right)f(x+y+z)\,dy\,dz$$
$$=h^{-n}\int\Psi_0\left(\frac{y}{h}\right)f(x+y)\,dy.$$

It is clear that

$$\frac{\partial}{\partial h} f_{hh}(x) = 2\frac{\partial}{\partial h}(\dot{f}_t)_h\big|_{t=k}.$$

Further, we have

$$\frac{\partial}{\partial h}\int h^{-n}\Psi\left(\frac{z}{h}\right) g(x+z)\,dz = \frac{\partial}{\partial h}\int A\sum_{k=1}^{M}\frac{(-1)^{M-k}}{k} C_M^k \eta(z)\, g(x+khz)\,dz$$

$$= \sum_{i=1}^{n}\int A\sum_{k=1}^{M}(-1)^{M-k} C_M^k z_i \eta(z)\, D_i g(x+khz)\,dz$$

$$= \sum_{i=1}^{n}\int A z_i \eta(z)\,\Delta^m(hz)\,D_i g(x)\,dz.$$

Substituting $g(x) = f_h(x)$ and carrying out differentiation on the kernel, we obtain

$$\frac{\partial}{\partial h} f_{hh}(x) = -2A\sum_{i=1}^{n}\iint h^{-2n-1}\frac{z_i}{h}\eta\left(\frac{z}{h}\right)\Psi_i'\left(\frac{y}{h}\right)\Delta^m(z) f(x+y+z)\,dydz.$$

By the choice of A,

$$\iint \Psi(y)\Psi(z)\,dydz = 1,$$

whence for almost all x we get

$$L^{\mathrm{loc}}\ f_{hh}(x) \to f(x) \quad \text{as} \quad h\to 0,$$

in the sense of L^{loc}.

By means of a basic lemma of integral calculus we now obtain the representation

$$f(x) = f_{HH}(x) + \int_0^H\iint h^{-2n-1}K\left(\frac{y}{h},\ \frac{z}{h}\right)\Delta^M(z) f(x+y)\,dydzdh \qquad (5)$$
$$= f_{HH}(x) + \tilde{f}(x),$$

where

$$K(y,\ z) = 2A\sum_{i=1}^{n}\Psi_i'(y)\, z_i\eta(z).$$

Clearly, for a suitable choice of support η any cone $x + 2V(H)$ can serve as the body of representation (5).

For $M > l$, let

$$N_T\left[T^{-l}\int_{\substack{|\zeta|=1\\ \zeta\in V}}\|\Delta^M(T\zeta) f\|_{\Gamma+V(H+bH)}\,d\zeta\right] < \infty,$$

$$\|f\|_{\Gamma+V(2H)} < \infty.$$

It is not hard to establish by means of (5) the existence of all generalized derivatives $f^{(\beta)}(x)$, $|\beta| \leqslant s$, with finite norm in $L_p(\widetilde{\Omega}(H))$ and $L_p(\Gamma)$ (see [4]).

Now we turn to estimation of the remainder of the Taylor formula

$$R_s^{(\beta)}(x-v;\ f(v)) = f^{(\beta)}(x) - P_s^{(\beta)}(x-v;\ f(v))$$
$$= \int_0^1 \frac{(1-t)^{s-|\beta|}}{(s-|\beta|)!} \frac{\partial^{s+1-|\beta|}}{\partial t^{s+1-|\beta|}} f^{(\beta)}[v+t(x-v)]\,dt. \quad (6)$$

Here we assume that $v \in \Gamma$, $x - v \in V(bH)$ and $|\beta| \leqslant S$.

Let us apply the operation of taking the remainder $R_s^{(\beta)}$ to the representation (5). Since f_{HH} is the convolution of f with a smooth finite kernel, the estimation of $R_s^{(\beta)}(x-v; f_{HH}(v))$ poses no problem. For estimation of $R_s^{(\beta)}(x-v; \overline{f}(v))$ we carry out the operation $R_s^{(\beta)}$ (by means of a translation of variables) on the kernel $K((y-x)/h, z/h)$.

By means of (6) we obtain

$$R_s^{(\beta)}\left(x-v;\ K\left(\frac{y-v}{h},\ \frac{z}{h}\right)\right) \Big| \leqslant C_b h^{-|\beta|}\left(\frac{|x-v|}{h}\right)^{s-|\beta|}$$
$$\times \begin{cases} \dfrac{|x-v|}{h} \displaystyle\int_0^1 \tilde{K}\left(\frac{y-v+t(x-v)}{h},\ \frac{z}{h}\right)dt & \text{if } bh \geqslant |x-v|, \\ \tilde{K}\left(\dfrac{y-v}{h},\ \dfrac{z}{h}\right) + \tilde{K}\left(\dfrac{y-x}{h},\ \dfrac{z}{h}\right) & \text{if } bh \leqslant |x-v|. \end{cases}$$

Here

$$\tilde{K}(y,\ z) = \max_{|\alpha| \leqslant s+1} |D_y^\alpha K(y,\ z)|.$$

Thus taking the remainder under the integral sign leads to the following estimate (for simplicity we restrict to the case $x - v \in V(bH)$):

$$|R_s^{(\beta)}(x-v;\ \bar{f}(v))|$$
$$\leqslant C \int_0^{b^{-1}|x-v|} \iint h^{-2n-1-|\beta|}\left(\frac{|x-v|}{h}\right)^{s-|\beta|} \tilde{K}\left(\frac{y}{h},\ \frac{z}{h}\right)\{|\Delta^M(z) f(v+y)|$$
$$+ |\Delta^M(z) f(x+y)|\}\,dy\,dz\,dh$$
$$+ C \int_{b^{-1}|x-v|}^{H} \iint h^{-2n-|\beta|-1}\left(\frac{|x-v|}{h}\right)^{s+1-|\beta|} \tilde{K}\left(\frac{y}{h},\ \frac{z}{h}\right)$$
$$\times \int_0^1 |\Delta^M(z) f(v+t(x-v)+y)|\,dt\,dy\,dz\,dh.$$

Let us estimate the integrals with respect to h of the kernels

$$A(y, z) = T^{s-|\beta|} \int_0^T h^{-2n-1-s} \tilde{K}\left(\frac{y}{h}, \frac{z}{h}\right) dh,$$

$$B(y, z) = T^{s+1-|\beta|} \int_T^H h^{-2n-2-s} \tilde{K}\left(\frac{y}{h}, \frac{z}{h}\right) dh.$$

From the form of the support of $\widetilde{K}(y, z)$ it follows that $A(y, z)$ can differ from zero only if $z \in V(T/M)$, and $B(y, z)$ can differ from zero only if $z \in V(H/M)$, $|z| > \epsilon T$. Hence, letting $\chi(t)$ denote the characteristic function of $(0, 1)$, for y, $z \in V(H)$ we obtain

$$A(y, z) \leqslant C_1 \chi\left(\frac{|z|}{T}\right) T^{s-|\beta|} \int_{h \geqslant |y|,\, h \geqslant |z|} h^{-2n-1-s} dh \leqslant C\chi\left(\frac{|z|}{T}\right) T^{s-|\beta|} (|y| + |z|)^{-2n-s},$$

$$B(y, z) \leqslant C \frac{\left[\chi\left(\frac{|z|}{H}\right) - \chi\left(\frac{|z|}{\varepsilon T}\right)\right] T^{s+1-|\beta|}}{(|y| + |z| + T)^{2n+s+1}}.$$

Combining the estimates, for $w = x - v \in V(bH)$, $v \in \Gamma$, we obtain

$$|R_s^{(\beta)}(w; \tilde{f}(v))| \leqslant$$

$$\leqslant C \int_V \int_{V(H)} \chi\left(\frac{b|z|}{|w|}\right) \frac{|w|^{s-|\beta|}}{(|y| + |z|)^{2n+s}} [\Delta^M(z) f(v + y)|$$

$$+ |\Delta^M(z) f(v + y + w)|] dy dz +$$

$$+ C \int_V \int_{V(H)} \frac{\left[\chi\left(\frac{|z|}{H}\right) - \chi\left(\frac{b|z|}{\varepsilon |w|}\right)\right] |w|^{s+1-|\beta|}}{(|y| + |z| + |w|)^{2n+s+1}} \int_0^1 |\Delta^M(z) f(v + tw + y)| dt dy dz.$$

We obtain an estimate of the $L_p(\Gamma)$ norm (in v) of the left side by applying to the right side first the Hölder inequality in y'' (with exponents p' and p), and then the known estimate in v' of the convolution $\|k * f\|_p \leqslant \|k\|_1 \|f\|_p$. We will have

$$\| R_s^{(\beta)}(w; \tilde{f}(\cdot))\|_\Gamma = \left\{ \int_\Gamma |R_s^{(\beta)}(w; \tilde{f}(v))|^p dv' \right\}$$

$$\leqslant C \int_V \chi\left(\frac{b|z|}{|w|}\right) \frac{|w|^{s-|\beta|}}{|z|^{n+s+\frac{n-m}{p}}} \|\Delta^M(z) f\|_{\Gamma+V(H+bH)} dz$$

$$+ C \int_V \frac{\left[\chi\left(\frac{|z|}{H}\right) - \chi\left(\frac{b|z|}{\varepsilon |w|}\right)\right] |w|^{s+1-|\beta|}}{(|z| + |w|)^{n+s+\frac{n-m}{p}+1}} \|\Delta^M(z) f\|_{\Gamma+V(H+bH)} dz.$$

Next, introducing the spherical coordinates $z = r\zeta$, $r \geqslant 0$, $|\zeta| = 1$, $dz = r^{n-1} dr d\zeta$, we obtain for $w \in V(bH)$

$$\| R_s^{(\beta)}(w;\ \bar f(\cdot))\|_\Gamma \leqslant C \int_0^H |w|^{s-|\beta|} r^{-s-\frac{n-m}{p}-1} \left[\frac{|w|}{r}\right]_1 \int_{\substack{|\zeta|=1\\ \zeta\in V}} \|\Delta^M(r\zeta) f\|_{\tilde\Omega(H+bH)} d\zeta dr, \tag{7}$$

where $[a]_1 = \min\{a, 1\}$ is a cut-off at unity. Adding and subtracting $\bar f^{(\beta)}(x)$ in (3) and estimating one of the integrals, we obtain the following estimate:

$$\omega_{s+1}(\delta;\ \bar f) \leqslant C \sum_{|\beta|\leqslant s} \delta^{-\frac{n}{p}+|\beta|} \left\{ \int_{\Omega(\delta)} \int_{\Gamma(x)} |\bar f^{(\beta)}(x) - P_s^{(\beta)}(x-y;\ \bar f(y)|^p\, dy' dx \right\}^{\frac{1}{p}}.$$

Changing the order of integration and making the change $x = w + y$, we obtain

$$\omega_{s+1}(\delta;\ \bar f) \leqslant C \sum_{|\beta|\leqslant s} \delta^{-\frac{n}{p}+|\beta|} \left\{ \int_{V(b\delta)} \| R_s^{(\beta)}(w;\ \bar f(\cdot))\|_\Gamma^p\, dw \right\}^{\frac{1}{p}}.$$

Drawing in the estimate (7) we obtain for $0 < \delta \leqslant H$ that

$$\omega_{s+1}(\delta;\ \bar f) \leqslant C\delta^s \int_0^H r^{-s-\frac{n-m}{p}-1} \left[\frac{\delta}{r}\right]_1 \int_{\substack{|\zeta|=1\\ \zeta\in V}} \|\Delta^M(r\xi) f\|_{\tilde\Omega(H+bH)}\, d\zeta dr.$$

Hence after the substitutions $\delta = T$ and $r = Tt$ we get

$$\frac{\omega_{s+1}(T;\ \bar f)}{T^{l-\frac{n-m}{p}}} \leqslant C \int_0^{\frac{H}{T}} t^{l-s-\frac{n-m}{p}-1} \left[\frac{1}{t}\right]_1 (Tt)^{-l} \int_{\substack{|\zeta|=1\\ \zeta\in V}} \|\Delta^M(Tt\zeta) f\|_{\tilde\Omega(H+bH)}\, d\zeta dt.$$

Applying N_T, we obtain, by virtue of (3), that

$$N_T\left[\frac{\omega_{s+1}(T;\ \bar f)}{T^{l-\frac{n-m}{p}}}\right] \leqslant CN_T\left[T^{-l} \int_{\substack{|\zeta|=1\\ \zeta\in V}} \|\Delta^M(T\zeta) f\|_{\tilde\Omega(H+bH)} d\zeta\right].$$

Since $f = f_{HH} + \bar f$ in (5) and f_{HH} is convolution of f with a smooth finite kernel, we conclude by means of very simple estimates of f_{HH} that

$$\|f\|_\Gamma + N_T\left[\frac{\omega_{s+1}(T;\ f)}{T^{l-\frac{n-m}{p}}}\right] \leqslant C\|f\|_{\tilde\Omega(2H)} + CN_T\left[T^{-l} \int_{\substack{|\zeta|=1\\ \zeta\in V}} \|\Delta^M(T\zeta) f\|_{\tilde\Omega(H+bH)} d\zeta\right]. \tag{8}$$

2. Throughout below we will impose a new meaning to the notation in §1. Namely, we now let

$$V(H) = \{x\colon |x''| > M_1|x'|,\ |x| < H\}, \qquad m < n-1,$$
$$V(H) = \{x\colon x_n > M_1|x'|,\ |x| < H\}, \qquad m = n-1,$$

and the other sets $\widetilde{\Omega}(H)$, $\Omega(H)$, $\Gamma(x)$, and so on, are constructed from the above by the same rules as in §1. Note that for $m = n - 1$ our geometric objects have not changed, rather have only taken on a special character. As to the case $m < n - 1$, $V(H)$ is not a cone now but the union of a finite number of cones for various orientations of their axes in E^{n-m} ($\perp E^m$), whence also the estimates $\omega_{s+1}(T; f)$ (in the new notation) are obtained as the union of a finite number of estimates in §1. Thus, for example, estimate (8) is true in the new notation.

3. Now consider the question of extending a system of functions $\{f_\alpha(y)\}$, $|\alpha| \leqslant s$, $y \in \Gamma$. For the polynomial $P_s(x - y; y)$ constructed by (2) and a point $x \in \Omega(H)$ (in the notation of §2) consider

$$f(x) = \int_\Gamma P_s(x-y;\ y)\frac{1}{r^m(x)}\,\omega\left(\frac{y'-x'}{r(x)}\right)dy'. \tag{9}$$

Here for $x \in \Omega(H)$ and any α we have

$$c_1\rho(x) < r(x) < c_2\rho(x), \quad |D^\alpha r(x)| \leqslant C_\alpha \rho(x)^{1-|\alpha|}.$$

The existence of such an $r(x) \in C^\infty(\Omega(H))$ is shown in [5].

The kernel $\omega(x') \in C_0^\infty(E^m)$, satisfies $\int \omega(x')dx' = 1$, and in addition its support is contained in a neighborhood of zero, $|x'| < \epsilon$, so small that for each $x \in \Omega(H)$ the support of the function $\kappa(y') = \omega((y' - x')/r(x))$ is contained in the projection G' of Γ onto E^m. Hence

$$f(x) = P_s(x-z;\ z) + \int [P_s(x-y;\ y) - P_s(x-z;\ z)]\frac{1}{r(x)^m}\,\omega\left(\frac{y'-x'}{r(x)}\right)dy'.$$

Further, let

$$l - 1 \leqslant |\alpha| < l, \qquad |\alpha| + k > l, \cdots, k \text{ an integer} \tag{10}$$

Then

$$\Delta^k(v)D^\alpha f(x) = \sum_{\beta\leqslant\alpha} C_\beta \int_\Gamma\int_\Gamma \Delta_x^k(v)\Big\{[P_s^{(\beta)}(x-y;\ y) - P_s^{(\beta)}(x-z;\ z)]$$
$$\times D_x^{\alpha-\beta}\left[\frac{1}{r^m(x)}\,\omega\left(\frac{y'-x'}{r(x)}\right)\right]\frac{1}{r^m(x)}\,\omega\left(\frac{z'-x'}{r(x)}\right)\Big\}\,dy'dz'.$$

We will assume that $[x, x + kv] \in \Omega(H)$ and $q > 0$. For $\rho(x) \leqslant q|v|$ let us estimate the difference under the integral by the sum of the moduli. For $q|v| \leqslant \rho(x)$ we replace the difference $\Delta_x^k(v)$ by the integral of the kth derivative.

In the first case we obtain that for some $q < a$ and all $|v| \leqslant H/a$,

$$I_1 = \|\Delta^k(\Omega(H),\ v)D^\alpha f\|_{\rho(x)<q|v|} \leqslant C\|D^\alpha f\|_{\Omega(a|v|)}$$
$$\leqslant C_1\sum_{\beta\leqslant\alpha}\left\{\int_{\Omega(b|v|)}\int_{\Gamma(x)}\int_{\Gamma(x)}\rho(x)^{-2m}\left|\frac{P_s^{(\beta)}(x-y;\ y) - P_s^{(\beta)}(x-z;\ z)}{\rho(x)^{|\alpha-\beta|}}\right|^p dy'dz'dx\right\}^{\frac{1}{p}}$$
$$\leqslant C_2\sum_{\beta\leqslant\alpha}\sum_{j=0}^{\infty}|2^{-j}v|^{-|\alpha-\beta|-\frac{2m}{p}}\left\{\int_{\Omega(2^{-j}a|v|)\setminus\Omega(2^{-j-1}a|v|)}\int_{\Gamma(x)}\int_{\Gamma(x)}|P_s^{(\beta)}(x-y;\ y)\right.$$

$$- P_s^{(\beta)}(x-z;\ z)|^p\,dy'dz'dx\Big\}^{\frac{1}{p}} \leqslant C_3 \sum_{j=0}^{\infty} |2^{-j}v|^{-|\alpha|+\frac{n-m}{p}}\,\omega_{s+1}(2^{-j}a\,|v|).$$

In the second case for some $q > k$ (the existence of such a q is geometrically obvious) and some $c \in (0, q-k)$, if $c|v| < q|v| \leqslant H$, then

$$I_2 = \|\Delta^k(\Omega(H),\ v)\,D^\alpha f\|_{\rho(x)>q|v|}$$

$$\leqslant C_1|v|^k \sum_{|\beta+\gamma|=|\alpha|+k} \Bigg\{ \int\limits_{\substack{\rho(x)>q|v|\\ [x,\,x+kv]\in\Omega(H)}} \Big\{\int\limits_\Gamma\int\limits_\Gamma |P_s^{(\beta)}(x+tv-y;\ y)$$

$$- P_s^{(\beta)}(x+tv-z;\ z)|\,\Big|D_x^\gamma\Big[\frac{1}{r^m(x+tv)}\,\omega\Big(\frac{y'-x'-tv'}{r(x+tv)}\Big)\Big]$$

$$\times \frac{1}{r^m(x+tv)}\,\omega\Big(\frac{z'-x'-tv'}{r(x+tv)}\Big)\Big|\,dtdy'dz'\Big\}^p dx\Bigg\}^{\frac{1}{p}}$$

$$\leqslant C_2|v|^k \sum_{|\beta|\leqslant|\alpha|} \Bigg\{ \int\limits_{\Omega(H)\setminus\Omega(c|v|)} \Big[\int\limits_{\Gamma(x)}\int\limits_{\Gamma(x)} \rho(x)^{-2m}$$

$$\times \Big|\frac{P_s^{(\beta)}(x-y;\ y) - P_s^{(\beta)}(x-z;\ z)}{\rho(x)^{|\alpha|-|\beta|+k}}\Big|\,dy'dz'\Big]^p dx\Bigg\}^{\frac{1}{p}}$$

$$\leqslant C_2|v|^k \sum_{|\beta|\leqslant|\alpha|}\sum_{j=0}^{\infty} \Bigg\{ \int\limits_{\Omega(H)\cap\Omega(2^{j+1}c|v|)\setminus\Omega(2^jc|v|)} \Big[\int\limits_{\Gamma(x)}\int\limits_{\Gamma(x)} \rho(x)^{-2m}$$

$$\times \Big|\frac{P_s^{(\beta)}(x-y;\ y) - P_s^{(\beta)}(x-z;\ z)}{\rho(x)^{|\alpha|-|\beta|+k}}\Big|\,dy'dz'\Big]^p dx\Bigg\}^{\frac{1}{p}}$$

$$\leqslant C_3|v|^k \Big\{ H^{-|\alpha|-k+\frac{n-m}{p}}\,\omega_{s+1}(H) + \sum_{\substack{j\geqslant 0\\ 2^{j+1}c|v|<H}} |2^jv|^{-|\alpha|-k+\frac{n-m}{p}}\,\omega_{s+1}(2^{j+1}c\,|v|)\Big\}.$$

In the final estimates for I_1 and I_2, replacing the sums in j by integrals, we obtain

$$\|\Delta^k(\Omega(H),\ v)\,D^\alpha f\| \leqslant J_1 + J_2 \leqslant C_1 \int\limits_{\frac{H}{|v|}}^{\min\{2a,\,\frac{H}{|v|}\}} |tv|^{-|\alpha|+\frac{n-m}{p}}\,\omega_{s+1}(t\,|v|)\,\frac{dt}{t}$$

$$+ C_2|v|^k \int\limits_{\min\{c,\,\frac{H}{|v|}\}}^{\frac{H}{|v|}} |tv|^{-|\alpha|+\frac{n-m}{p}-k}\,\omega_{s+1}(t\,|v|)\,\frac{dt}{t}.$$

Hence

$$\| \Delta^k (\Omega (H), v) D^\alpha f \| \leqslant C \int_0^{\frac{H}{|v|}} \frac{t^{l-|\alpha|}}{[t^k]_1} \frac{\omega_{s+1} (t |v|)}{|tv|^{l-\frac{n-m}{p}}} \frac{dt}{t},$$

and, after applying the N-functional,

$$N_T \left[\frac{\sup\limits_{|v|=T} \| \Delta^k (\Omega (H), v) D^\alpha f \|_\Omega}{T^{l-|\alpha|}} \right] \leqslant CN \left[\frac{\omega_{s+1} (T)}{T^{l-\frac{n-m}{p}}} \right]. \tag{11}$$

By similar but simpler estimates we obtain

$$\sum_{|\beta|<l} \| D^\beta f \|_{\Omega(H)} \leqslant C \sum_{|\beta|\leqslant s} \| f_\beta \|_\Gamma + CN_T \left[\frac{\omega_{s+1} (T)}{T^{l-\frac{n-m}{p}}} \right]. \tag{12}$$

Finally we need to show that in a certain sense

$$f_\beta (y) f = {}^{(\beta)} (y), \quad y \in \Gamma, \quad |\beta| \leqslant s,$$

where $f(x)$ is constructed from $\{f_\beta\}$ by means of (9) and (2). By the properties of $\omega(x')$ for $z \in \Gamma$,

$$f(x) - P_s (x - z;\ z) = \int_\Gamma [P_s (x - y;\ y) - P_s (x - z;\ z)] \frac{1}{r^m (x)} \omega \left(\frac{y' - x'}{r(x)} \right) dy'.$$

Hence for $|\alpha| \leqslant s$

$$| f^{(\alpha)} (x) - P_s^{(\alpha)} (x - z;\ z) |$$
$$\leqslant C \sum_{\beta \leqslant \alpha} r(x)^{-\frac{m}{p} - |\alpha - \beta|} \left\{ \int_{\Gamma(x)} | P_s^{(\beta)} (x - y;\ y) - P_s^{(\beta)} (x - z;\ z) |^p dy' \right\}^{\frac{1}{p}}.$$

Hence, if $\epsilon = l - |\alpha| - (n - m)/p > 0$, then

$$\delta^{-\frac{n}{p} - \varepsilon} \left\{ \int_{V(\delta)} \int_{\Gamma_\delta} | f^{(\alpha)} (z + w) - P_s^{(\alpha)} (w;\ z) |^p dz' dw \right\}^{\frac{1}{p}}$$

$$\leqslant \delta^{-\frac{n}{p} - \varepsilon} \left\{ \int_{\Gamma_\delta} \int_{z + V(\delta)} | f^{(\alpha)} (x) - P_s^{(\alpha)} (x - z;\ z) |^p dx dz' \right\}^{\frac{1}{p}}$$

$$\leqslant \delta^{-\frac{n}{p} - \varepsilon} \left\{ \int_{\Omega(\delta)} \int_{\Gamma(x)} | f^{(\alpha)} (x) - P_s^{(\alpha)} (x - z;\ z) |^p dz' dx \right\}^{\frac{1}{p}}$$

$$\leqslant C \delta^{-l + \frac{n-m}{p}} \sum_{\beta \leqslant \alpha} \delta^{-\frac{m+n}{p} + |\beta|} \left\{ \int_{\Omega(\delta)} \int_{\Gamma(x)} \int_{\Gamma(x)} | P_s^{(\beta)} (x - y;\ y) - \right.$$

$$- P_s^{(\beta)}(x-z;\ z)|^p\,dy'dz'dx\Big\}^{\frac{1}{p}} \leqslant C\delta^{-l+\frac{n-m}{p}}\omega_{s+1}(\delta).$$

Thus for $|\alpha| \leqslant s$ and $\epsilon = l - |\alpha| - (n-m)/p > 0$ we have

$$N_T\left[T^{-\frac{n}{p}-\varepsilon}\left\{\int\limits_{V(T)}\int\limits_{\Gamma_T}|f^{(\alpha)}(z+w)-P_s^{(\alpha)}(w;\ z)|^p\,dz'dw\right\}^{\frac{1}{p}}\right] \leqslant CN_T\left[T^{-l+\frac{n-m}{p}}\omega_{s+1}(T)\right]. \tag{13}$$

In our case of a finite right side this estimate indicates in what sense the derivatives $f^{(\alpha)}(x)$ of the function (9) take on the values $f_\alpha(x)$ on the boundary Γ.

REMARK. In the estimate of traces of functions (8) and its converse (11) the function $\omega_{s+1}(\delta)$ can be replaced (in fact without change in proof) by

$$\sum_{|\beta|\leqslant s}\delta^{-\frac{n+m}{p}+|\beta|}\left\{\int\limits_{\Omega(\delta)\setminus\Omega\left(\frac{\delta}{2}\right)}\int\limits_{\Gamma(x)}\int\limits_{\Gamma(x)}|P_s^{(\beta)}(x-y;\ y)-P_s^{(\beta)}(x-z;\ z)|^p\,dy'dz'dx\right\}^{\frac{1}{p}}$$

or for $\epsilon > 0$ by

$$\sum_{|\beta|\leqslant s}\delta^{s-\varepsilon}\left\{\int\limits_{\Omega(\delta)}\int\limits_{\Gamma(x)}\int\limits_{\Gamma(x)}\left|\frac{P_s^{(\beta)}(x-y;\ y)-P_s^{(\beta)}(x-z;\ z)}{\rho(x)^{s-|\beta|+\frac{n-m}{p}-\varepsilon}}\right|^p dy'dz'dx\right\}^{\frac{1}{p}}.$$

4. Let us apply our results to the case of a region $\Omega \subset E^n$ of a definite class.

Let the open set $\Omega_0 \subset \Omega$, $\partial\Omega_0 \subset \Omega$, and for $k \geqslant 1$, the regions $\widetilde{\Omega}_k \subset \Omega$, $\Gamma_k = \partial\Omega \cap \partial\widetilde{\Omega}_k$, have dimension m_k, and set

$$\Omega_k^{(\delta)} = \{x : x \in \Omega_k,\ \rho(x,\ \partial\Omega_k \setminus \partial\Omega) > \delta\},$$

where under a suitable rotation of the cartesian coordinates in E^n the sets Γ_k, Ω_k, $\widetilde{\Omega}_k$ and $\Gamma_k(x)$ have the forms (1), (1′), (1″) and (1‴) of §2. For the region $\Omega \subset E^n$ let there exist a $\delta > 0$ and a natural number K such that

$$\Omega = \bigcup_{k=0}^{K}\widetilde{\Omega}_k = \bigcup_{k=0}^{K}\Omega_k^{(\delta)}. \tag{14}$$

THEOREM. *Let the region Ω satisfy condition* (14) *for $s = s_k$ and $m = m_k$ satisfying condition* (4). *Then*

$$\sum_{k=1}^{K}\sum_{|\beta|\leqslant s_k}\|f^{(\beta)}\|_{\Gamma_k} + \sum_{k=1}^{K}N_T\left[\frac{\omega_{s_k+1}(T;\ f;\ \Gamma_k)}{T^{l-\frac{n-m_k}{p}}}\right] \leqslant C\sum_{|\beta|<l}\|D^\beta f\|_\Omega + CN_T\left[T^{-l}\int\limits_{|\zeta|=1}\|\Delta^M(T\zeta;\ \Omega)f\|\,d\zeta\right], \tag{15}$$

where $\omega_{s_k+1}(T; f; \Gamma_k)$ *is* $\omega_{s_k+1}(T; f)$ *for* $\Gamma = \Gamma_k$ *and is constructed in correspondence with* (3) *according to the Taylor polynomials*

$$P_{s_k}(w;\ f(y)),\ y \in \Gamma_k.$$

Conversely, on $\partial\Omega = \bigcup_1^k \Gamma_k$ *let there be given a system* $\{f_\alpha(y)\}$, $|\alpha| \leqslant s_k$ *if* $y \in \Gamma_k$. *Then there exists a function* $f(x)$, $x \in \Omega$, *such that*

$$\sum_{|\beta|<l} \|D^\beta f\|_\Omega + N_T\left[T^{-l} \sup_{|\zeta|=1} \|\Delta^M(T\zeta;\ \Omega) f\|\right]$$

$$\leqslant C \sum_{k=1}^K \sum_{|\beta|\leqslant s_k} \|f^{(\beta)}\|_{\Gamma_k} + \sum_{k=1}^K N_T\left[\frac{\omega_{s_k+1}(T;\ f;\ \Gamma_k)}{T^{l-\frac{n-m_k}{p}}}\right], \quad (16)$$

where the $\omega_{s_k+1}(T; f; \Gamma_k)$ *are constructed from polynomials of the form* (2). *Here* $f^{(\alpha)}(y) = f_\alpha(y)$, $y \in \partial\Omega$, *in the sense of the finiteness of the left side of* (13).

The proof follows from the previous estimates and use of a partition of unity.

BIBLIOGRAPHY

1. S. M. Nikol'skiĭ, *Approximation of functions of several variables and embedding theorems,* "Nauka", Moscow, 1969. (Russian)

2. V. P. Il'in, *The properties of some classes of differentiable functions of several variables defined in an n-dimensional region,* Trudy Mat. Inst. Steklov. **66** (1962), 227–363; English transl., Amer. Math. Soc. Transl. (2) **81** (1969), 91–256. MR #3750.

3. K. K. Golovkin, *Embedding theorems for fractional spaces,* Trudy Mat. Inst. Steklov. **70** (1964), 38–46; English transl., Amer. Math. Soc. Transl. (2) **91** (1970), 57–67. MR **28** #5282; **42** #2.

4. O. V. Besov, *Estimates of moduli of smoothness on domains and embedding theorems,* Trudy Mat. Inst. Steklov. **117** (1972), 22–46 = Proc. Steklov. Inst. Math. **117** (1972), 25–53.

5. ———, *The behavior of differentialbe functions on a nonsmooth surface,* Trudy Mat. Inst. Steklov. **117** (1972), 3–10 = Proc. Steklov Inst. Math. **117** (1972), 1–9.

Translated by J. ELLIS

Trudy Mat. Inst. Steklov.
117 (1972)

Proc. Steklov Inst. Math.
117 (1972)

ESTIMATES OF MODULI OF SMOOTHNESS ON DOMAINS, AND IMBEDDING THEOREMS

UDC 517.518.22

O. V. BESOV

Abstract. For a function defined in a region of a certain class, estimates are established relating moduli of smoothness of coordinate directions of the function and its derivatives in various (mixed) L_p norms. As a corollary, imbedding and extension theorems are obtained for classes of functions which generalize a series of known results.

Bibliography: 17 items.

In this article we study functions defined on a domain (of a certain class) of the Euclidean space E^n, and for these functions we establish estimates relating the moduli of smoothness (or of the difference) of functions and of their derivatives with respect to the L_p and L_q norms. The study is carried out for the nonisotropic case, when the functions have different differential properties in different directions; moreover, the L_p and L_q norms are mixed norms (i.e. p and q are vectors). As a corollary, theorems on imbedding and extension are obtained for certain classes of functions characterized, for instance, by the behavior of differences of coordinate directions. These results are more general than previously known, even for functions defined on the entire space E^n with isotropic properties and with respect to the usual (unmixed) norms. The basic tool in these considerations is a known integral representation of functions due to V. P. Il'in, and representations close to it.

We remark that for functions defined on the entire space E^n, results of various authors concerning estimates of moduli of smoothness in the usual L_p norms are generalized in the book [1] (see, for instance, §6.4 and others). Recently P. L. Ul'janov obtained inequalities for the modulus of continuity in various norms in the one-dimensional case [2]. These results can be considered complete (see [2] and the work by V. A. Andrienko [3]). Below we give, in particular, a generalization of such inequalities to the case of functions of several variables, defined in a domain and with respect to mixed norms. We also remark that a generalization of similar results of P. L. Ul'janov on imbedding into L_q was already used in [4].

In §1 we give an integral representation of functions and auxiliary estimates of integrals of potential type.

AMS (MOS) subject classifications (1970). Primary 26A54, 46E15; Secondary 46E35.

§ §2 and 3 are devoted to the basic estimates of moduli of smoothness of a function and of its extension in the case when relations between the steps in coordinate directions are those of a power.

In § §4 and 5 we present applications of these results to imbedding and extension of classes of functions.

An integral representation in terms of differences with a step of variable direction (§6) is used in §7 for finding relations between the modulus of smoothness of coordinate directions and the variable direction (for individual functions and for classes of functions).

The problem of constructing functions with the best properties of smoothness from the given trace is solved in §8.

Some remarks on the constructive method of proving imbedding theorems are contained in §9.

§10 is devoted to generalizations of imbedding theorems for the case when relations between smoothness in different coordinate directions are not those of a power.

§1. Auxiliary propositions and representation of functions

Let $p = (p_1, \cdots, p_n)$, $1 \leqslant p_i \leqslant \infty$, $x \in E^n$, $\varphi = \varphi(x)$ and $\psi = \psi(x, t)$. Set

$$\|\varphi\|_p = \left\{ \int_{-\infty}^{\infty} \left[\cdots \left(\int_{-\infty}^{\infty} |\varphi|^{p_1} dx_1 \right)^{\frac{1}{p_1}} \cdots \right]^{p_n} dx_n \right\}^{\frac{1}{p_n}},$$

$$\|\varphi\|_{p(\Omega)} = \|\varphi_\Omega\|_p,$$

where $\varphi_\Omega(x) = \varphi(x)$ for $x \in \Omega \subset E^n$, $\varphi_\Omega(x) = 0$ for $x \notin \Omega$, and

$$\|\psi\|_{p,\theta} = \left\{ \int_0^\infty \|\psi(\cdot, t)\|_p^\theta \, dt \right\}^{\frac{1}{\theta}}, \qquad 1 \leqslant \theta \leqslant \infty.$$

As usual, for $p_i = \infty$ under the partial norm we understand the essential supremum of the absolute value.

We also assume that $1/p = (1/p_1, \cdots, 1/p_n)$; $p \leqslant q$ $(p < q)$ if $p_i \leqslant q_i$ $(p_i < q_i)$ for all $i = 1, \cdots, n$; $\mathbf{1} = (1, \cdots, 1)$, $\mathbf{0} = (0, \cdots, 0)$, $\infty = (\infty, \cdots, \infty)$. For $h > 0$ and $\sigma \geqslant \mathbf{0}$ we set $x/h^\sigma = (x_1/h^{\sigma_1}, \cdots, x_n/h^{\sigma_n})$, $|\sigma| = \Sigma_1^n \sigma_i$ and $\sigma^\gamma = (\sigma_1^{\gamma_1}, \cdots, \sigma_n^{\gamma_n})$.

Set

$$\Phi(x) = \int_{E^n} \frac{\varphi(x+y) \, |y_n|^\varepsilon \, dy}{\rho(y)^{\left(\varkappa, 1 - \frac{1}{p} + \frac{1}{q}\right) + \varkappa_n \varepsilon}},$$

where $\mathbf{1} \leqslant p \leqslant q \leqslant \infty$, $q_n = \kappa > 0$, $\epsilon \geqslant 0$ and $\rho(y) = \Sigma_1^n |y_i|^{1/\kappa_i}$.

The following theorem is true.

THEOREM 1. *For* $\epsilon > 0$ *let either* $p_n = 1$, *or* $1 \leqslant p_n \leqslant q_{n-1}$ *and* $1 \leqslant p_{n-1} < q_{n-1} < \infty$, *and for* $\epsilon = 0$ *either* $1 < p_n \leqslant q_{n-1}$, $1 \leqslant p_{n-1} < q_{n-1} < \infty$, *or* $p_n = p_{n-1} = \cdots = p_{n-k+1} = 1, q_n = q_{n-1} = \cdots = q_{n-k+1} = \infty, 1 < p_{n-k} < q_{n-k} < \infty$, *or* $p_n = p_{n-1} = \cdots = p_{n-k} = 1$, $q_n = q_{n-1} = \cdots = q_{n-k} = \infty$, $1 < p_{n-k} \leqslant q_{n-k-1}$, $1 \leqslant p_{n-k-1} < q_{n-k-1} < \infty$. *Then*

$$\|\Phi\|_q \leqslant C\|\varphi\|_p \tag{1}$$

where the constant C *does not depend on* φ.

This theorem was established by V. P. Il'in [5] for $n = 2$ and afterwards was generalized in [4] to the multidimensional case.

We observe that in an interchange of two consecutive partial norms in construction of $\|\varphi\|_p$ the value of the latter changes, in general, and it is comparable with the initial one thanks to Minkowski's inequality for integrals

$$\|\varphi\|_{(p_1,\, p_2)} \leqslant \|\varphi\|_{(p_2,\, p_1)} \qquad (n=2,\ p_1 \leqslant p_2).$$

On the basis of such an interchange, Theorem 1 could be formulated for $n \geqslant 3$ under more general relations between p and q (compare [4]). However, we shall not do this here in order not to make formulations cumbersome.

We also note that the extension of estimate (1) to the case $n > 2$ is carried out with the help of a known estimate for convolution:

$$\|k * f\|_q \leqslant \|k\|_r \|f\|_p, \qquad 0 \leqslant \frac{1}{r} = 1 - \frac{1}{p} + \frac{1}{q} \leqslant 1, \tag{2}$$

whose one-dimensional case is successively applied to x_1, then to x_2, etc.

The region $R = R(\sigma)$ $(\sigma > 0)$

$$R(\sigma) = \bigcup_{0<h<H} \{x;\ 0 < a_i h^{\sigma_i} < x_i < b_i h^{\sigma_i} \quad (i = 1, \ldots, n)\}. \tag{3}$$

will be called a horn.

For a locally summable function $f(x)$, the following integral representation is valid (in the sense of L^{loc}-convergence).

$$\begin{aligned} f(x) = & \int_{E^n} H^{-|\sigma|}\Psi_0\left(\frac{y}{H^\sigma}\right) f(x+y)\,dy \\ & + \int_0^H \sum_{i=1}^n \int_{E^n}\int_0^\infty h^{-1-|\sigma|-\sigma_i}\Psi_i\left(\frac{y}{h^\sigma}\right)\xi_i\left(\frac{t}{h^{\sigma_i}}\right)\Delta_i^{m_i}(\delta_i t) f(x+y+te_i)\,dt\,dy\,dh. \end{aligned} \tag{4}$$

(see, for instance, [6], and also §10 of the present article). Here $\Delta_i^{m_i}(t)f(x)$ is the difference of order $m_i \geqslant 1$ with the step t in the direction of the ith unit vector e_i, $\sigma > 0$, $\delta_1 = \delta^{\sigma_i}$ and $\delta \in (0, 1]$.

The functions $\Psi_i(x) \in C_0^\infty(E^n)$ and $\xi_i(t) \in C_0^\infty(E^1)$, and they are concentrated in a preassigned subcube of $\{x | 0 < x < 1\}$ and a subinterval of $(0, 1)$, respectively. Their supports and also the number $\delta \in (0, 1]$ are such that the actual integration in (4) is carried out inside of a horn $R(\sigma)$ given in advance; and in representation (4) of the function $f(x)$ only its values in the points of the shifted horn $x + R(\sigma)$ are taken into account. We also emphasize that the parameter $H > 0$ can be arbitrary. Here and in what follows the sign $+$ will denote the arithmetic sum of sets.

We shall say that a region $\Omega \subset E^n$ satisfies a σ-horn condition if there exist a finite number K of open sets Ω_k and horns $R_k(\sigma)$ (of the form (3) or obtained from (3) by symmetric reflection with respect to one or more coordinate-hyperplanes) satisfying the following conditions.

1°. $\Omega_k + R_k(\sigma) \subset \Omega$ $(k = 1, \cdots, K)$.

2°. For some $\epsilon > 0$, $\Omega = \bigcup_{k=1}^K \Omega_k^{(\epsilon)}$, where

$$\Omega_k^{(\varepsilon)} = \{x;\ x \in \Omega_k,\ \rho(x,\ \partial\Omega_k \setminus \partial\Omega) > \varepsilon\}.$$

We shall say that Ω satisfies a strong σ-horn condition if instead of 2° the following stronger condition is fulfilled:

3°. For some $\epsilon > 0$, $\Omega = \bigcup_{k=1}^K \Omega_k^{[\epsilon]}$, where

$$\Omega_k^{[\varepsilon]} = \{x;\ x \in \Omega_k,\ \rho(x,\ \Omega \setminus \Omega_k) > \varepsilon\}.$$

For an open set $\Omega \subset E^n$ we set

$$\Delta_i^{m_i}(\Omega,\ t) f(x) = \Delta_i^{m_i}(t) f(x),$$

if in the construction of the difference the points and segments joining them belong to Ω, and $\Delta_i^{m_i}(\Omega, t) f(x) = 0$ otherwise.

Let an open set $U \subset E^n$ and a function $f(x)$ be given on $U + R(\sigma)$, $1 \leqslant \theta \leqslant \infty$. Set

$$\varphi_i(x,\ t) = t^{-\frac{1}{\theta_i} - l_i} \Delta_i^{m_i}(U + R,\ \delta t) f(x + te_i);$$

$f_0(x)$ is the extension of $f(x)$ equal to zero outside $U + R$. Then the function $\tilde{f}(x)$ defined on E^n,

$$\tilde{f}(x) = \int_{E^n} H^{-|\sigma|} \Psi_0\left(\frac{y}{H^\sigma}\right) f_0(x + y)\, dy$$
$$+ \int_0^H \sum_{i=1}^n \int_{E^n} \int_0^\infty h^{-1-|\sigma|-\sigma_i} \Psi_i\left(\frac{y}{h^\sigma}\right) \xi_i\left(\frac{t}{h^{\sigma_i}}\right) t^{\frac{1}{\theta_i} + l_i} \varphi_i(x + y,\ t)\, dt\, dy\, dh \tag{5}$$

coincides with $f(x)$ on U.

§2. Estimates of moduli of smoothness in a region

We shall estimate the difference $\Delta_j^M(\tau^{\sigma_j})D^\alpha\tilde{f}(x)$ in L_q. Denote by $\chi(z)$, $z \in E^k$, the characteristic function of the cube $\{0 < x_i < 1,\ i = 1, \cdots, k\}$. First we estimate the quantities

$$A(y,\ t) = \int_0^T h^{-1-|\sigma|-\sigma_i-(\alpha,\sigma)}\chi\left(\frac{y}{h^\sigma}\right)\left|\xi_i\left(\frac{t}{h^{\sigma_i}}\right)\right| dh,$$

$$B(y,\ t) = \tau^{\sigma_j M}\int_T^H h^{-1-|\sigma|-\sigma_i-(\alpha,\sigma)-\sigma_j M}\chi\left(\frac{y}{h^\sigma}\right)\left|\xi_i\left(\frac{t}{h^{\sigma_i}}\right)\right| dh$$

for $0 < \tau \leqslant T \leqslant H,\ y \geqslant 0$.

Since the support of ξ_i is bounded away from the origin, it follows that $A(y, t) = 0$ for $t > T^{\sigma_i}$, and $B(y, t) = 0$ for $t > H^{\sigma_i}$ and for $t < \epsilon T^{\sigma_i}$ $(\epsilon > 0)$. Hence

$$A(y,\ t) \leqslant c_1\chi\left(\frac{t}{T^{\sigma_i}}\right)\int_{\max\left[|y_1|^{\frac{1}{\sigma_1}},\ldots,|y_n|^{\frac{1}{\sigma_n}},\,t^{\frac{1}{\sigma_i}}\right]}^{\infty} h^{-1-|\sigma|-\sigma_i-(\alpha,\sigma)}\,dh \leqslant \frac{c_2\chi\left(\frac{t}{T^{\sigma_i}}\right)}{\left[\sum_{j=1}^n |y_j|^{\frac{1}{\sigma_j}} + t^{\frac{1}{\sigma_i}}\right]^{|\sigma|+\sigma_i+(\alpha,\sigma)}}.$$

Similarly

$$B(y,\ t) \leqslant c_3\left[\chi\left(\frac{t}{H^{\sigma_i}}\right) - \chi\left(\frac{t}{\varepsilon T^{\sigma_i}}\right)\right]\frac{\tau^{\sigma_j M}}{\left[\sum_{j=1}^n |y_j|^{\frac{1}{\sigma_j}} + t^{\frac{1}{\sigma_i}} + T\right]^{|\sigma|+\sigma_i+(\alpha,\sigma)+\sigma_j M}}.$$

From this for $M \geqslant 0$ and from (5) we obtain

$$\begin{aligned}|\Delta_j^M(\tau^{\sigma_j})D^\alpha\tilde{f}(x)| &\leqslant \int_{E^n} H^{-|\sigma|-(\alpha,\sigma)}\left|\Delta_{y_j}^M(-\tau^{\sigma_j})\Psi_0^{(\alpha)}\left(\frac{y}{H^\sigma}\right)f_0(x+y)\right|dy \\ &+ c_4\sum_{i=1}^n\int_0^{T^{\sigma_i}}\int_{E^n}\frac{t^{\frac{1}{\theta_i}+l_i}|\Delta_j^M(\tau^{\sigma_i})\varphi_i(x+y,\ t)|\,dydt}{\left[\sum_{j=1}^n|y_j|^{\frac{1}{\sigma_j}} + t^{\frac{1}{\sigma_i}}\right]^{\lambda_i}} \\ &+ c_4\sum_{i=1}^n\tau^{\sigma_j M}\int_{\varepsilon T^{\sigma_i}}^{H^{\sigma_i}}\int_{E^n}\frac{t^{\frac{1}{\theta_i}+l_i}\int_0^1|\varphi_i(x+y+M\tau^{\sigma_j}ue_j,\ t)|\,dudydt}{\left[\sum_{j=1}^n|y_j|^{\frac{1}{\sigma_j}} + t^{\frac{1}{\sigma_i}} + T\right]^{\lambda_i+\sigma_j M}} \\ &= A_0(x,\ T,\ H) + c_4\sum_{i=1}^n A_i(x,\ T,\ H) + c_4\sum_{i=1}^n B_i(x,\ T,\ H),\end{aligned} \tag{6}$$

where

$$\lambda_i=\left(\sigma,\ 1-\frac{1}{p^i}+\frac{1}{q}\right)+\sigma_i\left(1-\frac{1}{\theta_i}\right)+\sigma_i\left(\frac{1}{\theta_i}+l_i\right).$$

We assume here that

$$\sigma_i l_i=\left(\sigma,\ \frac{1}{p^i}-\frac{1}{q}+\alpha\right)\geqslant 0 \qquad (i=1,\ldots,n). \tag{7}$$

To estimate the L_q norm of the left side of inequality (6) we estimate A_0 and B_i with the help of (2), assuming that the suitable relations for the vectors $(p_1,\cdots,p_n,\theta_i)$ and $(q_1,\cdots,q_n,\infty)$ are fulfilled. To A_i we apply estimate (1). Evidently we obtain $(p^0\leqslant q)$

$$\|A_0\|_q\leqslant C_1T^{\sigma_j M}H^{-\sigma_j M-\left(\sigma,\frac{1}{p^0}-\frac{1}{q}+\alpha\right)}\|f_0\|_{p^0},$$

$$\|A_i\|_q+\|B_i\|_q\leqslant C_2\left\{\int_0^{T^{\sigma_i}}\|\varphi_i(\cdot,\ t)\|_{p^i}^{\theta_i}dt\right\}^{\frac{1}{\theta_i}}$$

$$+C_2T^{\sigma_j M}\int_{\varepsilon T^{\sigma_i}}^{H^{\sigma_i}}\|\varphi_i(\cdot,\ t)\|_{p^i}\frac{t^{\frac{1}{\theta_i}+l_i}dt}{(T^{\sigma_i}+t)^{1+\frac{\sigma_j}{\sigma_i}M+l_i}}$$

$$\leqslant C_3\left\{\int_0^{T^{\sigma_i}}\|\varphi_i(\cdot,\ t)\|_{p^i}^{\theta_i}dt\right\}^{\frac{1}{\theta_i}}+C_3T^{\sigma_j M}\int_{T^{\sigma_i}}^{H^{\sigma_i}}\|\varphi_i(\cdot,\ t)\|_{p^i}t^{\frac{1}{\theta_i}-1-\frac{\sigma_j}{\sigma_i}M}dt.$$

According to a well-known property of the norm of a difference,

$$\|\varphi_i(\cdot,\ 2^k t)\|_{p^i}\leqslant 2^{k\left(m_i-\frac{1}{\theta_i}-l_i\right)}\|\varphi_i(\cdot,\ t)\|_{p^i}.$$

Consequently for $\sigma_j M>\sigma_i m_i-\sigma_i l_i$ we have

$$T^{\sigma_j M}\int_{T^{\sigma_i}}^{H^{\sigma_i}}\|\varphi_i(\cdot,\ t)\|\,t^{\frac{1}{\theta_i}-1-\frac{\sigma_j}{\sigma_i}M}dt\leqslant T^{\sigma_j M}\sum_{k=0}^{\infty}\int_{2^kT^{\sigma_i}}^{2^{k+1}T^{\sigma_i}}\|\varphi_i(\cdot,\ t)\|_{p^i}\frac{dt}{t^{1-\frac{1}{\theta_i}+\frac{\sigma_j}{\sigma_i}M}}$$

$$\leqslant T^{\sigma_j M}\sum_{k=1}^{\infty}2^{k\left(m_i-l_i-\frac{\sigma_j}{\sigma_i}M\right)}\int_{\frac{1}{2}T^{\sigma_i}}^{T^{\sigma_i}}\|\varphi_i(\cdot,\ t)\|_{p^i}\frac{dt}{t^{1-\frac{1}{\theta_i}+\frac{\sigma_j}{\sigma_i}M}}$$

$$\leqslant C_4\left\{\int\limits_{\frac{1}{2}T^{\sigma_i}}^{T^{\sigma_i}}\|\varphi_i(\cdot,\ t)\|_{p^i}^{\theta_i}dt\right\}^{\frac{1}{\theta_i}}.$$

Collecting the estimates we have obtained, we arrive at the inequality

$$\begin{aligned}\|\Delta_j^M(\tau^{\sigma_j})D^\alpha\tilde{f}\|_q &\leqslant CT^{\sigma_j M}H^{-\sigma_j M-\left(\sigma,\frac{1}{p^0}-\frac{1}{q}+\alpha\right)}\|f\|_{p_0(U+R)}\\ &+C\sum_{i=1}^n\left\{\int\limits_0^T\|\Delta_i^{m_i}(U+R,\ \delta^{\sigma_i}t^{\sigma_i})f\|_{p^i}^{\theta_i}\frac{dt}{t^{1+\theta_i l_i\sigma_i}}\right\}^{\frac{1}{\theta_i}}+C\sum_{i=1}^n J_i(T,\ H)\end{aligned}\tag{8}$$

and (with a slight change in the proof) at the inequality

$$\begin{aligned}\|\Delta_j^M(U,\ \tau^{\sigma_j})D^\alpha f\|_{q(U)} &\leqslant CT^{\sigma_j N}H^{-\sigma_j N-\left(\sigma,\frac{1}{p_0}-\frac{1}{q}+\beta\right)}\|\Delta_j^{M-N}(U+R,\ \tau^{\sigma_j})D^{\alpha-\beta}f\|_{p^0}\\ &+C\sum_{i=1}^n\left\{\int\limits_0^T\|\Delta_i^{m_i}(U+R,\ \delta^{\sigma_i}t^{\sigma_i})f\|_{p^i}^{\theta_i}\frac{dt}{t^{1+\theta_i l_i\sigma_i}}\right\}^{\frac{1}{\theta_i}}+C\sum_{i=1}^n J_i(T,\ H),\end{aligned}\tag{9}$$

where in the general case

$$J_i(T,\ H)\leqslant T^{\sigma_j M}\int\limits_T^H\|\Delta_i^{m_i}(U+R,\ \delta^{\sigma_i}t^{\sigma_i})f\|_{p^i}t^{-1-l_i\sigma_i-\sigma_j M}dt,$$

and $J_i(T, H) = 0$ for $\sigma_j M > \sigma_i m_i - \sigma_i l_i$.

In inequalities (8) and (9) we have $0 \leqslant N \leqslant M$, $0 \leqslant \beta \leqslant \alpha$ and $0 < \tau \leqslant T \leqslant H$.

From estimate (9) we now obtain the following theorem.

THEOREM 2. *Let a region $\Omega \subset E^n$ satisfy the σ-horn condition, and let $1 \leqslant p^i \leqslant q \leqslant \infty$ $(i = 0, 1, \cdots, n)$. Let condition (7) be fulfilled, and also for every $i = 1, \cdots, n$ let either $\theta_i = 1$, or $1 \leqslant \theta_i \leqslant q_n$ and $1 \leqslant p_n^i < q_n < \infty$. Then, for $0 < \tau \leqslant T \leqslant H \leqslant H(\Omega)$, $0 \leqslant N \leqslant M$ and $0 \leqslant \beta \leqslant \alpha$,*

$$\begin{aligned}\|\Delta_j^M(\Omega,\ \tau^{\sigma_j})D^\alpha f\|_{q(\Omega)} &\leqslant CT^{\sigma_j N}H^{-\sigma_j N-\left(\sigma,\frac{1}{p_0}-\frac{1}{q}+\beta\right)}\|\Delta_j^{M-N}(\Omega,\ \tau^{\sigma_j})D^{\alpha-\beta}f\|_{p^0(\Omega)}\\ &+C\sum_{i=1}^n\left\{\int\limits_0^T\|\Delta_i^{m_i}(\Omega,\delta^{\sigma_i}t^{\sigma_i})f\|_{p^i}^{\theta_i}\frac{dt}{t^{1+\theta_i l_i\sigma_i}}\right\}^{\frac{1}{\theta_i}}+C\sum_{i=1}^n J_i(T,\ H),\end{aligned}\tag{10}$$

where in the general case

$$J_i(T,\ H)\leqslant T^{\sigma_j M}\int\limits_T^H\|\Delta_i^{m_i}(\Omega,\ \delta^{\sigma_i}t^{\sigma_i})f\|_{p^i}\frac{dt}{t^{1+l_i\sigma_i+\sigma_j M}},$$

and $J_i(T, H) = 0$ for $\sigma_j M > \sigma_i m_i - \sigma_i l_i$.

REMARK 1. Theorem 2 can be obtained, in a similar way, also in a more general form, assuming that to every $k = 1, \cdots, K$ there corresponds a vector $\sigma = \sigma^{(k)}$ (compare [4]).

REMARK 2. For $M = 0$ condition 2° can be relaxed, admitting in it $\epsilon = 0$ (see [4]).

REMARK 3. From (10) one may obtain the estimate

$$\sup_{0<\tau\leqslant T} \| \Delta_j^{m_j} (\Omega, \tau^{\sigma_j}) D^\alpha f \|_{q(\Omega)} \leqslant C \sum_{i=1}^{n} \left\{ \int_0^T \sup_{0<\tau\leqslant t} \| \Delta_i^{m_i} (\Omega, \delta^{\sigma_i}\tau^{\sigma_i}) f \|_{p^i}^{\theta_i} \frac{dt}{t^{1+\theta_i l_i \sigma_i}} \right\}^{\frac{1}{\theta_i}}$$

which, under conditions of Theorem 2, generalizes the corresponding estimate of P. L. Ul'janov [2] (p. 672=624). To this end it suffices to let $H = T$, $M = m_j$, $N = 0$ and $\alpha = \beta$, and to take the supremum with respect to τ on both sides.

By means of the same methods, generalizing Theorem 4 somewhat along the lines of Theorem 2, one may obtain a generalization of the estimate of P. L. Ul'janov, similar to the cited one, to the n-dimensional case for the full modulus of smoothness.

We also observe that for $J_i(T, H) = 0$ $(i = 1, \cdots, n)$ the generality of estimate (10) is not reduced if we assume that $\tau = T$.

REMARK 4. Estimates of type (8), (9) and (10) for $\theta_i = 1$ can be obtained in a simpler way by applying convolution estimate (2) with respect to h under the integral sign in (6) (see §10).

REMARK 5. In estimate (10) and its further corollaries, evidently one may increase l_i, i.e. assume that instead of condition (7) the condition

$$\sigma_i l_i \geqslant \left(\sigma, \frac{1}{p^0} - \frac{1}{q} + \alpha\right) \geqslant 0 \quad (i = 1, \ldots, n) \tag{7'}$$

is fulfilled.

§3. Difference properties of the extending function

We turn now to problems of extension of functions and to investigation of their properties. Formula (5) defines an extension of the function $f(x)$ outside the set U; estimates of such an extension are contained in (8). We restrict outselves to the case when $q = p^i = p \geqslant p^0$ and $\theta_i = 1$ $(i = 1, \cdots, n)$. We shall use a partition of unity for "piecing together" such extensions.

LEMMA 1. *Let $U \subset \widetilde{U}$ be open sets in E^n, let a function $\eta(x) \in C^\infty(E^n)$ have bounded derivatives up to a sufficiently high order, and let*

$$\operatorname{supp} \eta \subset \widetilde{U}^\varepsilon = \{x \in \widetilde{U},\ \rho(x, E^n \setminus \widetilde{U}) > \varepsilon\}, \quad \varepsilon > 0.$$

Let a function $f(x)$ *be given in* $U+R(\sigma)$, *and let the function* $\widetilde{f}(x)$ *be defined by fornula* (5). *Then for* $\sigma_i l_i = (\sigma, \alpha)$ $(i=1, \cdots, n)$

$$0 \leqslant \tau \leqslant T \leqslant H \leqslant H_0(\varepsilon), \qquad M \geqslant 0,$$

$$\| \Delta_j^M (\widetilde{U}, \tau^{\sigma_j}) D^\alpha (\eta \widetilde{f}) \|_p \leqslant C T^{\sigma_j M} H^{-\sigma_j M - \left(\sigma, \frac{1}{p^0} - \frac{1}{p} + \alpha\right)} \| f \|_{p^0(U+R)}$$
$$+ C \sum_{i=1}^n \int_0^T \| \Delta_i^{m_i} (U+R, \delta^{\sigma_i} t^{\sigma_i}) f \|_p t^{-1-(\sigma, \alpha)} dt + C \sum_{i=1}^n J_i(T, H),$$

where in the general case

$$J_i(T, H) \leqslant T^{\sigma_j M} \int_T^H \| \Delta_i^{m_i} (U+R, \delta^{\sigma_i} t^{\sigma_i}) f \|_p t^{-1-(\sigma, \alpha) - \sigma_j M} dt,$$

and $J_i(T, H) = 0$ *for* $\sigma_j M > \sigma_i m_i - \sigma_i l_i$.

PROOF. By virtue of the formula for a finite difference of the product of two functions (analogous to the Leibniz formula for differentiation of a product) we have

$$\| \Delta_j^M (\widetilde{U}, \tau^{\sigma_j}) \eta^{(\alpha-\beta)} D^\beta \widetilde{f} \|_p \leqslant C \sum_{0 \leqslant N \leqslant M} \tau^{\sigma_j (M-N)} \left\| \Delta_j^N \left(\widetilde{U}^{\frac{\varepsilon}{2}}, \tau^{\sigma_j} \right) D^\beta \widetilde{f} \right\|_{p(\widetilde{U})}.$$

Applying to each term of the sum estimate (8), we obtain the desired result.

THEOREM 3. *Let a region* $\Omega \subset E^n$ *satisfy the strong σ-horn condition, and let* $\sigma_i l_i = (\sigma, \alpha)$ $(i = 1, \cdots, n)$. *Then there exists a linear operation* $f(x) \to \widetilde{\widetilde{f}}(x)$ *of extension of functions on* E^n *such that for all functions with a finite right side*

$$\| \Delta_j^M (\tau^{\sigma_j}) D^\alpha \widetilde{\widetilde{f}} \|_p \leqslant C T^{\sigma_j M} H^{-\sigma_j M - \left(\sigma, \frac{1}{p^0} - \frac{1}{p} + \alpha\right)} \| f \|_{p^0(\Omega)}$$
$$+ C \sum_{i=1}^n \int_0^T \| \Delta_i^{m_i} (\Omega, \delta^{\sigma_i} t^{\sigma_i}) f \|_p t^{-1-(\sigma, \alpha)} dt + C \sum_{i=1}^n J_i(T, H), \tag{11}$$

where in the general case

$$J_i(T, H) \leqslant T^{\sigma_j M} \int_T^H \| \Delta_i^{m_i} (\Omega, \delta^{\sigma_i} t^{\sigma_i}) f \|_p t^{-1-(\sigma, \alpha) - \sigma_j M} dt,$$

and $J_i(T, H) = 0$ *for* $M\sigma_j > \sigma_i m_i - (\sigma, \alpha)$.

In this connection $\widetilde{f}(x) = f(x)$ $(x \in \Omega)$, $\widetilde{\widetilde{f}}(x) = 0$ *outside of an ϵ-neighborhood of* Ω *given in advance,* $0 \leqslant \tau \leqslant T \leqslant H \leqslant H(\Omega)$, $\sigma_i l_i = (\sigma, \alpha)$ $(i = 1, \cdots, n)$, $j = 1, \cdots, n$, $M = 0, 1, \cdots, M_0$ *and* $C = C(\Omega, \epsilon, M_0, p, p^0, \alpha)$.

PROOF. We construct the following system of infinitely differentiable functions $\{\eta_k(x)\}_1^K$, forming a partition of unity for the region Ω and corresponding to the covering $\{\Omega_k\}_1^K$:

$$0 \leqslant \eta_k(x) \leqslant 1 \quad \text{for} \quad x \in E^n,$$

$$\sum_{k=1}^{K} \eta_k(x) = 1 \quad \text{for} \quad x \in \Omega,$$

$$\eta_k(x) = 0 \text{ outside an } \frac{\epsilon}{2}\text{-neighborhood of } \Omega_k^{[\epsilon]},$$

$$|D^\alpha \eta_k(x)| \leqslant C \quad \text{for} \quad x \in E^n, \quad |\alpha| \leqslant N.$$

Set

$$\tilde{\tilde{f}}(x) = \sum_{k=1}^{K} \eta_k(x) \tilde{f}_k(x),$$

where $\tilde{f}_k(x)$ is the extension of the function $f(x), x \in \Omega_k + R_k(\sigma) \subset \Omega$, in accordance with Lemma 1 with Ω_k instead of U and $R_k(\sigma)$ instead of $R(\sigma)$.

By virtue of Lemma 1 and the properties of $\eta_k(x)$ we arrive at the proof of the theorem.

REMARK 6. By the same method one deduces an estimate for the extended function $\tilde{\tilde{f}}$ in terms of f, which is more general than (11) and which has the form of estimate (10) for $\beta = 0$ and $N = M$.

§4. Imbedding of classes of functions

As a corollary of Theorems 2 and 3 it is possible to obtain theorems on imbedding, extension and equivalence of norms of various kinds for the spaces H_p^l studied by Nikol'skiĭ [7], the spaces $B_{p,\theta}^l$ (see, for instance, [6]), and their generalizations connected with functionals of maximization type (see §9) introduced by K. K. Golovkin (see, for instance, [8]). Such theorems are obtained from Theorems 2 and 3 with the help of Minkowski's inequality. We state an imbedding theorem connected with a functional more general than a functional of maximization type.[1]

A nonnegative functional $N[\psi(T)] = N_T[\psi(T)]$, defined on nonnegative measurable functions $\psi(T), 0 < T < H < \infty$, and satisfying the following conditions for any $a \geqslant 0, b \geqslant 0, \delta > 0$:

1°. $N[a\psi_1(T) + b\psi_2(T)] \leqslant CaN[\psi_1(T)] + CbN[\psi_2(T)]$,

2°. $N[\psi_1(T)] \leqslant CN[\psi_2(T)]$ for $0 \leqslant \psi_1(T) \leqslant \psi_2(T)$,

3°. $N_T\left[\int_0^1 \psi(Tt)\, t^{-1+\delta} dt\right] + N_T\left[\int_1^{\frac{H}{T}} \psi(Tt)\, t^{-1-\delta} dt\right] \leqslant C(\delta) N[\psi(T)]$,

4°. $N[T^\delta] < \infty$,

5°. $\int_0^H \psi(t)\, t^{-1+\delta} dt \leqslant C(\delta) N[\psi(T)]$,

---

(1) We mention the work of A. S. Džafarov [9] on imbedding of classes of functions whose smoothness is characterized by a certain function with Lozinskiĭ's condition. Such a characterization is outside of the scope of a functional of maximization type.

will be called an *N-functional.*

Observe that condition 3° is fulfilled, for instance, if the following conditions hold:

3°_1. $N_T[\psi(Tt)] \leqslant CN_T[\psi(T)], \quad t \in (0, \infty)$.

3°_2. $N_T\left[\int_0^1 \psi(Tt)\, t^{-1+\delta} dt\right] \leqslant C\int_0^1 t^{-1+\delta} N_T[\psi(Tt)]\, dt$.

3°_3. $N_T\left[\int_1^\infty \psi(Tt)\, t^{-1-\delta} dt\right] \leqslant C\int_1^\infty t^{-1-\delta} N_T[\psi(Tt)]\, dt$.

The functionals $\sup_{0<T<H}\psi(T)$ and

$$J_\theta[\psi(T)] = \left\{\int_0^H \psi^\theta(T)\frac{dT}{T}\right\}^{\frac{1}{\theta}} \quad \text{for} \quad 1 \leqslant \theta < \infty$$

are examples of N-functionals which are functionals of maximization type.

An example of an N-functional which is not of maximization type is the functional

$$\left(\int_0^H |\psi(T)|^\theta \mu(T)\frac{dT}{T}\right)^{\frac{1}{\theta}}, \quad 1 \leqslant \theta < \infty,$$

with the condition that for any $\delta > 0$, $0 < t \leqslant 1$ and $0 < T \leqslant H$

$$0 < C_1(\delta)\, t^\delta \leqslant \frac{\mu(tT)}{\mu(T)} \leqslant C_2(\delta)\, t^{-\delta}.$$

In particular, $\mu(T) = \ln^s(2H/T)$, $-\infty < s < \infty$.

REMARK 7. In what follows the parameter $H > 0$ of an N-functional will be taken sufficiently small. We direct the reader's attention to the fact that the properties 1°, 2° and 3° of an N-functional are the most essential ones; they are defining properties and play the fundamental role. This will be seen below. One could define an N-functional for $H = +\infty$ with properties 1°, 2° and 3° preserved.

We rewrite inequality (10) for(2) $\theta_i = 1$ in the form (eliminating the parameter $\delta \in (0, 1)$ by changing the notation for T and H)

$$\sup_{0<\tau\leqslant T} \|\Delta_j^M(\Omega, \tau^{\sigma_j}) D^\alpha f\|_{q(\Omega)} \leqslant CT^{\sigma_j M} H^{-\sigma_j M - \left(\sigma, \frac{1}{p^0} - \frac{1}{q} + \alpha\right)} \|f\|_{p^0(\Omega)}$$
$$+ C\sum_{i=1}^n \left(\int_0^T + T^{\sigma_j M}\int_T^H t^{-\sigma_j M}\right) t^{-1-\sigma_i l_i} \|\Delta_i^{m_i}(\Omega, t^{\sigma_i}) f\|_{p_i} dt.$$

(2) Probably the utilization of estimate (10) for $\theta_i \geqslant 1$ would make it possible to somewhat enlarge the class of admissible functions.

We multiply both sides by $T^{-\sigma_j r_j}$ for $0 < r_j < M$ and apply $N_T[\ ]$:

$$N_T[T^{-\sigma_j r_j} \sup_{0<\tau\leqslant T} \|\Delta_j^M(\Omega, \tau^{\sigma_j}) D^\alpha f\|_q]$$

$$\leqslant CN_T\Big[T^{\sigma_j(M-r_j)} H^{-\sigma_j M-\left(\sigma, \frac{1}{p^0}-\frac{1}{q}+\alpha\right)} \|f\|_{p^0(\Omega)}\Big] + C\sum_{i=1}^n N_T\Bigg[\left(\int_0^1 + \int_1^{\frac{H}{T}} t^{-\sigma_j M}\right)$$

$$\times t^{\sigma_j r_j - 1}(Tt)^{-\sigma_i l_i - \sigma_j r_j}\|\Delta_i^{m_i}(\Omega, (Tt)^{\sigma_i}) f\|_{p_i} dt\Bigg] \tag{12}$$

$$\leqslant C_1 H^{-\sigma_j M-\left(\sigma, \frac{1}{p^0}-\frac{1}{q}+\sigma\right)} \|f\|_{p^0(\Omega)} + C_1\sum_{i=1}^n N_T\Big[T^{-\sigma_i l_i - \sigma_j r_j}\|\Delta_i^{m_i}(\Omega, T^{\sigma_i}) f\|_{p^i}\Big].$$

Estimation of $\|D^\alpha f\|_{q(\Omega)}$ is also carried out with the help of inequality (10) for $M = 0$ and of property 5° of an N-functional (which is used only here):

$$\|D^\alpha f\|_{q(\Omega)} \leqslant CH^{-\left(\sigma, \frac{1}{p^0}-\frac{1}{q}+\alpha\right)} \|f\|_{p^0(\Omega)} + C\sum_{i=1}^n \int_0^H t^{-1-l_i\sigma_i - r_j\sigma_j}\|\Delta_i^{m_i}(\Omega, t^{\sigma_i}) f\|_{p^i} t^{r_j\sigma_j} dt$$

$$\leqslant CH^{-\left(\sigma, \frac{1}{p^0}-\frac{1}{q}+\alpha\right)} \|f\|_{p^0(\Omega)} + C_1\sum_{i=1}^n N_T[T^{-l_i\sigma_i - r_j\sigma_j}\|\Delta_i^{m_i}(\Omega, T^{\sigma_i}) f\|_{p^i}]. \tag{13}$$

We summarize the results in the form of a theorem.

THEOREM 4. *Let a region $\Omega \subset E^n$ satisfy the σ-horn condition, let $1 \leqslant p^i \leqslant q \leqslant \infty$ $(i = 1, \cdots, n)$ and let conditions (7′) be fulfilled. Then for $0 < r_j < M$ and $0 < H \leqslant H(\Omega)$*

$$N_T[T^{-\sigma_j r_j} \sup_{0<\tau\leqslant T} \|\Delta_j^M(\Omega, \tau^{\sigma_j}) D^\alpha f\|_q \leqslant CH^{-\sigma_j M-\left(\sigma, \frac{1}{p^0}-\frac{1}{q}+\alpha\right)} \|f\|_{p^0(\Omega)}$$

$$+ C\sum_{i=1}^n N_T[T^{-\sigma_i l_i - \sigma_j r_j}\|\Delta_i^{m_i}(\Omega, T^{\sigma_i}) f\|_{p^i}],$$

$$\|D^\alpha f\|_{q(\Omega)} \leqslant CH^{-\left(\sigma, \frac{1}{p^0}-\frac{1}{q}+\alpha\right)} \|f\|_{p^0(\Omega)} + C\sum_{i=1}^n N_T[T^{-l_i\sigma_i - r_j\sigma_j}\|\Delta_i^{m_i}(\Omega, T^{\sigma_i}) f\|_{p^i}].$$

Theorem 4 shows that $D^\alpha f$ has in L_q "smoothness" r_j with respect to the variable x_j if f has in L_{p^i} "smoothness" $l_i + \sigma_j r_j/\sigma_i$ with respect to the variables x_i $(i = 1, \cdots, n)$, respectively.

COROLLARY. *For $p^0 \leqslant p^i = p = q$ $(i = 1, \cdots, n)$ and $\sigma_1 r_1 = \cdots = \sigma_n r_n > 0$, estimate* (12), *together with the inequality*

$$\|\Delta_i^{m_i+k_i}(\Omega, t) f\|_p \leqslant t^{k_i}\|\Delta_i^{m_i}(\Omega, t) D_i^{k_i} f\|_p$$

shows that for a fixed H the expressions

$$\|f\|_{p_0(\Omega)} + \sum_{i=1}^n N_T[T^{-\sigma_i(r_i-k_i)}\|\Delta_i^{m_i}(\Omega, T^{\sigma_i}) D_i^{k_i} f\|_p], \quad m_i + k_i > r_i > k_i \geqslant 0,$$

are equivalent for different m_i *and* k_i.

§5. Extension of classes of functions

We now study the problem of extension of functions $f(x), x \in \Omega \subset E^n$, to outside of Ω with preservation of properties. The basic part of this problem is estimation of the moduli of smoothness (11) for $q = p$. We shall assume that Ω satisfies the strong σ-horn condition.

As in derivation of estimates (12) and (13), we obtain the estimates

$$N_T\Big[T^{-\sigma_j r_j} \sup_{0<\tau\leqslant T} \|\Delta_j^M(\tau^{\sigma_j}) D^\alpha \tilde{\tilde{f}}\|_p\Big] \leqslant CH^{-\sigma_j M-\left(\sigma, \frac{1}{p^0}-\frac{1}{p}+\alpha\right)} \|f\|_{p^0(\Omega)} + C\sum_{i=1}^n N_T\Big[T^{-(\sigma,\alpha)-\sigma_j r_j}\|\Delta_i^{m_i}(\Omega, T^{\sigma_i})f\|_p\Big],$$

$$\|D^\alpha \tilde{\tilde{f}}\|_p \leqslant CH^{-\left(\sigma, \frac{1}{p^0}-\frac{1}{p}+\alpha\right)}\|f\|_{p^0(\Omega)} + C\sum_{i=1}^n N_T\Big[T^{-1-(\sigma,\alpha)-r_j\sigma_j}\|\Delta_i^{m_i}(\Omega, T^{\sigma_i})f\|_p\Big].$$

Estimates (14) and (15) imply, in particular,

THEOREM 5. *Let a region* $\Omega \subset E^n$ *satisfy the strong* σ*-horn condition, let* $m_i > r_i > 0$, *and let* $\sigma_i r_i = \text{const}$ $(i = 1, \cdots, n)$. *Then for functions* $f(x), x \in \Omega$, *with the finite functional* $(0 < H \leqslant H(\Omega))$

$$|||f|||_{N,r,\Omega} = \|f\|_{p(\Omega)} + \sum_{i=1}^n N_T\Big[T^{-\sigma_i r_i}\|\Delta_i^{m_i}(\Omega, T^{\sigma_i})f\|_p\Big]$$

there exists a linear extension $f \to \tilde{\tilde{f}}$ *to* E^n *for which*

$$|||\tilde{\tilde{f}}|||_{N,r,E^n} \leqslant C|||f|||_{N,r,\Omega}$$

with a constant C *independent of* f.

REMARK 8. From estimate (11) follow certain theorems on extension of classes of functions with prescribed majorants of moduli of smoothness and with extensions belonging to the same class. We state one of them.

Denote by $H^\kappa(\Omega)$, $\kappa = \kappa(t)$, the class of functions for which

$$\|f\|_{p(\Omega)} \leqslant K_f,$$

$$T^{-m_j\sigma_j}\|\Delta_j^{m_j}(\Omega, T^{\sigma_j})f\|_p \leqslant K_f\kappa(T), \quad T \in (0, H], \; j = 1, \ldots, n.$$

Moreover, let $m_i\sigma_i = a$ $(i = 1, \cdots, n)$ and

$$T^{-a}\int_0^T \kappa(t)t^{-1+a}dt + \int_T^H \kappa(t)t^{-1}dt \leqslant c\kappa(T),$$

$$T \in (0, H], \qquad i, j = 1, \ldots, n.$$

Then $\tilde{\tilde{f}}(x) \in H^\kappa(E^n)$ and $K_{\tilde{\tilde{f}}} \leqslant CK_f$, where the constant C does not depend on f.

§6. An integral representation in terms of differences with step of variable direction

Of some interest also are estimates in various metrics for differences with a step having the direction not only along the coordinates, but in an arbitrary (variable) direction. For such differences one could carry out investigations similar to those presented above. We shall restrict ourselves to some of those which are related to the problems on equivalence of classes of functions characterized by various types of properties of differences.

We state an integral representation of functions in terms of a finite difference with a vector-step of variable direction. The derivation of this representation in the one-dimensional case coincides with the one given in [6].

Let a function $\eta(x) \in C_0^\infty(E^n)$, $\operatorname{supp} \eta \subset \{\epsilon \mathbf{1} < x < \mathbf{1}, \epsilon > 0\}$, $\int \eta(x)dx = 1$, $\xi_i(x) = 2A\sigma_i x_i \eta(x)$, $0 < \delta \leqslant 1$, and let

$$\Psi(x) = A \sum_{k=0}^{m} \frac{(-1)^{m-k}}{(1+k\delta)^{n+1}} C_m^k \eta\left(\frac{x}{1+k\delta}\right),$$

$$\frac{1}{A} = \sum_{k=0}^{m} \frac{(-1)^k}{1+k\delta} C_m^k = \int_0^1 (1-t^\delta)^m \, dt > 0.$$

We set

$$f_{h^\sigma}(x) = h^{-|\sigma|} \int \Psi\left(\frac{y}{h^\sigma}\right) f(x+y)\, dy,$$

$$\hat{f}_{h^\sigma}(x) = (f_{h^\sigma})_{h^\sigma}(x) = h^{-2|\sigma|} \int \int \Psi\left(\frac{y}{h^\sigma}\right) \Psi\left(\frac{z}{h^\sigma}\right) f(x+y+z)\, dydz$$
$$= h^{-|\sigma|} \int \Psi_0\left(\frac{y}{h^\sigma}\right) f(x+y)\, dy.$$

It is clear that

$$\frac{\partial}{\partial h} \hat{f}_{h^\sigma}(x) = 2 \frac{\partial}{\partial h} (f_{t^\sigma})_{h^\sigma} \big|_{t=h}.$$

Further, we have

$$\frac{\partial}{\partial h} \int h^{-|\sigma|} \Psi\left(\frac{z}{h^\sigma}\right) g(x+z)\, dz$$

$$= \frac{\partial}{\partial h} \int A \sum_{k=0}^{m} \frac{(-1)^{m-k}}{1+k\delta} C_m^k \eta(z)\, g(x + (1+k\delta) h^\sigma z)\, dz$$

$$= \frac{1}{2} \sum_{i=1}^{n} \int h^{-|\sigma|-1} \xi_i \left(\frac{z}{h^\sigma}\right) h^{\sigma_i} \Delta^m(\delta z)\, D_i g(x+z)\, dz.$$

Substituting here $g(x) = f_{h^\sigma}(x)$ and applying the differentiation $D_i f_{h^\sigma}(x)$ to the kernel, we obtain

$$\frac{\partial}{\partial h}\hat{f}_{h^\sigma}(x) = -\sum_{i=1}^{n}\iint h^{-2|\sigma|-1}\xi_i\left(\frac{z}{h^\sigma}\right)\Psi_i\left(\frac{y}{h^\sigma}\right)\Delta^m(\delta z)\,f(x+y+z)\,dydz$$
$$= -\iint h^{-2|\sigma|-1}K\left(\frac{y}{h^\sigma},\frac{z}{h^\sigma}\right)\Delta^m(\delta z)\,f(x+y+z)\,dydz,$$

where $K(y, z) = \Sigma_1^n \Psi_i(y)\xi_i(z)$.

By virtue of the choice of A,

$$\iint \Psi(y)\Psi(z)\,dydz = 1,$$

from which it follows that $\hat{f}_{h^\sigma}(x) \to f(x)$ $(h \to 0)$ in the sense of L^{loc}.

With the help of the fundamental theorem of integral calculus, we now obtain the representation

$$f(x) = \hat{f}_{H^\sigma}(x) + \int_0^H\iint h^{-1-2|\sigma|}K\left(\frac{y}{h^\sigma},\frac{z}{h^\sigma}\right)\Delta^m(\delta z)\,f(x+y+z)\,dydzdh. \tag{16}$$

It is evident that with a suitable choice of the support of the function $\eta(x)$, any horn $x + R(\sigma)$ can be used as the body of the last representation.

Let $U \subset E^n$ be an open set and let a function $f(x)$ be given on $U + R(\sigma)$, $1 \leqslant \theta \leqslant \infty$. Set

$$\rho(z) = \left(\sum_1^n |z_i|^{\frac{1}{\sigma_i}}\right)^{\frac{|\sigma|}{n}}, \qquad \varphi(x, z) = \rho(z)^{-n-\mu}\Delta^m(U+R, \delta z)\,f(x+z), \quad \mu > 0,$$

let $f(x)$ be extended to outside $U + R$ by zero and denote the extended function by $f_0(x)$. Then the function $\tilde{f}(x)$ defined on E^n by

$$\begin{aligned}\tilde{f}(x) = &\int_{E_n} H^{-|\sigma|}\Psi_0\left(\frac{y}{H^\sigma}\right)f_0(x+y)\,dy + \\ &+\int_0^H\sum_{i=1}^n\iint_E h^{-1-2|\sigma|}\Psi_i\left(\frac{y}{h^\sigma}\right)\xi_i\left(\frac{z}{h^\sigma}\right)\rho(z)^{n+\mu}\varphi(x+y, z)\,dzdydh\end{aligned} \tag{17}$$

coincides with $f(x)$ on U.

§7. Estimates with differences of variable direction and the equivalence of classes of functions

We estimate the difference $\Delta^M(v^\sigma)D^\alpha\tilde{f}(x)$ in L_q for $0 \leqslant v_i \leqslant T \leqslant H$ $(i = 1, \cdots, n)$, where $v^\sigma = (v_1^{\sigma_1}, \cdots, v_n^{\sigma_n})$. First we estimate the quantities

$$A(y, z) = \int_0^T h^{-1-2|\sigma|-(\alpha,\sigma)}\chi\left(\frac{y}{h^\sigma}\right)\left|\xi_i\left(\frac{z}{h^\sigma}\right)\right|dh,$$

$$B(y, z) = \int_T^H h^{-1-2|\sigma|-(\alpha,\sigma)}\left|\frac{v^\sigma}{h^\sigma}\right|^M\chi\left(\frac{y}{h^\sigma}\right)\left|\xi_i\left(\frac{z}{h^\sigma}\right)\right|dh \leqslant$$

$$\leqslant |v_{\max}|^{M\sigma_{\min}} \int_T^H h^{-1-2|\sigma|-(\alpha,\sigma)-M\sigma_{\min}} \chi\left(\frac{y}{h^\sigma}\right) \left|\xi_i\left(\frac{z}{h^\sigma}\right)\right| dh,$$

for $0 \leqslant v_i \leqslant T \leqslant H$, where $\chi(x)$ is the characteristic function of the cube $\mathbf{0} < x < \mathbf{1}$, $v_{\max} = \max v_i$ and $\sigma_{\min} = \min \sigma_i$.

Since the support of ξ_i is bounded and it may be assumed that it is at a certain distance from the coordinate planes, it follows that $A(y, z)$ can differ from zero only for $\mathbf{0} \leqslant z \leqslant T^\sigma$, and $B(y, z)$ only for $\epsilon T^\sigma \leqslant z \leqslant H^\sigma$ $(\epsilon > 0)$. Hence $(y \geqslant \mathbf{0}, z \geqslant \mathbf{0})$

$$A(y, z) \leqslant C_1 \chi\left(\frac{z}{T^\sigma}\right) \int_{h^\sigma \geqslant y,\ h^\sigma \geqslant z} h^{-1-2|\sigma|-(\alpha,\sigma)} dh \leqslant C_2 \frac{\chi\left(\frac{z}{T^\sigma}\right)}{\left[\rho(y)^{\frac{n}{|\sigma|}} + \rho(z)^{\frac{n}{|\sigma|}}\right]^{2|\sigma|+(\alpha,\sigma)}}.$$

Similarly

$$B(y, z) \leqslant C_3 \left[\chi\left(\frac{z}{H^\sigma}\right) - \chi\left(\frac{z}{\varepsilon T^\sigma}\right)\right] \frac{|v|^{M\sigma_{\min}}}{\left[\rho(y)^{\frac{n}{|\sigma|}} + \rho(z)^{\frac{n}{|\sigma|}} + T\right]^{2|\sigma|+(\alpha,\sigma)+M\sigma_{\min}}}.$$

Hence for $M \geqslant 0$

$$|\Delta^M(v^\sigma) D^\alpha \tilde{f}(x)| \leqslant \int_{E^n} H^{-|\sigma|-(\alpha,\sigma)} \left|\Delta_y^M(-v^\sigma) \Psi_0^{(\alpha)}\left(\frac{y}{H^\sigma}\right) f_0(x+y)\right| dy$$

$$+ C_4 \int_0^{T^\sigma} \int_{E^n} \frac{\rho(z)^{n+\mu} |\Delta_x^M(v^\sigma) \varphi(x+y, z)| \, dy dz}{\left[\rho(y)^{\frac{n}{|\sigma|}} + \rho(z)^{\frac{n}{|\sigma|}}\right]^\nu}$$

$$+ C_4 |v_{\max}|^{M\sigma_{\min}} \int_{\varepsilon T^\sigma}^{H^\sigma} \int_{E^n} \frac{\rho(z)^{n+\mu} |\varphi^*(x+y, z)| \, dy dz}{\left[\rho(y)^{\frac{n}{|\sigma|}} + \rho(z)^{\frac{n}{|\sigma|}} + T\right]^{\nu+M\sigma_{\min}}},$$

$$\nu = \left(\sigma, \mathbf{1} - \frac{1}{p} + \frac{1}{q}\right) + |\sigma|\left(1 + \frac{\mu}{n}\right) = 2|\sigma| + (\alpha, \sigma), \quad \mu \geqslant 0, \quad \mathbf{1} \leqslant p \leqslant q \leqslant \infty,$$

$$\varphi^*(x, v^\sigma, z) = |\varphi(x, z)| \quad \text{for} \quad M = 0,$$

$$\varphi^*(x, v^\sigma, z) = \sum_{|\beta|=M} \int_0^1 \cdots \int_0^1 \left| c_\beta \varphi\left(x + \sum_{i=1}^n \gamma_i^{(\beta)} v_i^{\sigma_i} e_i + \sum_{i=1}^n \sum_{j=1}^{\beta_i} u_{ij} v_i^{\sigma_i} e_i\right)\right| \prod_{i=1}^n \prod_{j=1}^{\beta_i} du_{ij}.$$

To estimate the L_q norm of the left side of the last inequality, we apply (2) to each term of the right side. We evidently obtain $(p^0 \leqslant q)$

$$\|\Delta^M(v^\sigma) D^\alpha \tilde{f}(x)\|_q \leqslant CT^{M\sigma\min} H^{-M\sigma\min-\left(\sigma, \frac{1}{p^0}-\frac{1}{q}+\alpha\right)} \|f_0\|_{p^0}$$
$$+ C\int_0^{T^\sigma} \|\varphi(\cdot, z)\|_p\, dz + CT^{M\sigma\min} \int_{\varepsilon T^\sigma}^{H^\sigma} \|\varphi(\cdot, z)\|_p$$
$$\times \frac{\rho(z)^{n+\mu} dz}{\left(\rho(z)^{\frac{n}{|\sigma|}} + T\right)^{|\sigma|\left(1+\frac{\mu}{n}\right)+M\sigma\min}} \leqslant CT^{M\sigma\min} H^{-M\sigma\min-\left(\sigma, \frac{1}{p^0}-\frac{1}{q}+\alpha\right)} \|f_0\|_{p^0}.$$
$$+ C_1\int_0^{T^\sigma} \|\varphi(\cdot, z)\|_p\, dz + C_1 T^{M\sigma\min} \int_{T^\sigma}^{H^\sigma} \|\varphi(\cdot, z)\|_p\, \rho(z)^{-Mn\frac{\sigma\min}{|\sigma|}}\, dz.$$

THEOREM 6. *Let a region* $\Omega \subset E^n$ *satisfy the σ-horn condition, let* $1 \leqslant p \leqslant \infty$, *and let* $|\sigma|\mu/n = (\sigma, 1/p - 1/q + \alpha) \geqslant 0$. *Then for* $0 \leqslant v_i \leqslant T \leqslant H \leqslant H(\Omega)$

$$\|\Delta^M(\Omega, v^\sigma) D^\alpha f\|_{q(\Omega)} \leqslant CT^{M\sigma\min} H^{-M\sigma\min-\left(\sigma, \frac{1}{p_0}-\frac{1}{q}+\alpha\right)} \|f\|_{p^0}$$
$$+ C\left\{\int_{\rho(z)\leqslant T^{\frac{|\sigma|}{n}}} + T^{M\sigma\min} \int_{T^{\frac{|\sigma|}{n}} \leqslant \rho(z) \leqslant H^{\frac{|\sigma|}{n}}} \rho(z)^{-\frac{n}{|\sigma|}M\sigma\min}\right\} \tag{18}$$
$$\times \|\Delta^m(\Omega, \delta z) f\|_p \frac{dz}{\rho(z)^{n+\mu}}, \quad \rho(z) = \max |z_i|^{\frac{|\sigma|}{n\sigma_i}}.$$

In this connection $\sigma_{\min}$ *can be replaced by* σ_j, *if* $0 \leqslant v_j \leqslant T$ *and* $v_i = 0$ *for all* $i \neq j$.

Since $z = \rho(z)^{n\sigma/|\sigma|}\zeta$ and $\max |\zeta_i| = 1$ (see [10]), on changing to new variables in the last estimate (18) we obtain

$$\|\Delta_j^M(\Omega, v_j^{\sigma_j}) D^\alpha f\|_{q(\Omega)} \leqslant CT^{M\sigma_j} H^{-M\sigma_j-\left(\sigma, \frac{1}{p_0}-\frac{1}{q}+\alpha\right)} \|f\|_{p^0(\Omega)}$$
$$+ C\left(\int_{\rho\leqslant T^{\frac{|\sigma|}{n}}} + T^{M\sigma_j} \int_{T^{\frac{|\sigma|}{n}} \leqslant \rho \leqslant H^{\frac{|\sigma|}{n}}} \rho^{-\frac{n}{|\sigma|}M\sigma_j}\right) \tag{18'}$$
$$\times \int_{\max|\zeta_i|=1} \left\|\Delta^m\left(\Omega, \delta\rho^{\frac{n\sigma}{|\sigma|}}\zeta\right) f\right\|_p \frac{ds(\zeta)\, d\rho}{\rho^{\mu+1}}.$$

We multiply both sides by $T^{-\sigma_j r_j}$ and, after the substitution $\rho^{n/|\sigma|} = t$, we apply the functional $N_T[\ ]$, assuming $0 < r_j < M$:

$$
\begin{aligned}
&N_T\Big[T^{-\sigma_j r_j}\sup_{0<v_j\leqslant T}\|\Delta_j^M(\Omega,\ v_j^{\sigma_j})D^\alpha f\|_q\Big]\\
&\qquad\leqslant CN_T\Big[T^{\sigma_j(M-r_j)}H^{-M\sigma_j-\left(\sigma,\frac{1}{p^0}-\frac{1}{q}+\alpha\right)}\|f\|_{p^0(\Omega)}\Big]\\
&+CN_T\Bigg[(Tt)^{-\sigma_j r_j-\frac{|\sigma|}{n}\mu}\,t^{\sigma_j r_j-1}\Bigg(\int_0^1+\int_1^{\frac{H}{T}}t^{-M\sigma_j}\Bigg)\\
&\times\int_{\max|\zeta_i|=1}\|\Delta^m(\Omega,\ \delta(Tt)^\sigma\zeta)f\|_p\,ds(\zeta)\,dt\Bigg]\leqslant C_1H^{-\sigma_j M-\left(\sigma,\frac{1}{p_0}-\frac{1}{q}+\alpha\right)}\|f\|_{p^0(\Omega)} \qquad (19)\\
&+C_1N_T\Bigg[T^{-\sigma_j r_j-\frac{|\sigma|}{n}\mu}\int_{\max|\zeta_i|=1}\|\Delta^m(\Omega,\ \delta T^\sigma\zeta)f\|_p\,ds(\zeta)\Bigg].
\end{aligned}
$$

For $q=p$, $\alpha=\mathbf{0}$ and $\mu=0$ the last formula yields an estimate of coordinate differences in terms of differences with a step of variable direction in the same L_p norm relative to the N-functional.

We now turn to the derivation of the converse estimate. To this end we use representation (5) and results similar to those following (5). For $0\leqslant z_i\leqslant T\leqslant H$ and $\theta=1$ we obtain

$$
\begin{aligned}
|\Delta^M(z^\sigma)\tilde f(x)|&\leqslant\int_{E^n}H^{-|\sigma|}\left|\Delta_y^M(-z^\sigma)\Psi_0\left(\frac{y}{H^\sigma}\right)f_0(x+y)\right|dy\\
&+C\sum_{i=1}^n\int_0^{T^{\sigma_i}}\int_{E^n}\frac{t\,|\Delta^M(z^\sigma)\varphi_i(x+y,\ t)|\,dy\,dt}{\left[\sum_{j=1}^n|y_j|^{\frac{1}{\sigma_j}}+t^{\frac{1}{\sigma_i}}\right]^{|\sigma|+\sigma_i}}\\
&+C\sum_{i=1}^nT^{M\sigma_{\min}}\int_{\varepsilon T^{\sigma_i}}^{H^{\sigma_i}}\int_{E^n}\frac{\varphi^*(x+yv^\sigma,\ t)\,dy\,dt}{\left[\sum_{j=1}^n|y_j|^{\frac{1}{\sigma_j}}+t^{\frac{1}{\sigma_i}}+T\right]^{|\sigma|+\sigma_i+M\sigma_{\min}}}.
\end{aligned}
$$

Hence for $0\leqslant z_i\leqslant T\leqslant H\leqslant H(\Omega)$ we have

$$
\begin{aligned}
\|\Delta^M(\Omega,\ z^\sigma)f\|_p&\leqslant CT^{M\sigma_{\min}}H^{-M\sigma_{\min}-\left(\sigma,\frac{1}{p^0}-\frac{1}{p}\right)}\|f\|_{p^0(\Omega)}\\
&+C\sum_{i=1}^n\left(\int_0^T+T^{M\sigma_{\min}}\int_T^Ht^{-M\sigma_{\min}}\right)\|\Delta_i^{m_i}(\Omega,\ \delta^{\sigma_i}t^{\sigma_i})f\|_p\frac{dt}{t}.
\end{aligned}
\tag{20}
$$

Consequently the same estimate (with a different constant) holds also for

$$
\sup_{|\zeta_i|\leqslant 1}\|\Delta^M(\Omega,\ T^\sigma\zeta)f\|_{p(\Omega)}.
$$

We multiply both sides by $T^{-\sigma_j r_j}$ for $M\sigma_{\min}>r_j\sigma_j$ and apply the N-functional:

$$N_T\Big[T^{-\sigma_j r_j}\sup_{|\zeta_i|\leqslant 1}\|\Delta^M(\Omega,\ T^\sigma\zeta)f\|_p\Big]\leqslant C_1H^{-M\sigma\min-\left(\sigma,\frac{1}{p^0}-\frac{1}{p}\right)}\|f\|_{p^0(\Omega)}$$

$$+C_1\sum_{i=1}^n N_T\Bigg[\Bigg(\int_0^1+\int_1^{\frac{H}{T}}t^{-M\sigma\min}\Bigg)\|\Delta_i^{m_i}(\Omega,\ \delta^{\sigma_i}T^{\sigma_i}t^{\sigma_i})f\|_p$$

$$\times(Tt)^{-\sigma_j r_j}t^{\sigma_j r_j-1}dt\Bigg]\leqslant C_1H^{-M\sigma\min-\left(\sigma,\frac{1}{p^0}-\frac{1}{p}\right)}\|f\|_{p^0(\Omega)} \tag{21}$$

$$+C_2\sum_{i=1}^n N_T[\|\Delta_i^{m_i}(\Omega,\ \delta^{\sigma_i}T^{\sigma_i})f\|_p T^{-\sigma_j r_j}].$$

§8. Construction of functions from a given trace

In the left side of inequality (10) some coordinates q_{j_k} $(k=1,\cdots,m)$ of the vector q can be equal to infinity. In this case estimate (10) for $j\neq j_k$ yields an estimate of the trace of the function $f(x)$, $x\in\Omega$, on the intersection of Ω and the plane $x_{j_k}=x^0_{j_k}$ $(k=1,\cdots,m)$. We shall now consider the converse problem, i.e. the problem of constructing a best extension (in a certain sense) of a function, given in an m-dimensional space, to a function in an n-dimensional space $(n>m)$, and of establishing estimates of difference properties of this extension. A similar problem was first posed and solved by S. M. Nikol'skiĭ [7] for the H-classes. In our paper we shall use a modification of this method.

Let

$$x=(x_1,\ldots,x_n),\qquad x'=(x_{n+1},\ldots,x_{n+m}),\qquad \sigma=(\sigma_1,\ldots,\sigma_n),$$
$$\sigma'=(\sigma_{n+1},\ldots,\sigma_{n+m}),\ p=(p_1,\ldots,p_n),\ p'=(p_{n+1},\ldots,p_{n+m}).$$

Let a function $f(x)$ be defined on $U+R$ and let representation (4), (5) be valid on U. On the basis of this representation we construct the function

$$f(x,\ x')=\prod_{k=n+1}^{n+m}\frac{x_k^{\beta_k}}{\beta_k!}\mu\left(\frac{x_k}{H^{\sigma_k}}\right)\int_{E^n}H^{-|\sigma|}\Psi_0\left(\frac{y}{H^\sigma}\right)f_0(x+y)\,dy$$

$$+\int_0^H\sum_{i=1}^n\int_{E^n}\int_0^\infty\prod_{k=n+1}^{n+m}\frac{x_k^{\beta_k}}{\beta_k!}\mu\left(\frac{x_k}{h^{\sigma_k}}\right)h^{-1-|\sigma|-\sigma_i} \tag{22}$$

$$\times\Psi_i\left(\frac{y}{h^\sigma}\right)\xi_i\left(\frac{t}{h^{\sigma_i}}\right)t^{1+l_i}\varphi(x+y+te_i,\ t)\,dt\,dy\,dh,$$

where $\mu(t)\in C_0^\infty(E^1)$, $\mu(t)=1$ for $|t|<\epsilon$ and $\mu(t)=0$ for $|t|>2\epsilon$ $(\epsilon>0)$.

From the construction of the function $f(x,x')$ it is clear that $D^{\alpha'}f(x,0)=0$ for $\alpha'\neq\beta'$, and $D^{\beta'}f(x,0)=f(x)$, since in the last case the right side of (22) becomes representation (5). We estimate $D^{(\alpha,\alpha')}f(x,x')$ for $(x,x')\in\widetilde{U}=U\times E^m$ and $0<\tau\leqslant T\leqslant H\leqslant H_0$:

$$\|\Delta_j^M(\bar{U},\ \tau^{\sigma_j})D^{(\alpha,\alpha')}f\|_{(p,p')} \leqslant C(H)T^{\sigma_j M}\|f\|_{p_0(U+R)}$$

$$+C\sum_{i=1}^{n}\left\| \int_0^{T^{\sigma_i}}\int_{E^n} \frac{t^{1+l_i}|\varphi_i(x+y,\ t)|\,dydt}{\left[\sum_{j=1}^{n}|y_j|^{\frac{1}{\sigma_j}}+\sum_{j=n+1}^{n+m}|x_j|^{\frac{1}{\sigma_i}}+t^{\frac{1}{\sigma_i}}\right]^{\lambda_i}}\right\|_{(p,p')}$$

$$+C\sum_{i=1}^{n}\left\| T^{\sigma_j M}\int_{\varepsilon T^{\sigma_i}}^{H^{\sigma_i}}\int_{E^n} \frac{t^{1+l_i}|\varphi_i(x+y,\ t)|\,dydt}{\left[\sum_{j=1}^{n}|y_j|^{\frac{1}{\sigma_j}}+\sum_{j=n+1}^{n+m}|x_j|^{\frac{1}{\sigma_j}}+t^{\frac{1}{\sigma_i}}+T\right]^{\lambda_i+\sigma_i M}}\right\|_{(p,p')}$$

where $\lambda_i = |\sigma| + \sigma_i + \sigma_i l_i + (\sigma', 1/p')$, and l_i is defined by the condition

$$\sigma_i l_i = (\alpha,\ \sigma) + (\alpha',\ \sigma') - (\beta',\ \sigma') - \left(\sigma',\ \frac{1}{p'}\right) \geqslant 0, \qquad (i = 1, \ldots, n). \tag{23}$$

Consequently

$$\|\Delta_j^M(\bar{U},\ \tau^{\sigma_j})D^{(\alpha,\alpha')}f\|_{(p,p')} \leqslant C(H)T^{\sigma_j M}\|f\|_{p^0(U+R)}$$

$$+C_1\sum_{i=1}^{n}\int_0^{T^{\sigma_i}}\frac{t^{1+l_i}\|\varphi_i(\cdot,\ t)\|_p\,dt}{\left[\frac{1}{t^{\sigma_i}}\right]^{\sigma_i+\sigma_i l_i}} + C_1\sum_{i=1}^{n}T^{\sigma_j M}\int_{\varepsilon T^{\sigma_i}}^{H^{\sigma_i}}\frac{t^{1+l_i}\|\varphi_i(\cdot,\ t)\|_p\,dt}{\left[\frac{1}{t^{\sigma_i}}+T\right]^{\sigma_i+\sigma_i l_i+\sigma_j M}}$$

$$\leqslant C(H)T^{\sigma_j M}\|f\|_{p^0(U+R)} + C_2\sum_{i=1}^{n}\left(\int_0^T + T^{\sigma_j M}\int_T^H t^{-\sigma_j M}\right)$$

$$\times\|\Delta_i^{m_i}(U+R,\ \delta^{\sigma_i}t^{\sigma_i})f\|_p\frac{dt}{t^{1+\sigma_i l_i}}.$$

In the last estimate, after a change of notation for T and H, we may assume that $\delta = 1$.

Applying a partition of unity for $\Omega \subset E^n$, we arrive at the following theorem.

THEOREM 7. *Let a region $\Omega \subset E^n$ satisfy the σ-horn condition. If condition (23) is fulfilled, then with the function $f(x)$, $x \in \Omega$, a function $f_{\beta'}(x, x')$, $(x, x') \in \Omega \times E^m$, can be associated in a linear manner such that $D^{\alpha'}f_{\beta'}(x, 0) = 0$ for $\alpha' \neq \beta'$, $D^{\beta'}f_{\beta'}(x, 0) = f(x)$, and for $0 < T \leqslant H \leqslant H(\Omega)$, $j = 1, \cdots, n+m$*

$$\sup_{0<\tau\leqslant T}\|\Delta_j^M(\Omega\times E^m,\ \tau^{\sigma_j})D^{(\alpha,\alpha')}f_{\beta'}\|_{(p,p')} \leqslant C(H)T^{\sigma_j M}\|f\|_{p^0(\Omega)}$$

$$+C\sum_{i=1}^{n}\left(\int_0^T + T^{\sigma_j M}\int_T^H t^{-\sigma_j M}\right)\|\Delta_i^{m_i}(\Omega,\ t^{\sigma_i})f\|_p\frac{dt}{t^{1+\sigma_i l_i}}.$$

As a corollary we obtain a corresponding theorem on estimates of the extending function in terms of the N-functionals.

We multiply the last inequality by $T^{-\sigma_j r_j}$ for $0 < r_j < M$ $(1 \leqslant j \leqslant n+m)$, replace t by Tt and apply the N-functional:

$$N_T\Big[T^{-\sigma_j r_j} \sup_{0<\tau\leqslant T} \|\Delta_j^M(\Omega \times E^m, \tau^{\sigma_j}) D^{(\alpha, \alpha')} f_{\beta'}\|_{(p, p')}\Big] \leqslant C_1(H)\|f\|_{p^0(\Omega)} + C_1 \sum_{i=1}^n N_T[T^{-\sigma_i l_i - \sigma_j r_j}\|\Delta_i^{m_i}(\Omega, T^{\sigma_i}) f\|_p]. \tag{24}$$

From (23) for $M = 0$ we also obtain

$$\|D^{(\alpha, \alpha')} f_{\beta'}\|_{(p, p')\,(\Omega \times E^m)} \leqslant C_1(H)\|f\|_{p^0(\Omega)} + C_1 \sum_{i=1}^n N_T[T^{-\sigma_i l_i - \sigma_j r_j}\|\Delta_i^{m_i}(\Omega, T^{\sigma_i}) f\|_p]. \tag{25}$$

§9. On a constructive method of proving imbedding theorems

The general theory of imbedding of function spaces of differentiable functions of several real variables was initiated by S. L. Sobolev. He studied the function spaces $W_p^{(l)}$ with the norm $\Sigma_{|\alpha|\leqslant l}\|D^\alpha f\|_{p(\Omega)}$.

Imbedding theorems of spaces of differentiable functions with different properties of different variables and arbitrary positive indices of differentiability were first obtained by S. M. Nikol'skiĭ [7]. He proved imbedding theorems for the function spaces $H_p^r(E^n)$, $r = (r_1, \cdots, r_n)$, $r_i > 0$, with L_p-summable unmixed derivatives $D_i^{k_i} f$ $(r_i - 1 \leqslant k_i < r_i)$ satisfying the Hölder condition in L_p of degree $r_i - k_i$.

His method is based on the approximation of functions of several real variables by entire functions of finite degree (and in the periodic case by trigonometric polynomials). We shall point out two basic aspects of this method.

The first one is the equivalence of differential-difference properties of a class of functions and the rate of convergence to zero of the best approximations or approximations by a special linear method. In the simplest case, this is Jackson's Theorem and its converse obtained by S. N. Bernšteĭn.

Thus for functions $f(x)$ of one variable $x \in (-\infty, \infty)$ one constructs as approximating machinery entire functions (of exponential type) of degree n:

$$G_n(f; x) = n \int_{-\infty}^{\infty} g(ny)[f(x) + (-1)^{s+1}\Delta^s(y) f(x)]\, dy, \tag{26}$$

where $g(x)$ is an entire function of the first degree, bounded on the real line, decreasing polynomially for $x \to \pm\infty$ and such that

$$\int_{-\infty}^{\infty} g(x)\, dx = 1. \tag{27}$$

From (26) and (27) it evidently follows that for $0 < r < s$ the condition

$$\sup_y |y|^{-r} \| \Delta^s(y) f(x) \|_{L_p} \leqslant M \tag{28}$$

implies

$$\sup_{n=1,2\ldots} n^r \| f - G_n \|_{L_p} \leqslant CM. \tag{29}$$

With the help of the representation

$$f = G_1 + \sum_{n=0}^{\infty} (G_{2^{n+1}} - G_{2^n}) \tag{30}$$

one shows [7] that (29), in turn, implies (28).

The second central stage of the proof is estimating the norms of derivatives of the approximating functions in other metrics, and the norms of their traces on hyperplanes of lesser dimensions in terms of the norm of the function itself. These estimates [7] have the following form $(1 \leqslant p \leqslant q \leqslant \infty)$:

$$\| D^\alpha G_{\nu_1, \ldots, \nu_n} \|_{p(E^n)} \leqslant \nu_1^{\alpha_1} \ldots \nu_n^{\alpha_n} \| G_{\nu_1, \ldots, \nu_n} \|_{p(E^n)},$$

$$\| G_{\nu_1, \ldots, \nu_n} \|_{q(E^n)} \leqslant C \left(\prod_{i=1}^{n} \nu_i \right)^{\frac{1}{p} - \frac{1}{q}} \| G_{\nu_1, \ldots, \nu_n} \|_{p(E^n)},$$

$$\sup_{x_{m+1}, \ldots, x_n} \| G_{\nu_1, \ldots, \nu_n} \|_{p(E^m)} \leqslant \left(\prod_{i=m+1}^{n} \nu_i \right)^{\frac{1}{p}} \| G_{\nu_1, \ldots, \nu_n} \|_{p(E^n)}.$$

Estimates similar to the indicated ones for entire functions exist for trigonometric polynomials also in other cases, which makes it possible to construct a theory based on these estimates. Thus Ja. S. Bugrov in [11] obtained and used similar estimates for harmonic polynomials.

Estimates of this type exist also for *averaging functions* with the help of a kernel with compact support of finite or infinite smoothness. Specifically, if $h = (h_1, \cdots, h_n)$, $h_i > 0$, $y/h = (y_1/h_1, \cdots, y_n/h_n)$, and $K(x)$ is a function with compact support, differentiable sufficiently many times, $1 \leqslant p \leqslant q \leqslant \infty$ and $1 \leqslant m \leqslant n$, then for

$$\varphi_h(x) = \int \prod_{i=1}^{n} \frac{1}{h_i} K\left(\frac{y}{h}\right) \varphi(x+y)\, dy, \tag{31}$$

we have

$$\sup_{x_{m+1}, \ldots, x_n} \| D^\alpha \varphi_h \|_{q(E^m)} \leqslant C \left(\prod_1^n h_i^{-\alpha_i} \right) \left(\prod_1^n h_i^{-\frac{1}{p}} \right) \left(\prod_1^m h_i^{\frac{1}{q}} \right) \| \varphi \|_{p(E^n)}. \tag{32}$$

Estimate (32) is derived with the help of Hölder's inequality for three functions. It was actually used by V. P. Il'in [12], K. K. Golovkin [8] and others.

We take a closer look at the first problem. An estimate of the rate of approximation by differential-difference properties of a function can be obtained not only for the class of functions (28), but also for an arbitrary function in terms of its integral modulus of smoothness (see, for instance, [1], Russian p. 274)

$$\left\{\int_{\infty}^{\infty} |f(x) - G_n(f;\ x)|^p\,dx\right\}^{\frac{1}{p}} \leqslant C\omega_s\left(f;\ \frac{1}{n}\right)_p. \tag{33}$$

However, this estimate, in general, has no converse, i.e. the information on the rate of decrease of approximations to zero contained in (33) is not sufficient for estimation of the right side of (33) in terms of the left side. An attempt to obtain a two-sided estimate is unsuccessful without an estimate of the type

$$\int_0^h \frac{\omega(t)}{t}\,dt \leqslant C\omega(h), \qquad 0 < h < \varepsilon, \tag{34}$$

for $\omega(t) = \omega(f;\ t)$, which does not hold for all functions $\omega(t)$.(3) For instance, it is not true for the continuous increasing function $\omega(t) = \ln^{-2} t$.

In this connection one introduces a functional Φ for which, for instance, one now has

$$\Phi\left[h^{-r}\int_0^h \frac{\omega(f;\ t)}{t}\,dt\right] \leqslant C\Phi\,[h^{-r}\omega(f;\ h)], \tag{35}$$

whenever the right side is finite. This functional determines the class of functions for which a constructive method of proving imbedding theorems and theorems on extension of functions can be carried out without loss of differential properties. In [7] such a functional is

$$J_\infty[\psi(h)] = \sup_{h>0} \psi(h).$$

In [13] we studied with the help of approximation by entire functions the spaces $B^r_{p,\theta}(E^n)$, characterized with the help of the functional

$$J_\theta[\psi(h)] = \left\{\int_0^H |\psi(h)|^\theta \frac{dh}{h}\right\}^{\frac{1}{\theta}}, \qquad 1 \leqslant \theta < \infty.$$

For similar function spaces and their generalizations (pertaining to vector-values of the index θ and others), V. P. Il'in [12] obtained imbedding theorems for regions with the help of integral representation of functions which he established. Despite a

---

(3) We observe that the property $t^{-1}\omega(f;\ t)$ implies that the converse inequality of inequality (34) is always valid for $\omega(t) = \omega(f;\ t)$ ([1], §3.2.4).

different approach, his method fits into the same scheme of approximation of functions (by their averages), of representation of a function of type (30) (but in the form of an integral), and of application of estimates (32). In this connection for averaged functions one may use construction (26) with n replaced by $1/h$ and $g(y)$ by a sufficiently smooth function $\xi(y)$ with a compact support in $(0, 1)$ (see [8, 12, 14]).(4) We note that expansion into averaged functions to prove imbedding theorems was first used by A. A. Dezin [15].

We also remark that the method of approximation by averaged functions has the advantage that for the construction of an average at a point one uses the values of the function to be averaged on a part of a small neighborhood of the given point. This makes it possible to study classes of functions defined in a region.

K. K. Golovkin [8], using the method of approximation by averaging, obtained imbedding theorems on $\Omega = E^m \times (0, \infty)^{n-m}$ for more general classes of functions, defined with the help of a "special functional of maximization type", i.e. a functional $J = J[\psi(h)]$ satisfying the following conditions $(0 < h < \infty)$;

$$1^\circ.\ J[\psi_1(h) + \psi_2(h)] \leqslant J[\psi_1(h)] + J[\psi_2(h)].$$
$$2^\circ.\ J[a\psi(h)] = |a|\, J[\psi(h)] = |a|\, J[|\psi(h)|].$$
$$3^\circ.\ \psi(h) \geqslant \varphi(h) \geqslant 0 \text{ implies } J[\psi(h)] \geqslant J[\varphi(h)].$$
$$4^\circ.\ J[\psi(kh)] = J[\psi(h)], \qquad k > 0.$$
$$5^\circ.\ J[\psi(h^\alpha)] \sim J[\psi(h)].$$

Though the functionals J_∞ and J_θ are the most useful and interesting (see [7, 12, 13] and others), we have made an attempt in the present article to single out possibly a wider class of functions (N-functionals) which admit an application of the constructive method in obtaining theorems on imbedding, extension and equivalence. Without considering the rate of convergence of approximations (which is an intermediate link in the constructive method), we made use of a representation of functions in the form of integrals of differences to obtain corresponding estimates. (This approach was used before by V. P. Il'in [12].)(5)

§10. Possible generalizations of results

The preceding results cover classes of functions such that for each of these classes the majorants of the moduli of smoothness of coordinate directions are related to one another (roughly speaking) as powers of the step with different exponents. Many of them are also valid for more general relations between the moduli of smoothness of

---

(4) In connection with what has been pointed out, K. K. Golovkin's statement in the introduction to [14], asserting that convenience of approximating by averaging functions is comparable to that of approximation by entire functions, is surprising.

(5) *Added in Proof.* We mention the work by Ju. A. Brudnyĭ and V. K. Šalašov [17], in which in estimates of type (35) one takes different functionals on the left and on the right sides.

coordinate directions. We shall demonstrate this in one case generalizing the imbedding Theorems 2 and 4.

We remark that relations between smoothness of different coordinate directions which are not those of a power were first studied in articles by A. S. Džafarov (see, for instance, [16], where one considers power relations with logarithmic additional terms).

Let $\sigma(h) = (\sigma_1(h), \cdots, \sigma_n(h))$ be a vector-function defined for $h \in [0, \infty]$, and let $\sigma_i(0) = 0$ and $\sigma_i'(h) > 0$ $(h > 0; i = 1, \cdots, n)$.

Consider the average $f_{\sigma(h)}(x)$ with the step $\sigma_i(h)$ with respect to x_i:

$$f_{\sigma(h)}(x) = \int\int \prod_{i=1}^{n} \chi_i(y_i)\chi_i(z_i) f(x + \sigma(h)y + \sigma(h)z)\,dydz,$$

where $\sigma(h)y$ and $\sigma(h)z$ are coordinatewise products of the vectors,

$$\chi_i(t) = \frac{1}{A_i}\sum_{k=0}^{m_i} \frac{(-1)^{m_i-k}}{(1+k\delta_i)^2}\binom{m_i}{k}\eta_i\left(\frac{t}{1+k\delta_i}\right),$$

$0 < \delta < 1$, the functions $\eta_i(t) \in C_0^\infty$ and have support in a segment $[\gamma_i', \gamma_i''] \subset (0, 1)$ which is given in advance, $\xi_i(t) = t\eta_i(t)$, and

$$A_i = \int_0^1 (1 - t^{\delta_i})^{m_i}\,dt = \sum_{k=0}^{m_i} \frac{(-1)^k}{1+k\delta_i}\binom{m_i}{k} > 0.$$

For $f \in L^{\mathrm{loc}}$, $f_{\sigma(h)} \to f$ $(h \to 0)$, at least in the sense of L^{loc}. As in the derivation of (4), we obtain the representation

$$\begin{aligned} f(x) = f_{\sigma(H)}(x) &+ \int_0^H \sum_{i=1}^n \int_{E^n}\int_0^\infty \frac{\sigma_i'(h)}{\sigma_i^2(h)} \prod_1^n \frac{1}{\sigma_j(h)} \\ &\times \Psi_i\left(\frac{y}{\sigma(h)}\right)\xi_i\left(\frac{t}{\sigma_i(h)}\right)\Delta_i^{m_i}(\delta_i t) f(x + y + te_i)\,dtdydh. \end{aligned} \tag{36}$$

Here $\Psi_i(x) \in C_0^\infty(E^n)$ and they have their supports in a subcube of $\{x, 0 < x < 1\}$ given in advance. As a consequence of this, the body of the representation in (36) is the "$\sigma(h)$-horn"

$$\mathscr{R} = \mathscr{R}(\sigma) = \bigcup_{0<h<H} \{x;\ 0 < a_i\sigma_i(h) < x_i < b_i\sigma_i(h) \quad (i = 1, \ldots, n)\}$$

(with numbers $b_i > a_i > 0$ given in advance), generalizing the concept of the σ-horn (3). Of course, the concepts of a region satisfying the $\sigma(h)$-horn condition and the strong $\sigma(h)$-horn condition can also be generalized.

From (36) we obtain the following estimate for the difference for $0 < \tau \leqslant \sigma_j(T)$:

$$\begin{aligned}\Delta_j^M(\tau)\,D^\alpha f(x)| &\leqslant |\Delta_j^M(\tau)\,D^\alpha f_{\sigma(H)}(x)| \\ &+\sum_{i=1}^n\int_{E^n}\int_0^\infty\Bigg\{\int_0^T\left|\Delta_{y_j}^M(-\tau)\,\Psi_i^{(\alpha)}\left(\frac{y}{\sigma(h)}\right)\right|+\int_T^H\left[\frac{\tau}{\sigma_j(h)}\right]^M \\ &\times\int_0^M\left|\Psi_i^{(\alpha+Me_j)}\left(\frac{y-\tau u e_j}{\sigma(h)}\right)\right|du\Bigg\}\frac{\sigma_i'(h)}{\sigma_i^2(h)}\prod_{k=1}^n\sigma_k(h)^{-1-\alpha_k} \\ &\times\left|\xi_i\left(\frac{t}{\sigma_i(h)}\right)\right|\cdot|\Delta_i^{m_i}(\delta_i t)\,f(x+y+te_i)\,dh\,dt\,dy.\end{aligned}\tag{37}$$

Applying to (37) the estimate for convolution (2) with respect to t and h under the integrals, for $0<\tau\leqslant\sigma_j(T)$, $T\leqslant H$, we obtain

$$\begin{aligned}\|\Delta_j^M(U,\tau)\,D^\alpha f\|_{q(U)} &\leqslant C\left[\frac{\tau}{\sigma_j(H)}\right]^M\Lambda_0(H)\,\|f\|_{p^0(U+\mathscr{R})} \\ &+C\sum_{i=1}^n\left\{\int_0^T+\int_T^H\left[\frac{\tau}{\sigma_j(h)}\right]^M\right\}\Lambda_i(h)\,\frac{\sigma_i'(h)}{\sigma_i(h)}\int_{\delta_i\varepsilon}^{\delta_i}\|\Delta_i^{m_i}[U+\mathscr{R},\,t\sigma_i(h)]\,f\|_{p^i}\,dt\,dh,\end{aligned}\tag{38}$$

where it is natural to call the factors

$$\Lambda_i(h)=\prod_{k=1}^n\sigma_k(h)^{\frac{1}{q_k}-\frac{1}{p_k^i}-\alpha_k}$$

the imbedding factors.

Estimate (38) yields the following theorem.

THEOREM 8. *Let a region $\Omega\subset E^n$ satisfy the $\sigma(h)$-horn condition, $1\leqslant p^i\leqslant q\leqslant\infty$. Then for $0<\tau<\sigma_j(T)$ and $0<T\leqslant H\leqslant H(\Omega)$ the estimate*

$$\begin{aligned}\omega_j^M(\Omega,\sigma_j(T),D^\alpha f)_q &\leqslant C\left[\frac{\sigma_j(T)}{\sigma_j(H)}\right]^M\Lambda_0(H)\,\|f\|_{p^0(\Omega)} \\ &+C\sum_{i=1}^n\left\{\int_0^T+\int_T^H\left[\frac{\sigma_j(T)}{\sigma_j(h)}\right]^M\right\}\Lambda_i(h)\,\frac{\sigma_i'(h)}{\sigma_i(h)}\,\bar\omega_i^{m_i}(\Omega,\sigma_i(h),f)_{p^i}\,dh\end{aligned}\tag{39}$$

holds, where

$$\omega_i^{m_i}(\Omega,\tau,f)_p=\sup_{0<t\leqslant\tau}\|\Delta_i^{m_i}(\Omega,t)\,f\|_p,$$

$$\bar\omega_i^{m_i}(\Omega,\tau,f)_p=\int_0^1\|\Delta_i^{m_i}(\Omega,t\tau)\,f\|_p\,dt.$$

In certain cases under more accurate examination it is possible to obtain an estimate more precise than (38) or (39). Namely, in estimating the L_p norm of the integrals $\int_0^T\cdots dt$ in (37), we apply the estimate for convolution (2) with respect to the first $n-1$ variables only. Then, instead of the first term under the summation sign in (38), we shall have the L_{q_n} norm

$$g(x_n)=\int_0^T\int_0^\infty\int_0^\infty \psi_i\left(\frac{y_n}{\sigma_n(h)}\right)\frac{\sigma_i'(h)}{\sigma_i(h)^2}\Lambda_i(h)$$

$$\times\, \sigma_n(h)^{-1+\frac{1}{p_n^i}-\frac{1}{q_n}}\left|\xi_i\left(\frac{t}{\sigma_i(h)}\right)\right|\varphi_i(x_n+y_n,\ \delta_i t)\,dt\,dy\,dh,$$

where $\psi_i(x_n)$ is the characteristic function of the support

$$\sup_{x_1,\,\ldots,\,x_{n-1}}\Psi_i^{(\alpha)}(x),\ \|\varphi_i(\cdot,\ \delta_i t)\|_{p_n^i}\leqslant\|\Delta_i^{m_i}(U+\mathscr{R},\ \delta_i t)f\|_{p^i}.$$

We have, further,

$$g(x_n)\leqslant C\int_0^{\sigma_i(T)}\int_0^{\sigma_n(T)}\int_{\max[\sigma_n^{-1}(y_n),\,\sigma_i^{-1}(t)]}^{T}\frac{\sigma_i'(h)}{\sigma_i(h)^2}\Lambda_i(h)$$

$$\times\,\sigma_n(h)^{-1+\frac{1}{p_n^i}-\frac{1}{q_n}}\,dh\,\varphi_i(x_n+y_n,\ \delta_i t)\,dy_n\,dt.$$

Already in the case $\sigma_i(h)=h^{\sigma_i}$ $(i=1,\cdots,n)$ for $1\leqslant p_n^i<q_n<\infty$ and $1\leqslant\theta_i\leqslant q_n$, by virtue of Theorem 1 (with $n=2$) we obtain estimate (10).

We divide both sides of (39) by a nondecreasing function $r_j(T)$, $r_j(0)=0$, and apply the N^*-functional with respect to T. We set

$$a_{ij}(h)=\sup_{0<T\leqslant H}\frac{r_j(Th)}{r_j(T)}\frac{Th\sigma_i'(Th)}{\sigma_i(Th)}\qquad(0<h\leqslant 1),$$

$$b_{ij}(h)=\sup_{0<T\leqslant H}\left[\frac{\sigma_j(T)}{\sigma_j(Th)}\right]^M\frac{r_j(Th)}{r_j(T)}\frac{Th\sigma_i'(Th)}{\sigma_i(Th)}\qquad\left(1\leqslant h\leqslant\frac{H}{T}\right)$$

and require that conditions 1° and 2° for an N-functional be fulfilled for the N^*-functional, and also that the conditions

3*. $$N_T^*\left[\int_0^1\psi(Tt)a_{ij}(t)\frac{dt}{t}\right]+N_T^*\left[\int_0^{\frac{H}{T}}\psi(Tt)b_{ij}(t)\frac{dt}{t}\right]\leqslant C(a_{ij},\ b_{ij})N_T^*[\psi(T)].$$

4*. $$N_T^*\left[\frac{\sigma_j(T)^M}{r_j(T)}\right]<\infty.$$

be fulfilled for the N^*-functional and for the functions $r_j(T)$ and $\sigma_j(T)$.

The requirements 3* and 4* are fulfilled for $N_T^*\equiv N_T$ if the functions $r_j(T)$ and $\sigma_j(T)$ are powers.

Taking into account the properties of the N^*-functional, we obtain the following imbedding theorem (for the class of functions with a finite right side):

$$N_T^*\left[\frac{1}{r_j(T)}\omega_j^M(\Omega, \sigma_j(T), D^\alpha f)_q\right] \leqslant C\frac{\Lambda_0(H)}{\sigma_j(H)^M}\|f\|_{p^0(\Omega)} + C\sum_{i=1}^n N_T^*\left[\frac{\Lambda_i(T)}{r_j(T)}\bar{\omega}_i^{m_i}(\Omega, \sigma_i(T), f)_{p^i}\right]. \tag{40}$$

Analogous generalizations to smoothness relations in different coordinate directions, which are not those of a power, can be carried out also for other of the above results.

BIBLIOGRAPHY

1. A. F. Timan, *Theory of approximation of functions of a real variable,* Fizmatgiz, Moscow, 1960; English transl., Internat. Series of Monographs in Pure and Appl. Math., vol. 34, Macmillan, New York, 1963. MR **22** #8257; **33** #456.

2. P. L. Ul'janov, *The imbedding of certain function classes* H_p^W, Izv. Akad. Nauk SSSR Ser. Mat. 32 (1968), 649–686 = Math. USSR Izv. **2** (1968), 601–638. MR **37** #6749.

3. V. A. Andrienko, *Necessary conditions for imbedding the function classes* H_p^W, Mat. Sb. **78** (120) (1969), 280–300 = Math. USSR Sb. **7** (1969), 273–290. MR **38** #6347.

4. O. V. Besov and V. P. Il'in, *An imbedding theorem for the limit exponent,* Mat. Zametki **6** (1969), 129–138 = Math. Notes 6 (1969), 537–542. MR **41** #4225.

5. V. P. Il'in, *On an integral inequality,* Mat. Zametki **6** (1969), 139–148 = Math. Notes 6 (1969), 543–548. MR **41** #4309.

6. O. V. Besov and V. P. Il'in, *Natural extension of the class of regions in imbedding theorems,* Mat. Sb. **75** (117) (1968), 483–495 = Math. USSR Sb. **4** (1968), 445–456. MR **37** #1987.

7. S. M. Nikol'skiĭ, *Inequalities for entire functions of exponential type and their applications to the theory of differentiable functions of several variables,* Trudy Mat. Inst. Steklov. **38** (1951), 244–278; English transl., Amer. Math. Soc. Transl. (2) **80** (1969), 1–38. MR **14**, 32.

8. K. K. Golovkin, *Imbedding theorems for fractional spaces,* Trudy Mat. Inst. Steklov. **70** (1964), 38–46; English transl., Amer. Math. Soc. Transl. (2) **91** (1970), 57–67. MR **28** #5282; **42** #2.

9. A. S. Džafarov, *Imbedding theory of classes of functions with differentiable properties in norms of special spaces,* Akad. Nauk Azerbaĭdžan. SSR Dokl. **21** (1965), no. 2, 10–14. (Russian) MR **32** #4535.

10. O. V. Besov and P. L. Lizorkin, *Singular integral operators and sequences of convolutions in the* L_p*-spaces,* Mat. Sb. **73** (115), 65–88 = Math. USSR Sb. **2** (1967), 57–76. MR **36** #1936.

11. Ja. S. Bugrov, *Inequalities for harmonic polynomials,* Izv. Vysš. Učebn. Zaved. Matematika **1963**, no. 5 (36), 12–23. (Russian) MR **27** #5924.

12. V. P. Il'in, *The properties of some classes of differentiable functions of several variables defined in an n-dimensional region,* Trudy Mat. Inst. Steklov. **66** (1962), 227–363; English transl., Amer. Math. Soc. Transl. (2) **81** (1969), 91–256. MR **27** #3750.

13. O. V. Besov, *Investigation of a family of function spaces in connection with imbedding and extension theorems,* Trudy Mat. Inst. Steklov. **60** (1961), 42–81; English transl., Amer. Math. Soc. Transl. (2) **40** (1964), 85–126. MR **24** #A3501.

14. K. K. Gołovkin, *On the approximation of functions in arbitrary norms,* Trudy Mat. Inst. Steklov. **70** (1964), 26–37; English transl., Amer. Math. Soc. Transl. (2) **91** (1970), 43–56. MR **28** #5281; **42** #2.

15. A. A. Dezin, *On imbedding theorems and the problem of extension of a function,* Dokl. Akad. Nauk SSSR **88** (1953), 741–743. (Russian) MR **15**, 205.

16. A. S. Džafarov, *Theorems of weighted imbedding,* Dokl. Akad. Nauk SSSR **142** (1962), 514–517 = Soviet Math. Dokl. **3** (1962), 82–85. MR **24** #A3500.

17. Ju. A. Brudnyĭ and V. K. Šalašov, *Lipschitz spaces of functions,* Dokl. Akad. Nauk SSSR **197** (1971), 18–20 = Soviet Math. Dokl. **12** (1971), 383–386. MR **43** #3789.

Translated by G. BIRIUK

Trudy Mat. Inst. Steklov.
117 (1972)

Proc. Steklov Inst. Math.
117 (1972)

PROPERTIES OF SOLUTIONS OF DIFFERENTIAL EQUATIONS OF HIGHER ORDER IN TERMS OF WEIGHT CLASSES

UDC 517.51, 517.946

Ja. S. BUGROV

Abstract. The paper investigates the differential properties of the solutions of certain boundary value problems in the unit disk, depending on the differential properties of the boundary values in terms of weight classes. The equation

$$\frac{\partial^2 u}{\partial \rho^2}+\frac{1}{\rho}\frac{\partial u}{\partial \rho}+(-1)^{l+1}\frac{1}{\rho^2}\frac{\partial^{2l}u}{\partial x^{2l}}=0,$$

is considered, where $l \geqslant 1$ is a natural number.

Bibliography: 5 items.

Introduction

In [1] we considered the equation

$$A^{(l,1)}u \equiv \frac{\partial^2 u}{\partial \rho^2}+\frac{1}{\rho}\frac{\partial u}{\partial \rho}+\frac{(-1)^{l+1}}{\rho^2}\frac{\partial^{2l}u}{\partial x^{2l}}=0 \tag{1}$$

in the unit disk $\sigma = \{0 \leqslant \rho < 1, -\pi \leqslant x \leqslant \pi\}$, where $l \geqslant 1$ is a natural number. The boundary value problem for equation (1) is posed as follows: to find a solution $u(x, \rho; l)$ which is bounded in the unit disk σ and satisfies the condition

$$u(x, \rho; l)\big|_{\rho=1}=\varphi(x) \tag{2}$$

on the disk's boundary. Here $\varphi(x)$ is a function of period 2π, and equality (2) is understood in the sense that

$$\lim_{\rho\to 1-0}\int_{-\pi}^{\pi}|u(x, \rho; l)-\varphi(x)|^p\,dx=0 \quad (1\leqslant p\leqslant\infty). \tag{2'}$$

The formal solution of problem (1)–(2) has the form

$$u(x, \rho; l)=\frac{1}{\pi}\int_{-\pi}^{\pi}P(t, \rho; l)\,\varphi(x+t)\,dt, \tag{3}$$

where

$$P(t, \rho; l)=\frac{1}{2}+\sum_{1}^{\infty}\rho^{kl}\cos kt. \tag{4}$$

AMS (MOS) subject classifications (1970). Primary 35B30; Secondary 35J05, 35C15.

Let us note that for $l = 1$ the operator $A^{(1,1)}u \equiv \Delta u$, where Δ is a Laplace operator in polar coordinates and the kernel $P(t, \rho; 1) = P(t, \rho)$ is a Poisson kernel.

In [1] we investigated the properties of solution (3) in terms of the classes H, W and B. We show in the present article that solution (3) has derivatives of all orders which are summable with a certain weight. We study in addition the function

$$v(x, \rho; l) = -\frac{1}{\pi} \int_{-\pi}^{\pi} Q(t, \rho; l)\, \varphi(t + x)\, dt, \tag{5}$$

where

$$Q(t, \rho; l) = \sum_{1}^{\infty} \rho^{kl} \sin kt. \tag{6}$$

The function v is also a solution of equation (1), and we shall call it the l-conjugate of the function $u(x, \rho; l)$. For $l = 1$ it proves to be the harmonic function conjugate to the harmonic function $u(x, \rho; 1)$. We shall call the kernels $P(t, \rho; l)$ and $Q(t, \rho; l)$ generalized Poisson kernels.

§1. Notation and auxiliary data

For the function $\varphi(x)$ we shall denote the first and second difference as follows:

$$\Delta_h \varphi(x) = \varphi(x + h) - \varphi(x), \quad \Delta_h^2 \varphi(x) = \varphi(x + 2h) - 2\varphi(x + h) + \varphi(x).$$

We shall sometimes assume that

$$\Delta_t^2 \varphi(x) = \varphi(x + t) - 2\varphi(x) + \varphi(x - t).$$

DEFINITION 1. *Let $r > 0$, $r = \bar{r} + \alpha$, $0 < \alpha \leqslant 1$, where $\bar{r}$ is an integer $\geqslant 0$. We say that the periodic function $\varphi(x)$ belongs to the class $H_p^{(r)*}$ $(1 \leqslant p \leqslant \infty)$ if it, together with its generalized derivatives (in the Sobolev sense) $\partial^k \varphi / \partial x^k$ $(k = 1, \cdots, \bar{r})$, is pth power summable and the derivative $\varphi^{(\bar{r})}$ satisfies the condition*

$$\sup_{h>0} h^{-\alpha} \| \Delta_h^2 \varphi^{(\bar{r})}(x) \|_p = M_p^{(r)}(\varphi) < \infty,$$

where

$$\| \varphi \|_p = \left(\int_{-\pi}^{\pi} | \varphi(x) |^p dx \right)^{\frac{1}{p}}, \quad 1 \leqslant p < \infty,$$

$$\| \varphi \|_\infty = \operatorname*{ess\,sup}_{-\pi \leqslant x \leqslant \pi} | \varphi(x) |, \quad p = \infty.$$

A norm is introduced on this class as follows:

$$\| \varphi \|_{H_p^{(r)*}} = \| \varphi \|_p + M_p^{(r)}(\varphi).$$

DEFINITION 2. *Let* $r_1 = \bar{r}_1 + \alpha$ *and* $r_2 = \bar{r}_2 + \alpha_2$, *where* $0 < \alpha_i \leqslant 1$ *and the* $\bar{r}_i$ *are nonnegative integers. We say that a function* $f(x, \rho)$ *belongs to the class* $H_p^{(r_1, r_2)}(\sigma)$ *relative to polar coordinates if it has generalized derivatives*

$$\frac{\partial^k f}{\partial x^k}\,(k=0,\ 1,\ \ldots,\ \bar{r}_1),\ \frac{\partial^k f}{\partial \rho^k}\,(k=0,\ 1,\ \ldots,\ \bar{r}_2),$$

which are pth power summable $(1 \leqslant p \leqslant \infty)$ *and*

$$\sup_{h>0} h^{-\alpha_1}\| \Delta_{hx}^2 f_x^{(\bar{r}_1)} \|_{L_p(\sigma)} = M_p^{(r_1)}(f) < \infty,$$

$$\sup_{h>0} h^{-\alpha_2}\| \Delta_{h\rho}^2 f_\rho^{(\bar{r}_2)} \|_{L_p(\sigma)}^{*} = M_p^{(r_2)}(f) < \infty,$$

where

$$\| f \|_{L_p(\sigma)} = \left(\int_0^1 \int_{-\pi}^{\pi} | f |^p \rho\, d\rho\, dx \right)^{\frac{1}{p}},$$

$$\| f \|_{L_p(\sigma)}^{*} = \left(\int_0^{1-2h} \int_{-\pi}^{\pi} | f |^p \rho\, d\rho\, dx \right)^{\frac{1}{p}},$$

and $\Delta_{hx}^2 f$, *and* $\Delta_{h\rho}^2 f$ *are second h-differences in the variables* x *and* ρ, *respectively.*

A norm is introduced on this class in the following way:

$$\| f \|_{H_p^{(r_1, r_2)}(\sigma)} = \| f \|_{L_p(\sigma)} + M_p^{(r_1)}(f) + M_p^{(r_2)}(f).$$

S. M. Nikol'skiĭ [2] was the first to consider H classes.

DEFINITION 3. *We say that a periodic function* $\varphi(x) \in B_p^{(r)*}$ *if* $\varphi^{(k)}(x) \in L_p$ $(k = 0, \cdots, \bar{r})$, *and* $\varphi^{(\bar{r})}(x)$ *satisfies the condition*

$$\| \varphi^{(\bar{r})} \|_{B_p^{(\alpha)}*} = \int_{-\pi}^{\pi} | h |^{-1-p\alpha} \| \Delta_h^2 \varphi^{(\bar{r})}(x) \|_p^p \, dh < \infty \qquad (1 \leqslant p \leqslant \infty),$$

$$\| \varphi \|_{B_{p*}^{(r)}} = \| \varphi \|_p + \| \varphi^{(\bar{r})} \|_{B_p^{(\alpha)}*}.$$

In O. V. Besov's paper [3] (Theorem 1.1) it is shown that the equivalent norm

$$\| \varphi \|_{B_{p*}^{(r)}} = \| \varphi \|_p + \left\{ \sum_{k=0}^{\infty} b^{kp} A_k^p (\varphi)_p \right\}^{\frac{1}{p}},$$

can be introduced on this class, where $b = a^r > 1$, $A_k(\varphi)_p = E_{a^k}(\varphi)_p$; here $E_\nu(\varphi)_p = \inf_{T_\nu} \| \varphi - T_\nu \|_p$ is the best approximation of the function $\varphi \in L_p$ by means of trigonometric polynomials of degree ν.

The class $B_p^{(r_1, r_2)}(\sigma)$ with norm

$$\| f \|_{B_p^{(r_1, r_2)}(\sigma)} = \| f \|_{L_p(\sigma)} + \| f_x^{(\bar{r}_1)} \|_{B_{px}^{(\alpha_1)}(\sigma)} + \| f_\rho^{(\bar{r}_2)} \|_{B_{p\rho}^{(\alpha_2)}(\sigma)}.$$

can be introduced quite analogously. For $\alpha < 1$ the first differences $\Delta_t \varphi(x)$, $\Delta_{h\rho} f$, etc. can be taken everywhere. In the sequel we shall often make use of the so-called Abel transformation ([4], Chap. I, formulas (2.1))

$$\sum_{k=0}^{n} \mu_k b_k = \sum_{k=0}^{n-1} v_k \Delta\mu_k + v_n \mu_n,$$

where

$$\Delta\mu_k = \mu_k - \mu_{k+1}, \qquad v_k = \sum_{j=0}^{k} b_j.$$

By applying the Abel transformation to various trigonometric sums we shall obtain Dirichlet and Fejér kernels and their conjugates. Therefore it makes sense at this point to note a number of properties of these kernels.

For the Dirichlet kernel

$$\mathscr{D}_n(t) = \frac{1}{2} + \sum_{1}^{n} \cos kt = \frac{\sin \frac{2n+1}{2} t}{2 \sin \frac{t}{2}}$$

the estimate

$$|\mathscr{D}_n(t)| \leqslant c(n+1)^{\delta} t^{\delta-1} \qquad (0 < t \leqslant \pi, \quad 0 \leqslant \delta \leqslant 1) \tag{7}$$

holds, where c is a constant. A similar estimate holds for the conjugate kernel

$$\widetilde{\mathscr{D}}_n(t) = \sum_{1}^{n} \sin kt.$$

For the Fejér kernel

$$\mathscr{F}_n(t) = \frac{2}{n+1} \left(\frac{\sin \frac{n+1}{2} t}{2 \sin \frac{t}{2}} \right)^2$$

the estimate

$$|\mathscr{F}_n(t)| \leqslant c(n+1)^{\delta-1} t^{\delta-2} \tag{8}$$

$(0 < t \leqslant \pi)$ holds, where $0 \leqslant \delta \leqslant 1$. The latter estimate follows from the inequality $|\sin t| \leqslant |t|^{\delta}$, $0 \leqslant \delta \leqslant 1$.

We shall often take the coefficient $\mu_0 = 0$ for a sum of the form $\Sigma_1^\infty \mu_k \sin kt$; then

$$\sum_{1}^{\infty} \mu_k \sin kt = \mu_0 \left(-\frac{1}{2} \operatorname{ctg} \frac{t}{2} \right) + \sum_{1}^{\infty} \mu_k \sin kt.$$

Applying the Abel transformation twice, we obtain

$$\sum_{1}^{\infty} \mu_k \sin kt = \sum_{0}^{\infty} (k+1) \Delta^2 \mu_k \widetilde{\mathscr{F}}_k(t) \qquad (\mu_k \to 0, \ k \to \infty),$$

where

$$\Delta^2\mu_k = \mu_k - 2\mu_{k+1} + \mu_{k+2}, \qquad (n+1)\tilde{\mathcal{F}}_n(t) = \sum_0^n \tilde{\tilde{Q}}_k(t),$$

$$\tilde{\tilde{Q}}_0(t) = -\frac{1}{2}\operatorname{ctg}\frac{t}{2}, \qquad \tilde{\tilde{Q}}_k(t) = \tilde{\tilde{Q}}_0(t) + \sum_1^n \sin kt.$$

It is easy to show that

$$\tilde{\tilde{Q}}_k(t) = -\frac{\cos\frac{k+1}{2}t}{2\sin\frac{t}{2}}, \qquad \tilde{\mathcal{F}}_n(t) = \frac{-\sin(n+1)t}{4(n+1)\sin^2\frac{t}{2}}.$$

Hence it is clear that $(0 < t \leqslant \pi;\ 0 \leqslant \delta \leqslant 1)$

$$|\tilde{\mathcal{F}}_n(t)| \leqslant c\ \ (n+1)^{\delta-1} t^{\delta-2}. \tag{9}$$

§2. Properties of generalized Poisson kernels

Let us consider the trigonometric series

$$\psi_c(t,\ \rho;\ l,\ m) = \sum_{k=0}^{\infty} k^m \rho^{kl} \cos kt, \tag{10}$$

$$\psi_s(t,\ \rho;\ l,\ m) = \sum_{k=0}^{\infty} k^m \rho^{kl} \sin kt, \tag{11}$$

where $m \geqslant 0$ and $l \geqslant 1$; we shall take the first term of series (10) equal to ½ for $m = 0$ and $k = 0$.

It is clear that for some m the functions ψ_c and ψ_s coincide with generalized Poisson kernels or their derivatives. Therefore, having obtained estimates for these functions, by the same token we obtain estimates of the kernels $P(t, \rho; l)$ and $Q(t, \rho; l)$. For example,

$$\frac{\partial^{\nu+s} P}{\partial \rho^\nu \partial t^s} = \frac{1}{\rho^\nu} \sum_{j=1}^{\nu} \lambda_j \sum_0^\infty k^{jl+s} \rho^{kl} \begin{Bmatrix} \cos kt \\ \sin kt \end{Bmatrix} = \rho^{-\nu} \sum_{j=1}^{\nu} \lambda_j \begin{Bmatrix} \psi_c(t,\ \rho;\ l,\ jl+s) \\ \psi_s(t,\ \rho;\ l,\ jl+s) \end{Bmatrix},$$

where the λ_j are constants.

LEMMA 1. *With* $0 < t \leqslant \pi$ *and* $½ \leqslant \rho < 1$ *the estimates*

$$\left|\begin{Bmatrix} \psi_c \\ \psi_s \end{Bmatrix}\right| \leqslant c t^{\delta-1} (1-\rho)^{-\frac{m+\delta}{l}}, \qquad m \geqslant 0, \quad 0 \leqslant \delta \leqslant 1, \tag{12}$$

$$\left|\begin{Bmatrix} \psi_c \\ \psi_s \end{Bmatrix}\right| \leqslant c t^{\delta-2} (1-\rho)^{\frac{1-m-\delta}{l}} \tag{13}$$

hold for the functions ψ_c *and* ψ_s, *where* $\delta = 0$ *for* $m \geqslant 1$; $1 - m < \delta < 1$ *for* $0 < m < 1$;

$$|\psi_c| \leqslant c t^{-2} (1-\rho)^{\frac{1}{l}} \qquad m = 0. \tag{14}$$

PROOF. Applying the Abel transformation, we obtain

$$\psi_c(t, \rho; l, m)=\sum_0^\infty \Delta k^m\rho^{kl}\mathscr{D}_k(t).$$

for $l \geqslant 1, m \geqslant 0$ and $\rho < 1$.

It is easy to see that $(m > 0)$

$$\begin{aligned}|\Delta k^m\rho^{kl}| &= |k^m\rho^{kl}-(k+1)^m\rho^{(k+1)^l}| = \left|\int_0^1 [(k+u)^m\rho^{(k+u)^l}]'_u\,du\right| \\ &\leqslant c\int_0^1\left[(k+u)^{m-1}+(k+u)^{m+l-1}\ln\frac{1}{\rho}\right]\rho^{(k+u)^l}du \\ &\leqslant c[(k+1)^{m-1}+(k+1)^{m+l-1}(1-\rho)]\rho^{kl},\end{aligned} \tag{15}$$

since $\ln_p^{-1} = \ln(1+(1-\rho)/\rho) \sim (1-\rho)$ for $½ \leqslant \rho \leqslant 1$. For $m = 0$

$$|\Delta\rho^{kl}| \leqslant c(1-\rho)k^{l-1}\rho^{kl}. \tag{15'}$$

Hence, by virtue of (7),

$$\begin{aligned}|\psi_c| &\leqslant c\sum_0^\infty[(k+1)^{m-1+\delta}+(k+1)^{m+l-1+\delta}(1-\rho)]\,t^{\delta-1}\rho^{kl} \\ &\leqslant \left(\rho=e^{-z},\ z=\ln\frac{1}{\rho},\ z\sim 1-\rho\right) \\ &\leqslant ct^{\delta-1}\sum_0^\infty[(k+1)^{m-1+\delta}+z(k+1)^{m+l-1+\delta}]\,e^{-k^lz}.\end{aligned}$$

Let us fix z and choose a natural number N so that

$$(N+1)^{-l} \leqslant z < N^{-l}. \tag{16}$$

Then

$$\begin{aligned}|\psi_c| &\leqslant ct^{\delta-1}\left\{\sum_0^N\cdots+\sum_{N+1}^\infty\cdots\right\} \\ &\leqslant ct^{\delta-1}\left\{\sum_0^N(k^{m-1+\delta}+zk^{m+l-1+\delta})+\sum_{N+1}^\infty(k^{m-1+\delta}+zk^{m+l-1+\delta})\,e^{-(N+1)^lz\left(\frac{k}{N+1}\right)^l}\right\} \\ &\leqslant ct^{\delta-1}\left\{N^{m+\delta}+zN^{m+l+\delta}+\sum_{N+1}^\infty(k^{m+\delta-1}+zk^{m+l-1+\delta})\,e^{-\left(\frac{k}{N+1}\right)^l}\right\},\end{aligned}$$

since $(N+1)^lz \geqslant 1$. Furthermore,

$$\begin{aligned}&\sum_{N+1}^\infty(k^{m-1+\delta}+zk^{m+l-1+\delta}\,e^{-\left(\frac{k}{N+1}\right)^l} \leqslant c\int_{N+1}^\infty(x^{m-1+\delta}+zx^{m+l-1+\delta})\,e^{-\left(\frac{x}{N+1}\right)^l}dx \\ &\leqslant (x=(N+1)\,t) \leqslant cN^{m+\delta}\int_1^\infty(t^{m-1+\delta}+N^lzt^{m+l-1+\delta})\,e^{-t^l}dt \\ &\leqslant c\,(N^{m+\delta}+zN^{m+l+\delta}).\end{aligned}$$

Thus, by virtue of (16),

$$|\psi_c| \leqslant ct^{\delta-1}\{N^{m+\delta} + zN^{m+l+\delta}\} \leqslant cz^{-\frac{m+\delta}{l}}t^{\delta-1} \leqslant ct^{\delta-1}(1-\rho)^{-\frac{m+\delta}{l}}$$

for $m > 0$ and $0 \leqslant \delta \leqslant 1$. If $m = 0$, then, utilizing estimate (15′), we obtain the result required in this case.

Quite analogously, applying estimate (7) for $\widetilde{\mathfrak{D}}_k(t)$, we obtain

$$|\psi_s| \leqslant \sum_1^\infty |\Delta k^m \rho^{kl}|\,|\widetilde{\mathscr{D}}_k(t)| \leqslant ct^{\delta-1}\sum_1^\infty [(k+1)^{m-1+\delta} + (1-\rho)k^{m+l+\delta-1}]\rho^{kl}$$

$$\leqslant ct^{\delta-1}(1-\rho)^{-\frac{m+\delta}{l}} \qquad (m \geqslant 0,\ 0 \leqslant \delta \leqslant 1),$$

and (12) is proved.

Applying the Abel transformation once more, we shall have

$$\psi_c = \sum_0^\infty (k+1)\,\Delta^2 k^m \rho^{kl} \mathscr{F}_k(t).$$

Utilizing estimate (8), we obtain

$$|\psi_c| \leqslant c\sum_0^\infty |\Delta^2 k^m \rho^{kl}|\,(k+1)^\delta\, t^{\delta-2}.$$

It is easy to verify that the second differences $\Delta^2 k^m \rho^{kl}$ can be evaluated by means of an integral (cf. [1], p. 9)

$$|\Delta^2 k^m \rho^{kl}| = \left|\int_0^1 (1-u)\,[(k+u+1)^m \rho^{(k+u+1)^l}]''_{u^2}\,du + \int_0^1 u\,[(k+u)\,\rho^{(k+u)^l}]''_{u^2}\,du\right| \tag{17}$$

$$\leqslant c\,[(k+1)^{m-2} + (1-\rho)k^{m+l-2} + (1-\rho)^2 k^{m+2l-2}]\rho^{kl},$$

if $m > 0, m \neq 1$.

In the same way we obtain the estimates

$$|\Delta^2 k\rho^{kl}| \leqslant c\,[(k+1)^{l-1}(1-\rho) + k^{2l-1}(1-\rho)^2]\rho^{kl} \qquad (m=1), \tag{17′}$$

$$|\Delta^2 \rho^{kl}| \leqslant c\,[(k+1)^{l-2}(1-\rho) + k^{2l-2}(1-\rho)^2]\rho^{kl} \qquad (m=0) \tag{17″}$$

for $m = 0$ and $m = 1$. Hence

$$|\psi_c| \leqslant c\sum_0^\infty [(k+1)^{m-2+\delta} + (1-\rho)k^{m+l-2+\delta} + (1-\rho)^2 k^{m+2l-2+\delta}]\rho^{kl} t^{\delta-2}.$$

If $m > 1$, then, setting $\delta = 0$, we obtain

$$|\psi_c| \leqslant ct^{-2}\sum_1^\infty (k^{m-2}+(1-\rho)k^{m+l-2}+(1-\rho)^2k^{m+2l-2})\rho^{kl}$$

$$\leqslant ct^{-2}(N^{m-1}+zN^{m+l-1}+z^2N^{m+2l-1}) \leqslant ct^{-2}(1-\rho)^{-\frac{m-1}{l}}.$$

For $m = 1$ we obtain the same result by utilizing (17′). If, however, $m = 0$, we get

$$|\psi_c| \leqslant ct^{-2}\sum_0^\infty [(k+1)^{l-2}(1-\rho)+k^{2l-2}(1-\rho)^2]\rho^{kl}$$

$$\leqslant ct^{-2}(N^{l-1}(1-\rho)+N^{2l-1}(1-\rho)^2) \leqslant ct^{-2}(1-\rho)^{\frac{1}{l}},$$

by virtue of (17″), and inequality (14) is proved.

If, however, $0 < m < 1$, we obtain

$$|\psi_c| \leqslant ct^{\delta-2}\sum_0^\infty [(k+1)^{m-2+\delta}+(1-\rho)k^{m+l-2+\delta}+(1-\rho)^2k^{m+2l-2+\delta}]\rho^{kl}$$

$$\leqslant ct^{\delta-2}(N^{\delta-1+m}+zN^{m+l-1+\delta}+z^2N^{m+2l-2+\delta}) \leqslant ct^{\delta-2}(1-\rho)^{\frac{1-\delta-m}{l}}$$

for $1 - m < \delta < 1$. For $m > 0$ we obtain, by virtue of (9),

$$|\psi_s| = |\tilde{Q}_0(t) + \sum_1^\infty k^m\rho^{kl}\sin kt| = \left|\sum_0^\infty (k+1)\,\Delta^2k^m\rho^{kl}\tilde{\mathcal{F}}_k(t)\right|$$

$$\leqslant ct^{\delta-2}\sum_0^\infty [(k+1)^{m-2+\delta}+(1-\rho)k^{m+l-2+\delta}+(1-\rho)^2k^{m+2l-2+\delta}]\rho^{kl} \leqslant ct^{\delta-2}(1-\rho)^{\frac{1-\delta-m}{l}}$$

for the function ψ_s, where $\delta = 0$ if $m \geqslant 1$; $1 - m < \delta < 1$ if $0 < m < 1$. The lemma is completely proved.

REMARK 1. If $m = \nu l + s$, the lemma affords estimates of the kernels $P(t, \rho; l)$, $Q(t, \rho; l)$ and their derivatives of order ν with respect to the variable ρ and of order s with respect to the variable t.

REMARK 2. For $l = 1$ there are estimates of the derivatives of the Poisson kernel in our paper [5], where

$$\left|\frac{\partial^m P(t,\rho;1)}{\partial\rho^{m-s}\partial t^s}\right| \leqslant c[(1-\rho)^2+t^2]^{-\frac{m+1}{2}} \qquad (m = 0, 1, \ldots,)$$

for $0 < t \leqslant \pi$ and $\frac{1}{2} \leqslant \rho < 1$.

REMARK 3. If $t^l \leqslant 1 - \rho$, then

$$\frac{1}{2(1-\rho)^2} \leqslant \frac{1}{(1-\rho)^2+t^{2l}} \leqslant \frac{1}{(1-\rho)^2},$$

and therefore

$$\left|\begin{Bmatrix}\psi_c\\ \psi_s\end{Bmatrix}\right| \leqslant c[(1-\rho)^2+t^{2l}]^{-\frac{m+1}{2}}.$$

If, however, $t^l \geqslant 1 - \rho$ and $m = 0$ or 1, we obtain

$$\left|\begin{Bmatrix}\psi_c\\ \psi_s\end{Bmatrix}\right| \leqslant c[(1-\rho)^2+t^{2l}]^{-\frac{m+1}{2l}}$$

from Lemma 1. Thus the estimate

$$\left|\begin{Bmatrix}\psi_c\\ \psi_s\end{Bmatrix}\right| \leqslant c\,[(1-\rho)^2+t^{2l}]^{-\frac{m+1}{2l}} \qquad \left(0\leqslant t\leqslant \pi,\ \ \tfrac{1}{2}\leqslant \rho<1,\ \ m=0,\ 1\right) \tag{18}$$

holds for $m = 0, 1$.

REMARK 4. Since the functions ψ_c and ψ_s are even or odd, Lemma 1 holds also for $-\pi \leqslant t < 0$; it is necessary only to replace t by $|t|$ in the respective estimates.

LEMMA 2. *The inequality*

$$I=\int_{-\pi}^{\pi}\left|\begin{Bmatrix}\psi_c\\ \psi_s\end{Bmatrix}\right|^{\varepsilon q}|t|^{\gamma}\,dt \leqslant c\,(1-\rho)^{\frac{1+\gamma-\varepsilon q(m+1)}{l}}$$

holds, where $1 \leqslant q < \infty$ *and* $\gamma > -1$; $1 > \epsilon > (1+\gamma)/2q$ *for* $m \geqslant 1$; $\epsilon > (1+\gamma)/(2-\delta)q$ *for* $0 < m < 1$, $1-m < \delta < 1$; $\epsilon > (1+\gamma)/2q$ *for* $m = 0$ *for the function* ψ_c; $\epsilon > (1+\gamma)/q$ *for* $m = 0$ *for the function* ψ_s.

PROOF. By virtue of (12) we have

$$I=\int_{|t|^l\leqslant 1-\rho}\cdots+\int_{|t|^l\geqslant 1-\rho}\cdots\leqslant c\int_{t^l\leqslant 1-\rho}(1-\rho)^{-\frac{m+1}{l}\varepsilon q}t^{\gamma}\,dt+\int_{|t|^l>1-\rho}\cdots$$

$$\leqslant c\,(1-\rho)^{\frac{1+\gamma-\varepsilon q(m+1)}{l}}+\int_{|t|^l>1-\rho}\cdots \qquad (\gamma>-1)$$

for $\delta = 1$. Now let us evaluate the second integral. If $m \geqslant 1$, we have from (13) for $\delta = 0$

$$\int_{|t|^l>1-\rho}\left|\begin{Bmatrix}\psi_c\\ \psi_s\end{Bmatrix}\right|^{\varepsilon q}|t|^{\gamma}\,dt\leqslant c\int_{t^l>1-\rho}t^{-2\varepsilon q}(1-\rho)^{\frac{1-m}{l}\varepsilon q}t^{\gamma}dt\leqslant (t^l=(1-\rho)v)$$

$$\leqslant c\,(1-\rho)^{\frac{1+\gamma-\varepsilon q(m+1)}{l}}\int_1^{\infty}v^{\frac{1+\gamma-l-2\varepsilon q}{l}}\,dv\leqslant c\,(1-\rho)^{\frac{1+\gamma-\varepsilon q(m+1)}{l}} \qquad \left(\varepsilon>\frac{1+\gamma}{2q}\right).$$

If $0 < m < 1$, we obtain analogously from (13) for $1-m < \delta < 1$

$$\int_{|t|^l>1-\rho}\left|\begin{Bmatrix}\psi_c\\ \psi_s\end{Bmatrix}\right|^{\varepsilon q}|t|^{\gamma}\,dt\leqslant c\,(1-\rho)^{\frac{1+\gamma-\varepsilon q(m+1)}{l}}\int_1^{\infty}v^{\frac{1+\gamma-l-2\varepsilon q+\delta\varepsilon q}{l}}\,dv$$

$$\leqslant c\,(1-\rho)^{\frac{1+\gamma-\varepsilon q(m+1)}{l}} \qquad \left(\varepsilon>\frac{1+\gamma}{(2-\delta)\,q}\right).$$

For $m = 0$ for the function ψ_c we obtain the required estimate for $\epsilon > (1+\gamma)/2q$ by utilizing (14). The same estimate (for $m = 0$) is valid for the function ψ_s for

$\epsilon > (1+\gamma)/q$. It is necessary to utilize (12) for $\delta = m = 0$.

COROLLARY.

$$\left(\int_{-\pi}^{\pi}\left|\begin{Bmatrix}\psi_c\\ \psi_s\end{Bmatrix}\right|^p dt\right)^{\frac{1}{p}} \leqslant c\,(1-\rho)^{\frac{1}{lp}-\frac{m+1}{l}},$$

where $1 \leqslant p < \infty$ *for the function* ψ_c; $1 < p < \infty$ *for the function* ψ_s.

PROOF. Setting $\epsilon q = p$ and $\gamma = 0$, we obtain the fact that the estimate holds for $p > 1/2$ if $m > 0$. The same is also true for the function ψ_c for $m = 0$. For the function ψ_s, the condition $p > 1$ must be fulfilled for $m = 0$.

LEMMA 3. *The inequalities*

$$I = \int_{1/2}^{1}\left|\begin{Bmatrix}\psi_c\\ \psi_s\end{Bmatrix}\right|^{(1-\varepsilon)p}(1-\rho)^{\gamma}\,d\rho \leqslant ct^{(1+\gamma)l-(1-\varepsilon)(m+1)p}$$

$(0 < t \leqslant \pi)$ *hold, where* $m \geqslant 0, \gamma \geqslant -1, 1 \leqslant p < \infty$, *and* $\epsilon < 1-(1+\gamma)l/(m+1)p$;

$$\varepsilon > 1 - \frac{(1+\gamma)\,l}{p\,(m-1)} \quad \text{for} \quad m \geqslant 1;$$

$$\varepsilon > 1 - \frac{(1+\gamma)\,l}{p\,(m+\delta-1)} \quad \text{for} \quad 0 < m < 1, \quad 1-m < \delta < 1,$$

and $0 < \epsilon < 1$ *for* $m = 0$.

PROOF. By virtue of Remark 3 on Lemma 1, we have

$$\begin{aligned} I &\leqslant c\int_{1-\rho>t^l}[(1-\rho)^2+t^{2l}]^{\frac{m+1}{2l}(\varepsilon-1)p}(1-\rho)^{\gamma}\,d\rho + \int_{1-\rho<t^l}\cdots \\ &\leqslant ct^{(1+\gamma)l+(m+1)(\varepsilon-1)p}\int_1^{\infty}u^{\gamma}(u^2+1)^{\frac{m+1}{2l}(\varepsilon-1)p}\,du \\ &+ \int_{1-\rho<t^l}\cdots \leqslant ct^{(1+\gamma)l+(m+1)(\varepsilon-1)p} + \int_{1-\rho<t^l}\cdots \end{aligned}$$

for $\epsilon < 1-(1+\gamma)l/p(m+1)$.

Now let us evaluate the second integral. It is necessary to consider the cases $m \geqslant 1$, $0 < m < 1$ and $m = 0$. Since the estimates are the same in all cases, let us carry out the proof for the case $0 < m < 1$ as an example. By virtue of (13),

$$\begin{aligned} \int_{1-\rho<t^l}\left|\begin{Bmatrix}\psi_c\\ \psi_s\end{Bmatrix}\right|^{(1-\varepsilon)p}(1-\rho)^{\gamma}\,d\rho &\leqslant ct^{(\delta-2)(1-\varepsilon)p}\int_{1-\rho<t^l}(1-\rho)^{\gamma+\frac{1-m-\delta}{l}(1-\varepsilon)p}\,d\rho \\ &\leqslant (1-\rho=ut^l) \leqslant ct^{(1+\gamma)l+(\delta-2)(1-\varepsilon)p+(1-m-\delta)(1-\varepsilon)p}\int_0^1 u^{\gamma+(1-\varepsilon)p\frac{1-m-\delta}{l}}\,du \\ &\leqslant ct^{(1+\gamma)l+(m+1)(\varepsilon-1)p} \end{aligned}$$

for $\epsilon > 1 - (1+\gamma)l/p(\delta + m - 1)$, $1 - m < \delta < 1$.

The estimates are carried out in the same way for the function ψ_c for $m = 0$.

We use (12) for the estimate of the function ψ_s for $\delta = m = 0$. The lemma is proved.

COROLLARY.

$$\left(\int_{\frac{1}{2}}^{1} \left| \begin{Bmatrix} \psi_c \\ \psi_s \end{Bmatrix} \right|^q d\rho \right)^{\frac{1}{q}} \leqslant c t^{\frac{l}{q} - m - 1},$$

where $l/(m+1) < q < l/(m-1)$ *for* $m \geqslant 1$; $l/(m+1) < q < l/(\delta + m - 1)$ *for* $0 < m < 1$, $1 - m < \delta < 1$; *and* $q > l$ *for* $m = 0$.

The proof follows from Lemma 3 for $\gamma = 0$ and $(1-\epsilon)p = q$.

LEMMA 4. *For fixed* $\delta > 0$,

$$I_\delta = \int_{|t| \geqslant \delta} |P(t, \rho; l)|\, dt \leqslant c(1-\rho)^{\frac{1}{l}},$$

where the constant c *depends on* δ.

PROOF. The kernel $P(t, \rho; l)$ coincides with the function ψ_c for $m = 0$; therefore

$$I_\delta \leqslant c \int_\delta^\pi t^{-2} (1-\rho)^{\frac{1}{l}}\, dt \leqslant c(1-\rho)^{\frac{1}{l}},$$

by virtue of (14).

LEMMA 5. *The equalities*

$$\frac{1}{\pi} \int_{-\pi}^{\pi} P(t, \rho; l)\, dt = \frac{1}{\pi} \int_{-\pi}^{\pi} P(t, \rho + h; l)\, dt = 1,$$

$$\int_{-\pi}^{\pi} \frac{\partial^{\nu+s} P}{\partial \rho^\nu \partial t^s}\, dt = 0, \qquad \nu + s > 0,$$

hold for the kernel $P(t, \rho; l)$.

This follows from the explicit form of the kernel $P(t, \rho; l)$ and the orthogonality of the trigonometric system.

§3. Differential properties of the functions $u(x, \rho; l)$ and $v(x, \rho; l)$

If $0 \leqslant \rho \leqslant 1/2$, the trigonometric series of $P(t, \rho; l)$ and $Q(t, \rho; l)$ converge uniformly; therefore their sums are bounded. The derivatives of the kernels are also bounded for $0 \leqslant \rho \leqslant 1/2$.

THEOREM 1. *If* $0 \leqslant \rho \leqslant 1/2, \varphi(t) \in L_p(-\pi, \pi)$ *and* $1 \leqslant p \leqslant \infty$, *then*

$$\left\| \frac{\partial^{\nu+s} u}{\partial \rho^\nu \partial x^s} \right\|_{L_p\left(\sigma_{\frac{1}{2}}\right)} \leqslant c \| \varphi \|_p \qquad (\nu + s \geqslant 0),$$

where c *is a constant and* $\sigma_{1/2} = \{0 \leqslant \rho \leqslant 1/2; -\pi \leqslant t \leqslant \pi\}$. *A similar estimate holds for the function* $v(x, \rho; l)$.

PROOF. According to (3) we have

$$\frac{\partial^{\nu+s} u}{\partial \rho^\nu \partial x^s} = c \int_{-\pi}^{\pi} \frac{\partial^{\nu+s} P(t, \rho; l)}{\partial \rho^\nu \partial t^s} \varphi(x + t)\, dt.$$

Applying the generalized Minkowski inequality, we obtain

$$\left\| \frac{\partial^{\nu+s} u}{\partial \rho^\nu \partial x^s} \right\|_{L_p\left(\sigma_{\frac{1}{2}}\right)} \leqslant c \int_{-\pi}^{\pi} \left(\int_0^{\frac{1}{2}} \int_{-\pi}^{\pi} \left| \frac{\partial^{\nu+s} P(t, \rho; l)}{\partial \rho^\nu \partial t^s} \right|^p | \varphi(x + t)|^p \, dx d\rho \right)^{\frac{1}{p}} dt$$

$$\leqslant c \| \varphi \|_p \int_{-\pi}^{\pi} \left(\int_0^{\frac{1}{2}} \left| \frac{\partial^{\nu+s} P}{\partial \rho^\nu \partial t^s} \right|^p d\rho \right)^{\frac{1}{p}} dt \leqslant c \| \varphi \|_p,$$

where the constant c depends on ν, s and l. The theorem is proved.

REMARK. The theorem just proved implies that the functions u and v are infinitely differentiable in $\sigma_{1/2}$ and their differential properties in $\sigma_{1/2}$ are determined entirely by the properties of the kernels P and Q. Near the boundary of the circle σ (i.e. for $1/2 \leqslant \rho < 1$) the differential properties of the functions u and v depend wholly on the properties of the function $\varphi(x)$.

THEOREM 2. *If the functions* $\varphi(x) \in H_p^{(\alpha)}$ *, *where* $1 \leqslant p \leqslant \infty$ *and* $0 < \alpha < 1$, *then*

$$I = \left(\int_{-\pi}^{\pi} | u(x, \rho; l) - \varphi(x)|^p \, dx \right)^{\frac{1}{p}} \leqslant c (1 - \rho)^{\frac{\alpha}{l}} \to 0 \qquad as \quad \rho \to 1 - 0;$$

$$\left(\int_{-\pi}^{\pi} | v(x, \rho; l) - \tilde{\varphi}(x)|^p \, dx \right)^{\frac{1}{p}} \leqslant c (1 - \rho)^{\frac{\alpha}{l}} \to 0 \qquad as \quad \rho \to 1 - 0,$$

where $\tilde{\varphi}(x)$ *is the function trigonometrically conjugate to* $\varphi(x)$.

PROOF. By virtue of Lemma 5,

$$u(x, \rho; l) - \varphi(x) = \frac{1}{\pi} \int_{-\pi}^{\pi} P(t, \rho; l) [\varphi(x + t) - \varphi(x)]\, dt.$$

Since $\varphi \in H_p^{(\alpha)}*$, it follows that $\|\Delta_t \varphi\|_p \leqslant c|t|^\alpha$; therefore, applying the generalized Minokowski inequality, we obtain

$$I \leqslant c \int_{-\pi}^{\pi} |P(t, \rho; l)| \cdot \|\Delta_t \varphi(x)\|_p \, dt$$

$$\leqslant c \int_{-\pi}^{\pi} |P(t, \rho; l)| \cdot |t|^\alpha \, dt \leqslant c \int_0^{\pi} |P(t, \rho; l)| \, t^\alpha dt.$$

By virtue of Lemma 2 with $\epsilon q = 1, \gamma = \alpha$ and $m = 0$ we have $I \leqslant c(1-\rho)^{\alpha/l} \to 0$ for $0 < \alpha < 1$ as $\rho \to 1 - 0$.

Let us prove the second assertion. Let the Fourier series of the function $\varphi(x)$ have the form

$$\frac{a_0}{2} + \sum_1^\infty (a_k \cos kx + b_k \sin kx).$$

Then

$$\tilde{\varphi}(x) = \sum_1^\infty (a_k \sin kx - b_k \cos kx).$$

If $\varphi(x) \in L_p(-\pi, \pi)$, $1 < p < \infty$, then $\widetilde{\varphi}(x) \in L_p(-\pi, \pi)$ (cf. [4], Chap. VII, Theorems 2.4 and 6.4) and $S(\widetilde{\varphi}) = \widetilde{S}(\varphi)$, i.e. the Fourier series of $\widetilde{\varphi}(x)$ coincides with the conjugate Fourier series of the function $\varphi(x)$. If $\widetilde{a}_k$ and $\widetilde{b}_k$ are Fourier coefficients of $\widetilde{\varphi}(x)$, then $\widetilde{a}_k = -b_k$ and $\widetilde{b}_k = a_k$. Hence, taking into account that $\int_{-\pi}^{\pi} \widetilde{\varphi}(x) dx = 0$, we obtain

$$v(x, \rho; l) = -\frac{1}{\pi} \int_{-\pi}^{\pi} Q(t, \rho; l) \varphi(x+t) \, dt = -\sum_1^\infty \rho^{k^l} (b_k \cos kx - a_k \sin kx)$$

$$= \sum_1^\infty \rho^{k^l} (\tilde{a}_k \cos kx + \tilde{b}_k \sin kx) = \frac{1}{\pi} \int_{-\pi}^{\pi} \sum_1^\infty \rho^{k^l} \cos kt \tilde{\varphi}(x+t) \, dt$$

$$= \frac{1}{\pi} \int_{-\pi}^{\pi} P(t, \rho; l) \tilde{\varphi}(x+t) \, dt.$$

Thus

$$v(x, \rho; l) = \frac{1}{\pi} \int_{-\pi}^{\pi} P(t, \rho; l) \tilde{\varphi}(x+t) \, dt. \tag{19}$$

Since $\varphi \in H_p^{(\alpha)}*$, we have $\widetilde{\varphi} \in H_p^{(\alpha)}*$ for $1 \leqslant p \leqslant \infty$ (cf. [4], Chap. III, Theorem 13.29 and Chap. VII, Theorem 6.14). Therefore, taking (19) into account, it is easy to obtain the speed of the convergence of the function $v(x, \rho; l)$ towards the boundary values of $\widetilde{\varphi}(x)$:

$$\left(\int_{-\pi}^{\pi} |v(x, \rho; l) - \tilde{\varphi}(x)|^p \, dx\right)^{\frac{1}{p}} \leqslant c(1-\rho)^{\frac{\alpha}{l}} \to 0$$

as $\rho \to 1-0$, and the theorem is proved.

REMARK. If $\varphi(x) \in L_p(-\pi, \pi)$, $1 < p < \infty$, it can also be proved that the function u assumes the values of $\varphi(x)$ in the sense of convergence in the p-mean, while the function v assumes the values of $\widetilde{\varphi}(x)$.

As a matter of fact, in this case

$$I \leqslant c \int_{-\pi}^{\pi} |P(t, \rho; l)| \cdot \|\Delta_t \varphi(x)\|_p \, dt.$$

Since $\varphi \in L_p$, it is continuous in the large, i.e. $\|\Delta_t \varphi(x)\|_p < \epsilon$ as soon as $|t| < \delta$. Therefore

$$I \leqslant c\varepsilon \int_{|t| \leqslant \delta} |P(t, \rho; l)| \, dt + c\|\varphi\|_p \int_{|t| > \delta} |P(t, \rho; l)| \, dt \leqslant c\varepsilon + c\|\varphi\|_p (1-\rho)^{\frac{1}{l}}$$

by virtue of Lemma 4 and the corollary of Lemma 2 (for $p = 1$ and $m = 0$).

By virtue of the arbitrariness of $\epsilon > 0$ we conclude from this that $I \to 0$ as $\rho \to 1-0$. The estimates for the function v are carried out in the same way. If $p = 1$, the function u takes the values of $\varphi(x)$. The function $\widetilde{\varphi}(x)$ can be nonexistent in this case. If, however, it is known beforehand that $\widetilde{\varphi}(x)$ is also Lebesgue intergrable, then $v(x, \rho; l) \to \widetilde{\varphi}(x)$ in the L metric (cf. [4], Chap. VII, Theorem 4.4).

THEOREM 3. *Let* $\varphi(x) \in H_p^{(r)}*$ $(1 \leqslant p \leqslant \infty; r > 0)$. *Then* $u(x, \rho; l)$ *and* $v(x, \rho; l)$, *as functions of polar coordinates, belong to the class* $H_p^{(r_1, r_2)}(\sigma)$, *and*

$$\|u\|_{H_p^{(r_1, r_2)}(\sigma)} \leqslant c\|\varphi\|_{H_p^{(r)}*},$$

$$\|v\|_{H_p^{(r_1, r_2)}(\sigma)} \leqslant c\|\tilde{\varphi}\|_{H_p^{(r)}*},$$

where $r_1 = r + l/p$ *and* $r_2 l = r_1$.

PROOF. The assertion in regard to the function $u(x, \rho; l)$ is proved in our paper [1]. The proof is the very same for the function $v(x, \rho; l)$, since (19) holds in this case and $\widetilde{\varphi}(x)$ belongs to $H_p^{(r)}*$.

THEOREM 4. *If* $\varphi(x) \in B_p^{(r)}*$ $(1 \leqslant p \leqslant \infty; r < 0)$, *then*

$$\left\|\begin{Bmatrix} u \\ v \end{Bmatrix}\right\|_{B_p^{(r_1, r_2)}(\sigma)} \leqslant c \begin{cases} \|\varphi\|_{B_p^{(r)}*}, \\ \|\tilde{\varphi}\|_{B_p^{(r)}*}, \end{cases}$$

where r_1 *and* r_2 *are defined in Theorem* 3.

PROOF. The membership of the conjugate function of $\widetilde{\varphi}(x)$ in the class $B_p^{(r)}*$ follows from the fact that the best approximation $E_\nu(\varphi)_p$ is of the same order as $E_\nu(\widetilde{\varphi})_p$ (cf. [4], Chap. III, Theorem 13.27); consequently

$$\|\tilde{\varphi}\|_{B_p^{(r)}*} = \|\tilde{\varphi}\|_p + \left\{\sum_{k=1}^{\infty} b^{kp} A_k^p(\tilde{\varphi})_p\right\}^{\frac{1}{p}} < \infty.$$

For the rest, the theorem is proved the same as in our paper [1] in regard to the function $u(x, \rho; l)$.

REMARK. The membership of the function $v(x, \rho; l)$ in the classes $H_p^{(r_1, r_2)}(\sigma)$ and $B_p^{(r_1, r_2)}(\sigma)$ can be expressed in the language of approximating the function v by the polynomials $\Phi_n(x, \rho; l)$ (necessary and sufficient conditions), as this is done for the function u in [1].

THEOREM 5. *Let $\varphi \in B_p^{(r)}*$, where $r = \bar{r} + \alpha$, $0 < \alpha < 1$, and $\bar{r}$ is a nonnegative integer. Then the functions $u(x, \rho; l)$ and $v(x, \rho; l)$ have derivatives (of order higher than admitted by Theorem 4) which are summable with the weight*

$$\left\|(1-\rho)^{\varkappa} \frac{\partial^{s_1+s_2} u}{\partial x^{s_1} \partial \rho^{s_2}}\right\|_{L_p\left(\sigma \setminus \sigma_{\frac{1}{2}}\right)} \leqslant c \|\varphi\|_{B_p^{(r)}*},$$

where $\kappa = s_2 + s_1/l - r/l - 1/p$, s_1 and s_2 are integers $\geqslant 0$, and $1 \leqslant p < \infty$.

Similar inequalities hold for $v(x, \rho; l)$, it being necessary to change the norm of φ to the norm of $\widetilde{\varphi}$.

PROOF. We find from (3) that

$$\frac{\partial^{s_1+s_2} u}{\partial x^{s_1} \partial \rho^{s_2}} = \frac{1}{\pi} \int_{-\pi}^{\pi} \frac{\partial^{s_1+s_2} P}{\partial t^{s_1} \partial \rho^{s_2}} \varphi(x+t)\, dt,$$

where

$$\frac{\partial^{s_1+s_2} P}{\partial t^{s_1} \partial \rho^{s_2}} = \frac{1}{\rho^{s_2}} \sum_{j=1}^{s_2} \lambda_j \sum_{0}^{\infty} k^{jl+s_1} \rho^{k^l} \begin{Bmatrix} \cos kt \\ \sin kt \end{Bmatrix}.$$

Then, integrating by parts $\bar{r}$ times (if $s_1 > \bar{r}$, $\bar{r}$ derivatives could be immediately transferred to the function $\varphi(x)$) and utilizing Lemma 5, we obtain

$$\frac{\partial^{s_1+s_2} u}{\partial x^{s_1} \partial \rho^{s_2}} = \frac{c}{\rho^{s_2}} \sum_{j=1}^{s_2} \lambda_j \int_{-\pi}^{\pi} \begin{Bmatrix} \psi_c(t, \rho;\, l,\, m_j) \\ \psi_s(t, \rho;\, l,\, m_j) \end{Bmatrix} \varphi^{(\bar{r})}(x+t)\, dt$$

$$= \frac{c}{\rho^{s_2}} \sum_{j=1}^{s_2} \lambda_j \int_{-\pi}^{\pi} \begin{Bmatrix} \psi_c \\ \psi_s \end{Bmatrix} \Delta_t \varphi^{(\bar{r})}(x)\, dt,$$

where the $m_j = jl + s_1 - \bar{r}$ are integers; $j = 1, \cdots, s_2$.

It is sufficient for proving the theorem to evaluate the addend for $j = s_2$, i.e. for $m_{s_2} = m = s_2 l + s_1 - \bar{r}$. Applying Hölder's inequality $(1/p + 1/q = 1; 0 < \epsilon < 1)$, we obtain

$$I^p = c \int_{\frac{1}{2}}^{1} \int_{-\pi}^{\pi} (1-\rho)^{\varkappa p} \left| \int_{-\pi}^{\pi} \begin{Bmatrix} \psi_c(t, \rho; l, m) \\ \psi_s(t, \rho; l, m) \end{Bmatrix} \Delta_t \varphi^{(\bar{r})}(x)\, dt \right|^p \rho d\rho dx$$

$$\leqslant c \int_{\frac{1}{2}}^{1} \int_{-\pi}^{\pi} (1-\rho)^{\varkappa p} \left[\int_{-\pi}^{\pi} \left| \begin{Bmatrix} \psi_c \\ \psi_s \end{Bmatrix} \right|^{\varepsilon} \cdot \left| \begin{Bmatrix} \psi_c \\ \psi_s \end{Bmatrix} \right|^{1-\varepsilon} \cdot | \Delta_t \varphi^{(\bar{r})}(x) |\, dt \right]^p d\rho dx$$

$$\leqslant c \int_{\frac{1}{2}}^{1} \int_{-\pi}^{\pi} (1-\rho)^{\varkappa p} \left(\int_{-\pi}^{\pi} \left| \begin{Bmatrix} \psi_c \\ \psi_s \end{Bmatrix} \right|^{\varepsilon q} dt \right)^{\frac{p}{q}} \int_{-\pi}^{\pi} \left| \begin{Bmatrix} \psi_c \\ \psi_s \end{Bmatrix} \right|^{(1-\varepsilon)p} | \Delta_t \varphi^{(\bar{r})}(x) |^p\, dt d\rho dx.$$

According to the theorem's condition, $\kappa \geqslant 0$; therefore $m = s_2 l + s_1 - \bar{r} \geqslant \alpha + l/p$; i.e. $m \geqslant 1$. Now, applying Lemma 2, we obtain for $\epsilon > 1/2q$

$$I^p \leqslant c \int_{\frac{1}{2}}^{1} \int_{-\pi}^{\pi} (1-\rho)^{\varkappa p + \frac{p-1-\varepsilon p(m+1)}{l}} \int_{-\pi}^{\pi} \left| \begin{Bmatrix} \psi_c \\ \psi_s \end{Bmatrix} \right|^{(1-\varepsilon)p} | \Delta_t \varphi^{(\bar{r})}(x) |^p\, dt d\rho dx$$

$$\leqslant c \int_{-\pi}^{\pi} \| \Delta_t \varphi^{(\bar{r})}(x) \|_p^p \int_{\frac{1}{2}}^{1} \left| \begin{Bmatrix} \psi_c \\ \psi_s \end{Bmatrix} \right|^{(1-\varepsilon)p} (1-\rho)^{\varkappa p + \frac{p-1-\varepsilon p(m+1)}{l}} d\rho dt$$

According to Lemma 3, we obtain

$$\varepsilon < 1 - \frac{1}{2}\left(\alpha + \frac{1}{p}\right) = \frac{1}{2q} + \frac{1-\alpha}{2} \quad (0 < \alpha < 1)$$

for

$$I^p \leqslant c \int_{-\pi}^{\pi} \| \Delta_t \varphi^{(\bar{r})}(x) \|_p^p \cdot | t |^{l + \varkappa p l + p - 1 - \varepsilon p(m+1) + (\varepsilon - 1)(m+1)p} dt.$$

But

$$l + \varkappa p l + p - 1 - \varepsilon p(m+1) + (\varepsilon - 1)(m+1)p = l + \varkappa p l - mp - 1$$

$$= l + \left(s_2 + \frac{s_1}{l} - \frac{r}{l} - \frac{1}{p}\right) pl - p(s_2 l + s_1 - \bar{r}) - 1 = -1 - p\alpha;$$

therefore

$$I^p \leqslant c \int_{-\pi}^{\pi} | t |^{-1-p\alpha} \| \Delta_t \varphi^{(\bar{r})}(x) \|_p^p\, dt \leqslant c \| \varphi \|_{B_p^{(r)*}}^p.$$

This implies the theorem's assertion. The estimates for the function v are carried out in the same way using formula (19). The theorem is proved.

BIBLIOGRAPHY

1. Ja. S. Bugrov, *Inequalities of the type of Bernšteĭn inequalities and their application to the investigation of the differential properties of the solutions of differential equations of higher order,* Mathematica (Cluj) **5** (28) (1963), 5–25. (Russian) MR **31** #2514.

2. S. M. Nikol'skiĭ, *Properties of certain classes of functions of several variables on differentiable manifolds,* Mat. Sb. **33** (75) (1953), 261–326; English transl., Amer. Math. Soc. Transl. (2) **80** (1969), 39–118. MR **16**, 453.

3. O. V. Besov, *Investigation of a family of function spaces in connection with imbedding and extension theorems,* Trudy Mat. Inst. Steklov. **60** (1961), 42–81; English transl., Amer. Math. Soc. Transl. (2) **40** (1964), 85–126. MR **24** #A3501.

4. A. Zygmund, *Trigonometric series.* Vol. I, 2nd rev. ed., Cambridge Univ. Press, New York, 1959; Russian transl., "Mir", Moscow, 1965. MR **21** #6498; **31** #2554.

5. Ja. S. Bugrov, *Properties of polyharmonic functions,* Izv. Akad. Nauk SSSR Ser. Mat. **22** (1958), 491–514. (Russian) MR **20** #5372.

Translated by F. M. GOLDWARE

Trudy Mat. Inst. Steklov.
117 (1972)

Proc. Steklov Inst. Math.
117 (1972)

ON APPROXIMATING FUNCTIONS OF THE SPACE $C^r(\Omega)$ BY FUNCTIONS WITH COMPACT SUPPORT, FOR AN ARBITRARY OPEN SET Ω (1)

V. I. BURENKOV

UDC 517.518.22

Abstract. It is proved that for a sequence $\varphi_s(x)$ such that

$$\varphi_s(x) \in C_0^\infty(\Omega), \qquad \lim_{s\to\infty} \|f - \varphi_s\|_{C^r(\Omega)} = 0, \tag{1}$$

to exist for a function $f \in C^r(\Omega)$, where Ω is an arbitrary open set, it is necessary and sufficient that for any $x \in \Gamma(\Omega)$,

$$\lim_{\substack{y\to x\\ y\in\Omega}} D^k f(y) = 0$$

and in the case of an unbounded open set,

$$\lim_{\substack{y\to\infty\\ y\in\Omega}} D^k f(y) = 0.$$

If a bounded open set Ω satisfies the condition $\Gamma(\Omega) = \Gamma(\overline{\Omega})$, then for (1) to be fulfilled for it it is necessary and sufficient that

$$\Phi(x) = \begin{cases} f(x), & x\in\Omega \\ 0, & x\notin\Omega \end{cases} \in C^r(E_n).$$

Bibliography: 7 items.

Let E_n be the n-dimensional euclidean space of the points $x = (x_1, \cdots, x_n)$, and let $\Omega \subset E_n$ be an arbitrary open set. The function $f(x) \in C^r(\Omega)$, $r = 0, 1, \cdots$, if both the continuous partial derivatives

$$D^k f = \frac{\partial^{k_1+\ldots k_n} f}{\partial x_1^{k_1} \ldots \partial x_n^{k_n}}, \quad |k| = k_1 + \ldots + k_n \leqslant r,$$

and the finite norm

$$\|f\|_{C^r(\Omega)} = \max_{|k|\leqslant r} \sup_{x\in\Omega} |D^k f(x)| \tag{1}$$

exist in Ω. The function $f(x) \in C^\infty(\Omega)$ if $f(x) \in C^r(\Omega)$ for any $r = 1, 2, \cdots$, and $f(x) \in C_0^\infty(\Omega)$ if $f \in C^\infty(\Omega)$ and $f(x) \equiv 0$ outside of $\Omega - K_f$, where $K_f \subset \Omega$ is some compactum.

AMS (MOS) subject classifications (1970). Primary 41A30, 41A63, 46E15.

(1) The results of the present article were presented without proof in the author's note [1].

In many problems it is necessary to consider, along with $C^r(\Omega)$, the spaces which are the closure of the set $C_0^\infty(\Omega)$ in the metric of $C^r(\Omega)$.

It is well known that if Ω is a bounded region with sufficiently "good" boundaries, these spaces coincide with the spaces $C^r(\Omega)$ of functions of $C^r(\Omega)$ which together with all of their derivatives up through order r vanish on the boundary Γ of the region Ω.

More precisely, $f \in C^r(\Omega)$ (Ω is any open set) if $f \in C^r(\Omega)$ and for any $x \in \Gamma$

$$\lim_{\substack{y \to x \\ y \in \Omega}} D^k f(y) = 0, \qquad |k| \leqslant r. \tag{2}$$

By following the argument of G. G. Kazarjan's article [2] a similar(2) assertion can be obtained for a very broad class of regions (which admit "*local shift*"). The method of approximating a function by means of its average with shift, which was utilized earlier in E. Gagliardo's article [3], is modified in [2].

In the present article it will be proved in particular that the closure in the metric of $C^r(\Omega)$ of the set $C_0^\infty(\Omega)$ of functions infinitely differentiable in Ω and having compact support coincides with the space $\dot{C}^r(\Omega)$, for an arbitrary bounded open set Ω.

As usual, let us denote by Ω_δ the totality of those $x \in \Omega$ which are at a distance of more than δ from the boundary.

THEOREM 1. *Let Ω be an arbitrary open set. For any $\delta > 0$, the following inequality is valid for $f \in C^r(\Omega) \cap \dot{C}^{r-1}(\Omega)$:*

$$\| f \|_{C(\Omega - \Omega_\delta)} \leqslant \frac{n^{\frac{r}{2}} \delta^r}{r!} \max_{|k| = r} \| D^k f \|_{C(\Omega - \Omega_\delta)}. \tag{3}$$

The constant in inequality (3) *is sharp, in the sense that for any r, n and δ*

$$\sup_{\Omega} \sup_{f \neq 0} \frac{\| f \|_{C(\Omega - \Omega_\delta)}}{\max_{|k| = r} \| D^k f \|_{C(\Omega - \Omega_\delta)}} = \frac{n^{\frac{r}{2}} \delta^r}{r!}. \tag{4}$$

PROOF. Let $y \in \Omega - \Omega_\delta$. Let us denote by x some point of the boundary $\Gamma(\Omega)$ for which $|x - y| = \rho(y, \Gamma) = \rho \leqslant \delta$. By Taylor's formula for the function

---

(2) The space $C^r(\Omega)$ is defined as the set of functions $f \in C^r(\Omega)$ which preserve differential properties under continuation as zero on $E_n - \Omega$, i.e.

$$\Phi(x) = \begin{cases} f(x), & x \in \Omega, \\ 0, & x \notin \Omega, \end{cases} \in C^r(E_n)$$

(cf. the Remark below and Theorems 3 and 5).

$\varphi(t) = f(x + t(y - x))$ we have

$$f(y) = \varphi(1) = \varphi(0) + \cdots + \frac{\varphi^{(r-1)}(0)}{(r-1)!} + \frac{1}{(r-1)!}\int_0^1 (1-t)^{r-1}\varphi^{(r)}(t)\,dt.$$

By $\varphi(0), \cdots, \varphi^{(r-1)}(0)$ we understand the limits of the functions $\varphi(t), \cdots, \varphi^{(r-1)}(t)$ as $t \to +0$, which equal zero since $f \in \dot{C}^{r-1}(\Omega)$; therefore

$$\begin{aligned} f(y) &= \frac{1}{(r-1)!}\int_0^1 (1-t)^{r-1}\varphi^{(r)}(t)\,dt \\ &= \frac{1}{(r-1)!}\int_0^1 (1-t)^{r-1}\left[\sum_{|k|=r}\frac{r!}{k!}D^k f(x + t(y-x))(y-x)^k\right]dt. \end{aligned} \tag{5}$$

Here we have written $(y - x)^k = (y_1 - x_1)^{k_1} \cdots (y_n - x_n)^{k_n}$ and $k! = k_1! \cdots k_n!$. Since $|y_1 - x_1| + \cdots + |y_n - x_n| \leqslant \delta\sqrt{n}$, we have

$$\begin{aligned} |f(y)| &\leqslant \frac{1}{r!}\max_{|k|=r}\sup_{z\in(x,y]}|D^k f(z)|\sum_{|k|=r}\frac{r!}{k_1!\ldots k_n!}|y_1 - x_1|^{k_1}\ldots|y_n - x_n|^{k_n} \\ &\leqslant \frac{1}{r!}\max_{|k|=r}\|D^k f\|_{C(\Omega-\Omega_\delta)}(|y_1 - x_1| + \ldots + |y_n - x_n|)^r \\ &\leqslant \frac{n^{\frac{r}{2}}\delta^r}{r!}\max_{|k|=r}\|D^k f\|_{C(\Omega-\Omega_\delta)}, \end{aligned} \tag{6}$$

for $y \in \Omega - \Omega_\delta$, which was to be proved.

Now let Ω be the half-space defined by the inequality $x_1 + \cdots + x_n > 0$. Let us set $f(x) = (x_1 + \cdots + x_n)^r$. This function satisfies the theorem's conditions; both $\|f\|_{C(\Omega-\Omega_\delta)} = n^{r/2}\delta^r$ for it and $\|D^k f\|_{C(\Omega-\Omega_\delta)} = r!$ for $|k| = r$; therefore for this function inequality (3) turns into equality.

COROLLARY 1. *Let Ω be a bounded open set, and let d be its diameter. If $f(x) \in C^r(\Omega) \cap \dot{C}^{r-1}(\Omega)$, then*

$$\|f\|_{C(\Omega)} \leqslant \frac{n^{\frac{r}{2}}}{r!}\left(\frac{d}{2}\right)^r \max_{|k|=r}\|D^k f\|_{C(\Omega)}. \tag{7}$$

COROLLARY 2. *Let K_δ^x be a sphere of radius δ, with center at the point $x \in \Gamma$. If $f(x) \in C^r(\Omega) \cap \dot{C}^{r-1}(\Omega)$, then*

$$\|f\|_{C(\Omega K_\delta^x)} \leqslant \frac{n^{\frac{r}{2}}\delta^r}{r!}\max_{|k|=r}\|D^k f\|_{C(\Omega K_\delta^x)} \tag{8}$$

for any $\delta > 0$ and $x \in \Gamma$.

This is proved just as inequality (3). The constant in inequality (8) is also sharp.

LEMMA 1. *Let* $f \in \dot{C}^r(\Omega)$, *where* Ω *is an arbitrary open set. Then*

$$\Phi(x) = \begin{cases} f(x), & x \in \Omega, \\ 0, & x \notin \Omega, \end{cases} \in C^r(E_n). \tag{9}$$

PROOF. If $x \in \Omega$ or $x \notin \overline{\Omega}$, the existence and continuity of the derivatives $D^k\Phi$, $|k| \leqslant r$, are obvious. Let $x \in \Gamma$; then $\Phi(x) = 0$, and

$$\frac{\partial \Phi}{\partial x_1}(x) = \lim_{\Delta x_1 \to 0} \frac{\Phi(x_1 + \Delta x_1, x_2, \ldots, x_n)}{\Delta x_1} = 0,$$

since, by virtue of (8),

$$\Phi(x_1 + \Delta x_1, x_2, \ldots, x_n) = f(x_1 + \Delta x_1, x_2, \ldots, x_n) = o(|\Delta x_1|^r)$$

for $(x_1 + \Delta x_1, x_2, \cdots, x_n) \in \Omega$. The function $\partial\Phi/\partial x_1$ is continuous at the points $x \in \Gamma$. In fact, $\lim_{y \to x}(\partial\Phi/\partial x_1)(y) = 0$, since $\partial\Phi/\partial x_1 = 0$ for $y \notin \Omega$, while, by virtue of (8),

$$\frac{\partial \Phi}{\partial x_1}(y) = \frac{\partial f}{\partial x_1}(y) = o(|y - x|^{r-1})$$

for $y \in \Omega$. The existence and continuity of the derivatives $\partial\Phi/\partial x_i$, $i = 2, \cdots, n$, are proved in just the same way, and then also the derivatives $D^k\Phi$, $1 \leqslant |k| \leqslant r$.

LEMMA 2. *Let* $f \in \dot{C}^r(\Omega)$, *where* Ω *is an arbitrary bounded open set; then, as* $\delta \to 0$,

$$\|f\|_{C^l(\Omega - \Omega_\delta)} = o(\delta^{r-l}); \qquad l = 0, 1, \ldots, r. \tag{10}$$

PROOF. It is sufficient to prove (10) for the case $r = l = 0$. In the general case, (10) follows from this special case and inequality (3).

So let $f \in \dot{C}(\Omega)$. This means that for any $x \in \Gamma$

$$\|f\|_{C(\Omega K_\delta^x)} = o(1), \qquad \delta \to 0. \tag{11}$$

We shall prove that (11) is fulfilled uniformly with respect to $x \in \Gamma$. Let us assume the contrary: there is ϵ_0 such that for any n there exist $x_n \in \Gamma$ and δ_n, $0 \leqslant \delta_n < n^{-1}$, such that

$$\|f\|_{C(\Omega K_{\delta_n}^{x_n})} \geqslant \varepsilon_0. \tag{12}$$

Since Γ is a compact set, a subsequence (let us again denote it by x_n) can be picked out which converges to the point $x_0 \in \Gamma$. But $\|f\|_{C(\Omega K_\delta^x)} = o(1)$ as $\delta \to 0$; hence for sufficiently small δ_0

$$\|f\|_{C\left(\Omega K^{x_0}_{\delta_0}\right)} < \varepsilon_0,$$

which contradicts (12) since $\Omega K^{x_n}_{\delta_n} \subset \Omega K^{x_0}_{\delta_0}$ for sufficiently large n.

Since condition (11) is fulfilled uniformly with respect to $x \in \Gamma$, it implies (10):

$$\|f\|_{C(\Omega-\Omega_\delta)} \leqslant \|f\|_{C\left(\bigcup\limits_{x\in\Gamma} \Omega K^x_{2\delta}\right)} \leqslant \sup_{x\in\Gamma} \|f\|_{C\left(\Omega K^x_{2\delta}\right)} = o(1), \quad \delta \to 0.$$

THEOREM 2. *Let $f \in C^r(\Omega)$, where Ω is an arbitrary open set in E_n. For a sequence $\varphi_s \in C_0^\infty(\Omega)$ to exist such that*

$$\lim_{s\to\infty} \|f - \varphi_s\|_{C^r(\Omega)} = 0, \tag{13}$$

it is necessary and sufficient that $f \in \dot{C}^r(\Omega)$ and, in the case of an unbounded open set Ω,

$$\lim_{\substack{y\to\infty\\ y\in\Omega}} D^k f(y) = 0, \qquad |k| \leqslant r. \tag{14}$$

PROOF. NECESSITY. For given $\epsilon > 0$, let us choose s_0 so that $\|f - \varphi_{s_0}\|_{C^r(\Omega)} < \epsilon$; then

$$|D^k f(y)| \leqslant |D^k f(y) - D^k_{\varphi_{s_0}}(y)| + |D^k \varphi_{s_0}(y)| < \varepsilon + |D^k \varphi_{s_0}(y)|$$

for $y \in \Omega$ and $|k| \leqslant r$.

Let $x \in \Gamma$ and $|y - x| \dot{<} \rho$, where ρ is the distance from x to the support of the function $\varphi_{s_0}(x)$; then $|D^k f(y)| < \epsilon$, i.e.

$$\lim_{\substack{y\to x\\ y\in\Omega}} D^k f(y) = 0, \qquad |k| \leqslant r,$$

and $f \in \dot{C}^r(\Omega)$. If Ω is an unbounded open set, let us choose l so that $K_l \supset \operatorname{supp} \varphi_{s_0}(x)$ (K_l is a sphere of radius l, with center at the coordinate origin); then

$$\|f\|_{C^r(\Omega-K_l)} \leqslant \|f - \varphi_{s_0}\|_{C^r(\Omega-K_l)} \leqslant \|f - \varphi_{s_0}\|_{C^r(\Omega)} < \varepsilon,$$

from which (14) follows.

SUFFICIENCY. First let Ω be a bounded open set. Let us set

$$f_\delta(x) = \frac{1}{\delta^n} \int\limits_{\Omega_{2\delta}} \omega\left(\frac{x-y}{\delta}\right) f(y)\, dy, \tag{15}$$

where $\omega(z) \in C^\infty(E_n)$; $\omega(z) = 0$ if $|z| \geqslant 1$; $\int_{E_n} \omega(z) dz = 1$. The function $f_\delta(x)$ is an average, with averaging radius equal to the δ of the function

$$\Phi_{2\delta}(x)=\begin{cases} f(x), & x\in\Omega_{2\delta},\\ 0, & x\notin\Omega_{2\delta},\end{cases}$$

i.e. $f_\delta(x)=[\Phi_{2\delta}(x)]_\delta$. The function $f_\delta(x)\in C_0^\infty(\Omega)$. Let us prove that

$$\lim_{\delta\to 0}\|f(x)-f_\delta(x)\|_{C^r(\Omega)}=0. \tag{16}$$

First of all,

$$\|f-f_\delta\|_{C^r(\Omega)}\leqslant\|f-f_\delta\|_{C^r(\Omega_{3\delta})}+\|f\|_{C^r(\Omega-\Omega_{3\delta})}+\|f_\delta\|_{C^r(\Omega-\Omega_{3\delta})}.$$

Since the equality $f_\delta(x)=[f(x)]_\delta=[\Phi(x)]_\delta$ is valid for $x\in\Omega_{3\delta}$ and $\Phi(x)\in C^r(E_n)$ (cf. Lemma 1), we have

$$\|f-f_\delta\|_{C^r(\Omega_{3\delta})}=\|\Phi-[\Phi]_\delta\|_{C^r(\Omega_{3\delta})}\leqslant\|\Phi-[\Phi]_\delta\|_{C^r(E_n)}\to 0$$

as $\delta\to 0$ because of the properties of average functions. More explicitly:

$$\|\Phi-[\Phi]_\delta\|_{C(E_n)}=\sup_{x\in E_n}\left|\Phi(x)-\frac{1}{\delta^n}\int\limits_{|x-y|\leqslant\delta}\omega\left(\frac{x-y}{\delta}\right)\Phi(y)\,dy\right|$$

$$=\sup_{x\in E_n}\left|\frac{1}{\delta^n}\int\limits_{|x-y|\leqslant\delta}\omega\left(\frac{x-y}{\delta}\right)(\Phi(x)-\Phi(y))\,dy\right|\leqslant\sup_{|x-y|\leqslant\delta}|\Phi(x)-\Phi(y)|\to 0$$

as $\delta\to 0$, by virtue of the uniform continuity of the function $\Phi(x)$ on E_n. That $\|D^k\Phi-D^k[\Phi]_\delta\|_{C(E_n)}\to 0$ as $\delta\to 0$, $|k|\leqslant r$, is proved analogously. It must be taken into account that $D^k[\Phi]_\delta=[D^k\Phi]_\delta$ for all $x\in E_n$, since $\Phi(x)\in C^r(E_n)$. Furthermore, for $l=r$, (10) implies that as $\delta\to 0$,

$$\|f\|_{C^r(\Omega-\Omega_{3\delta})}\to 0.$$

Taking into account that $\omega((x-y)/\delta)=0$ for $|x-y|\geqslant\delta$, we obtain for $x\in\Omega-\Omega_{3\delta}$

$$|D^kf_\delta(x)|\leqslant\sup_{\substack{y\in\Omega_{2\delta}\\|y-x|\leqslant\delta}}|f(y)|\frac{1}{\delta^{n+|k|}}\int\limits_{|x-y|\leqslant\delta}\left|D^k\omega\left(\frac{x-y}{\delta}\right)\right|dy$$

$$=\frac{1}{\delta^{|k|}}\|f\|_{C(\Omega_{2\delta}K_\delta^x)}\int\limits_{|z|\leqslant 1}|D^k\omega(z)|\,dz\leqslant\frac{c_k}{\delta^{|k|}}\|f\|_{C(\Omega K_\delta^x)}$$

since

$$D^kf_\delta(x)=\frac{1}{\delta^{n+|k|}}\int\limits_{\Omega_{2\delta}}D^k\omega\left(\frac{x-y}{\delta}\right)f(y)\,dy. \tag{17}$$

Thus

$$\|D^kf_\delta\|_{C(\Omega-\Omega_{3\delta})}\leqslant\frac{c_k}{\delta^{|k|}}\|f\|_{C(\Omega-\Omega_{4\delta})}. \tag{18}$$

Lemma 2 and (18) imply that as $\delta \to 0$

$$\| f_\delta \|_{C^r(\Omega-\Omega_{3\delta})} \leqslant \frac{c_r}{\delta^r} \| f \|_{C(\Omega-\Omega_{4\delta})} \leqslant c_r \frac{o(\delta^r)}{\delta^r} \to 0.$$

By the same token (16) is proved.

If Ω is an unbounded open set, let us consider first the sequence of functions $f_s(x) = f(x)\psi(|x|/s)$, where $\psi(y) \in C^\infty(E_1)$, $\psi(y) = 1$ for $|y| \leqslant 1$, and $\psi(y) = 0$ for $|y| \geqslant 2$. It is not difficult to see that $f_s(x) \in \dot{C}^r(\Omega K_{2s})$ and that (14) implies that $\lim_{s\to\infty} \|f - f_s\|_{C^r(\Omega)} = 0$. This and what is proved above for bounded open sets imply the validity of Theorem 2 in the general case also.

REMARK. The question of the possibility of approximating the function $f \in C^r(\Omega)$ with any degree of accuracy by means of infinitely differentiable functions with compact support is closely connected with the question of the possibility of continuing f as zero on $E_n - \Omega$ with preservation of the class. If condition (13) is fulfilled, the function $\Phi(x)$ defined by equality (9) belongs to the space $C^r(E_n)$, which follows from Theorem 2 and Lemma 1, already proved. This result can also be obtained immediately. Indeed, if (13) is fulfilled, then

$$\| \varphi_s - \varphi_t \|_{C^r(E_n)} = \| \varphi_s - \varphi_t \|_{C^r(\Omega)} \to 0$$

as $s, t \to \infty$. The completeness of the space $C^r(E_n)$ implies the existence of a function $F(x) \in C^r(E_n)$ such that $\|F - \varphi_s\|_{C^r(E_n)} \to 0$ as $s \to \infty$. If $x \in \Omega$, (13) implies that $F(x) = f(x)$. If $x \notin \Omega$, then $F(x) = 0$, since $\varphi_s(x) = 0$ for any s and, as $s \to \infty$,

$$|F(x)| = |F(x) - \varphi_s(x)| \leqslant \| F(x) - \varphi_s(x) \|_{C^r(E_n)} \to 0.$$

Thus $F(x) = \Phi(x)$.

Let us note that in general the converse is untrue. For example, let Ω be the cube $\{|x_i| \leqslant 1\}$, from which part of the plane $x_n = 0, 0 \leqslant x_i \leqslant 1, i = 1, \cdots, n-1$, has been removed, while $f(x) = \rho(x) \sin x_n$, where $\rho(x)$ is an infinitely differentiable function equal to unity for $|x_i| < ½, i = 1, \cdots, n$, and equal to zero if $|x_i| > ¾$ for at least one i. It is obvious that $\Phi(x) \in C^r(E_n)$, but one cannot approximate the function $f(x)$ by means of the functions $\varphi_s \in C_0^\infty(\Omega)$ as accurately as one would wish, since $\partial f/\partial x_n = 1$ for $x_n = 0$, $0 < x_i < ½, i = 1, \cdots, n-1$; i.e. $f \notin \dot{C}^r(\Omega)$.

However, under certain restrictions on Ω, condition (9) is sufficient for the validity of (13).

THEOREM 3. *Let Ω be a bounded open set satisfying the following condition:*

$$\Gamma(\Omega) = \Gamma(\bar{\Omega}), \tag{19}$$

and $f \in C^r(\Omega)$. Then for a sequence $\varphi_s \in C_0^\infty(\Omega)$ to exist such that

$$\lim_{s\to\infty}\|f-\varphi_s\|_{C^r(\Omega)}=0, \tag{13}$$

it is necessary and sufficient that

$$\Phi(x)=\begin{cases} f(x), & x\in\Omega,\\ 0, & x\notin\Omega,\end{cases}\quad \in C^r(E_n). \tag{9}$$

PROOF. First of all, condition (19) is equivalent to the following condition:

$$\begin{array}{l}\text{for any } x\in\Gamma(\Omega) \text{ and any neighborhood } U_k \text{ of } x,\\ \text{a sphere } K \text{ exists such that } K\subset U_x \text{ and } K\Omega=\phi.\end{array} \tag{20}$$

As a matter of fact, if (19) is fulfilled and $x\in\Gamma(\Omega)$, then there is $y\notin\overline{\Omega}$ in U_x; hence $y\in E_n-\overline{\Omega}$, and y enters into U_x together with some neighborhood of points not belonging to Ω. Conversely, if (20) is fulfilled and $x\in\Gamma(\Omega)$, there is a point $y\notin\overline{\Omega}$ in K (since in the opposite case all points $y\in K$ would be points of the boundary $\Gamma(\Omega)$, which is impossible); consequently $\Gamma(\Omega)\subset\Gamma(\overline{\Omega})$, from which we obtain (19) (since $\Gamma(\overline{\Omega})\subset\Gamma(\Omega)$ for any Ω).

Then let (9) be fulfilled and let $x\in\Gamma(\Omega)$; then by virtue of (20) a sequence $y_n\to x$ exists such that $D^k\Phi(y_n)=0$.

Consequently for $|k|\leqslant r$

$$\lim_{\substack{y\to x\\ y\in\Omega}} D^k f(y)=\lim_{y\to x} D^k\Phi(y)=\lim_{n\to\infty} D^k\Phi(y_n)=0,$$

i.e. $f\in\dot{C}^r(\Omega)$. It is now sufficient to use Theorem 2. (Necessity is proved in the Remark, for any open set Ω.)

The necessary and sufficient conditions on $f\in C^r(\Omega)$ under which (13) is fulfilled can be formulated in terms of continuation on $E_n-\Omega$ also in the case of an arbitrary open bounded set.

THEOREM 4. *Let Ω be an arbitrary open bounded set, and let $f\in C^r(\Omega)$. For condition* (13) *to be fulfilled, it is necessary and sufficient that for any k, $|k|\leqslant r$,*

$$\Phi_k(x)=\begin{cases} D^k f(x), & x\in\Omega,\\ 0, & x\notin\Omega,\end{cases}\quad \in C(E_n). \tag{21}$$

PROOF. Since $D^k f\in C(\Omega)$, necessity is proved in the Remark. Furthermore, (21) implies that for $x\in\Gamma$

$$\lim_{\substack{y\to x\\ y\in\Omega}} D^k f(y)=0,\quad |k|\leqslant r.$$

Thus Theorem 2 implies that condition (21) is sufficient.

Below we deduce the necessary and sufficient condition on Ω under which conditions (13) and (9) are equivalent for any $f \in C^r(\Omega)$. To this end we must introduce the following definitions.

Let G be some nonempty set in E_n. For $x \in G$, let us denote by K_x the set of partial limits of the function $(y - z)/|y - z|$ as $y, z \to x$, where $y, z \in G$, i.e. the set of vectors τ such that in G sequences $y_k \to x$ exist for them for which

$$\tau = \lim_{k\to\infty} \frac{y_k - z_k}{|y_k - z_k|}. \tag{22}$$

In addition, let L_x be the subspace spanned by the vectors of K_x,

$$L_x^{(s)} = L_x^{(s-1)} + \overline{\lim_{\substack{y\to x\\ y\in G}}} L_y^{(s-1)}, \qquad s = 1, 2, \ldots,\text{(3)} \quad L_x^{(0)} \equiv L_x, \qquad M_x = \sum_{s=0}^{\infty} L_x^{(s)} \tag{23}$$

(an error was made at this point in [1]). Let us call M_k the *tangent subspace* of the set G at the point x, and let

$$n_x = \dim M_x. \tag{24}$$

If $n_x = m$ for any $x \in G$, we shall say that the set G has the "uniform dimension" m: $\dim^* G = m$ (this definition is not equivalent to the conventional definitions of dimension which are cited, for example, in [4]; cf. the following examples).

EXAMPLES. To write $\dim^* G = 0$ for $G \subset E_1$ means that the set G consists of isolated points; to write $\dim^* G = 1$ means that G does not contain isolated points. If G is an $(n - 1)$-dimensional differentiable surface, then $\dim^* G = n - 1$, and $M_x = L_x$ represents the set of vectors lying in the tangent plane of the surface at the point x. If G is a straight line, $\dim^* G = 1$. If, however, G is a configuration consisting of m straight lines which do not lie in the same $(m - 1)$-dimensional plane and intersect at the single point x_0, $\dim^* G$ is not defined: for $x \in G, x \neq x_0$,

---

(3) Let us recall the respective definitions. Let the L_α be subspaces. The subspace consisting of vectors each of which is the limit of some function l_α as $\alpha \to \alpha_0$ (for any $l_\alpha \in L$) is called the lower limit $\underline{\lim}_{\alpha\to\alpha_0} L_\alpha$. The subspace which is the sum of the subspaces $\underline{\lim}_{s\to\infty} L_{\alpha_s}$ corresponding to all possible sequences α_s converging to α_0 is called the upper limit $\overline{\lim}_{\alpha\to\alpha_0} L_\alpha$. If the lower and upper limits coincide, then

$$\lim_{\alpha\to\alpha_0} L_\alpha = \underline{\lim}_{\alpha\to\alpha_0} L_\alpha = \overline{\lim}_{\alpha\to\alpha_0} L_\alpha.$$

A vector $h \in \overline{\lim}_{\alpha\to\alpha_0} L_\alpha$ if and only if it is representable as a finite linear combination of the vectors $l_i = \lim_{k\to\infty} l_i^{(k)}$, where $l_i^{(k)}$ is some convergent sequence of vectors $l_i^{(k)} \in L_{\alpha_k}$ $(\alpha_k \to \alpha_0)$ (apropos of these definitions, see, for example, [5], pp. 224–225).

we have $n_x = 1$, but $n_{x_0} = m$; in the given case $M_{x_0} = L_{x_0}$ is the set of vectors lying in the m-dimensional plane spanned by the given straight lines. Now let $G \subset E_3$ consist of the point $x_0 = (0, 0, 0)$ and segments $[x_k, x_{k+1}]$, $k = 1, 2, \cdots$, where $x_k = (1/k, 0, 0)$ for $k = 2l - 1$, $x_k = (1/k, 1/k^3, 0)$ for $k = 4l - 2$, and $x_k = (1/k, 0, 1/k^3)$ for $k = 4l$ (everywhere $l = 1, 2, \cdots$). In this case $n_x = 1$ for $x \in G$, $x \neq x_0, x_1, \cdots$; $n_{x_k} = 2$ for $k = 1, 2, \cdots$, and $n_{x_0} = 3$, where $M_{x_0} = L_{x_0}^{(1)} = L_{x_0}^{(2)} = \cdots$; $L_{x_0}^{(1)} \neq L_{x_0}$ (here $M_{x_0} = E_3$ and L_{x_0} coincides with the axis of x_1).

THEOREM 5. *Let $\Omega \subset E_n$ be an open set. For condition* (13) *to be equivalent to condition* (9) *for any function $f \in C^r(\Omega)$, it is necessary and sufficient that the set Ω be bounded and*

$$\dim^*(E_n - \Omega) = n. \tag{25}$$

(25) follows easily from condition (19)–(20), since in this case a sequence $x_k \in E_n - \Omega$, $x_k \to x$, exists for any $x \in \Gamma$ such that $L_{x_k} = E_n$. If $\Omega \subset E_1$, condition (25) is equivalent to the fact that every point of the set $\Gamma(\Omega) - \Gamma(\overline{\Omega})$ is a limit point. Thus the open Cantor set is a very simple example of a set $\Omega \subset E_1$ for which $\Gamma(\Omega) \neq \Gamma(\overline{\Omega})$ but $\dim^*(E_1 - \Omega) = 1$.

We need a number of lemmas for the proof of Theorem 5.

LEMMA 3. *Let $f \in C^1(E_n)$ and $f(x) = 0$ on $G \subset E_n$. If $x \in G$ and the unit vector $l \in M_x$, then $(\partial f/\partial l)(x) = 0$.*

PROOF. First let $l \in K_x$, i.e. let sequences $y_k, z_k \in G$, $y_k, z_k \to x$, exist such that

$$l = \lim_{k\to\infty} l_k = \lim_{k\to\infty} \frac{y_k - z_k}{|y_k - z_k|}.$$

Since $f(y_k) = f(z_k) = 0$, a point $\xi_k \in (y_k, z_k)$ exists such that $(\partial f/\partial l_k)(\xi_k) = 0$. Since $l_k \to l$, $\xi_k \to x$ and $f \in C^1(E_n)$, by passing to the limit we obtain $(\partial f/\partial l)(x) = 0$.

If a unit vector $l \in M_x$, then $l \in L_x^{(s)}$ for some s. Let us argue by induction. If $l \in L_x^{(0)}$, then $l = \sum_1^m \alpha_i l_i$, where $l_1, \cdots, l_m$ are a normalized basis in $L_x^{(0)}$, $l_i \in K_x$. Consequently $(\partial f/\partial l_i)(x) = 0$ also. Let us assume that $(\partial f/\partial l)(x) = 0$ is proved for $l \in L_x^{(s-1)}$. Then, if $l \in L_x^{(s)}$, we have $l = \alpha\mu + \beta\nu$, where μ and ν are unit vectors, $\mu \in L_x^{(s-1)}$ and $\nu \in \overline{\lim}_{y\to x} L_y^{(s-1)}$.(4)

(4) We shall write $\overline{\lim}_{y\to x} L_y$ instead of $\overline{\lim}_{y\to x, y\in G} L_y$ here and everywhere in the sequel where ambiguity is not possible.

In turn, the vector ν is representable as a linear combination of unit vectors $\nu_i = \lim_{k\to\infty}\nu_i^{(k)}$, where $|\nu_i^{(k)}| = 1$, $\nu_i^{(k)} \in L_{y_i^{(k)}}^{(s-1)}$, $y_i^{(k)} \in G$ and $y_i^{(k)} \to x$ as $k \to \infty$. By virtue of the induction hypothesis, $(\partial f/\partial\mu)(x)$ and $(\partial f/\partial\nu_i^{(k)})(y_i^{(k)}) \to 0$. Passing to the limit as $k \to \infty$ and taking into account that $f \in C^1(E_n)$, we obtain $(\partial f/\partial\nu_i)(x) = 0$. Consequently $(\partial f/\partial\nu)(x) = 0$, and hence $(\partial f/\partial l)(x) = 0$ also, which was to be proved.

LEMMA 4. *Let $x \in G$ be a limit point of the set $G \subset E_n$; then*

$$\overline{\lim_{\substack{y\to x\\ y\in G}}} M_y \subset M_x. \tag{26}$$

PROOF. Since $L_x^{(i)} \subset L_x^{(i+1)}$, $i = 0, 1, \cdots$, and $\overline{\lim}_{y\to x} L_y^{(i)} \subset L_x^{(i+1)}$, we have

$$\overline{\lim_{y\to x}} M_y = \overline{\lim_{y\to x}} \sum_{s=0}^{\infty} L_y^{(s)} \subset \sum_{s=0}^{\infty} \overline{\lim_{y\to x}} L_y^{(s)} \subset \sum_{s=0}^{\infty} L_y^{(s+1)} \subset M_x, \tag{27}$$

which was to be proved.

LEMMA 5. *Let $x \in G$ be a limit point of the set G. A neighborhood U_x of the point x exists such that*

$$n_y \leqslant n_x \tag{28}$$

for all $y \in U_x G$.

PROOF. First of all, if $\dim M_k = m$ for $k = 1, 2, \cdots$, then $\dim \overline{\lim}_{k\to\infty} M_k \geqslant m$. Indeed, let us choose an orthonormal basis $\tau_1^{(k)}, \cdots, \tau_m^{(k)}$ in M_k and pick out a sequence k_p such that the $\lim_{p\to\infty}\tau_i^{(k_p)} = \tau_i$ exist. The vectors τ_i, $i = 1, \cdots, m$, are linearly independent; consequently $\dim \overline{\lim}_{k\to\infty} M_k \geqslant m$.

Now let us assume the opposite of (28): a sequence $y_k \to x$ exists such that $n_{y_k} \geqslant n_x + 1$. Since, by virtue of Lemma 4,

$$\overline{\lim_{k\to\infty}} M_{y_k} \subset \overline{\lim_{y\to x}} M_y \subset M_x,$$

it follows that $\dim M_x \geqslant n_x + 1$. Contradiction.

LEMMA 6. *Let $G \subset E_n$, and let $\dim^* G = m$; then the function M_x is continuous on G, i.e.*

$$\lim_{\substack{y\to x\\ y\in G}} M_y = M_x \tag{29}$$

for any limit point $x \in G$.

PROOF. Considering Lemma 4, it is sufficient to prove that $\underline{M}_x = \underline{\lim}_{y\to x} M_y \supset M_x$. Let us set $s = \dim \underline{M}_x$ and assume the opposite. Since, by virtue of Lemma 4, $\underline{M}_x \subset \overline{M}_x \subset M_x$, this means that $s < m$. Let us denote by $\tau_1, \cdots, \tau_s$ an orthonormal basis in $\underline{M}_x$. Then the functions $\tau_i^{(y)} \in M_y$ exist such that $\lim_{y\to x}\tau_i^{(y)} = \tau_i$, $i = 1, \cdots, s$. For y sufficiently close to x the vectors $\tau_1^{(y)}, \cdots, \tau_s^{(y)}$ are linearly independent. Let us fill them out with the unit vectors $e_{s+1}^{(y)}, \cdots, e_m^{(y)}$ to a basis in M_y, letting the vectors $e_{s+1}^{(y)}, \cdots, e_m^{(y)}$ be orthogonal among themselves and to $\tau_1^{(y)}, \cdots, \tau_s^{(y)}$. The limit $\lim_{y\to x} e_{s+1}^{(y)}$ does not exist (in other words, $\dim \underline{M}_x > s$); hence two noncollinear vectors $e_{s+1}^{(1)}$ and $e_{s+1}^{(2)}$ and sequences $y_k \to x$ and $z_k \to x$ exist such that $e_{s+1}^{(1)} = \lim_{k\to\infty} e_{s+1}^{(y_k)}$ and $e_{s+1}^{(2)} = \lim_{k\to\infty} e_{s+1}^{(z_k)}$. We obtain the vectors $e_{s+2}^{(1)}, e_{s+2}^{(2)}, \cdots, e_m^{(1)}, e_m^{(2)}$ analogously. By virtue of the construction, the vectors $\tau_1, \cdots \tau_s, e_{s+1}^{(1)}, \cdots, e_m^{(1)}$ are mutually orthogonal. The vector $e_{s+1}^{(2)}$ is, by virtue of the construction, orthogonal to $\tau_1, \cdots, \tau_s, e_{s+2}^{(1)}, \cdots, e_m^{(1)}$ and not collinear with $e_{s+1}^{(1)}$, i.e. the vectors $\tau_1, \cdots, \tau_s, e_{s+1}^{(1)}, \cdots, e_m^{(1)}, e_{s+1}^{(2)}$ are linearly independent. Since these vectors are contained in $\overline{\lim}_{y\to x} M_y \subset M_x$, it follows that $\dim M_x \geqslant m + 1$. Contradiction.

LEMMA 7. *Let $G \subset E_n$, $\dim^* G = m < n$ and $x_0 \in G_0$. A neighborhood U of the point x_0 and a vector function $\tau(x)$ exist such that*

1) $|\tau(x)| = 1$,

2) $\tau(x) \perp M_x$ *for* $x \in UG$,

3) $\tau(x)$ *is continuous in* UG.

PROOF. First let $m = 0$. Then a neighborhood U of the point x_0 exists such that $UG = \{x_0\}$; in addition, $M_{x_0} = \{0\}$. Therefore $\tau(x)$ is defined only at the point x_0; any unit vector may be taken as $\tau(x_0)$.

Now let $m = n - 1$. Let us take a unit basis vector in $M_{x_0}^\perp$ as $\tau(x_0)$. Let us choose a neighborhood U of the point x_0 small enough for the angle between $M_{x_0}^\perp$ and $M_x^\perp$ not to exceed $\pi/4$. (This is possible since $\lim_{x\to x_0, x\in G} M_x^\perp = M_{x_0}^\perp$, by virtue of Lemma 6.) Of two possible unit basis vectors in $M_x^\perp$, let us take as $\tau(x)$ the one for which the angle between $\tau(x_0)$ and $\tau(x)$ is acute. By virtue of the construction, $\tau(x) \perp M_x$, and by virtue of Lemma 6 the function $\tau(x)$ is continuous in UG.

In the general case let $f_1(x), \cdots, f_m(x)$ be an orthonormal basis in M_x. We fill it out for $x = x_0$ to an orthonormal basis in E_n: $f_1(x_0), \cdots, f_m(x_0), e_{m+1}, \cdots, e_n$. We set $\tau(x_0) = e_n$. We choose the vector $\tau(x)$ for $x \in UG$ so that $\tau(x) \perp f_1(x), \cdots, f_m(x), e_{m+1}, \cdots, e_{n-1}$, and so that the angle between $\tau(x_0)$ and $\tau(x)$ is acute.

LEMMA 8. *Under the assumptions of Lemma 7 for $x \in UG$,*

$$\lim_{\substack{y,\, z\to x \\ y,\, z\in UG}} \left(\tau(y), \frac{y-z}{|y-z|}\right) = 0 \tag{30}$$

PROOF. Let $f_1, \cdots, f_m$ be an orthonormal basis in M_x. Let us fill it out to an orthonormal basis in E_n so that the vector $\tau \equiv \tau(x)$: $f_1, \cdots, f_m, f_{m+1}, \cdots, f_n$; $f_n \equiv \tau$ enters into the complement of basis vectors.

Let us expand the vectors $(y-z)/|y-z|$ and $\tau(y)$ in this basis:

$$\frac{y-z}{|y-z|} = \sum_{i=1}^{n} \alpha_i(y, z) f_i, \qquad \tau(y) = \sum_{i=1}^{n} \beta_i(y) f_i.$$

Since all partial limits of the function $(y-z)/|y-z|$ belong to M_x, we have

$$\lim_{\substack{y, z \to x \\ y, z \in \complement G}} \alpha_i(y, z) = 0, \quad i = m+1, \ldots, n. \tag{31}$$

On the other hand,

$$\lim_{\substack{y \to x \\ y \in \complement G}} \beta_i(y) = 0, \quad i = 1, \ldots, n-1; \quad \lim_{\substack{y \to x \\ y \in \complement G}} \beta_n(y) = 1, \tag{32}$$

by virtue of the continuity of $\tau(y)$. (31) and (32) imply (30), which was to be proved.

PROOF OF THEOREM 5. SUFFICIENCY. Let $x \in \Gamma$. Since $n_x = n$, $\dim M_x = n$; let us choose a basis $l_1, \cdots, l_n$ in M_x. By virtue of Lemma 3, $(\partial f/\partial l_1)(x) = \cdots = (\partial f/\partial l_n)(x) = 0$, and hence also $(\partial f/\partial x_i)(x) = 0$ for $i = 1, \cdots, n$. By applying Lemma 3 repeatedly we obtain the fact that $D^k f(x) = 0$ for all $|k| \leqslant r$. Thus $f \in \dot{C}^r(\Omega)$, and hence, according to Theorem 2, (13) is fulfilled.

NECESSITY. 1. Let us first prove the necessity of condition (25). To this end in the case when $\dim^*(E_n - \Omega) \neq n$ we construct a function $\Phi(x) \in C^r(E_n)$ such that $\Phi(x) = 0$ outside of Ω and such that $(\partial^r \Phi/\partial x_i^r)(x_0) \neq 0$ for at least one i and some $x_0 \in \Gamma$. Let us denote by $f(x)$ the restriction of the function $\Phi(x)$ to Ω: the function $f \notin \dot{C}^r(\Omega)$, since

$$\lim_{\substack{y \to x \\ y \in \Omega}} \frac{\partial^r f}{\partial x_i^r}(y) \neq 0,$$

i.e. it cannot be approximated with the accuracy desired by means of functions of $C_0^\infty(\Omega)$, while

$$\Phi(x) = \begin{cases} f(x), & x \in \Omega \\ 0, & x \notin \Omega \end{cases} \in C^r(E_n).$$

The construction of the indicated function $\Phi(x)$ will be realized by means of the Whitney Theorem[(5)] (cf. [6, 7]).

Let $m = \min_{x\in E_n - \Omega} n_x$. Let us choose a point $x_0 \in G = E_n - \Omega$ such that $n_{x_0} = m$. By virtue of Lemma 5, a sphere K_R of radius R with center at the point x_0 exists such that $n_x = m$ for any $x \in K_R G$. According to Lemma 7 there exists a vector function $\tau(x) \perp M_x$ for $x \in K_R G$ which is continuous in $K_R G$: $\tau(x) = (\tau_1(x), \cdots \tau_n(x))$. Let us complete the definition of the functions $\tau_i(x)$ as zero outside of $K_R G$.

Let us define the functions Φ_k on the closed set G as follows:

$$\Phi_k \equiv 0, \qquad |k| < r,$$

$$\Phi_k = \tau_1^{k_1}(x)\,\tau_2^{k_2}(x)\ldots\tau_n^{k_n}(x)\,\rho(x), \qquad |k| = r$$

(or, in vector notation, $\Phi_k(x) = \tau^k(x)\rho(x)$), where $\rho(x) \in C_0^\infty(K_{R/2})$, $\rho(x) \geqslant 0$ and $\rho(x_0) = 1$). Let us show that the function $\Phi(x) \equiv 0$ on G belongs to the class $C^r(G)$ relative to Φ_k, $|k| \leqslant r$. If $|k| = r$, then

$$R_k(x, y) = \Phi_k(y) - \Phi_k(x). \tag{34}$$

If $|k| < r$, then

$$\begin{aligned} R_k(x, y) &= -\sum_{|k+l|\leqslant r} \frac{\Phi_{k+l}(x)}{l!}(y-x)^l = -\sum_{|k+l|=r} \frac{\tau^{k+l}(x)}{l!}(y-x)^l \rho(x) \\ &= -\frac{\tau^k(x)\rho(x)}{(r-|k|)!} \sum_{|l|=r-|k|} \frac{(r-|k|)!}{l!}\tau^l(x)(y-x)^l \\ &= -\frac{\tau^k(x)\rho(x)}{(r-|k|)!}(\tau_1(x)(y_1-x_1)+\ldots+\tau_n(x)(y_n-x_n))^{r-|k|}, \end{aligned}$$

whence

$$|R_k(x, y)| \leqslant \rho(x)\left|\tau_1(x)\frac{y_1-x_1}{|y-x|}+\ldots+\tau_n(x)\frac{y_n-x_n}{|y-x|}\right|^{r-|k|}|y-x|^{r-|k|}. \tag{35}$$

---

(5) Let us cite a formulation of this theorem. Let G be a closed set; let Φ and Φ_k, $|k| \leqslant r$, be real functions defined on G. The function $\Phi \in C^r(G)$ relative to Φ_k, $|k| \leqslant r$, if for any $x, y \in G$,

$$\Phi_k(y) = \sum_{|k+l|\leqslant r} \frac{\Phi_{k+l}(x)}{l!}(y-x)^l + R_k(x, y),$$

where $R_k(x, y)$ satisfies the condition that for any $z \in G$ and for any $\eta > 0$ there exists $\delta > 0$ such that for all $x, y \in G$ for which $|z - x| < \delta$ and $|z - y| < \delta$ the inequality

$$|R_k(x, y)| \leqslant \eta |y-x|^{r-|k|} \tag{33}$$

is valid.

WHITNEY THEOREM. *Let G be an arbitrary closed set in E_n, and let $\Phi \in C^r(G)$ relative to Φ_k, $|k| \leqslant r$. Then there exists $F(x) \in C^r(E_n)$ such that*

1) *$F(x) = \Phi(x)$ on G,*
2) *$D^k F(x) = \Phi_k(x)$ on G ($|k| \leqslant r$), and*
3) *$F(x)$ is analytic on $E_n - G$.*

By virtue of the continuity of the functions $\Phi_k(x)$, and by Lemma 8, (34) and (35) imply (33) for any $z \in G$, i.e. $\Phi \in C^r(G)$ relative to Φ_k, $|k| \leqslant r$.

2. If the set Ω is unbounded, a sequence of nonintersecting cubes $R_s = \{a_i^{(s)} \leqslant x_i \leqslant a_i^{(s)} + \rho_s\} \subset \Omega$ exists such that $\rho(R_s, \Gamma) > 0$ and $\lim_{s\to\infty} a^{(s)} = \infty$. Let $g(u)$ be a function of one variable, $g(u) \in C_0^\infty(0, 1)$, $g(½) = 1$. Let us consider the function

$$f(x) = \begin{cases} \rho_s^r g\left(\frac{x_1 - a_1}{\rho_s}\right) \cdots g\left(\frac{x_n - a_n}{\rho_s}\right), & x \in R_s, \\ 0, & x \in \Omega - \bigcup_{s=1}^{\infty} R_s. \end{cases}$$

The limit

$$\lim_{\substack{x\to\infty \\ x\in\Omega}} \frac{\partial^r f}{\partial x_1^r}(x)$$

does not exist for it; therefore, by virtue of Theorem 2, condition (13) is not fulfilled. At the same time,

$$\Phi(x) = \begin{cases} f(x), & x \in \Omega, \\ 0, & x \in E_n - \Omega \in C^r(E_n). \end{cases}$$

Thus the condition of the boundedness of Ω is necessary. Theorem 5 is completely proved.

BIBLIOGRAPHY

1. V. I. Burenkov, *On the approximation of functions in $C^r(\Omega)$, where Ω is an arbitrary open set,* Dokl. Akad. Nauk SSSR **202** (1972), 16–18 = Soviet Math. Dokl. **13** (1972), 10–12. MR **45** #2462.

2. G. G. Kazarjan, *On the density of smooth functions in $\overset{\circ}{W}{}_p^r(\Omega)$ with compact support,* Mat. Zametki 2 (1967), 45–52 = Math. Notes 2 (1967), 514–517. MR **35** #3431.

3. E. Gagliardo, *Proprietà di alcune classi di funzioni in più variabili,* Richerche Mat. **7** (1958), 102–137. MR **21** #1526.

4. W. Hurewicz and H. Wallman, *Dimension theory,* Princeton Math. Series, vol. 4, Princeton Univ. Press, Princeton, N. J., 1941; Russian transl., IL, Moscow, 1948. MR **3**, 312.

5. I. M. Glazman and Ju. I. Ljubič, *Finite-dimensional linear analysis in problems,* "Nauka", Moscow, 1969. (Russian)

6. H. Whitney, *Analytic extensions of differentiable functions defined in closed sets,* Trans. Amer. Math. Soc. **36** (1934), 36–89.

7. A. Goullet de Rugy, *Le théorème de Whitney sur les extensions différentiables,* Séminaire Choquet: 1965/66, Initiation à l'analyse, fasc. 1, Exposé 5, Secrétariat mathématique, Paris, 1968. MR **38** #4629.

Translated by F. M. GOLDWARE

Trudy Mat. Inst. Steklov.
117 (1972)

Proc. Steklov Inst. Math.
117 (1972)

BOUNDARIES OF SUBDOMAINS, HÖLDER WEIGHT CLASSES AND SOLUTIONS IN THESE CLASSES OF THE POISSON EQUATION

E. A. VOLKOV

UDC 517. 518. 22

Abstract. Let Ω be a bounded region of n-dimensional space, $n \geqslant 2$, γ the boundary of Ω, Ω_δ the subregion consisting of all points of Ω which are separated from γ by a distance exceeding δ, and γ_δ the boundary of Ω_δ. It is established that if γ is Lipschitz or has everywhere continuous kth derivatives satisfying a Hölder condition, $k \geqslant 2$, then the property is retained by γ_δ for $\delta > \delta_0$, where δ_0 is some positive number. On the other hand, it is proved that if γ has a tangent hyperplane whose rotation angle satisfies a Hölder condition with a positive index $\lambda < 1$ ($k = 1$), then this condition cannot be retained by γ_δ no matter how small $\delta > 0$ is.

On a region with Lipschitz boundary, we introduce functions whose derivatives of a given order are continuous on an open region and satisfy a Hölder condition with a weight depending on the distance from the boundary of the region. An imbedding theorem for various weight metrics is proved. In the conclusion, we analyse the solution of the Dirichlet problem for the Poisson equation with right side from a weight class. A priori estimates of solutions are obtained for corresponding weight norms.

Bibliography: 16 items.

Introduction

Let Ω be a bounded region of n-dimensional Euclidean space, $n \geqslant 2$, γ the boundary of Ω, Ω_δ the subregion consisting of all points of Ω which are separated from γ by a distance larger than δ, and γ_δ the boundary of Ω_δ. If γ possesses some "supply" of smoothness, then the question of smoothness of γ_δ arises. For example, let $n = 2$ and let a non-self-intersecting boundary γ be given parametrically:

$$x = \varphi(s), \qquad y = \psi(s) \qquad (s \text{ is an arc of } \gamma),$$

where the functions $\varphi(s)$ and $\psi(s)$ are periodic with period equal to the length of γ, and everywhere k times differentiable,

$$k \geqslant 2, \qquad \varphi(0) = \psi(0) = \psi'(0) = 0, \qquad \varphi'(0) = 1.$$

For sufficiently small $\delta > 0$, by Lemma 2.2, the curve γ_δ can be given parametrically along the arc s of the boundary γ:

AMS (MOS) subject classifications (1970). Primary 26A16, 35J05, 46E35.

$$x_\delta = \varphi(s) - \delta\psi'(s), \qquad y_\delta = \psi(s) + \delta\varphi'(s). \tag{1}$$

Hence it is clear that for sufficiently small $\delta > 0$, the curve γ_δ is differentiable in some neighborhood of the point $x_\delta = 0$ ($s = 0$), where

$$\frac{dy_\delta}{dx_\delta} = \frac{\psi'(s) + \delta\varphi''(s)}{\varphi'(s) - \delta\psi''(s)}. \tag{2}$$

Considering x_δ given parametrically through s by formula (1), one can obviously differentiate expression (2) $k - 2$ times with respect to x_δ in a neighborhood of the point $x_\delta = 0$. In this way one can establish the existence of the $(k - 1)$th derivative of γ_δ, in the expression for which $\varphi^{(k)}(s)$ and $\psi^{(k)}(s)$ clearly appear. Further differentiation is formally inadmissible, and the question of existence of the kth derivative of γ_δ remains open.

However, one can treat this somewhat differently. Having differentiated the equality $\varphi'^2(s) + \psi'^2(s) \equiv 1$ with respect to s, we find that $\varphi''(s) = -\psi'(s)\psi''(s)/\varphi'(s)$, and, having put the resulting expression for $\varphi''(s)$ in (2), we obtain a simpler expression for dy_δ/dx_δ:

$$\frac{dy_\delta}{dx_\delta} = \frac{\psi'(s)}{\varphi'(s)}. \tag{3}$$

The existence of the kth derivative of γ_δ follows from (3) for sufficiently small $\delta > 0$.

Below, in the n-dimensional case ($n \geqslant 2$) it is established in §2 (Theorem 2.1) that if the boundary γ has everywhere continuous kth derivatives satisfying a Hölder condition ($k \geqslant 2$), then this property is retained by γ_δ for $0 < \delta < d_0$, where d_0 is the radius of a maximal ball, the boundary of which touches some point of γ, while the ball lies entirely in Ω. Analogously (Theorem 1.1), if the boundary γ is Lipschitz, then γ_δ is Lipschitz for $0 < \delta < \delta_0$, where δ_0 is some value depending on γ.

However, if one supposes that γ has everywhere a tangent, the rotation angle of which satisfies Hölder's condition with index λ, $0 < \lambda < 1$ (γ is the Ljapunov surface), then this property cannot be retained by γ_δ, no matter how small $\delta > 0$ is, since γ_δ can have a corner (conic) point for all δ, $0 < \delta < \delta_0$. The corresponding example is constructed at the end of §2.

Therefore, it was not by accident that we succeeded in obtaining the simple expression (3) for dy_δ/dx_δ, not containing $\varphi''(s)$ and $\psi''(s)$, since we had supposed the existence of $\varphi''(s)$ and $\psi''(s)$. If γ is a Ljapunov curve, then (3) is, in general, a priori false, since the left side cannot exist for $x_\delta = 0$ and $\delta > 0$ (cf. §2).

We introduce in §3, for consideration on a bounded region Ω with Lipschitz boundary, classes $C^{m,\mu}_{k,\lambda}(\Omega)$ ($m \geqslant \max\{0, k\}$, $m + \mu \geqslant k + \lambda$) of differentiable functions, all mth order derivatives of which are either ($\mu = 0$) continuous on Ω and bounded with weight $\rho^{m-k-\lambda}$, where ρ is the distance from the boundary of the region, or ($\mu > 0$) satisfy a generalized Hölder condition with index μ and weight factor $\rho_*^{m+\mu-k-\lambda}$, where ρ_* is the distance from the nearest point to the boundary of the region. An inclusion theorem for various weight metrics is proved (Theorem 3.1), and completeness of the Hölder weight classes (spaces) is established (Lemma 3.1). In particular, for $k \geqslant 0$ and $0 < \lambda < 1$, the Hölder weight classes $C_{k,\lambda}^{m;\mu}(\Omega)$ are imbedded in the usual classes $C_{k,\lambda}(\Omega)$ of functions having kth derivatives satisfying a Hölder condition with index λ.

A systematic study of weight classes was begun by L. D. Kudrjavcev [1]. A large number of articles by many authors have been devoted to an analysis of various weight spaces, to the fundamentals of integral metrics, and to applications of the imbedding theorems for these spaces to the solution of boundary-value problems. For reviews and literature in this question, cf. S. M. Nikol'skiĭ [2], Kudrjavcev [3], and J. Nečas [4]. Theorem 3.1, established below, on the imbedding $C_{k,\lambda}^{m;\mu}(\Omega) \to C_{k,\lambda}^{m',\mu'}(\Omega)$ of classes of differentiable functions with "nonintegral" weight metrics on a region with Lipschitz boundary relates to the special case $p = \infty$, and it is not contained in the known results in the literature on weight spaces.

In §4 we consider the Dirichlet problem for the Poisson equation with principal part f from a weight class $C_{k,\lambda}^{m;\mu}(\Omega)$, $k \geqslant -1$, $0 < \lambda < 1$, and $0 < \mu < 1$. In particular, for $k = -1$ the function f can increase indefinitely as $\rho^{\lambda-1}$ in a neighborhood of the boundary. Under the additional assumptions that the boundary $\gamma \in C_{k^*,\lambda}$, $k^* = \max\{2, k+2\}$, and the given boundary values $\varphi \in C_{k+2,\lambda}(\gamma)$, we prove the existence of a unique solution $u \in C_{k+2,\lambda}^{m+2;\mu}(\Omega)$ of the Dirichlet problem, and we establish an estimate of the solution in the corresponding weight norms (Theorem 4.1) which is a natural generalization of the a priori estimates of Schauder in the classes $C_{k,\lambda}$.

In terms of the Hölder weight classes, the author has found conditions for uniform convergence of the network method of solution of the Poisson equation up to the second order relative to the grid of the network, and he has also found conditions for convergence with some weight. The resulting conditions of uniform convergence of the network method is essentially more precise than the known conditions in terms of the classes $C_{k,\lambda}$ (cf. [5, 6]).

The author would like to thank P. I. Lizorkin for discussion and valuable advice.

§1. Regions with Lipschitz boundary

Let Ω be a bounded region of n-dimensional Euclidean space, $n \geqslant 2$, and γ the boundary of Ω.

DEFINITION 1.1.(1) We shall say that $\gamma \in C_{0,1}(a, b)$, $a > 0$, $b > 0$, or, in other

(1) One can prove that Definition 1.1 is equivalent to the definition of Lipschitz boundary given by E. Gagliardo [7]. In particular, if γ is Lipschitz in the sense of [7], then there exist $a > 0$, $b > 0$, for which $\gamma \in C_{0,1}(a, b)$.

words, that the boundary γ of the region Ω is Lipschitz, if to each point $M \in \gamma$ one can associate a local rectangular system with coordinates ξ, η $(\xi = \{\xi_1, \cdots, \xi_{n-1}\})$ with origin at M such that the part of γ situated inside the cylinder $Q = Q\{|\eta| < H, |\xi| < a\}$ $(H = a(1 + \sqrt{1 + b^2}), |\xi|^2 = \xi_1^2 + \cdots + \xi_{n-1}^2)$ admits the representation

$$\eta = \psi(\xi), \quad |\xi| < a, \tag{1.1}$$

where $\psi(\xi)$ is a univalent function, $\psi(0) = 0$, and, for any $|\xi| < a$ and $|\zeta| < a$,

$$|\psi(\xi) - \psi(\zeta)| \leqslant b|\xi - \zeta| \tag{1.2}$$

and, moreover,

$$\begin{gathered} Q^* = Q^*\{\psi(\xi) < \eta < H, \quad |\xi| < a\} \subset \Omega, \\ \Omega \cap Q \setminus Q^* = \varnothing. \end{gathered} \tag{1.3}$$

Let us denote by $\Omega_\delta(\gamma_\delta)$ the subregion of Ω which consists of all points of Ω separated from γ by a distance exceeding δ (the boundary of Ω_δ).

THEOREM 1.1. *If $\gamma \in C_{0,1}(a, b)$ and $\delta < a/2$, then $\gamma_\delta \in C_{0,1}(a_\delta, b)$, where $a_\delta = a - 2\delta$.*

PROOF. We shall consider an arbitrary point $M \in \gamma$ and construct a local rectangular system of coordinates ξ, η with origin at m and the cylinder Q according to Definition 1.1. It is obvious that the intersection of Q and the cone $K = K\{\eta > b|\xi|\}$ is entirely situated in Ω. Consequently the figure consisting of these points satisfying the inequalities

$$\delta\sqrt{1 + b^2} + b|\xi| < \eta < H - \delta, \qquad |\xi| < a - \delta,$$

where $0 < \delta < a$, whose projection onto the hyperplane $\eta = 0$ coincides with the disk $\sigma = \sigma\{|\xi| < a - \delta\}$, is situated in $Q \cap \Omega_\delta$. Therefore, by the continuity of the distance from γ, each line which is parallel to the η axis and is separated from it by at least $a - \delta$ intersects at least once the part of γ_δ which is situated inside the cylinder $Q_1\{|\eta| < H - \delta, |\xi| < a - \delta\}$.

Let us show that the lines parallel to the η axis intersect γ_δ, $0 < \delta < a$, inside Q only once. Let us take an arbitrary point $M_\delta \in \gamma_\delta \cap Q_1$ with coordinates ξ_δ, η_δ. By virtue of the fact that M_δ is separated from γ by the distance δ, $0 < \delta < a$ and $\gamma \in C_{0,1}(a, b)$, the points of γ_δ are missing in the intersection of the cone $K_1\{\eta - \eta_\delta > b|\xi - \xi_\delta|\}$ and the cylinder Q_1, i.e. there are no points of γ_δ inside Q_1 other than M_δ.

Let us consider one of the points $M' \in \gamma$ with coordinates ξ', η' closest to M_δ.

Obviously $M' \in Q$. By the Lipschitz nature of γ, the intersection of the cone $K_2\{\eta - \eta' < -b|\xi - \xi'|\}$ and the cylinder Q does not have common points with Ω. Therefore the intersection of the cylinder Q and the figure A which is the locus of points satisfying the conditions

$$\eta < \eta' + \begin{cases} \sqrt{\delta^2 - |\xi|^2}, & |\xi| \leqslant \delta b/\sqrt{1+b^2}, \\ \delta\sqrt{1+b^2} - b|\xi|, & |\xi| > \delta b/\sqrt{1+b^2} \end{cases}$$

(a cone whose top is rounded off to a sphere of radius δ), does not contain points of Ω_δ. Consequently the points belonging to γ_δ are missing in the intersection of Q and the cone $K_3\{\eta - \eta_\delta < -b|\xi - \xi_\delta|\}$, which is entirely situated in A.

Thus $\gamma_\delta \cap Q_1 \cap (K_1 \cup K_3) = \phi$, i.e. $\gamma_\delta \cap Q_1$ admits the representation

$$\eta = \psi_\delta(\xi), \qquad |\xi| < a - \delta,$$

where $\psi_\delta(\xi)$ is a univalent function satisfying a Lipschitz condition with constant b. Moreover,

$$Q^*_\delta\{\psi_\delta(\xi) < \eta < H - \delta,\ |\xi| < a - \delta\} \subset \Omega_\delta,$$
$$\Omega_\delta \cap Q_1 \setminus Q^*_\delta = \varnothing.$$

Hence the proof of Theorem 1.1 is easily completed. Let $0 < \delta < a/2$, and let M_δ be an arbitrary point in γ_δ. We shall take one of the points $M \in \gamma$, separated from M_δ by a distance δ, and construct a local rectangular system of coordinates ξ, η with origin at M by Definition 1.1. The cylinder $Q_2\{|\eta - \eta_\delta| < (a - 2\delta) \times (1 + \sqrt{1+b^2}),\ |\xi - \xi_\delta| < a - 2\delta\} \subset Q_1$. Therefore, if one considers the rectangular system of coordinates with origin at M_δ and axes parallel to the axes of the chosen system ξ, η, then, by the arbitrary nature of M_δ, all requirements of γ_δ are satisfied in accordance with Definition 1.1 with the replacement of a by $a - 2\delta$. Theorem 1.1 is proved.

Let R be the diameter of Ω, $S(P, r)$ the open n-dimensional ball of radius r with center at the point P, and $\rho(\sigma, \kappa)$ the distance between the sets σ and κ. Let us prove a lemma which will be used in the sequel.

LEMMA 1.1. *If* $\gamma \in C_{0,1}(a, b)$, *then there exist two numbers* L *and* $\theta > 0$ *such that*: 1) *any two points* P_1 *and* P_2 *from* Ω *can be joined by a polygonal line, situated in* Ω, *the length of which does not exceed* $L\rho(P_1, P_2)$; 2) *for any ball* $S(P, r), P \in \Omega, 0 < r \leqslant R$, *there exists an open ball of radius* θr, *entirely situated in* $\Omega \cap S(P, r)$.

PROOF. Let us introduce the following notation: $\nu = \{\nu_1, \cdots, \nu_n\}$ is an integral multi-index; $g_\nu = g_\nu\{\nu_i h < x_i \leqslant (\nu_i + 1)h, i = 1, \cdots, n\}$ is a semi-open cube, where $h = a/4\sqrt{n}$; $\{x_1, \cdots, x_n\}$ is a point in n-dimensional Euclidean space; $E =$

$E\{\nu; g_\nu \cap \gamma \neq \phi\}$. Let us put in correspondence to each $\nu \in E$ an arbitrary point $M_\nu \in \gamma \cap g$. Let M_ν^1 and M_ν^2 be arbitrary points from $\Omega \cap S(M_\nu, a)$ which have coordinates ξ^1, η^1 and ξ^2, η^2 in the local rectangular system of coordinates ξ, η with origin at M_ν, constructed in accordance with Definition 1.1. These points can be joined by a polygonal line consisting of two linear segments $M_\nu^1 M^3$ and $M^3 M_\nu^2$, where the point M^3 has coordinates $\xi^3 = (\xi^1 + \xi^2)/2$, $\eta^3 = \max\{\eta^1, \eta^2\} + b|\xi^1 - \xi^2|/2 < a(1 + b)$. By the Lipschitz nature of γ, this polygonal line is situated in $Q \cap \Omega$, and its length does not exceed the quantity

$$2(\eta^3 - \min\{\eta^1, \eta^2\}) + |\xi^1 - \xi^2| \leqslant (3+b)\rho(M_\nu^1, M_\nu^2) < 2a(3+b). \quad (1.4)$$

Hence, if one fixes points M_ν^1, $\nu \in E$, by the finiteness of the set E, where $\gamma \subset \bigcup_{\nu\in E} S(M, a)$, it easily follows that any two points P_1 and P_2 belonging to Ω can be joined by a polygonal line whose length is bounded by some quantity $R_0 < \infty$ not depending on the choice of these points. Let $\epsilon_0 = a/4\sqrt{n}$. Three cases are a priori possible: a) $\rho(P_1, P_2) \geqslant \epsilon_0$; b) $\rho(P_1, P_2) < \epsilon_0$, $\rho(P_1 \cup P_2, \gamma) \geqslant \epsilon_0$; c) $\rho(P_1, P_2) < \epsilon_0$, $\rho(P_1 \cup P_2, \gamma) < \epsilon_0$. In case c), as simple calculations show, the points P_1 and P_2 lie simultaneously inside some single sphere $S(M_{\nu^*}, a)$, $\nu^* \in E$. Consequently, taking (1.4) into account, one can put $L = \max\{R_0/\epsilon_0, 3 + b\}$ in the general case.

Let us prove the second assertion of the lemma. Two cases are possible: a) $\rho(P, \gamma) \geqslant a/2$; b) $\rho(P, \gamma) < a/2$. In case a) $S(P, r^*) \subset \Omega \cap S(P, r)$, where $r^* = \min\{r, a/2\}$. In case b) we shall consider a point $M \in \gamma$, situated a distance $\rho(P, \gamma)$ from P, and construct a local rectangular system of coordinates ξ, η with origin at M, and the cylinder Q in accordance with Definition 1.1. We have $K \cap S(P, r^*) \subset \Omega \cap Q \cap S(P, r)$, $K = K\{\eta - \eta_P > b|\xi - \xi_P|\}$ is a circular cone, and ξ_P, η_P are the coordinates of P. As elementary calculations show, one can place a ball of radius $r_0 = r^*/2\sqrt{1 + b^2}$ in the figure $K \cap S(P, r^*)$ and, consequently, in $\Omega \cap S(P, r)$. One can put $\theta = a/2(a + R)\sqrt{1 + b^2}$ in case b) just as in case a). Lemma 1.1 is proved.

§2. Regions with a differentiable boundary

Definition 2.1. The boundary $\gamma \in C_{k,\lambda}(a, b)$, $a > 0$, $b > 0$, $k \geqslant 1$, $0 \leqslant \lambda \leqslant 1$, if the following requirements are satisfied: 1) γ has everywhere a tangent hyperplane; 2) for no matter what point $M \in \gamma$ and rectangular system of coordinates ξ, η with origin at M and positive semi-axis of η directed along the interior normal to γ, the part of γ situated inside the cylinder $Q = Q\{|\eta| < H, |\xi| < a\}$, $H = a(1 + \sqrt{1 + b^2})$, admits the representation (1.1), where conditions (1.2) and (1.3) are satisfied; 3) the function $\psi(\xi)$ is k times continuously differentiable in the disk $\sigma = \sigma\{|\xi| < a\}$ and, moreover, for $\lambda = 0$

$$\max_{\|\nu\|=k}\ \sup_{|\xi|<a} |\psi^{(\nu)}(\xi)| \leqslant b,$$

while for $\lambda > 0$

$$\max_{\|\nu\|=k}\ \sup_{\substack{\xi,\zeta\in\sigma\\ \xi\neq\zeta}} \frac{|\psi^{(\nu)}(\xi)-\psi^{(\nu)}(\zeta)|}{|\xi-\zeta|^{\lambda}} \leqslant b,$$

where $\nu = \{\nu_1, \cdots, \nu_{n-1}\}$, $\nu_i \geqslant 0$, $i = 1, \cdots, n-1$, are integers, $\|\nu\| = \nu_1 + \cdots + \nu_{n-1}$, and

$$\psi^{(\nu)}(\xi) = \partial^{\|\nu\|}\psi(\xi)/\partial\xi_1^{\nu_1}\partial\xi_2^{\nu_2}\dots\partial\xi_{n-1}^{\nu_{n-1}}.$$

REMARK 2.1. Since

$$\sup_{|\xi|<a} |\psi(\xi)| \leqslant ab,$$

it follows by the estimate of [8], §33, that

$$\max_{\|\nu\|=m}\ \max_{|\xi|\leqslant a-\varepsilon} |\psi^{(\nu)}(\xi)| \leqslant cb, \qquad m=0,\ 1,\ \dots,\ k, \tag{2.1}$$

where $0 < \epsilon \leqslant a/2$ and c is a constant not depending on ϵ, the point $M \in \gamma$ or the choice of orthogonal axes $\xi_1, \cdots, \xi_{n-1}$ in the tangent hyperplane to γ at M.

LEMMA 2.1. *If $\gamma \in C_{k,\lambda}(a, b)$, $k \geqslant 2$, and $0 \leqslant \lambda \leqslant 1$, then there exists a ball of positive radius, the boundary of which can touch an arbitrary point $M \in \gamma$, such that this ball lies entirely in Ω (or has no points in common with Ω).*

PROOF. By (2.1),

$$|\partial^2\psi(\xi)/\partial l^2| \leqslant c_1 b, \qquad |\xi| < a,$$

where $\partial/\partial l$ is a differentiation operator along an arbitrary direction in the hyperplane $\eta = 0$, and c_1 is a constant not depending on the choice of M. Consequently the ball of radius $d = \min\{a, 1/c_1 b\}$ with center on the η axis at the point $\eta = d$ ($\eta = -d$) is entirely situated in $\Omega \cap Q$ (in $Q\backslash\Omega$), and its boundary touches $M \in \gamma$, where d does not depend on the choice of M.

In the sequel, if the values of the parameters a and b are unessential, we shall allow the abbreviated notation $\gamma \in C_{k,\lambda}$ instead of $\gamma \in C_{k,\lambda}(a, b)$.

Let d_0 be the radius of a maximal sphere whose boundary can touch an arbitrary point $M \in \gamma$ such that this sphere lies entirely in Ω.

LEMMA 2.2. *If $\gamma \in C_{k,\lambda}$, $k \geqslant 2$, $\lambda \geqslant 0$, and $0 < \delta < d_0$, then* 1) *γ_δ has everywhere a tangent hyperplane*; *and* 2) *there is a one-to-one correspondence between points of the boundaries γ and γ_δ, separated by the distance δ, where the correspondence lies along the common normals to γ and γ_δ.*

PROOF. Let $0 < \delta < d_0$, M_δ an arbitrary point in γ_δ, and $M \in \gamma$, where the distance between M and M_δ is $\rho(M, M_\delta) = \delta$. Obviously the segment $M_\delta M$ lies on the normal to γ, since otherwise there would be a closer point to M_δ on γ, which cannot be. Let the point P be situated on the extension of the segment MM_δ within the region, with $\rho(P, M) = d_0$.

We have $S(P, d_0) \subset \Omega$; consequently $\gamma_\delta \cap S(P, d_0 - \delta) = \phi$ (each point of the ball $S(P, d_0 - \delta)$ is separated from γ by a distance exceeding δ). Moreover, $\gamma_\delta \cap S(M, \delta) = \phi$ (each point of the ball $S(M, \delta)$ is situated closer than δ to $M \in \gamma$). Hence it follows that each linear segment $\bar{\tau}$, parallel to MM_δ and separated from M_δ by a distance not exceeding the quantity $\min\{d_0 - \delta, \delta\}$, whose ends lie on the boundaries of the balls $S(M, \delta) \cup S(P, d_0 - \delta)$, but do not themselves have points in common with $S(M, \delta) \cup S(P, d_0 - \delta)$, intersects γ_δ at least once, by the continuity of the distance from γ.

Since the boundaries of the balls $S(M, \delta)$ and $S(P, d_0 - \delta)$ have a common tangent hyperplane at M_δ, normal to $M_\delta M$, it follows that the length of the segment $\bar{\tau}$ is a quantity of a higher order of smallness than the distance from it to M_δ. Consequently γ_δ has at M_δ a tangent coinciding with the tangent hyperplane to the boundaries of the balls $S(M, \delta)$ and $S(P, d_0 - \delta)$, where $M_\delta M$ is normal to γ_δ at M_δ. There cannot be a second point on γ besides M which is separated from M_δ by the distance δ, since even on the boundary of the ball $S(P, d_0) \subset \Omega, M$ is the unique point separated by a distance δ from M_δ, and $M_\delta \in S(P, d_0)$.

A single-valued correspondence is also valid in the other direction, i.e. for each $M \in \gamma$ there exists a unique $M_\delta \in \gamma_\delta$, separated from M by a distance δ, $0 < \delta < d_0$, where M and M_δ lie on a common normal to γ and γ_δ. Indeed, if one considers a point M' on the interior normal to γ at M and separated from M by a distance $\delta < d_0$, then it will belong to γ_δ, since $S(M', \delta) \subset \Omega$, where the segment $M'M$ is normal to γ_δ at M' by what was proved above. On an arbitrary line going through M but not normal to γ at M, points which are separated by a distance δ from M obviously could not belong to γ_δ, since the distance from them to γ is less than δ. Finally, let P' be situated on the outer normal to γ at M, and $\rho(M, P') = \delta$. Then either P', in general, lies outside Ω, and therefore $P' \notin \gamma_\delta$, or between M and P' there is an intersection point of the outer normal with γ, and consequently $\rho(P', \gamma) < \delta$ and $P' \notin \gamma_\delta$. Lemma 2.2 is proved.

LEMMA 2.3. *Let $\gamma \in C_{2,0}(a, b)$; ξ, η and ξ^*, η^* rectangular systems of coordinates with origin at $M \in \gamma$, where the positive semi-axis of η is directed along the interior normal to γ, and the η^* axis forms the angle $\beta < \arctan(1/b)$ with the η axis; γ^M the piece of γ which is situated in the cylinder Q and represented in the system of coordinates ξ, η by equation* (1.1) *in the closed disk $\bar{\sigma} = \bar{\sigma}\{|\xi| \leqslant a/2\}$; $\bar{\sigma}^*$ the projection of γ^M onto the hyperplane $\eta^* = 0$;*

$$\eta^* = \psi^*(\xi^*), \qquad \xi^* \in \bar{\sigma}^*,$$

the equation of γ^M in the system of coordinates ξ^*, η^*; and $\omega^*(t, \varphi, \bar{\sigma}^*)$ the modulus *of the continuous function* $\varphi(\xi^*)$ *on* $\bar{\sigma}^*$. *Then there exists a positive constant* $\beta_0 <$ $\arctan(1/b)$, *not depending on the choice of* $M \in \gamma$, *and a continuous function* $\omega_M(t)$, $t \geqslant 0$, $\omega(0) = 0$, *such that* $\beta \leqslant \beta_0$ *implies*

$$\max_{1 \leqslant i \leqslant j \leqslant n-1} \omega^*\left(t, \frac{\partial^2 \psi^*}{\partial \xi_i^* \partial \xi_j^*}, \bar{\sigma}^*\right) \leqslant \omega_M(t), \qquad t \geqslant 0. \tag{2.2}$$

PROOF. First of all, note that the function ψ^* is univalent by (1.2). The following parametric representation for γ^M is valid:

$$\xi_i^* = F_i(\xi) \equiv \sum_{j=1}^{n-1} a_{ij}\xi_j + a_{in}\psi(\xi), \qquad i = 1, 2, \ldots, n-1, \tag{2.3}$$

$$\eta^* = \sum_{j=1}^{n-1} a_{nj}\xi_j + a_{nn}\psi(\xi), \tag{2.4}$$

where $|\xi| \leqslant a/2$, and a_{ij}, $1 \leqslant i, j \leqslant n$, are elements of an orthogonal transformation from ξ, η to ξ^*, η^*, where, by (2.1), one can obviously choose a sufficiently small number $\beta_0 > 0$, not depending on $M \in \gamma$, such that the Jacobian transformation (2.3) will satisfy the condition

$$J = \frac{\mathscr{D}(F_1, F_2, \ldots, F_{n-1})}{\mathscr{D}(\xi_1, \xi_2, \ldots, \xi_{n-1})} \geqslant \frac{1}{2}, \qquad |\xi| < \frac{3a}{4}, \tag{2.5}$$

if only $\beta \leqslant \beta_0$. By the circumstance that lines which are parallel to the η and η^* axes cannot intersect γ^M more than once, the one-to-one correspondence between ξ and ξ^*, connected by (2.3), follows. By this, (2.5), and the implicit function theorem (cf. [9], Chapter VI, §2), $\xi_1, \cdots, \xi_{n-1}$ and consequently η^* are twice continuously differentiable in $\xi_1^*, \cdots, \xi_{n-1}^*$ on $\bar{\sigma}^*$. Since the lengths of the projections of segments whose ends are situated in γ^M onto the hyperplanes $\eta = 0$ and $\eta^* = 0$ are equivalent, it follows that if one applies known formulas of differentiation of implicit functions (cf. [9], Chapter VI, §2), by (2.3)–(2.5), by the uniform continuity of the second derivatives of $\psi(\xi)$ for $|\xi| \leqslant a/2$, and by the boundedness of the unit elements of the matrix $\|a_{ij}\|_1^n$, one easily obtains (2.2), where

$$\omega_M(t) = \alpha_M \max_{1 \leqslant i \leqslant j \leqslant n-1} \omega_M\left(\beta_M t, \frac{\partial^2 \psi}{\partial \xi_i \partial \xi_j}, \bar{\sigma}\right),$$

$\omega_M(t, \varphi, \bar{\sigma})$ is the modulus of the continuous function $\varphi(\xi)$ on the closed disk $\bar{\sigma} = \bar{\sigma}\{|\xi| \leqslant a/2\}$, and α_M and β_M are positive constants.

THEOREM 2.1. *Let* $\gamma \in C_{k,\lambda}(a, b)$, $k \geqslant 2$, *and* $0 < \lambda < 1$. *Then* $\gamma_\delta \in C_{k,\lambda}(a_{\delta_0}, b_{\delta_0})$ *for* $0 < \delta < \delta_0 < d_0$, *where* a_{δ_0} *and* b_{δ_0} *are positive constants depending on* δ_0 *but not on* δ.

PROOF. Let us construct a local rectangular system of coordinates ξ, η with origin at an arbitrary point $M \in \gamma$, the positive η semi-axis directed along the interior normal to γ, and the $\xi_1, \cdots, \xi_{n-1}$ axes chosen so that

$$\left.\frac{\partial^2 \psi}{\partial \xi_i \partial \xi_j}\right|_{\xi=0} = 0, \qquad 1 \leqslant i < j \leqslant n-1 \quad (n \geqslant 3), \tag{2.6}$$

where $\psi(\xi)$ is the function (1.1). The realization of (2.6) can be achieved by virtue of the continuity of the second derivatives, which are analogously changed by an orthogonal transformation of independent variables to coefficients of a quadratic form, reduced to diagonal form in some system of coordinates by means of an orthogonal transformation. In this connection,

$$\left.\frac{\partial^2 \psi}{\partial \xi_i^2}\right|_{\xi=0} \leqslant \frac{1}{d_0}, \qquad i = 1, 2, \ldots, n-1 \quad (n \geqslant 2), \tag{2.7}$$

since otherwise the point M could not be touched by the boundary of a ball of radius d_0 entirely situated in Ω.

The points $(\xi_\delta, \eta_\delta)$, where $0 < \delta < d_0$,

$$\xi_{\delta i} = \xi_i - \delta \frac{\partial \psi(\xi)}{\partial \xi_i} \Big/ \left(1 + \sum_{p=1}^{n-1} \left(\frac{\partial \psi(\xi)}{\partial \xi_p}\right)^2\right)^{1/2} \quad (i = 1, 2, \ldots, n-1), \tag{2.8}$$

$$\eta_\delta = \psi(\xi) + \delta \Big/ \left(1 + \sum_{p=1}^{n-1} \left(\frac{\partial \psi(\xi)}{\partial \xi_p}\right)^2\right)^{1/2}, \tag{2.9}$$

for $|\xi| < a$, being situated on the interior normal to γ a distance δ from γ, belong to γ_δ by Lemma 2.2.

Relations (2.8) and (2.9) can be considered as parametric equations of some piece γ_δ^M of the boundary γ_δ, which is of the form of a piece γ^M, given by (1.1) for $|\xi| \leqslant a/2$. By Lemma 2.2, a one-to-one correspondence takes place between the points $(\xi, \eta) \in \gamma^M$ and $(\xi_\delta, \eta_\delta) \in \gamma_\delta^M$, connected by equations (2.8) and (2.9) for $|\xi| \leqslant a/2$ $(0 < \delta < d_0)$. A one-to-one correspondence between the points (ξ, η) and $(\xi_\delta, \eta_\delta)$, connected by expressions (2.8) and (2.9), is established for $0 < \delta < \delta^0$, where δ^0 is some positive quantity, in an article of S. M. Nikol'skiĭ ([10], §4).

A one-to-one correspondence takes place not only between (ξ, η) and $(\xi_\delta, \eta_\delta)$, but also between ξ and ξ_δ, connected by relations (2.8) for $|\xi| \leqslant a/2$. The single-valued correspondence from ξ to ξ_δ is obvious. Let us suppose that to some value $\xi_\delta = \xi_\delta^0$ there corresponds more than one value ξ, i.e. one has two points $P_\delta^i = P_\delta^i(\xi_\delta^0, \eta_\delta^i) \in \gamma_\delta^M$, $i = 1, 2$, where $\eta_\delta^2 > \eta_\delta^1$ (equality between η_δ^1 and η_δ^2 cannot be, or else Lemma 2.2 would be contradicted). Let $P^1 \in \gamma$ and $\rho(P^1, P_\delta^1) = \delta$. Let us extend the segment $P^1 P_\delta^1$ inside Ω to the point $P_{d_0}^1$, situated a distance d_0 from γ.

The point P_δ^1 coincides with the lower point of intersection of the line $\xi = \xi_\delta^0$ with the boundary of the ball $S(P_{d_0}^1, d_0 - \delta)$. Since $S(P_{d_0}^1, d_0 - \delta) \cap \gamma_\delta = \phi$, it follows that P_δ^2 must be situated on the line $\xi = \xi_\delta^0$ no lower than the upper point P^3 of intersection of the indicated line with the boundary of the ball $S(P_{d_0}^1, d_0 - \delta)$. But the point P^3 and all points of the line $\xi = \xi_\delta^0$ situated higher than it are separated from γ^M by a distance exceeding δ, since the set of points satisfying the inequality

$$\eta > \eta^0 - \sqrt{d_0^2 - |\xi - \xi^0|^2}, \qquad |\xi - \xi^0| < d_0,$$

(ξ^0, η^0 are the coordinates of $P_{d_0}^1$), whose boundary is separated from the half-line $\xi = \xi_\delta^0, \eta \geqslant \eta_\delta^2$ ($|\xi_\delta^0 - \xi^0| < d_0 - \delta, \eta_\delta^2 > \eta^0$) by more than δ, does not intersect with γ^M. Consequently lines parallel to the η axis intersect γ_δ^M, $0 < \delta < d_0$, not more than once, and a single-valued correspondence takes place from ξ_δ to ξ, connected with ξ by (2.8) for $|\xi| \leqslant a/2$.

Let us denote by σ_δ^M the projection of γ_δ^M onto the hyperplane to γ at M. We shall now establish that there exists a positive constant α_{δ_0}, depending on $\delta_0 < d_0$ but not on the choice of $M \in \gamma$, such that for each positive $r \leqslant a/2$ and for all δ, $0 < \delta < \delta_0$, the disk $\sigma_\delta\{|\xi_\delta| < \alpha_{\delta_0} r\} \subset \sigma_\delta^M$, where the preimage of the disk σ_δ is entirely contained in the disk $\sigma = \sigma\{|\xi| < r\}$.

Without loss of generality we shall suppose that $\delta_0 \geqslant d_0/2$. Let $Q_\tau = Q_\tau\{|\eta| < H, |\xi| < \tau\}$, where $H = a(1 + \sqrt{1 + b^2})$, be a cylinder. Obviously $\gamma \cap (Q \cup S(P_0, d_0)) = \gamma^M \cap Q_\tau$ for $0 < \tau \leqslant a/2$, where $P_0 = P_0(0, d_0)$.

Let us consider the cylinder with curvilinear bases $Q_\mu^* = Q_\mu^*\{\psi(\xi) < \eta < d_0 - \sqrt{(d_0 - \delta_0)^2 - |\xi|^2}, |\xi| < \mu\}$, where $0 < \mu \leqslant \min\{a/2, d_0 - \delta_0\}$. The upper base of Q_μ^* is situated on the boundary of the ball $S(P_0, d_0 - \delta_0)$, which does not contain the points of γ_δ, $0 < \delta < \delta_0$. The lower base of Q_μ^* is located on γ. Therefore, by continuity of the distance from γ, each line which is parallel to the η axis and separated from it by a distance less than μ intersects γ_δ at least once, $0 < \delta < \delta_0$.

Let $\gamma_{\delta\mu} = \gamma_\delta \cap Q_\mu^*$, $d_0/2 \leqslant \delta_0 < d_0$, $0 < \tau \leqslant \min\{a, d_0\}/2$, and $\mu = (d_0 - \delta_0)\tau/2d_0$. We have $Q_\mu^* \subset Q_\tau \cup S(P_0, d_0)$. As elementary calculations show, the distance from any point $P \in Q_\mu^*$ to $\gamma \cap Q_\tau$ is less than to the boundary of the figure $Q_\tau \cup S(P_0, d_0)$, where $\gamma \cap Q_\tau = \gamma \cap (Q_\tau \cup S(P_0, d_0)) \subset \gamma^M$. Consequently $\gamma_{\delta\mu} \subset \gamma_\delta^M$, and therefore, by what was proved above, lines parallel to the η axis intersect the piece $\gamma_{\delta\mu} \subset \gamma_\delta$ inside Q_μ^* exactly once, where the preimage of $\gamma_{\delta\mu}$ belongs to $\gamma \cap Q_\tau \subset \gamma^M$ and is projected entirely into the disk $\sigma\{|\xi| < \tau\}$. Hence it is clear that one can take for α_{δ_0} the quantity

$$\alpha_{\delta_0} = (d_0 - \delta_0)/2d_0.$$

Let

$$F_i(\xi, \xi_\delta) = \xi_i - \delta \frac{\partial\psi(\xi)}{\partial\xi_i} \Bigg/ \left(1 + \sum_{p=1}^{n-1} \left(\frac{\partial\psi(\xi)}{\partial\xi_p}\right)^2\right)^{1/2} - \xi_{\delta i}.$$

It follows from (2.6) and (2.7) that for $0<\delta<\delta_0<d_0$ the Jacobian

$$\left.\frac{\mathscr{D}(F_1, F_2, \ldots, F_{n-1})}{\mathscr{D}(\xi_1, \xi_2, \ldots, \xi_{n-1})}\right|_{\xi=0} \geqslant \left(1-\frac{\delta_0}{d_0}\right)^{n-1}>0. \tag{2.10}$$

Having realized a covering of γ, analogously as in the proof of Lemma 1.1, with a finite number of balls $S(M_\nu, a^*)$ ($a^* = \min\{a/2, \beta_0/4cbn\}$, where c is the constant from (2.1)) for $h = a^*/9\sqrt{n}$, considering (2.1), (2.6), (2.7), and (2.10), we establish with the aid of Lemma 2.3 the existence of a positive constant $a_{\delta_0} \leqslant a/2$, not depending on the choice of $M\in\gamma$, but depending, in general, on $\delta_0<d_0$, such that for $0<\delta<\delta_0$ the Jacobian

$$\frac{\mathscr{D}(F_1, F_2, \ldots, F_{n-1})}{\mathscr{D}(\xi_1, \xi_2, \ldots, \xi_{n-1})} \geqslant \frac{1}{2}\left(1-\frac{\delta_0}{d_0}\right)^{n-1}, \qquad |\xi|<a'_{\delta_0}. \tag{2.11}$$

By the implicit function theorem ([9], Chapter VI, §2), by (2.11) and by the existence of the one-to-one correspondence established above between ξ and ξ_δ, connected by (2.8), there exist continuous derivatives $\partial\xi_i/\partial\xi_{\delta j}$ $(i, j = 1, \cdots, n-1)$ for $\xi_\delta\in\sigma'_\delta$ (σ'_δ is the image of the disk $\sigma\{|\xi|<a'_{\delta_0}\}$), where

$$\frac{\partial\xi_i}{\partial\xi_{\delta j}} = -\frac{\dfrac{\mathscr{D}(F_1, F_2, \ldots, F_{n-1})}{\mathscr{D}(\xi_1, \ldots, \xi_{i-1}, \xi_{\delta j}, \xi_{i+1}, \ldots, \xi_{n-1})}}{\dfrac{\mathscr{D}(F_1, F_2, \ldots, F_{n-1})}{\mathscr{D}(\xi_1, \xi_2, \ldots, \xi_{n-1})}}. \tag{2.12}$$

By this and by (2.9) and (2.11) the existence of continuous derivatives $\partial\eta_\delta/\partial\xi_{\delta q}$, $q = 1, \cdots, n-1$, formally follows, at least for $\xi_\delta\in\sigma'_\delta$, where by Lemma 2.2 for $0<\delta<\delta_0<d_0$ we have

$$\frac{\partial\eta_\delta}{\partial\xi_{\delta q}} = \frac{\partial\psi(\xi)}{\partial\xi_q}, \qquad |\xi|<a'_{\delta_0} \quad (q=1, 2, \ldots, n-1). \tag{2.13}$$

By the assumptions of Theorem 2.1, one can continuously differentiate relations (2.13) in $\xi_{\delta 1}, \cdots, \xi_{\delta, n-1}$ at least $k-1$ times by applying to the right side the formula for differentiation of composite functions with regard to (2.11) and (2.12) (considering ξ as the intermediate variable). By (2.1) and (2.11), uniform estimates of all these derivatives of η_δ in $\xi_{\delta 1}, \cdots, \xi_{\delta, n-1}$ up to kth order inclusively in the disk $\sigma'\{|\xi_\delta|<\alpha_{\delta_0} a'_{\delta_0}\}$ easily follow (α_{δ_0} is the quantity chosen above). In general, they depend on δ_0, $\frac{1}{2}\leqslant\delta_0<d_0$, but not on δ, $0<\delta<\delta_0$, or the choice of $M\in\gamma$. The equivalence of the distance between any points in $\sigma'\{|\xi_\delta|<\alpha_{\delta_0}a'_{\delta_0}\}$ and their preimages in $\sigma\{|\xi|<a'_{\delta_0}\}$ follows from the Lagrange formula of finite increments, (2.8), (2.11), and the convexity of σ'. Therefore, if $\lambda>0$ in the conditions of the theorem, then for the derivatives of η_δ of order k, with regard to (2.11), it follows that they satisfy a Hölder condition with index λ on the disk σ', and the Hölder constants are bounded by certain quantities which depend on δ_0 but not on δ and $M\in\gamma$.

Let $d_0/2 \leqslant \delta_0 < d_0$ and let nb_{δ_0} be the minimum constant bounding for $\lambda = 0$ the modulus of all first and kth derivatives of η_δ in $\xi_{\delta 1}, \cdots, \xi_{\delta, n-1}$ in the disk $\sigma' \{|\xi_\delta| < \alpha_{\delta_0} a'_{\delta_0}\}$, and for $\lambda > 0$ the modulus of the first derivatives and the Hölder constants of kth derivatives of η_δ in the same disk for all δ, $0 < \delta < \delta_0$, and points $M \in \gamma$ for all possible methods of choice of the axes $\xi_1, \cdots, \xi_{n-1}$ of a rectangular system of coordinates in the tangent hyperplane to γ at M. If $0 < \delta_0 < d_0/2$, let $b_{\delta_0} = b_{d_0/2}$. Let us put

$$a_{\delta_0} = \min \{\alpha_{\delta_0} a'_{\delta_0}, \ \min \{H, \ d_0 - \delta_0\}/(2 + b_{\delta_0})\},$$

where $H = a(1 + \sqrt{1 + b^2})$. Then the cylinder

$$Q^{\delta_0} = Q^{\delta_0} \{|\eta - \delta| < H_{\delta_0}, \ |\xi| < a_{\delta_0}\} \subset Q \cup \bar{S}(P_0, \ d_0 - \delta_0) \cup Q^*_{a_{\delta_0}},$$

where $H_{\delta_0} = a_{\delta_0}(1 + \sqrt{1 + b^2_{\delta_0}})$ and $\bar{S}$ is the closure of S.

It remains to note that an arbitrary choice of rectangular system of coordinates ξ, η with origin in γ and positive semi-axis of η, directed along the interior normal to γ, is equivalent (to within a parallel translation along the common η axis), by Lemma 2.2, to an arbitrary choice of an analogous rectangular system of coordinates with origin in γ_δ.

By what was stated above for γ_δ, $0 < \delta < \delta_0$, all requirements of Definition 2.1 are satisfied for the given k, λ and $a = a_{\delta_0}$, $b = b_{\delta_0}$, i.e. $\gamma_\delta \in C_{k,\lambda}(a_{\delta_0}, b_{\delta_0})$, $0 < \delta < \delta_0$, and $0 < \delta < d_0$. Theorem 2.1 is proved.

In conclusion of the present section, we shall introduce an example from which it follows that Theorem 2.1 is false for $k = 1$ and $0 < \lambda < 1$.

Let us consider the infinite hyperplane γ given by the equation

$$\eta = |\xi|^{1+\lambda}, \tag{2.14}$$

where $0 < \lambda < 1$. Let γ_δ be the set of points separated from γ by a distance δ and situated above γ; $P_0 = P_0(0, \eta_0)$, $0 < \eta_0 < 1$, a point on the η axis; and

$$P_1 = P_1\left(\xi_0, \ \eta_0^{\frac{1+\lambda}{1-\lambda}}\right), \text{ where } |\xi_0| = \eta_0^{\frac{1}{1-\lambda}},$$

a point on γ. We have $P_0 \in \gamma_{\delta_*}$ and

$$\delta_*^2 = \rho^2(P_0, \ \gamma) \leqslant \rho^2(P_0, \ P_1) = \eta_0^2 - \eta_0^{\frac{2}{1-\lambda}} + \eta_0^{\frac{2(1+\lambda)}{1-\lambda}} < \eta_0^2.$$

Let the point $P_*(\xi_*, \eta_*) \in \gamma$, where $\rho(P_0, P_*) = \delta_*$ and consequently $\xi_* \neq 0$. Obviously all points (ξ, η_*), where $|\xi| = |\xi_*|$, are situated on γ and separated from P_0 by a distance δ_*. It is easy to see that the cone $K = K\{\eta > (1 + \lambda)|\xi_*|^\lambda |\xi| - \lambda |\xi_*|^{1+\lambda}\}$ touches γ at the point (ξ, η_*), where $|\xi| = |\xi_*|$, and completely contains an infinite region situated over γ. Each point of the boundary of the cone $K_1 =$

K_1 $\{\eta > (1 + \lambda)|\xi_*|^\lambda|\xi| + \eta_0\}$ is separated from the boundary of K by a distance equal to the distance from its top $P_0(0, \eta_0)$ to the boundary of K, i.e. by δ_*. Consequently the points of the hyperplane γ_{δ_*} are missing in all regions situated between the boundaries of K and K_1. Since the top of K_1 belongs to γ_{δ_*}, it follows from this that γ_{δ_*} cannot have a tangent at P_0. Since δ_* depends monotonically and continuously on η_0, where $\delta_* \to 0$ as $\eta_0 \to 0$, it follows that for all $\delta, 0 < \delta < \delta_0$ (δ_0 is the value of δ_* for $\eta_0 = ½$), γ_δ does not have a tangent on the η axis.

Obviously one could consider a bounded closed hyperplane $\gamma \in C_{1,\lambda}$, $0 < \lambda < 1$, which includes a piece described in some coordinate neighborhood of the origin by (2.14), to arrive at an analogous conclusion that $\gamma_\delta \notin C_{1,\lambda}$, $0 < \delta < \delta_0$.

It also follows from this example that if $\gamma \in C_{1,\lambda}$, $0 < \lambda < 1$, then the principle of a one-to-one correspondence between the points of γ and γ_δ, established by Lemma 2.2 in the case $\gamma \in C_{k,\lambda}$, $k \geqslant 2$, can be violated for all sufficiently small δ.

§3. Hölder weight classes

Let $\mathbf{x} = \{x_1, \cdots, x_n\}$ be a point of n-dimensional Euclidean space, $|\mathbf{x}|^2 = x_1^2 + \cdots + x_n^2$, $\nu = \{\nu_1, \cdots, \nu_n\}$, where $\nu_i \geqslant 0$, $i = 1, \cdots, n$ are integers, $\|\nu\| = \nu_1 + \cdots + \nu_n$, and $f^{(\nu)}(\mathbf{x}) = \partial^{\|\nu\|} f(\mathbf{x})/\partial x_1^{\nu_1} \cdots \partial x_n^{\nu_n}$.

DEFINITION 3.1. A function $f(\mathbf{x})$ belongs to the Hölder weight class $C_{k,\lambda}^{m,\mu}(\Omega)$ ($f(\mathbf{x}) \in C_{k,\lambda}^{m,\mu}(\Omega)$), where m and k are integers, $m \geqslant \max\{k, 0\}$, $0 \leqslant \mu \leqslant 1$, $0 \leqslant \lambda \leqslant 1$, and $k + \lambda \leqslant m + \mu$, if f is m times continuously differentiable on Ω, and also for $\mu = 0$

$$\max_{\|\nu\|=m} \sup_{\mathbf{x}\in\Omega} \rho_{\mathbf{x}}^{\alpha} |f^{(\nu)}(\mathbf{x})| = F_{k,\lambda}^{m,0} < \infty,$$

and for $\mu > 0$

$$\max_{\|\nu\|=m} \sup_{\substack{\mathbf{x},\mathbf{y}\in\Omega\\ \mathbf{x}\neq\mathbf{y}}} \rho_{\mathbf{x}\cup\mathbf{y}}^{\alpha} \frac{|f^{(\nu)}(\mathbf{x}) - f^{(\nu)}(\mathbf{y})|}{|\mathbf{x}-\mathbf{y}|^{\mu}} = F_{k,\lambda}^{m,\mu} < \infty,$$

where $\alpha = m + \mu - k - \lambda$ and ρ_σ is the distance from the set σ to the boundary γ of the region Ω.

Let us introduce a norm for $f \in C_{k,\lambda}^{m,\mu}(\Omega)$:

$$\|f\|_{C_{k,\lambda}^{m,\mu}(\Omega)} = F_{k,\lambda}^{m,\mu} + \sup_{\mathbf{x}\in\Omega} \rho_{\mathbf{x}}^{\beta} |f(\mathbf{x})|, \tag{3.1}$$

where $\beta = -\min\{0, k + \lambda\}$.

REMARK 3.1. In the special case when $m + \mu = k + \lambda$, these weight classes coincide with the usual classes $C_{k,\lambda}(\Omega)$ which consist of the functions having continuous kth derivatives satisfying Hölder's condition with index λ on Ω. We shall write

$C_{k,\lambda}(\Omega)$ instead of $C_{k,\lambda}^{k,\lambda}(\Omega)$ and, correspondingly, $\|f\|_{C_{k,\lambda}(\Omega)}$ instead of $\|f\|_{C_{k,\lambda}^{k,\lambda}(\Omega)}$.

REMARK 3.2. If $k \geqslant 0$, $0 < \lambda < 1$, and $\gamma \in C_{0,1}$, one has the imbedding $C_{k,\lambda}^{m,\mu} \to C_{k,\lambda}$ by Theorem 3.1 (below). Having in mind this property, and also in consideration of convenience for some applications, we took the notation of Hölder weight classes introduced above.

THEOREM 3.1. *Let* $\gamma \in C_{0,1}$, k, m, m' *integers*, $0 \leqslant \lambda < 1$, $0 \leqslant \mu \leqslant 1$, $0 \leqslant \mu' \leqslant 1$, $m \geqslant m' \geqslant \max\{0, k\}$, $m + \mu \geqslant m' + \mu' \geqslant k + \lambda$, *and* $\mu' + \alpha' > 0$, *where* $\alpha' = m' + \mu' - k - \lambda$. *Then one has the imbedding*

$$C_{k,\lambda}^{m,\mu}(\Omega) \to C_{k,\lambda}^{m',\mu'}(\Omega),$$

i.e. if $f \in C_{k,\lambda}^{m,\mu}(\Omega)$, *then* $f \in C_{k,\lambda}^{m',\mu'}(\Omega)$, *where*

$$\|f\|_{C_{k,\lambda}^{m',\mu'}(\Omega)} \leqslant c\|f\|_{C_{k,\lambda}^{m,\mu}(\Omega)},$$

where the constant c *does not depend on* f.

PROOF. Let $a > 0$ and $b > 0$, where $\gamma \in C_{0,1}(a, b)$. Obviously, if $0 < \delta < a$, then $f \in C_{m,\mu}(\overline{\Omega}_\delta)$, and in addition

$$\|f\|_{C_{m,\mu}(\overline{\Omega}_\delta)} \leqslant c_\delta' \|f\|_{C_{k,\lambda}^{m,\mu}(\Omega)}, \tag{3.2}$$

where the constant c_δ' does not depend on f. If $0 < \delta < a/2$, then on the basis of (3.2), Theorem 1.1, Lemma 1.1 (by which condition A of [8], §33, is satisfied), and the estimates of [8], loc. cit., we have

$$\max_{0 \leqslant q \leqslant m} \max_{\|\nu\|=q} \max_{\mathbf{x} \in \overline{\Omega}_\delta} |f^{(\nu)}(\mathbf{x})| \leqslant c_\delta^* \|f\|_{C_{k,\lambda}^{m,\mu}(\Omega)}, \tag{3.3}$$

where the constant c_δ^* does not depend on f.

Let $\mathbf{x} \in \Omega \backslash \overline{\Omega}_\delta$, $0 < \delta \leqslant a/3$, $M \in \gamma$, $\rho(\mathbf{x}, M) = \rho(\mathbf{x}, \gamma) = \rho_{\mathbf{x}}$, ξ, η a local rectangular system of coordinates with origin at M which satisfies the requirements according to Definition 1.1, and $\xi_{\mathbf{x}}$, $\eta_{\mathbf{x}}$ the coordinates of $\mathbf{x}$ in the local system of coordinates. Let us consider the point $\mathbf{u} \in Q \cap \Omega \backslash \Omega_\delta$ with coordinates $\xi_{\mathbf{u}} = \xi_{\mathbf{x}}$, $\eta_{\mathbf{u}} > \eta_{\mathbf{x}}$ (Q is the cylinder constructed in accordance with Definition 1.1). By the Lipschitz nature of γ, we have

$$0 < |\mathbf{u} - \mathbf{x}| < \delta\sqrt{1 + b^2}, \tag{3.4}$$

$$\delta \geqslant \rho\left(\mathbf{x} + \frac{\tau}{|\mathbf{u} - \mathbf{x}|}(\mathbf{u} - \mathbf{x}),\ \gamma\right) > \frac{\rho_{\mathbf{x}} + \tau}{\sqrt{1 + b^2}}, \tag{3.5}$$

where $0 \leqslant \tau \leqslant |\mathbf{u} - \mathbf{x}|$.

Let us denote $N = 1 + \max\{0, [\log_2(|\mathbf{u} - \mathbf{x}|/\rho_{\mathbf{x}})]\}$; [ ] is the integral part symbol, $\tau_0 = 0$, $\tau_i = |\mathbf{u} - \mathbf{x}|2^{i-N}$, $i > 0$, and $\mathbf{x}_j = \mathbf{x} + \tau_j(\mathbf{u} - \mathbf{x})/|\mathbf{u} - \mathbf{x}|$.

Let us examine the following six special cases possible for the conditions of Theorem 3.1:

$$\begin{array}{llll}
1)\ \mu=0, & k+\lambda<m'=m-1, & 0\leqslant\mu'\leqslant 1,\\
2)\ \mu=0, & k+\lambda=m'=m-1, & 0<\mu'\leqslant 1,\\
3)\ \mu=0, & k+\lambda>m'=m-1, & 0<\lambda\leqslant\mu'\leqslant 1,\\
4)\ 0<\mu\leqslant 1, & k+\lambda<m'=m, & 0\leqslant\mu'\leqslant\mu,\\
5)\ 0<\mu\leqslant 1, & k+\lambda=m'=m, & 0<\mu'\leqslant\mu,\\
6)\ 0<\mu\leqslant 1, & k+\lambda>m'=m, & 0<\lambda\leqslant\mu'\leqslant\mu.
\end{array}$$

Let $\|\nu\| = m'$, $\beta = m + \mu - m' - \mu'$, and $\alpha' = m' + \mu' - k - \lambda$. Beginning with Definition 3.1, we find, in all six distinguished special cases, with regard to (3.5), that

$$\begin{aligned}|f^{(\nu)}(\mathbf{u})-f^{(\nu)}(\mathbf{x})| &\leqslant \sum_{j=1}^{N}|f^{(\nu)}(\mathbf{x}_j)-f^{(\nu)}(\mathbf{x}_{j-1})|\\ &\leqslant \psi_N|\mathbf{u}-\mathbf{x}|^{\mu'}\rho_{\mathbf{x}}^{-\alpha'}\|f\|_{C^{m,\mu}_{k,\lambda}(\Omega)},\end{aligned} \tag{3.6}$$

where

$$\psi_N=\frac{n\,(2\sqrt{1+b2})^{\alpha'+\beta}}{2^{N(\beta+\mu')}}\left(\frac{|\mathbf{u}-\mathbf{x}|}{\rho_{\mathbf{x}}}\right)^{\beta}\sum_{j=1}^{N}2^{j(\mu'-\alpha')},$$

and n is the dimension of the space.

Since $\alpha' \geqslant 0$, $\mu' \geqslant 0$, and $\mu' + \alpha' > 0$, where $|\mathbf{u} - \mathbf{x}| < \rho_{\mathbf{x}} 2^N$, in all six cases under consideration it follows that there exists a constant c^*, not depending on the chosen points $\mathbf{x}$ and $\mathbf{u}$, or on ν ($\|\nu\| = m'$) or on f, such that $\psi_N \leqslant c^*$, and consequently, by (3.6),

$$\max_{\|\nu\|=m'}|f^{(\nu)}(\mathbf{x})-f^{(\nu)}(\mathbf{u})|\leqslant c^*|\mathbf{x}-\mathbf{u}|^{\mu'}\rho_{\mathbf{x}}^{-\alpha'}\|f\|_{C^{m,\mu}_{k,\lambda}(\Omega)}, \tag{3.7}$$

where $\mathbf{x} \in \Omega\backslash\overline{\Omega}_\delta$ and, in a corresponding manner, the chosen point $\mathbf{u} \in \Omega\backslash\Omega_\delta$, $0 < \delta \leqslant a/3$.

In the special case when $\mathbf{u} \in \gamma_\delta \cap Q$, by the boundedness of Ω and by (3.7), (3.4), and (3.3) for $\delta = a/3$, one has the inequality

$$\max_{\|\nu\|=m'}|f^{(\nu)}(\mathbf{x})|\leqslant c_1\rho_{\mathbf{x}}^{-\alpha'}\|f\|_{C^{m,\mu}_{k,\lambda}(\Omega)}, \tag{3.8}$$

where the constant c_1 does not depend on $\mathbf{x} \in \Omega$ or f.

If $\mu' = 0$, which is possible in cases 1) and 4), then the assertion of Theorem 3.1 already follows from estimate (3.8).

For $\mu' > 0$, let us consider two arbitrary points: $\mathbf{x}, \mathbf{y} \in \Omega$, $\mathbf{x} \neq \mathbf{y}$. If $\rho_{\mathbf{x}\cup\mathbf{y}} \geqslant 2|\mathbf{x} - \mathbf{y}|$, then, obviously, in all six cases

$$\max_{\|\nu\|=m'} |f^{(\nu)}(\mathbf{x}) - f^{(\nu)}(\mathbf{y})| \leqslant n\,|\mathbf{x}-\mathbf{y}|^{m+\mu-m'}(\rho_{\mathbf{x}\cup\mathbf{y}} - |\mathbf{x}-\mathbf{y}|)^{-\alpha}\|f\|_{C^{m,\mu}_{k,\lambda}(\Omega)} \leqslant 2^{\alpha} n\,|\mathbf{x}-\mathbf{y}|^{\mu'}\rho_{\mathbf{x}\cup\mathbf{y}}^{-\alpha'}\|f\|_{C^{m,\mu}_{k,\lambda}(\Omega)}, \tag{3.9}$$

where $\alpha' = m' + \mu' - k - \lambda$. If $\rho_{\mathbf{x}\cup\mathbf{y}} < 2|\mathbf{x} - \mathbf{y}|$, where $|\mathbf{x} - \mathbf{y}| \geqslant a/36\sqrt{1 + b^2}$, then from (3.8) one has the inequality

$$\max_{\|\nu\|=m'} |f^{(\nu)}(\mathbf{x}) - f^{(\nu)}(\mathbf{y})| \leqslant c_2 |\mathbf{x}-\mathbf{y}|^{\mu'}\rho_{\mathbf{x}\cup\mathbf{y}}^{-\alpha'}\|f\|_{C^{m,\mu}_{k,\lambda}(\Omega)}, \tag{3.10}$$

where the constant c_2 does not depend on $\mathbf{x}$, $\mathbf{y}$, or f.

Finally, if $\rho_{\mathbf{x}\cup\mathbf{y}} < 2|\mathbf{x} - \mathbf{y}| < a/18\sqrt{1 + b^2}$, then we shall construct a local rectangular system of coordinates ξ, η with origin at some point $M \in \gamma$, $\rho(\mathbf{x} \cup \mathbf{y}, M) = \rho_{\mathbf{x}\cup\mathbf{y}}$, in accordance with Definition 1.1. Let us consider two auxiliary points $\mathbf{u}$ and $\mathbf{v}$ with local coordinates $\xi_{\mathbf{u}} = \xi_{\mathbf{x}}$, $\xi_{\mathbf{v}} = \xi_{\mathbf{y}}$, $\eta_{\mathbf{u}} = \eta_{\mathbf{v}} = \max\{\eta_{\mathbf{x}}, \eta_{\mathbf{y}}\} + 3\sqrt{1 + b^2}|\mathbf{x} - \mathbf{y}|$. Simple calculations show that $\mathbf{u}, \mathbf{v} \in Q \cap \Omega\backslash\bar{\Omega}_{a/4}$, where, by (3.5),

$$\rho(\mathbf{u} + \tau(\mathbf{v} - \mathbf{u})/|\mathbf{v} - \mathbf{u}|, \gamma) > 2|\mathbf{x} - \mathbf{y}|, \qquad 0 \leqslant \tau \leqslant |\mathbf{v} - \mathbf{u}|. \tag{3.11}$$

We have

$$|f^{(\nu)}(\mathbf{x}) - f^{(\nu)}(\mathbf{y})| \leqslant |f^{(\nu)}(\mathbf{x}) - f^{(\nu)}(\mathbf{u})| + |f^{(\nu)}(\mathbf{u}) - f^{(\nu)}(\mathbf{v})| + |f^{(\nu)}(\mathbf{v}) - f^{(\nu)}(\mathbf{y})|. \tag{3.12}$$

Taking into consideration that $|\mathbf{u} - \mathbf{v}| \leqslant |\mathbf{x} - \mathbf{y}|$ and $\max\{|\mathbf{x} - \mathbf{u}|, |\mathbf{v} - \mathbf{y}|\} < (1 + 3\sqrt{1 + b^2})|\mathbf{x} - \mathbf{y}|$, we estimate the first and third terms on the right side of (3.12) with the aid of (3.7) and the second term by means of (3.11). Analogously, we also arrive at (3.9) and (3.10), possibly for another constant c_2, for $\rho_{\mathbf{x}\cup\mathbf{y}} < 2|\mathbf{x} - \mathbf{y}| < a/18\sqrt{1 + b^2}$. From this, (3.9), and (3.10), the assertion of Theorem 3.1 follows in all six cases under consideration for $\mu' > 0$.

The proof of Theorem 3.1 in the general case is completed in the following manner. If the values m' and μ' do not fall under any of the six considered cases, we apply Theorem 3.1 several times consecutively in the proved cases: first, if $\mu > 0$, for the fourth case (for $\mu' = 0$), then several times in the first case (for $\mu' = 0$) and one more time in conclusion for one of the first three special cases, i.e.

$$C^{m,\mu}_{k,\lambda} \to C^{m,0}_{k,\lambda} \to C^{m-1,0}_{k,\lambda} \to \ldots \to C^{m'+1,0}_{k,\lambda} \to C^{m',\mu'}_{k,\lambda}.$$

Theorem 3.1 is proved.

REMARK 3.3. Theorem 3.1 is transitive.

REMARK 3.4. It is impossible to put $\mu' + \alpha' = 0$ in the conditions of Theorem 3.1, except in two trival cases: 1) $k = m' = \lambda = \mu' = 0$; 2) $k = m' = m > 0$ and

$\lambda = \mu' = \mu = 0$. In the remaining cases for $\mu' + \alpha' = 0$, a counterexample is the function $f(\mathbf{x}) = x_n^k \ln |\mathbf{x}|$, if $\gamma \in C_{0,1}$ contains the origin of coordinates, where Ω is entirely situated in the half-space $x_n > 0$.

REMARK 3.5. If $f \in C_{k,\lambda}^{m,\mu}(\Omega)$, $k \geqslant 0$, and $0 < \lambda < 1$, then $f \in C_{k,\lambda}(\Omega)$ by Theorem 3.1, but, in general, $f \notin C_{k,\lambda}(\Omega)$ for any $\epsilon > 0$. The function $f(\mathbf{x}) = |\mathbf{x}|^{k+\lambda}$ serves as an example, if the origin of coordinates is situated on the boundary $\gamma \in C_{0,1}$.

REMARK 3.6. If $\gamma \in C_{0,1}$, then the boundedness of the second seminorm in (3.1) follows from the boundedness of the first seminorm in all cases except $k + \lambda = 0$. This can be established by starting with the fact that $f \in C_{m,\mu}(\bar\Omega_{a/3})$, and consequently f is continuous and bounded on $\bar\Omega_{a/3}$, with the aid of estimates of the type (3.3) and (3.6) for $\|\nu\| = m, m-1, \cdots, 0$ with the replacement of $\|f\|_{C_{k,\lambda}^{m,\mu}(\Omega)}$ by $\|f\|_{C_{m,\mu}(\bar\Omega_{a/3})} + F_{k,\lambda}^{m,\mu}$. For $k + \lambda = 0$, boundedness of the function f does not follow, in general, from the boundedness of $F_{0,0}^{m,\mu} = F_{-1,1}^{m,\mu}$, $m + \mu > 0$, for example, for $f(\mathbf{x}) = \ln |\mathbf{x}|$, if the origin of coordinates is situated on γ.

LEMMA 3.1. *For no matter what region Ω, the Hölder weight classes (spaces) $C_{k,\lambda}^{m,\mu}(\Omega)$ are complete*, $m \geqslant \max\{0, k\}$, $m + \mu > k + \lambda$, *and* $0 < \mu < 1$.

We shall develop the proof according to the scheme set forth in §4.7 of the monograph [11]. We shall suppose that $\Omega' \subset \Omega$ and $\Omega_* \subset \bar\Omega_* \subset \Omega$ are open subregions of Ω.

Let us denote

$$\|f\|^*_{C_{k,\lambda}^{m,\mu}(\Omega')} = \sup_{\mathbf{x}\in\Omega'} \rho_{\mathbf{x}}^{\beta} |f(\mathbf{x})| + \max_{\|\nu\|=m} \sup_{\substack{\mathbf{x},\mathbf{y}\in\Omega' \\ \mathbf{x}\neq\mathbf{y}}} \rho_{\mathbf{x}\cup\mathbf{y}}^{\alpha} \frac{|f^{(\nu)}(\mathbf{x}) - f^{(\nu)}(\mathbf{y})|}{|\mathbf{x}-\mathbf{y}|^{\mu}},$$

where $\alpha = m + \mu - k - \lambda$, $\beta = -\min\{0, k + \lambda\}$, and ρ_σ is the distance from the set σ to the boundary γ of Ω. Obviously

$$\|f\|^*_{C_{k,\lambda}^{m,\mu}(\Omega')} \leqslant \|f\|^*_{C_{k,\lambda}^{m,\mu}(\Omega)} = \|f\|_{C_{k,\lambda}^{m,\mu}(\Omega)}, \tag{3.13}$$

$$\|f\|_{C_{m,\mu}(\Omega_*)} \leqslant c_{\Omega_*} \|f\|^*_{C_{k,\lambda}^{m,\mu}(\Omega_*)}, \tag{3.14}$$

where the constant c_{Ω_*} does not depend on f.

Let there be given a sequence of functions $f_i \in C_{k,\lambda}^{m,\mu}(\Omega)$ $(i = 1, 2, \cdots)$ satisfying Cauchy's condition in the metric of $C_{k,\lambda}^{m,\mu}(\Omega)$. Then, by (3.13), (3.14), and the completeness of $C_{m,\mu}(\Omega_*)$, which follows from the theorem of §4.7 of [11], there exists a function $f \in C_{m,\mu}(\Omega_*)$ such that

$$\|f_i - f\|_{C_{m,\mu}(\Omega_*)} \to 0 \quad (i \to \infty). \tag{3.15}$$

Given arbitrary $\epsilon > 0$, there exists an $N > 0$ such that for $i, j > N$, by (3.13),

$$\|f_i - f_j\|^*_{C^{m,\mu}_{k,\lambda}(\Omega_*)} \leqslant \|f_i - f_j\|_{C^{m,\mu}_{k,\lambda}(\Omega)} < \varepsilon. \tag{3.16}$$

By (3.15) the passage to the limit of the first term as $j \to \infty$ leads to an expression in which one must replace f_j by f, i.e.

$$\|f_i - f\|^*_{C^{m,\mu}_{k,\lambda}(\Omega_*)} \leqslant \varepsilon \qquad (i > N)$$

for any subregion Ω_*, and consequently

$$\|f_i - f\|_{C^{m,\mu}_{k,\lambda}(\Omega)} = \|f_i - f\|^*_{C^{m,\mu}_{k,\lambda}(\Omega)} \leqslant \varepsilon \qquad (i > N).$$

In addition f belongs (together with f_i) to $C^{m,\mu}_{k,\lambda}(\Omega)$.

REMARK 3.7. Relying on the completeness of $C_{\overline{m},0}(\Omega)$, one can analogously prove Lemma 3.1 for $\mu = 0$.

§4. On solution of the Poisson equation with right side from a Hölder weight class

DEFINITION 4.1. Let $k \geqslant 0$, $k^* = \max\{2, k\}$, $0 \leqslant \lambda \leqslant 1$, $\gamma \in C_{k^*,\lambda}$ for some $a > 0$ and $b > 0$, according to Definition 2.1; let φ be a function given on γ. We shall say that $\varphi \in C_{k,\lambda}(\gamma)$ if for every local rectangular system of coordinates ξ, η with origin at an arbitrary point $M \in \gamma$ and η axis directed along the normal to γ, the function φ, expressed in terms of the variable ξ, belongs to the class $C_{k,\lambda}(\sigma)$ on $\gamma \cap Q$, where $\sigma = \sigma\{|\xi| < a\}$ is a disk and Q is the cylinder constructed according to Definition 1.1.

Let us introduce a norm for $\varphi \in C_{k,\lambda}(\Omega)$:

$$\|\varphi\|_{C_{k,\lambda}(\gamma)} = \sup_{M\in\gamma} \inf_{\omega} \|\varphi\|_{C_{k,\lambda}(\sigma)}, \tag{4.1}$$

where inf is taken over all possible rotations ω in the disk σ.(2)

DEFINITION 4.2. The function $f \in C^{\infty}_{k,\lambda}(\Omega)$, k an integer and $0 \leqslant \lambda < 1$, if f is infinitely differentiable on Ω, and, moreover,

$$\max_{\|\nu\|=m} \sup_{\mathbf{x}\in\Omega} \rho_{\mathbf{x}}^{m-k-\lambda} \left| f^{(\nu)}(\mathbf{x}) \right| = F^m_{k,\lambda} < \infty,$$

where $\rho_{\mathbf{x}}$ is the distance from $\mathbf{x}$ to γ and $m = m_0, m_0 + 1, \cdots$ $(m_0 = \max\{0, k + 1\})$.

Let us consider the Dirichlet problem for the Laplace equation

$$\Delta u = 0 \text{ on } \Omega, \qquad u = \varphi \quad \text{on } \gamma, \tag{4.2}$$

where $\Delta \equiv \partial^2/\partial x_1^2 + \cdots + \partial^2/\partial x_n^2$.

---

(2) The norm (4.1) is introduced for some fixed a, associated with γ. If one changes the value of a, the new norm will be equivalent to the given one.

LEMMA 4.1. *If* $\gamma \in C_{k^*,\lambda}$, $\varphi \in C_{k,\lambda}(\gamma)$, $k \geqslant 0$, $k^* = \max\{2, k\}$, *and* $0 < \lambda < 1$, *then the solution* u *to problem* (4.2) *belongs to* $C^{\infty}_{k,\lambda}(\Omega)$, *where*

$$\max_{\|\nu\|=m} \sup_{\mathbf{x}\in\Omega} \rho_{\mathbf{x}}^{m-k-\lambda} |u^{(\nu)}(\mathbf{x})| \leqslant c_m \|\varphi\|_{C_{k,\lambda}(\gamma)}, \tag{4.3}$$

with $m = k+1, k+2, \cdots$ *and the constant* c_m *does not depend on* φ.

PROOF. We have $u \in C_{k,\lambda}(\Omega)$ (cf. [12], [8], and [13], Chapter III, §§1 and 2),(3) and also

$$\|u\|_{C_{k,\lambda}(\Omega)} \leqslant c\|\varphi\|_{C_{k,\lambda}(\gamma)},^{(4)} \tag{4.4}$$

where the constant c does not depend on φ.

Let $S = S\{|\mathbf{y}| < 1\}$ be the unit ball, Γ the boundary of S, ψ a continuous function on Γ, and $\Psi = \max_{\Gamma}|\psi|$. A harmonic function v on S, taking the given values of ψ on Γ, is representable (cf. [15]) by the Poisson integral

$$v(\mathbf{y}) = \frac{1}{\sigma_{n-1}} \int_{\Gamma} \psi(\theta) \frac{1-|\mathbf{y}|^2}{|\theta - \mathbf{y}|^n}\, d\theta,$$

where $\mathbf{y} \in S$, $\theta \in \Gamma$, and σ_{n-1} is the measure of Γ. From this we get

$$\max_{\|\nu\|=m} |v^{(\nu)}(\mathbf{0})| \leqslant c'_m \Psi, \qquad m = 0, 1, \ldots, \tag{4.5}$$

where the constant c'_m does not depend on ψ.

Let us consider an arbitrary point $\mathbf{z} \in \Omega$ and represent u in the form

$$u(\mathbf{x}) = P_k(\mathbf{x};\, u,\, \mathbf{z}) + v(\mathbf{x}), \tag{4.6}$$

where $P_k(\mathbf{x}; u, \mathbf{z})$ is the kth degree Taylor polynomial for u at $\mathbf{z}$, i.e.

$$\max_{0\leqslant q\leqslant k} \max_{\|\nu\|=q} \left| u^{(\nu)}(\mathbf{z}) - P_k^{(\nu)}(\mathbf{z};\, u,\, \mathbf{z}) \right| = 0,$$

where the differentiation of P_k is with respect to $\mathbf{x}$.

Since for $\|\nu\| \leqslant k$,

$$P_k^{(\nu)}(\mathbf{x};\, f,\, \mathbf{z}) \equiv P_{k-\|\nu\|}(\mathbf{x};\, f^{(\nu)},\, \mathbf{z}),$$

it follows by the harmonic nature of u that

$$\Delta P_k(\mathbf{x};\, u,\, \mathbf{z}) \equiv 0 \qquad (k \geqslant 0).$$

Consequently v is harmonic on Ω, and, moreover, by (4.4) and (4.6)

---

(3) The norms in $C_{k,\lambda}(\Omega)$ and $C_{k,\lambda}(\gamma)$ introduced in [13] and in this paper are equivalent for the cases under consideration by the estimates of [8], Russian p. 33.

(4) For $n = 2$ inequality (4.3) follows from this and the estimates of [14], §3.

$$\|P_k(\mathbf{x};\ u,\ \mathbf{z})\|_{C_{k,\ 0}(\Omega)} \leqslant \tilde{c}\,\|\varphi\|_{C_{k,\ \lambda}(\gamma)}, \tag{4.7}$$

$$\max_{\mathbf{x}\in\bar{S}(\mathbf{z},\ \rho_{\mathbf{z}})} |v(\mathbf{x})| \leqslant \tilde{c}\rho_{\mathbf{z}}^{k+\lambda}\|\varphi\|_{C_{k,\ \lambda}(\gamma)}, \tag{4.8}$$

where $\rho_{\mathbf{z}}$ is the distance from $\mathbf{z}$ to γ, $\bar{S}(\mathbf{z}, \rho_{\mathbf{z}})$ is the closed ball of radius $\rho_{\mathbf{z}}$ with center at $\mathbf{z}$, and $\tilde{c}$ is a constant not depending on φ or $\mathbf{z}$.

From (4.8) and (4.5) with the aid of uniform dilatation and contraction mappings, we obtain the estimate

$$\max_{\|\nu\|=m} |v^{(\nu)}(\mathbf{z})| \leqslant a_m \rho_{\mathbf{z}}^{k+\lambda-m}\|\varphi\|_{C_{k,\ \lambda}(\gamma)}, \tag{4.9}$$

where $m \geqslant 0$ and a_m is a constant not depending on φ or $\mathbf{z}$.

The assertion of Lemma 4.1 follows from (4.6), (4.7) and (4.9) by the boundedness of Ω and the analyticity of u on Ω.

Let us consider the Dirichlet problem for the Poisson equation:

$$\Delta u = f \text{ on } \Omega, \qquad u = \varphi \text{ on } \gamma. \tag{4.10}$$

LEMMA 4.2. *If* $\gamma \in C_{2,\lambda}$, $\varphi \in C_{1,\lambda}(\gamma)$, $f \in C^{0,\mu}_{-1,\lambda}(\Omega)$, *and* $0 < \mu \leqslant \lambda < 1$, *then there exists a unique continuous (on* $\overline{\Omega}$*) solution* $u \in C^{2,0}_{1,\lambda}(\Omega)$ *to problem* (4.10), *where*

$$\|u\|_{C^{2,\ 0}_{1,\ \lambda}(\Omega)} \leqslant c\left(\|f\|_{C^{0,\ \mu}_{-1,\ \lambda}(\Omega)} + \|\varphi\|_{C_{1,\ \lambda}(\gamma)}\right), \tag{4.11}$$

the constant c *not depending on* f *or* φ.

PROOF. Let us consider the function

$$w(\mathbf{x}) = \int_{\Omega} L(\mathbf{x},\ \mathbf{y}) f(\mathbf{y})\, d\mathbf{y}, \qquad \mathbf{x} \in \Omega,$$

where

$$L(\mathbf{x},\ \mathbf{y}) = \begin{cases} \dfrac{1}{2\pi} \ln |\mathbf{x} - \mathbf{y}|, & n = 2, \\ \dfrac{-1}{(n-2)\,\omega_{n-1}} |\mathbf{x} - \mathbf{y}|^{2-n}, & n > 2, \end{cases}$$

and ω_{n-1} is the measure of the surface of the n-dimensional unit sphere.

For $n \geqslant 2$ we have

$$|w(\mathbf{x})| \leqslant c_1 \|f\|_{C^{0,\ \mu}_{-1,\ \lambda}(\Omega)} \int_{\Omega} |\mathbf{x} - \mathbf{y}|^{1-n} \rho_{\mathbf{y}}^{\lambda-1}\, d\mathbf{y}, \tag{4.12}$$

where the constant c_1 does not depend on f.

Let us denote by κ_λ the integral

$$\varkappa_\lambda = \int_\Omega \rho_{\mathbf{y}}^{\lambda-1} d\mathbf{y}, \quad 0<\lambda<1,$$

and by $S(\mathbf{x})$ the sphere of radius $\rho_{\mathbf{x}}/2$ with center at $\mathbf{x}$.

Obviously the integral κ_λ exists, and

$$\int_\Omega |\mathbf{x}-\mathbf{y}|^{1-n} \rho_{\mathbf{y}}^{\lambda-1} d\mathbf{y} = I_1 + I_2, \tag{4.13}$$

where $\mathbf{x} \in \Omega$,

$$I_1 = \int_{S(\mathbf{x})} |\mathbf{x}-\mathbf{y}|^{1-n} \rho_{\mathbf{y}}^{\lambda-1} d\mathbf{y} < \omega_{n-1} \left(\frac{\rho_{\mathbf{x}}}{2}\right)^\lambda, \tag{4.14}$$

$$I_2 = \int_{\Omega \setminus S(x)} |\mathbf{x}-\mathbf{y}|^{1-n} \rho_{\mathbf{y}}^{\lambda-1} d\mathbf{y} < \varkappa_\lambda \left(\frac{\rho_{\mathbf{x}}}{2}\right)^{1-n}. \tag{4.15}$$

Let $a, b>0$ and $0<\delta<a/2$, where $\gamma \in C_{2,\lambda}(a, b)$. Let us represent the function $w(\mathbf{x})$ on Ω_δ in the form

$$w(\mathbf{x}) = \int_{\Omega_{\delta/2}} L(\mathbf{x}, \mathbf{y}) f(\mathbf{y})\, d\mathbf{y} + \int_{\Omega \setminus \Omega_{\delta/2}} L(\mathbf{x}, \mathbf{y}) f(\mathbf{y})\, d\mathbf{y}. \tag{4.16}$$

One can differentiate the second term in the right side of (4.16) inside the integral sign an arbitrary number of times on Ω_δ, since the corresponding integrals converge uniformly on $\bar{\Omega}_\delta$. By Theorem 12 of [8], Chapter VIII, one can differentiate the first term once inside the integral sign on Ω_δ. Moreover, by Theorem 12 of [8], Chapter II, the first term in the right side of (4.16), and consequently also the function $w(\mathbf{x})$, have continuous second derivatives on Ω_δ, where

$$\frac{\partial^2 w(\mathbf{x})}{\partial x_i \partial x_j} = \frac{\delta_{ij}}{n} f(\mathbf{x}) + \overset{*}{\int_{\Omega_{\delta/2}}} \frac{\partial^2 L(\mathbf{x}, \mathbf{y})}{\partial x_i \partial x_j} f(\mathbf{y})\, d\mathbf{y} + \int_{\Omega \setminus \Omega_{\delta/2}} \frac{\partial^2 L(\mathbf{x}, \mathbf{y})}{\partial x_i \partial x_j} f(\mathbf{y})\, d\mathbf{y}, \tag{4.17}$$

δ_{ij} being the Kronecker symbol, and the integral distinguished by an asterisk being understood in the sense of principal value.

Since

$$\overset{*}{\int_{\Omega_{\delta/2}}} \frac{\partial^2 L(\mathbf{x}, \mathbf{y})}{\partial x_i \partial x_j} f(\mathbf{y})\, d\mathbf{y} = \int_{S(\mathbf{x})} \frac{\partial^2 L(\mathbf{x}, \mathbf{y})}{\partial x_i \partial x_j} (f(\mathbf{y}) - f(\mathbf{x}))\, d\mathbf{y} + \int_{\Omega_{\delta/2} \setminus S(\mathbf{x})} \frac{\partial^2 L(\mathbf{x}, \mathbf{y})}{\partial x_i \partial x_j} f(\mathbf{y})\, d\mathbf{y},$$

it follows by (4.17) and the arbitrariness of δ, $0<\delta<a/2$, that $w(\mathbf{x})$ has continuous second derivatives on Ω, and also

$$\max_{\|\nu\|=2} |w^{(\nu)}(\mathbf{x})| \leqslant c_2 \rho_{\mathbf{x}}^{\lambda-1} \|f\|_{C^{0,\mu}_{-1,\lambda}(\Omega)} (1 + I_3 + \rho_{\mathbf{x}}^{1-\lambda} I_4), \tag{4.18}$$

where

$$I_3 = \rho_{\mathbf{x}}^{-\mu} \int_{S(\mathbf{x})} |\mathbf{x}-\mathbf{y}|^{\mu-n} d\mathbf{y} = \text{const} < \infty, \tag{4.19}$$

$$I_4 = \int_{\Omega \setminus S(\mathbf{x})} |\mathbf{x}-\mathbf{y}|^{-n} \rho_{\mathbf{y}}^{\lambda-1} d\mathbf{y} < \varkappa_\lambda \left(\frac{\rho_{\mathbf{x}}}{2}\right)^{-n}. \tag{4.20}$$

We shall make an additional estimate of the integrals I_2 and I_4 for the case $\rho_{\mathbf{x}} < a/4$. Let $\mathbf{x} \in \Omega \backslash \bar{\Omega}_{a/4}$ and $M_{\mathbf{x}} \in \gamma$, where $\rho(\mathbf{x}, M_{\mathbf{x}}) = \rho_{\mathbf{x}}$. Consequently $\mathbf{x}$ is situated on the normal to γ constructed at $M_{\mathbf{x}}$. Let us introduce a local rectangular system of coordinates ξ, η with origin at $M_{\mathbf{x}}$ and positive η semi-axis directed along the interior normal to γ.

Let $G(r)$, $0 < r \leqslant a$, be the cylinder $\{|\xi| < r,\ \psi(\xi) < \eta < r(1 + \sqrt{1+b^2})\}$, and $\psi(\xi)$ the function giving equation (1.1) of the piece $\gamma \cap Q$. Obviously $G(r) \subset Q \cap \Omega$, $0 < r \leqslant a$, $S(\mathbf{x}) \subset G(\rho_{\mathbf{x}})$, and by the Lipschitz nature of γ one has on $G(a/2)$ the inequality

$$\rho_{\mathbf{y}} \geqslant (\eta_{\mathbf{y}} - \psi(\xi_{\mathbf{y}}))/\sqrt{1+b^2}, \tag{4.21}$$

where $\xi_{\mathbf{y}}, \eta_{\mathbf{y}}$ are the coordinates of $\mathbf{y}$ in the local system of coordinates. Consequently, by (4.21) and the Lipschitz nature of γ, for $m = n-1$ or n

$$\begin{aligned}\int_{G(\rho_{\mathbf{x}}) \setminus S(\mathbf{x})} |\mathbf{x}-\mathbf{y}|^{-m} \rho_{\mathbf{y}}^{\lambda-1} d\mathbf{y} &\leqslant c_3 \rho_{\mathbf{x}}^{-m} \int_{|\xi|<\rho_{\mathbf{x}}} \int_{\psi(\xi)}^{\rho_{\mathbf{x}}(1+\sqrt{1+b^2})} (\eta - \psi(\xi))^{\lambda-1} d\eta\, d\xi \\ &\leqslant c_3 \rho_{\mathbf{x}}^{-m} \int_{|\xi|<\rho_{\mathbf{x}}} \int_0^{3\rho_{\mathbf{x}}(1+b)} \tau^{\lambda-1} d\tau\, d\xi = c_4 \rho_{\mathbf{x}}^{n-m-1+\lambda}\end{aligned} \tag{4.22}$$

and analogously for $\rho_{\mathbf{x}} \leqslant r \leqslant a/4$:

$$\int_{G(2r) \setminus G(r)} |\mathbf{x}-\mathbf{y}|^{-m} \rho_{\mathbf{y}}^{\lambda-1} d\mathbf{y} \leqslant c_5 r^{n-m-1+\lambda}, \tag{4.23}$$

where the constants c_3, c_4, and c_5 do not depend on $\mathbf{x}$. Let

$$N = [\log_2(a/4\rho_{\mathbf{x}})] + 1,\ G_0 = G(\rho_{\mathbf{x}}) \setminus S(\mathbf{x}),\ G_k = G(2^k \rho_{\mathbf{x}}) \setminus G(2^{k-1}\rho_{\mathbf{x}}),$$

$k \geqslant 1$. By (4.22) and (4.23), for $\rho_{\mathbf{x}} < a/4$

$$\begin{aligned}\int_{G(a/2) \setminus S(\mathbf{x})} |\mathbf{x}-\mathbf{y}|^{-m} \rho_{\mathbf{y}}^{\lambda-1} d\mathbf{y} &\leqslant \sum_{k=0}^N \int_{G_k} |\mathbf{x}-\mathbf{y}|^{-m} \rho_{\mathbf{y}}^{\lambda-1} d\mathbf{y} \\ &\leqslant c_6 \rho_{\mathbf{x}}^{n-m-1+\lambda} \sum_{k=1}^N 2^{k(n-m-1+\lambda)} \leqslant c_7 \begin{cases} 1, & m = n-1, \\ \rho_{\mathbf{x}}^{\lambda-1}, & m = n, \end{cases}\end{aligned} \tag{4.24}$$

where the constants c_6 and c_7 do not depend on $\mathbf{x} \in \Omega \backslash \bar{\Omega}_{a/4}$.

It follows from (4.12)–(4.15), (4.18)–(4.20), (4.24), and the boundedness of Ω that $w \in C^{2,0}_{1,\lambda}(\Omega)$, and also

$$\|w\|_{C^{2,0}_{1,\lambda}(\Omega)} \leqslant c_8 \|f\|_{C^{0,\mu}_{-1,\lambda}(\Omega)}, \tag{4.25}$$

where the constant c_8 does not depend on f. Moreover, by Theorem 3.1 we have $w \in C_{1,\lambda}(\Omega)$, and consequently (cf. [16], §2) $w \in C_{1,\lambda}(\Omega)$ and also

$$\|w\|_{C_{1,\lambda}(\bar{\Omega})} \leqslant c_9 \|f\|_{C^{0,\mu}_{-1,\lambda}(\Omega)}, \tag{4.26}$$

where the constant c_9 does not depend on f.

Let v be a continuous (on $\bar{\Omega}$) solution to the boundary problem

$$\Delta v = 0 \text{ on } \Omega, \qquad v = \varphi - w \text{ on } \gamma; \tag{4.27}$$

then by (4.26) and estimates of the type (4.4) we have $v \in C_{1,\lambda}(\Omega)$ and $v \in C^{2,0}_{1,\lambda}(\Omega)$, where

$$\|v\|_{C^{2,0}_{1,\lambda}(\Omega)} \leqslant c_{10} \left(\|f\|_{C^{0,\mu}_{-1,\lambda}(\Omega)} + \|\varphi\|_{C_{1,\lambda}(\gamma)} \right), \tag{4.28}$$

and the constant c_{10} does not depend on f or φ.

Since $L(\mathbf{x}, \mathbf{y})$ is harmonic for $\mathbf{x} \neq \mathbf{y}$, it follows on the basis of (4.17) that $\Delta w = f$ on Ω, and consequently by (4.27), (4.25) and (4.28), the function $u = v + w$ is the desired solution to problem (4.10).

Uniqueness of the solution follows specifically from the maximum principle for harmonic functions. Lemma 4.2 is proved.

THEOREM 4.1. *If* $\gamma \in C_{k^*,\lambda}$, $\varphi \in C_{k+2,\lambda}(\gamma)$, $f \in C^{m,\mu}_{k,\lambda}(\Omega)$, $k \geqslant -1$, $k^* = \max\{2, k+2\}$, $m \geqslant \max\{0, k\}$, $m + \mu \geqslant k + \lambda$, $0 < \lambda < 1$, *and* $0 < \mu < 1$, *then there exists a unique solution* $u \in C^{m+2,\mu}_{k+2,\lambda}(\Omega)$ *to problem* (4.10), *and also*

$$\|u\|_{C^{m+2,\mu}_{k+2,\lambda}(\Omega)} \leqslant c \left(\|f\|_{C^{m,\mu}_{k,\lambda}(\Omega)} + \|\varphi\|_{C_{k+2,\lambda}(\gamma)} \right), \tag{4.29}$$

where the constant c *does not depend on* f *or* φ.

PROOF. For brevity, let us introduce the notation

$$\|f, \varphi\| = \|f\|_{C^{m,\mu}_{k,\lambda}(\Omega)} + \|\varphi\|_{C_{k+2,\lambda}(\gamma)}. \tag{4.30}$$

For $k \geqslant 0$, by Theorem 3.1 for $m' = k$ and $\mu' = \lambda$ we have $f \in C_{k,\lambda}(\Omega)$, and consequently (cf. [13], Chapter III, §§1 and 2) there exists a unique solution $u \in C_{k+2,\lambda}(\Omega)$, and also

$$\|u\|_{C_{k+2,\lambda}(\Omega)} \leqslant c_1 \|f, \varphi\|, \tag{4.31}$$

where the constant c_1 does not depend on f or φ. For $k = -1$ the existence of a unique solution $u \in C^{2,0}_{1,\lambda}(\Omega)$ to (4.10) satisfying (4.31) is a consequence of Theorem 3.1 and Lemma 4.2.

Let d_0 be the radius of a maximal sphere whose boundary can touch an arbitrary point $M \in \gamma$ such that this sphere belongs entirely to Ω, $d = \min\{1, d_0/2\}$, let $S_{\mathbf{z}}$ be the open ball of radius d with center at $\mathbf{z}$, and $\Gamma_{\mathbf{z}}$ the boundary of $S_{\mathbf{z}}$.

We shall suppose that $\mathbf{z} \in \Omega$, $\rho_{\mathbf{z}} > d$, and we shall represent u on $S_{\mathbf{z}}$ in the form

$$u = v + w, \tag{4.32}$$

where

$$\Delta v = 0 \text{ on } S_{\mathbf{z}}, \qquad v = u \text{ on } \Gamma_{\mathbf{z}}, \tag{4.33}$$

$$\Delta w = f \text{ on } S_{\mathbf{z}}, \qquad w = 0 \text{ on } \Gamma_{\mathbf{z}}. \tag{4.34}$$

By (4.31),

$$\| u \|_{C_{k+2,\lambda}(\Gamma_{\mathbf{z}})} \leqslant c_2 \| f, \varphi \|, \tag{4.35}$$

where the constant c_2 does not depend on f, φ, or $\mathbf{z}$.

Let $\mathbf{y} \in \Omega$. If $\rho_{\mathbf{y}} \geqslant 2d$, then we shall take $\mathbf{z} = \mathbf{y}$. With this, $\rho_{\mathbf{z}} - d \geqslant d$ and $\rho(\mathbf{y}, \Gamma_{\mathbf{z}}) = d$. If $\rho_{\mathbf{y}} < 2d$, then let us draw through $\mathbf{y}$ the normal to γ at the point $M_{\mathbf{y}} \in \gamma$, $\rho(\mathbf{y}, M_{\mathbf{y}}) = \rho_{\mathbf{y}}$, and let us fix on this normal the point $\mathbf{z} \in \Omega$, situated a distance $\rho_{\mathbf{y}}/2 + d$ from $M_{\mathbf{y}}$. In this connection, $\mathbf{y} \in S_{\mathbf{z}}$ and $\rho(\mathbf{y}, \Gamma_{\mathbf{z}}) = \rho_{\mathbf{z}} - d = \rho_{\mathbf{y}}/2$. Hence by Lemma 4.1 and (4.35) one has the inequality

$$\max_{\| \nu \| = m+2} | v^{(\nu)}(\mathbf{y}) | \leqslant c_3 \rho_{\mathbf{y}}^{k+\lambda-m} \| f, \varphi \|, \tag{4.36}$$

where the constant c_3 does not depend on f, φ, or $\mathbf{y}$.

In the general case ($m \geqslant \max\{0, k\}$), by Definition 3.1

$$\| f \|_{C_{m,\mu}(S_{\mathbf{z}})} \leqslant c_4 (\rho_{\mathbf{z}} - d)^{k+\lambda-m-\mu} \| f \|_{C^{m,\mu}_{k,\lambda}(\Omega)},$$

and, if $m > \max\{0, k\}$, then, moreover, by Theorem 3.1 we have $C^{m,\mu}_{k,\lambda} \to C^{m-1,\lambda}_{k,\lambda}$, and consequently

$$\| f \|_{C_{m-1,\lambda}(S_{\mathbf{z}})} \leqslant c_5 (\rho_{\mathbf{z}} - d)^{k-m+1} \| f \|_{C^{m,\mu}_{k,\lambda}(\Omega)},$$

where the constants c_4 and c_5 do not depend on f or $\mathbf{z}$.

Hence (cf. [13], loc. cit.) in the general case ($m \geqslant \max\{0, k\}$) we have for a solution to problem (4.34)

$$\| w \|_{C_{m+2,\mu}(S_{\mathbf{z}})} \leqslant c_6 (\rho_{\mathbf{z}} - d)^{k+\lambda-m-\mu} \| f \|_{C^{m,\mu}_{k,\lambda}(\Omega)} \tag{4.37}$$

and, moreover, for $m > \max\{0, k\}$

$$\|w\|_{C_{m+1,\lambda}(S_{\mathbf{z}})} \leqslant c_7(\rho_{\mathbf{z}} - d)^{k-m+1} \|f\|_{C^{m,\mu}_{k,\lambda}(\Omega)}, \tag{4.38}$$

where the constants c_6 and c_7 do not depend on f or $\mathbf{z}$.

For $k < m$, let us estimate $(m+2)$th order derivatives of w on the closed sphere $\bar{S}_{\mathbf{z}}$. For $m > \max\{0, k\}$, let $w^{(\nu)}(\mathbf{x}) \equiv \partial w^{(\mathbf{p})}(\mathbf{x})/\partial x_1$ $(\|\nu\| = \|\mathbf{p}\| + 1 = m + 2)$ and attain the value of maximum modulus at a point $\mathbf{x}_0 \in \bar{S}_{\mathbf{z}}$. Let us define the points

$$\mathbf{x}_1 = \mathbf{x}_0 + R^{-1}(\rho_{\mathbf{z}} - d)(\mathbf{z} - \mathbf{x}_0) \in S_{\mathbf{z}},$$
$$\mathbf{x}_1^+ = \mathbf{x}_1 + \tau(\rho_{\mathbf{z}} - d)\mathbf{e}_1 \in S_{\mathbf{z}},$$
$$\mathbf{x}_1^- = \mathbf{x}_1 - \tau(\rho_{\mathbf{z}} - d)\mathbf{e}_1 \in S_{\mathbf{z}},$$

where $\tau = d/R$, R is the diameter of Ω, and $\mathbf{e}_1 = \{1, 0, 0, \cdots, 0\}$.

We have

$$\frac{|w^{(\mathbf{p})}(\mathbf{x}_1^+) - w^{(\mathbf{p})}(\mathbf{x}_1^-)|}{2\tau(\rho_{\mathbf{z}} - d)} = |w^{(\nu)}(\mathbf{x}_1')| \geqslant |w^{(\nu)}(\mathbf{x}_0)| - (2\tau(\rho_{\mathbf{z}} - d))^{\mu} \|w\|_{C_{m+2,\mu}(S_{\mathbf{z}})} \tag{4.39}$$

where $\mathbf{x}_1'$ is some point on the segment joining the points $\mathbf{x}_1^-$ and $\mathbf{x}_1^+$.

On the other hand,

$$|w^{(\mathbf{p})}(\mathbf{x}_1^+) - w^{(\mathbf{p})}(\mathbf{x}_1^-)| \leqslant (2\tau(\rho_{\mathbf{z}} - d))^{\lambda} \|w\|_{C_{m+1,\lambda}(S_{\mathbf{z}})}.$$

By this, (4.39), (4.37), (4.38) and the boundedness of Ω for $m > \max\{0, k\}$, one has

$$|w^{(\nu)}(\mathbf{y})| \leqslant \max_{\mathbf{x} \in \bar{S}_{\mathbf{z}}} |w^{(\nu)}(\mathbf{x})| = |w^{(\nu)}(\mathbf{x}_0)| \leqslant c_8(\rho_{\mathbf{z}} - d)^{k+\lambda-m} \|f\|_{C^{m,\mu}_{k,\lambda}(\Omega)}$$
$$\leqslant c_9 \rho_{\mathbf{y}}^{k+\lambda-m} \|f\|_{C^{m,\mu}_{k,\lambda}(\Omega)},$$

where the constants c_8 and c_9 do not depend on f or $\mathbf{y}$ $(\mathbf{z} = \mathbf{z}(\mathbf{y}))$.

The remaining derivatives of order $m + 2 > \max\{2, k+2\}$ are estimated analogously. Thus for $m > \max\{0, k\}$

$$\max_{\|\nu\| = m+2} |w^{(\nu)}(\mathbf{y})| \leqslant c_{10} \rho_{\mathbf{y}}^{k+\lambda-m} \|f\|_{C^{m,\mu}_{k,\lambda}(\Omega)},$$

where the constant c_{10} does not depend on f or $\mathbf{y}$.

Hence from (4.32), (4.36) for $m > \max\{0, k\}$, and on the basis of Lemma 4.2 for $k = -1$ and $m = 0$, we have

$$\max_{\|\nu\| = m+2} |u^{(\nu)}(\mathbf{x})| \leqslant c_{11} \rho_{\mathbf{x}}^{k+\lambda-m} \|f, \varphi\|, \qquad k < m, \tag{4.40}$$

where the constant c_{11} does not depend on f, φ, or $\mathbf{x}$.

Let $\mathbf{x}, \mathbf{y} \in \Omega$ be two arbitrary points. If $0 < \rho_{\mathbf{x} \cup \mathbf{y}} \leqslant 4|\mathbf{x} - \mathbf{y}|$ or $\rho_{\mathbf{x} \cup \mathbf{y}} > 4|\mathbf{x} - \mathbf{y}| \geqslant d$, then by (4.31) for $k = m$ we have

$$\max_{\|\nu\|=m+2} |u^{(\nu)}(\mathbf{x}) - u^{(\nu)}(\mathbf{y})| \leqslant |\mathbf{x}-\mathbf{y}|^{\lambda} \|u\|_{C_{m+2,\lambda}(\Omega)}$$

and consequently

$$\max_{\|\nu\|=m+2} |u^{(\nu)}(\mathbf{x}) - u^{(\nu)}(\mathbf{y})| \leqslant c_{12} |\mathbf{x}-\mathbf{y}|^{\mu} \rho_{\mathbf{x}\cup\mathbf{y}}^{-\alpha} \|f, \varphi\|, \tag{4.41}$$

where $\alpha = m + \mu - k - \lambda$ and the constant c_{12} does not depend on f, φ, $\mathbf{x}$, or $\mathbf{y}$. For $k < m$, inequality (4.41) (possibly with another constant c_{12}) follows directly from (4.40) in the case under consideration.

If $\rho_{\mathbf{x}\cup\mathbf{y}} > 4|\mathbf{x}-\mathbf{y}|$ but $|\mathbf{x}-\mathbf{y}| < d/4$, we shall choose a point $\mathbf{z}$ just as in the derivation of (4.36). In this connection, $\mathbf{x} \cup \mathbf{y} \in S_{\mathbf{z}} \subset \Omega$ and $\rho_{\mathbf{z}} - d \geqslant \rho_{\mathbf{y}}/2 > \rho_{\mathbf{x}\cup\mathbf{y}}/4$. Then by (4.37)

$$\max_{\|\nu\|=m+2} |w^{(\nu)}(\mathbf{x}) - w^{(\nu)}(\mathbf{y})| \leqslant c_{13} |\mathbf{x}-\mathbf{y}|^{\mu} \rho_{\mathbf{x}\cup\mathbf{y}}^{-\alpha} |f, \varphi\|, \tag{4.42}$$

where $\alpha = m + \mu - k - \lambda$ and c_{13} is a constant not depending on f, φ, $\mathbf{x}$ or $\mathbf{y}$.

By Lemma 4.1, (4.33), (4.35) and Theorem 3.1 we have $v \in C_{k+2,\lambda}^{m+2,\mu}(S_{\mathbf{z}})$, and also

$$\|v\|_{C_{k+2,\lambda}^{m+2,\mu}(S_{\mathbf{z}})} \leqslant c_{14} \|f, \varphi\|,$$

where the constant c_{14} does not depend on f, φ, or $\mathbf{z}$.

Since $4\rho(\mathbf{x} \cup \mathbf{y}, \Gamma_{\mathbf{z}}) > \rho_{\mathbf{x}\cup\mathbf{y}}$, it follows by Definition 3.1 that

$$\max_{\|\nu\|=m+2} |v^{(\nu)}(\mathbf{x}) - v^{(\nu)}(\mathbf{y})| \leqslant c_{15} |\mathbf{x}-\mathbf{y}|^{\mu} \rho_{\mathbf{x}\cup\mathbf{y}}^{-\alpha} \|f, \varphi\|, \tag{4.43}$$

where $\alpha = m + \mu - k - \lambda$ and c_{15} is a constant not depending on f, φ, $\mathbf{x}$, or $\mathbf{y}$.

The assertion of Theorem 4.1 easily follows from (4.31), (4.41)–(4.43), and (4.32), with the notation of (4.30).

REMARK 4.1. Theorems 3.1 and 4.1 remain valid if one weakens the requirements on the boundary $\gamma \in C_{k,\lambda}(a, b)$ and allows the existence of multiple points on the boundary, by introducing the following changes in Definitions 1.1 and 2.1. 1) Requirement (1.3) is removed in Definitions 1.1 and 2.1. 2) For no matter what point $M \in \gamma$, the intersection of Ω and the open ball of radius $a(3 + b)$ with center at M can be decomposed into a finite or infinite number of open connected nonintersecting sets, where a finite number of them have M as a limit point (we shall denote such sets by ω_i^M, $i = 1, \cdots, n_M$). 3) If $k \geqslant 1$, then the boundary γ_i^M of the set ω_i^M has a tangent hyperplane at M for $i = 1, \cdots, n_M$. 4) For each boundary γ_i^M, $i = 1, \cdots, n_M$, one can choose its own rectangular system of coordinates ξ, η with origin at M (for $k \geqslant 1$, with the η axis directed along the normal to γ_i^M at M), in which the piece $\gamma_i^M \cap \gamma \cap Q$ (Q is the cylinder constructed according to Definition 1.1) is represented in the form (1.2), (1.3) is satisfied, and, moreover, the requirements

on the function $\psi(\xi)$ in accordance with Definition 2.1 are satisfied for $k \geqslant 1$.
5) Each line which is parallel to the η axis and intersects Q also intersects both $\gamma_i^M \cap \Omega$ and $\gamma_i^M \cap \gamma$ exactly once, and also all points situated in Q above $\gamma_i^M \cap \gamma$ belong to Ω, $i = 1, \cdots, n_M$.

In this connection, by $|\mathbf{x} - \mathbf{y}|$ in a generalized Hölder condition in Definition 3.1 we understand not the Euclidean distance but the greatest lower bound of the length of curves joining $\mathbf{x}$ and $\mathbf{y}$ and situated entirely in Ω. Moreover, the function φ is considered to be given on the "interior side" of the boundary in Definition 4.1, and if M is a double point, then φ must satisfy the requirements prescribed by Definition 4.1 on $\gamma_i^M \cap \gamma$, $i = 1, 2$, on the side containing ω_i^M, respectively. Finally, if M is a double point, then the symbol sup in (4.1) means the maximum over the two sides of γ $(i = 1, 2)$.

REMARK 4.2. An approximate solution to the Poisson equation in Hölder weight spaces was considered in the two-dimensional case in [17].

BIBLIOGRAPHY

1. L. D. Kudrjavcev, *Direct and inverse imbedding theorems, Applications to the solution of elliptic equations by variational methods*, Trudy Mat. Inst. Steklov. **55** (1959); English transl., Transl. Math. Monographs, Amer. Math. Soc., Providence, R. I. (to appear). MR **33** #7838.

2. S. M. Nikol'skiĭ, *Embedding, continuation, and approximation of differentiable functions of several variables*, Uspehi Mat. Nauk **16** (1961), no. 5 (101), 63–114 = Russian Math. Surveys **16** (1961), no. 5, 55–104. MR **26** #6757.

3. L. D. Kudrjavcev, *Imbedding theorems for weight spaces and their application to the solution of a Dirichlet problem*, Theory of Functions (Proc. Second All-Union Conf., Baku, 1962), Izdat. Akad. Nauk Azerbaĭdžan,. SSR, Baku, 1965, pp. 493–501. (Russian) MR **34** #6521.

4. J. Nečas, *Les méthodes directes en théorie des équations elliptiques*, Masson, Paris; Academia, Prague, 1967. MR **37** #3168.

5. E. A. Volkov, *A posteriori and weighted error estimates for solutions of Poisson's equation and their derivatives computed by the net method*, Dokl. Akad. Nauk SSSR **192** (1970), 717–720 = Soviet Math. Dokl. **11** (1970), 699–703. MR **42** #7085.

6. ———, *Weighted error estimates for the mesh method of solving the Laplace and Poisson equations*, Trudy Mat. Inst. Steklov. **117** (1972), 100–112 = Proc. Steklov. Inst. Math. **117** (1972), 119–134.

7. E. Gagliardo, *Proprietà di alcune classi di funzioni in più variabili*, Ricerche Mat. **7** (1958), 102–137. MR **21** #1526.

8. C. Miranda, *Equazioni alle derivate parziali di tipo ellittico*, Ergebnisse der Mathematik und ihrer Grenzgebiete, Heft 2, Springer-Verlag, Berlin, 1955; English transl., Springer-Verlag, Berlin, 1970; Russian transl., IL, Moscow, 1957. MR **19**, 421.

9. G. M. Fihtengol'c, *A course in differential and integral calculus*. Vol. 1, Fizmatgiz, Moscow, 1958; German transl., VEB Deutscher Verlag, Berlin, 1968. MR **39** #1a.

10. S. M. Nikol'skiĭ, *Properties of certain classes of functions of several variables on differentiable manifolds*, Mat. Sb. **33** (75) (1953), 261–326; English transl., Amer. Math. Soc. Transl. (2) **80** (1969), 39–118. MR **16**, 453.

11. S. M. Nikol'skiĭ, *Approximation of functions of several variables and imbedding theorems*, "Nauka", Moscow, 1969. (Russian)

12. O. D. Kellogg, *On the derivatives of harmonic functions on the boundary*, Trans. Amer. Math. Soc. **33** (1931), 486–510.

13. O. A. Ladyženskaja and N. N. Ural'ceva, *Linear and quasi-linear equations of elliptic type*, "Nauka", Moscow, 1964; English transl., Academic Press, New York, 1968. MR **35** #1955; **39** #5941.

14. E. A. Volkov, *On the solution by the method of grids of the inner Dirichlet problem for the Laplace equation*, Vyčisl. Mat. **1** (1957), 34–61; English transl., Amer. Math. Soc. Transl. (2) **24** (1963), 279–307. MR **23** #B1665.

15. G. Gasper, *On the Littlewood-Paley g-function and the Lusin s-function*, Trans. Amer. Math. Soc. **134** (1968), 385–403. MR **38** #3450.

16. N. M. Gunther, *La théorie du potentiel et ses applications aux problèmes fondamentaux de la physique mathématique*, Gauthier-Villars, Paris, 1934; Russian transl., GITTL, Moscow, 1953. MR **16**, 357.

17. E. A. Volkov, *Approximate solution of the Laplace and Poisson equations in weighted Hölder spaces*, Trudy Mat. Inst. Steklov. **128** (1972), 76–112 = Proc. Steklov. Inst. Math. **128** (1972). (to appear)

Translated by J. D. FABREY

Trudy Mat. Inst. Steklov.
117 (1972)

Proc. Steklov Inst. Math.
117 (1972)

WEIGHTED ERROR ESTIMATES FOR THE MESH METHOD OF SOLVING THE LAPLACE AND POISSON EQUATIONS

E. A. VOLKOV

UDC 517.954.215

Abstract. Weighted error estimates for approximate solutions by the mesh method of the Dirichlet problem for the Laplace and Poisson equations are obtained in the n-dimensional case, $n \geqslant 2$. These estimates are more accurate than the uniform estimate of order h^2 (h is the mesh step), since they reveal an additional decrease in the error close to the boundary. While the smoothness conditions on the boundary of the domain, the boundary values and the right side of the equation, under which the estimates are obtained, may be in fact weakened on a few occasions, such weakening is only very slight. The right side is taken from a weighted class which, in particular, need not necessarily be imbedded in the space of functions satisfying a Lipschitz condition. The paper concludes by finding a lower bound for the error of order h^2 in a uniform metric, for individual solutions of the Laplace and Poisson equations in the case of arbitrary boundary value problems. This bound reflects the error resulting from approximation of the equation itself, and is not connected with the boundary conditions.

Bibliography: 20 items.

Introduction

Weighted error estimates for approximate solutions by the mesh method of the Dirichlet problem for the Laplace and Poisson equations are obtained below in the n-dimensional case. They are obtained under natural conditions on the data of the problem and prove to be more accurate than the familiar uniform estimates of order h^2 (h is the mesh step), since they reveal an additional decrease of the error close to the boundary. The author previously published such estimates in the two-dimensional case [1].

To obtain the weighted estimates, special error majorants are devised; these take account of the fact that the discrepancies of the difference equations for the solution of the initial problem increase quite rapidly on approaching the boundary; at the same time, the majorants themselves decrease close to the boundary, at the same rate as $(\rho + h)^\sigma h^2$, where ρ is the distance from the boundary and σ is a quantity which depends on the data of the problem, $0 < \sigma \leqslant 1$. While the smoothness conditions on the boundary, the boundary values and the right side of the equation, under which the estimates are obtained, may be weakened on a few occasions, such weakening is only very slight (see §3, and

---

AMS (MOS) subject classifications (1970). Primary 65N15; Secondary 35J05.

also [2–4]). The right side is here taken from a weighted Hölder class $C_{k,\lambda}^{m;\mu}(\Omega)$ (see [5] and §1 below).

When conditions are imposed on the right side of the equation in terms of weighted classes, they prove to be much weaker than when imposed in terms of the usual classes $C_{k,\lambda}$ of functions possessing continuous kth derivatives and satisfying a Hölder condition with index λ. To ensure that the approximate solution is uniformly convergent at a second order rate relative to the mesh step, it is usually required that the right side $f \in C_{2,\lambda}(\Omega)$, $\lambda > 0$ (see Lemma 5.4, $\nu = 2$, of [6], and Theorem 5.5 of [7]). In fact, under suitable conditions on the boundary and the boundary values, it is sufficient for this that $f \in C_{0,\lambda}^{2;\lambda}(\Omega)$, $\lambda > 0$; and if $\lambda < 1$, it is possible that $f \notin C_{0,\lambda+\epsilon}(\Omega)$ no matter what the $\epsilon > 0$ (see Remark 3.5 of [5]). While a function $f \in C_{0,\lambda}^{2;\lambda}(\Omega)$ has second derivatives in Ω which satisfy a generalized Hölder condition, these second derivatives may increase indefinitely in the neighborhood of the boundary, at the rate of $\rho^{\lambda-2}$.

On the other hand, conditions on the right side in terms of classes $C_{k,\lambda}$ cannot be significantly weakened. For, recalling N. S. Bahvalov's results of [8] (see also [9]), for the class of solutions of Poisson's equation with right side f belonging to $C_{1,\mu}(\Omega)$, $0 < \mu < 1$, defined by the condition $\|f\|_{C_{1,\mu}(\Omega)} \leqslant 1$, the mesh method cannot be uniformly convergent at a rate better than $h^{1+\mu}$. These circumstances suggest that, in the present context, $C_{k,\lambda}$ are not the natural classes to use when stating conditions on the right side f.

Under the assumptions of the present paper regarding the boundary, the boundary values and the right side of the equation, it is possible to use inequalities (17) and (18), based on Theorem 4.1 of [5], in conjunction with Bahvalov's lemma (see §1 of [3], and also [10]), to obtain a uniform error estimate of order h^2 for the mesh method; but such a procedure does not reveal the extra decrease of the error close to the boundary.

The weighted error estimates (2.1) ($k = 0, 1$) obtained in §2, under the conditions of Theorem 2.1 for the right side of the equation, appear to be new and to hold in a uniform metric.

A lower bound for the error of order h^2 in a uniform metric is obtained in §4, applicable to individual solutions of the Laplace and Poisson equations in the case of arbitrary boundary value problems. This bound reflects the error resulting from approximation of the equation itself, and is not connected with the boundary conditions.

§1. Definitions

Let Ω be a bounded domain of n-dimensional Euclidean space, $n \geqslant 2$, γ the boundary of Ω, $\mathbf{x} = \{x_1, \cdots, x_n\}$, $|\mathbf{x}|^2 = x_1^2 + \cdots + x_n^2$, and $\rho_{\mathbf{x}}$ the distance from the point $\mathbf{x} \in \Omega$ to γ.

Take the Dirichlet problem for Poisson's equation

$$\Delta u = f \text{ in } \Omega, \qquad u = \varphi \text{ on } \gamma, \tag{1.1}$$

where $\Delta \equiv \partial^2/\partial x_1^2 + \cdots + \partial^2/\partial x_n^2$.

It will be said that $\gamma \in C_{k,\lambda}$, $k \geqslant 1$, if γ has no self-intersections, has a tangent hyperplane everywhere, and in addition is k times continuously differentiable in a neighborhood of fixed radius of every point, if we take a local rectangular coordinate system with origin at this point and axes of the independent variables located in the tangent hyperplane, with the proviso, moreover, that these kth derivatives satisfy a Hölder condition with index λ. It will be said that the function $\varphi \in C_{k,\lambda}(\gamma)$ if, taking the same local coordinate system and expressing φ in terms of the independent variables on the tangent hyperplane, the function possesses kth derivatives in a fixed neighborhood of the origin, satisfying a Hölder condition with index λ. More explicit definitions, and in particular a definition of $\|\varphi\|_{C_{k,\lambda}(\gamma)}$, will be found in [5].

The function $f(\mathbf{x})$ belongs to the weighted Hölder class $C_{k,\lambda}^{m;\mu}(\Omega)$ ($f \in C_{k,\lambda}^{m;\mu}(\Omega)$), where k and m are integers, $0 \leqslant k \leqslant m$, $0 \leqslant \mu \leqslant 1$, $0 \leqslant \lambda < 1$ and $m + \mu \geqslant k + \lambda$, if f is m times continuously differentiable in Ω, and with $\mu = 0$

$$\max_{\|\nu\|=m} \sup_{\mathbf{x}\in\Omega} \rho_{\mathbf{x}}^{\alpha} \, | f^{(\nu)}(\mathbf{x}) | = F_{k,\lambda}^{m,0} < \infty,$$

while with $\mu > 0$

$$\max_{\|\nu\|=m} \sup_{\substack{\mathbf{x},\,\mathbf{y}\in\Omega \\ \mathbf{x}\neq\mathbf{y}}} \rho_{\mathbf{x}\cup\mathbf{y}}^{\alpha} \frac{| f^{(\nu)}(\mathbf{x}) - f^{(\nu)}(\mathbf{y}) |}{| \mathbf{x} - \mathbf{y} |^{\mu}} = F_{k,\lambda}^{m,\mu} < \infty,$$

where $\alpha = m + \mu - k - \lambda$, $\rho_{\mathbf{x}\cup\mathbf{y}} = \min\{\rho_{\mathbf{x}}, \rho_{\mathbf{y}}\}$, $\nu = \{\nu_1, \cdots, \nu_n\}$, the $\nu_i \geqslant 0$ are integers, $\|\nu\| = \nu_1 + \cdots + \nu_n$, and $f^{(\nu)}(\mathbf{x}) = \partial^{\|\nu\|} f(\mathbf{x})/\partial x_1^{\nu_1} \cdots \partial x_n^{\nu_n}$.

Introduce the norm

$$\| f \|_{C_{k,\lambda}^{m,\mu}(\Omega)} = F_{k,\lambda}^{m,\mu} + \sup_{\mathbf{x}\in\Omega} | f(\mathbf{x}) |. \tag{1.2}$$

If $m + \mu \geqslant k + \lambda$, $0 < \lambda < 1$, $\gamma \in C_{1,\lambda}$, and $f \in C_{k,\lambda}^{m;\mu}(\Omega)$, then, by Theorem 3.1 of [5], $f \in C_{k,\lambda}(\Omega)$, and a function $f^* \in C_{k,\lambda}^{m,\mu}(\Omega)$ exists such that $f^* \notin C_{k,\lambda+\epsilon}(\Omega)$ whatever the $\epsilon > 0$.

Construct a cubic mesh by means of the families of hyperplanes $x_i = 0, \pm h, \pm 2h, \cdots$ $(i = 1, \cdots, n)$. The points of intersection of the hyperplanes of these families will be called base points of the mesh. Denote by Ω_h the set of base points located in Ω. Introduce into Ω_h the difference operator Δ_h approximating the Laplace operator:

$$\Delta_h = \sum_{i=1}^{n} \delta_i^2,$$

where

$$\delta_i^2 u \equiv 2\frac{l_{i1}u_{i2}-(l_{i1}+l_{i2})\,u+l_{i2}u_{i1}}{l_{i1}l_{i2}\,(l_{i1}+l_{i2})}, \tag{1.4}$$

u is the value of the function at a point $\mathbf{x}\in\Omega_h$, and u_{ik} is the value of u at a point belonging to $\Omega_h\cup\gamma$ and situated at a minimum distance l_{ik} from the point $\mathbf{x}$, measured in the direction of positive variation of the ith coordinate if $k=2$ and in the direction of negative variation of the ith coordinate if $k=1$.

Denote by γ_h the set of all points of γ at which the value of the function is used when evaluating the operator Δ_h in Ω_h.

The following system of difference equations, approximating the problem (1.1), will be considered:

$$\Delta_h u_h = f \text{ in } \Omega_h, \qquad u_h=\varphi \text{ on } \gamma_h. \tag{1.5}$$

It is easily shown, by means of the maximum principle, that system (1.5) has a nonvanishing determinant.

§2. Weighted error estimates

THEOREM 2.1. *If* $\gamma\in C_{2+k,\lambda}$, $\varphi\in C_{2+k,\lambda}(\gamma)$, $f\in C^{m,\mu}_{k,\lambda}(\Omega)$, $0\leqslant k\leqslant 1$, $0<\lambda<1$, *and* $m+\mu>2$, *then*

$$|u_h-u|\leqslant c\,(\rho_{\mathbf{x}}+h)^{\sigma}\,h^2\,(\|f\|_{C^{m,\mu}_{k,\lambda}(\Omega)}+\|\varphi\|_{C_{2+k,\lambda}(\gamma)}), \tag{2.1}$$

in Ω_h, *where* u *is the solution of problem* (1.1), u_h *is the solution of system* (1.5), $\rho_{\mathbf{x}}$ *is the distance from the current point of* Ω_h *to* γ, $\sigma=\max\{k,\lambda\}$, *and* c *is a constant independent of* f, φ, h *and* $\rho_{\mathbf{x}}$.

PROOF. Let w be the solution of the boundary value problem

$$\Delta w=-2 \text{ in } \Omega, \qquad w=0 \text{ on } \gamma.$$

We have $w\in C_{2+k,\lambda}(\bar\Omega)$ (see [11], Chapter III, §§1 and 2). Hence it can be shown by means of Taylor's formula that an $h_0>0$ exists such that, with $0<h\leqslant h_0$,

$$\Delta_h w<-1 \text{ in } \Omega_h, \qquad w=0 \text{ on } \gamma_h. \tag{2.2}$$

In addition,

$$w(x)\leqslant c_1\rho_{\mathbf{x}} \text{ in } \bar\Omega, \tag{2.3}$$

where the constant c_1 is independent of $\mathbf{x}$.

Denote by Ω_h^1 the subset of Ω_h consisting of all base points at which at least one boundary value is used in the expression for the operator Δ_h, and denote

$\Omega_h^0 = \Omega_h \backslash \Omega_h^1$. Define a function z in $\Omega_h \cup \gamma_h$, representing the solution of the system

$$\Delta_h z = -h^{-2} \text{ in } \Omega_h^1, \qquad \Delta_h z = 0 \text{ in } \Omega_h^0, \qquad z = 0 \text{ on } \gamma_h. \tag{2.4}$$

It is easily shown, by the method described by L. Collatz in [12], that

$$z \leqslant 1 \text{ in } \Omega_h. \tag{2.5}$$

Let d_0 be the radius of the maximum sphere whose surface can touch the boundary at an arbitrary point $M \in \gamma$ and which belongs entirely to Ω (by Lemma 2.2 of [5], $d_0 > 0$). Consider in $\bar{\Omega}$ the function

$$v(\mathbf{x}) = g(\rho_{\mathbf{x}}), \tag{2.6}$$

where $g(t)$, $t \geqslant 0$, is a twice continuously differentiable function, and $g(t) = 0$ for $t \geqslant 2d_0/3$.

The function $v(\mathbf{x})$ is equal to a constant on any hypersurface γ_δ specified by the equation $\rho_{\mathbf{x}} = \delta$ ($\mathbf{x} \in \bar{\Omega}$, $\delta \geqslant 0$). By Theorem 2.1 of [5], $\gamma_\delta \in C_{2+k,\lambda}$, $0 < \delta \leqslant 5d_0/6$. In the neighborhood of every point $M \in \gamma_\delta$ the hypersurface γ_δ can be written in terms of rectangular coordinates ξ, η, where $\xi = \{\xi_1, \cdots, \xi_{n-1}\}$, and the η axis is directed along the normal to γ_δ at the point M, by means of the equation

$$\eta = \psi_\delta(\xi), \qquad |\xi| < a_0,$$

($|\xi|^2 = \xi_1^2 + \xi_2^2 + \cdots + \xi_{n-1}^2$), where

$$\max_{1 \leqslant i \leqslant j \leqslant n-1} \sup_{|\xi| < a_0} \left| \frac{\partial^2 \psi_\delta(\xi)}{\partial \xi_i \partial \xi_j} \right| \leqslant c_2,$$

and the constants a_0 and c_2 are independent of $M \in \gamma_\delta$ and δ, $0 \leqslant \delta \leqslant 5d_0/6$. Moreover, by Lemma 2.2 of [5] the normals are common for the family of hypersurfaces $\{\gamma_\delta\}$, $0 \leqslant \delta \leqslant 5d_0/6$.

Hence, with $0 < h \leqslant h_1$, where $h_1 = \min\{h_0, a_0, d_0/24\}$, on recalling the properties of the function $g(t)$ and performing elementary transformations on the expression for $\Delta_h v(\mathbf{x})$, the inequalities

$$\Delta_h v(\mathbf{x}) \leqslant \max_{t \in \tau(\mathbf{x}, h)} g''(t) + c_0 \max_{t \in \tau(\mathbf{x}, h)} |g'(t)|, \tag{2.7}$$

$$|\Delta_h v(\mathbf{x})| \leqslant \max_{t \in \tau(\mathbf{x}, h)} |g''(t)| + c_0 \max_{t \in \tau(\mathbf{x}, h)} |g'(t)|, \tag{2.8}$$

are obtained in Ω_h, where $\tau(\mathbf{x}, g)$ is the interval $[\max\{0, \rho_{\mathbf{x}} - h\}, \rho_{\mathbf{x}} + 2h]$, and c_0 is a constant independent of $\mathbf{x} \in \Omega_h$ and of the function g with the properties in question.

Let $\psi(t)$, $t \geqslant 0$, be a fixed, twice continuously differentiable function, where $\psi(t) = 1$ if $0 \leqslant t \leqslant d_0/2$ and $\psi(t) = 0$ if $t \geqslant 2d_0/3$. Take two particular cases:

1) $k = 0, 0 < \lambda < 1$,

$$g(t) = (t+h)^\lambda \psi(t);$$

2) $k = 1, 0 < \lambda < 1$,

$$g(t) = (t+h)(1-(t+h)^\lambda)\psi(t). \tag{2.10}$$

Using simple calculations, confined to the operations of the differential calculus, and taking inequalities (2.7) and (2.8) as basis, positive constants r $(r \leqslant d_0/2)$, κ, c_3 and h_* $(h_* \leqslant h_1)$ can be found such that, with $0 < h \leqslant h_*$,

$$\Delta_h v(\mathbf{x}) \leqslant -\kappa(\rho_{\mathbf{x}} + h)^{k+\lambda-2} \quad (\mathbf{x} \in \Omega_h^0,\ \rho_{\mathbf{x}} \leqslant r),$$
$$|\Delta_h v(\mathbf{x})| \leqslant c_3 \quad (\mathbf{x} \in \Omega_h^0,\ \rho_{\mathbf{x}} > r),$$
$$|\Delta_h v(\mathbf{x})| \leqslant c_3 h^{\sigma-2} \quad (\mathbf{x} \in \Omega_h^1),$$

where $v(\mathbf{x})$ is the function (2.6) and $\sigma = \max\{k, \lambda\}$, $k = 0, 1$.

Bounds will be obtained for the discrepancies of the difference equations for the solution of problem (1.1). Denote for brevity

$$\|f, \varphi\| = \|f\|_{C^{m,\mu}_{k,\lambda}(\Omega)} + \|\varphi\|_{C_{2+k,\lambda}(\gamma)}.$$

By Theorems 3.1 and 4.1 of [5], the solution of problem (1.1) is $u \in C^{4,0}_{2+k,\lambda}(\Omega)$, where

$$\|u\|_{C_{2+k,\lambda}(\Omega)} \leqslant c_4 \|f, \varphi\|, \qquad \max_{\|\nu\|=4} |u^{(\nu)}(\mathbf{x})| \leqslant c_5 \rho_{\mathbf{x}}^{k+\lambda-2} \|f, \varphi\|,$$

and the constants c_4 and c_5 are independent of f, φ and $\mathbf{x}$. Hence, using Taylor's formula,

$$|\Delta_h u - f| \leqslant c_6 h^\sigma \|f, \varphi\| \text{ in } \Omega_h^1,$$

$$|\Delta_h u - f| \leqslant c_7 h^2 (\rho_{\mathbf{x}} + h)^{k+\lambda-2} \|f, \varphi\| \text{ in } \Omega_h^0,$$

where the constants c_6 and c_7 are independent of f, φ, $\mathbf{x}$ and h.

Take the function

$$\Psi = (c_8 v + c_9 w + c_{10} h^\sigma z) h^2 \|f, \varphi\|, \tag{2.17}$$

specified in $\Omega_h \cup \gamma_h$, where $c_8 = c_7/\kappa$, $c_9 = c_3 c_8 + c_7 r^{k+\lambda-2}$, and $c_{10} = c_3 c_8 + c_6$. From (1.5), (2.2), (2.4), (2.11)–(2.13), (2.15) and (2.16), with $0 < h \leqslant h_*$, we get

$$0=|u_h-u|\leqslant\Psi \text{ on } \gamma_h, \qquad |\Delta_h(u_h-u)|\leqslant-\Delta_h\Psi \text{ in } \Omega_h.$$

Hence, by the comparison theorem (see [13], Russian 1966 ed., p. 594),

$$|u_h-u|\leqslant\Psi \text{ in } \Omega_h\cup\gamma_h, \qquad 0<h\leqslant h_*.$$

Consequently, using (2.3), (2.5), (2.6), (2.9) and (2.10), we see that the inequality (2.1) is satisfied for $0<h\leqslant h_*$. The same inequality holds for $h>h_*$ by virtue of (2.14) and the fact that the solution of system (1.5) is bounded by $c_{11}\|f,\varphi\|$, where the constant c_{11} is independent of f, φ and $h>h_*$, and may easily be determined, e.g., by the method described on Russian pp. 1073–1074 (English pp. 837–838) of [14].

Theorem 2.1 is now proved.

The theorem may be extended to the domains defined in Remark 4.1 of [5], whose boundaries may have multiple points.

Inequalities (2.1) and (2.14) show that quadratic interpolation may be used to compute the approximate solution of problem (1.1) at any point, with an error of order $(\rho_x+h)^\sigma h^2$, by means of the values of u_h in Ω_h.

Under conditions imposed directly on the required solution u, instead of on the data of the problem (1.1), the inequality (2.1), with $\sigma=1$ and with $(\|f\|_{C^{m,\mu}_{k,\lambda}(\Omega)}+\|\varphi\|_{C_{2+k,\lambda}(\gamma)})$ replaced by a constant explicitly dependent on u, follows from V. I. Lebedev's paper [15] in the case $f\equiv 0$, or from Bramble and Hubbard's paper [16] in the case $f\not\equiv 0$. It is assumed in [15] that the fourth derivatives of the required solution can only have singularities at a finite number of boundary points, while it is required in [16] that the required solution at least have fourth derivatives satisfying a Lipschitz condition.

The requirements of [15] and [16] are not in general satisfied under the hypotheses of Theorem 2.1.

§3. On the impossibility of weakening the conditions on the data of the problem

First, taking the case $k=0$, we shall show that the value $\sigma=\lambda$ of the power in (2.1) is exact, in the sense that, even if $\gamma\notin C_{2,\lambda+\epsilon}$, or $\varphi\notin C_{2,\lambda+\epsilon}(\gamma)$, or $f\notin C^{m,\mu}_{0,\lambda+\epsilon}(\Omega)$ $(m+\mu>2)$ for arbitrary $\epsilon>0$, the value of σ cannot in general be increased. For simplicity, the two-dimensional case will be considered.

The following example will show that the conditions on the boundary, with $k=0$ and $0<\lambda<1$, cannot be weakened. Take a finite two-dimensional domain Ω whose boundary $\gamma\in C_{2,\lambda}$ contains a piece described by the equation

$$x_2=|x_1|^{2+\lambda}, \qquad |x_1|\leqslant 1,$$

and is infinitely differentiable everywhere, except for the point located at the origin. Assume, in addition, that the whole of Ω lies above the curve $x_2=|x_1|^{2+\lambda}$

($|x_1| < \infty$), and that the whole of the sector $\{|x_1| < x_2 < \sqrt{1 - x_1^2}\}$ lies inside the domain.

Take in Ω the harmonic function

$$u = x_2 \cos \frac{\pi (2 + \lambda)}{2} - |\mathbf{x}|^{2+\lambda} \cos (2 + \lambda)\, \theta, \tag{3.1}$$

where $\theta = \operatorname{arc\,tan} (x_1/x_2)$. Simple computations show that the boundary values of u are thrice continuously differentiable.

With $x_2 > 0$ we have

$$\left.\frac{\partial^4 u}{\partial x_1^4}\right|_{x_1=0} = \left.\frac{\partial^4 u}{\partial x_2^4}\right|_{x_1=0} > \lambda (1 - \lambda)\, x_2^{\lambda-2}. \tag{3.2}$$

Also, in Ω,

$$\max_{\|\nu\|=5} |u^{(\nu)}(\mathbf{x})| < c_1 |\mathbf{x}|^{\lambda-3}, \tag{3.3}$$

where the constant c_1 is independent of $\mathbf{x}$.

Using these last two expressions and Taylor's formula, it can easily be seen that positive constants h_0, κ and τ (where τ is an integer) exist such that, provided only that $0 < h \leqslant h_0$, the point $\mathbf{x} = \{0, \tau h\} \in \Omega_h^0$, $h < \rho_{\mathbf{x}} \leqslant \tau h$, and also, at the point $\mathbf{x}$,

$$|\Delta_h u| > \kappa h^{\lambda}. \tag{3.4}$$

Since $\Delta_h(u_h - u) = -\Delta_h u$ in Ω_h, it follows from (3.4) that

$$|u_h - u| > \kappa h^{2+\lambda}/8. \tag{3.5}$$

at at least the base point $\{0, \tau h\}$, or at one of the four base points adjacent to it.

It follows immediately from this that the exponent $\sigma = \lambda$ in (2.1) cannot be increased in the present example. And it has also been shown that with $\sigma = \lambda$, $0 < \lambda < 1$ and $k = 0$ the conditions on γ in Theorem 2.1 cannot be weakened in terms of $C_{k,\lambda}$.

To show that in Theorem 2.1 with $k = 0$ the conditions on the boundary values in terms of $C_{k,\lambda}(\gamma)$, or on the right side in terms of $C_{k,\lambda}^{m;\mu}(\Omega)$ (for the subscripts) cannot be weakened, exactly the same procedure may be adopted as in the above example; we only have to take as Ω the circle $\{x_1^2 + (x_2 - 1)^2 < 1\}$, and as u the function (3.1) in the first case, and the function $u = x_2^{2+\lambda}$ in the second. In the case of $u = x_2^{2+\lambda}$, the boundary values are easily seen to be four times continuously differentiable.

Similar examples can be devised for a greater number of dimensions. For example, the lower bound (6.9) obtained in [17], §6, for the three-dimensional case shows that in a domain with a smooth boundary and $\epsilon = 2 + \lambda$ the conditions of Theorem 2.1 on the boundary values cannot be weakened with $n = 3$.

It will now be shown that with $\sigma = 1$ it is not sufficient that $f \in C^{m,\mu}_{1,0}(\Omega)$ for arbitrary $m + \mu > 2$ in the bound (2.1), and also that it is not possible to put $\sigma > 1$ under the conditions of Theorem 2.1 with arbitrary k and m, $0 < k + \lambda \leqslant m + \mu < \infty$.

Let $\Pi = \Pi\{0 < x_2 < 1\}$ be an infinite strip, $R = R\{|x_1| < 1, 0 < x_2 < 1\}$ a rectangle, Γ the boundary of R, and Ω a bounded domain, $R \subset \Omega \subset \Pi$, the boundary γ of which is infinitely differentiable ($\gamma \in C_\infty$); at every point of γ lying on the straight lines $x_2 = 0$ and $x_2 = 1$, the curvature and all the derivatives of γ vanish. (An example of an infinitely differentiable curve such that its curvature and all its derivatives vanish at every point of the curve lying on the line $x_2 = 0$ is provided by the curve $x_2 = 0$ for $x_1 \leqslant 0$, $x_2 = \exp\{-1/x_1\}$ for $x_1 > 0$.)

It will be assumed below that $h = 1/N$, where $N \geqslant 2$ is an integer. Also, let R_h (Γ_h) be the set of base points of a square mesh in R (on Γ), and $\tilde{\Omega}^i_h = \Omega^i_h \setminus R_h$, $i = 0, 1$.

Take the auxiliary function

$$v = (x_1^2 - 1)\, x_2 \ln x_2. \tag{3.6}$$

We have

$$\Delta_h v > -\frac{3}{x_2} \quad \text{in } R_h, \qquad v = 0 \quad \text{on } \Gamma_h. \tag{3.7}$$

In Ω, let

$$u = -24 x_2^3 \ln x_2. \tag{3.8}$$

The boundary values of u are easily shown to be infinitely differentiable ($u \in C_\infty(\gamma)$) under our assumptions regarding γ. Moreover, with $N \geqslant 2$,

$$f - \Delta_h u < -\frac{3h^2}{x_2} \quad \text{in } \Omega^0_h \cup R_h, \tag{3.9}$$

$$|f - \Delta_h u| \leqslant c_0 h \ln h^{-1} \quad \text{in } \tilde{\Omega}^1_h, \tag{3.10}$$

where $f = \Delta u$, and the constant c_0 is independent of $h \leqslant \frac{1}{2}$.

In Ω_h, let $r = f - \Delta_h u$. The error

$$\varepsilon_h = u_h - u, \tag{3.11}$$

where u_h is the solution of system (1.5) with $\varphi = u$ on γ_h, will be written as the sum of three terms:

$$\varepsilon_h = \varepsilon^0_h + \varepsilon^1_h + \varepsilon^2_h, \tag{3.12}$$

where $\epsilon^i_h = 0$ on γ_h, $i = 0, 1, 2$,

$$\begin{aligned} &\Delta_h \varepsilon^q_h = r \ \text{ in } \tilde{\Omega}^q_h, \quad \Delta_h \varepsilon^q_h = 0 \ \text{ in } \Omega_h \setminus \tilde{\Omega}^q_h, \quad q = 0, 1, \\ &\Delta_h \varepsilon^2_h = r \ \text{ in } R_h, \quad \Delta_h \varepsilon^2_h = 0 \ \text{ in } \Omega_h \setminus R_h. \end{aligned} \tag{3.13}$$

Using the method of [12] and noting (3.10), we find that

$$|\varepsilon_h^1| \leqslant c_0 h^3 \ln h^{-1} \text{ in } \Omega_h. \tag{3.14}$$

Using (3.9) and the maximum principle (see [13], Russian p. 593),

$$\varepsilon_h^p \geqslant 0 \text{ in } \Omega_h, \quad p = 0,\ 2. \tag{3.15}$$

Hence, using (3.7), (3.9), (3.13) and the comparison theorem ([13], Russian p. 594),

$$h^2 v \leqslant \varepsilon_h^3 \text{ in } R_h. \tag{3.16}$$

It follows from (3.6), (3.12) and (3.14)–(3.16) that, at points of intersection of the x_2 axis with Ω_h,

$$\varepsilon_h(0,\ x_2) \geqslant h^2 x_2 \ln x_2^{-1} - c_0 h^3 \ln h^{-1}$$

or, with $h \ln h^{-1} \leqslant x_2 \leqslant \exp\{-2c_0\}$,

$$\varepsilon_h(0,\ x_2) \geqslant h^2 x_2 \ln x_2^{-1}/2.$$

In view of this and (3.11), the inequality (2.1) cannot be satisfied with $\sigma = 1$ for all $h = 1/N$, $N = 2, 3, \cdots$, no matter what the constant c, if the solution of problem (1.1) is the function (3.8), and instead of $\|f\|_{C^{m,\mu}_{1,\lambda}(\Omega)}$ $(\lambda > 0)$ we have $\|f\|_{C^{m,\mu}_{1,0}(\Omega)}$ in the right side of (2.1). The additional conditions are required here that $f = \Delta u \in C^{m,\mu}_{1,0}(\Omega)$ for arbitrary $m + \mu > 2$, $\gamma \in C_\infty$, $u \in C_\infty(\gamma)$.

If we take $u = -x_2^4$ and $v = x_2(1 - x_2)(1 - x_1^2)$, it can be shown similarly that it is not possible to take $\sigma > 1$ in (2.1). Here f is analytic, $\gamma \in C_\infty$ and $u \in C_\infty(\gamma)$.

The question of the minimum conditions on the boundary and the boundary values, under which (2.1) holds with $\sigma = 1$, will now be discussed. First, it follows from the examples considered for $k = 0$ that, if $\gamma \in C_{2,1-\epsilon}$ and $\varphi \in C_{2,1-\epsilon}(\gamma)$, $0 < \epsilon < 1$, then it is not in general possible to put $\sigma > 1 - \epsilon$, although, by Theorem 2.1, it is sufficient that $\gamma \in C_{3,\epsilon}$ and $\varphi \in C_{3,\epsilon}(\gamma)$, with arbitrary $\epsilon > 0$, in order to obtain $\sigma = 1$. Hence the conditions in Theorem 2.1 on γ and φ cannot be significantly weakened with $\sigma = 1$. But stronger assertions than this can be made.

Let Ω be the same domain as in the previous example $(\Omega \subset \Pi)$, but let the origin be given a downwards shift of unity along the x_2 axis. Take the family of functions

$$u^\alpha(\mathbf{x}) = x_2 - 1 + (x_1^3 - 3x_1(x_2 - 1 + \alpha)^2) \operatorname{arc\,tan} \frac{x_1}{x_2 - 1 + \alpha}$$
$$+ ((x_2 - 1 + \alpha)^3 - 3x_1^2(x_2 - 1 + \alpha)) \ln \sqrt{x_1^2 + (x_2 - 1 + \alpha)^2} \tag{3.17}$$

harmonic in Ω, where α is a parameter, $0 \leqslant \alpha \leqslant 1$. With $0 < \alpha \leqslant 1$ the boundary values of u^α are infinitely differentiable; while simple computations show that all the derivatives of u along the boundary, up to and including the third order, are uniformly bounded with respect to α, i.e.

$$\|u^\alpha\|_{C_{3,0}(\gamma)} \leqslant c_0, \tag{3.18}$$

where the constant c_0 is independent of α, $0 < \alpha \leqslant 1$.

Assume that $h = 2/(2N + 1)$, $N = 1, 2, \cdots$. At the point $\{0, 1 + h/2\} \in \Omega_h^1$ we have

$$|\Delta_h u^h| > c_1 h \ln h^{-1} - c_2 h,$$

where the positive constants c_1 and c_2 are independent of h.

In view of this, and recalling the definition (1.3), (1.4) of the operator Δ_h, we arrive at

$$\sup_{0<\alpha\leqslant 1} \max_{\omega_h} |u_h^\alpha - u^\alpha| > \frac{c_1}{11} h^3 \ln h^{-1}, \tag{3.19}$$

where $h = 2/(2N + 1)$, $N > \exp\{2c_2/c_1\}$, ω_h is the set of four points: $\{0, 1 + h/2\}$, $\{0, 1 + 3h/2\}$, $\{- h, 1 + h/2\}$, $\{h, 1 + h/2\}$, at distances not exceeding $3h/2$ from γ, and u_h^α is the solution of system (1.5) with $f \equiv 0$ and $\varphi = u^\alpha$ on γ_h.

In view of (3.19) and (3.18), with the present domain and the present position of the origin, no constant c, independent of h, $\rho_{\mathbf{x}}$ and $\varphi \in C_{3,0}(\gamma)$, can be found such that (2.1) holds with $\sigma = 1$, when the $\|\varphi\|_{C_{3,\lambda}(\gamma)}$ $(\lambda > 0)$ on the right side is correspondingly replaced by $\|\varphi\|_{C_{3,0}(\gamma)}$, although $f \equiv 0$ and $\gamma \in C_\infty$.

If the function $v^\alpha(\mathbf{x}) = u^\alpha(\mathbf{x}) + u^\alpha(\mathbf{y})$, where $\mathbf{y} = \mathbf{y}_0 - \mathbf{x}$, $\mathbf{y}_0 = \{0, 3\}$, is considered instead of (3.17), a similar conclusion can be arrived at, regardless of the position of the origin and the direction of the axes in the coordinate system in which the square mesh is constructed.

Finally, consider in the neighborhood of the point $\{0, 1\}$ the curve whose implicit equation is

$$u^0(\mathbf{x}) = 0,$$

where $u^0(\mathbf{x})$ is the function (3.17) with $\alpha = 0$. Elementary computations show that the connected piece of this curve, located in a neighborhood of positive radius r_0 of the point $\{0, 1\}$, is given by the equation

$$x_2 = \psi(x_1), \tag{3.20}$$

where the single-valued function $\psi(x_1)$ has a second derivative satisfying a Lipschitz condition. Hence a closed curve γ with no self-intersections can be constructed as in [4], §4, such that it contains a piece of the curve described by (3.20), having $\{0, 1\}$ as an interior point. The curve γ in question is infinitely differentiable everywhere except at the point $\{0, 1\}$, while the domain Ω lying inside it contains points $\{0, 1 + \epsilon\}$ when $\epsilon > 0$ is sufficiently small, and is located entirely in the half-plane $x_2 > 0$.

Consider in this domain the harmonic function

$$u(\mathbf{x}) = u^0(\mathbf{x}),$$

where $u^0(\mathbf{x})$ is the function (3.17) with $\alpha = 0$. The boundary values of u are here infinitely differentiable along γ. Proceeding as in the previous example, it can be shown that

$$\max_{\omega_h} |u_h - u| \geqslant \kappa h^3 \ln h^{-1},$$

where $h = 2/(2N+1)$, $N = N_0, N_0 + 1, \cdots$; κ and N_0 are well defined positive numbers, and ω_h is the set introduced above.

The condition $\gamma \in C_{2,1}$ is therefore not in general sufficient for (2.1) to hold with $\sigma = 1$, even for individual solutions, with $f \equiv 0$ and $\varphi \in C_\infty(\gamma)$. The question remains open, whether the condition $\gamma \in C_{3,0}$ is sufficient for (2.1) to hold with $\sigma = 1$, while retaining the other conditions of Theorem 2.1.

To conclude this section, note the further point that it is not possible to put $m < 2$ in the conditions of Theorem 2.1 with arbitrary k, $0 \leqslant k \leqslant 1$, λ, $0 < \lambda < 1$, and μ, $0 \leqslant \mu < 1$, since, as we mentioned in the Introduction, N. S. Bahvalov's results of [8] (see also [9]) imply that, for the class of solutions of Poisson's equation with right side belonging to $C_{1,\mu}(\Omega)$, the mesh method cannot be uniformly convergent at a rate better than $h^{1+\mu}$. It is an open question whether (2.1) holds with $m = 2$, $\mu = 0$, or with $m = \mu = 1$.

§4. A lower bound for the error in a uniform metric for individual solutions of arbitrary boundary value problems

Let $R = R\{|x_i| < 1, i = 1, \cdots, n\}$ be a cube with boundary Γ, and u a function specified in an arbitrary n-dimensional domain Ω; assume that u has continuous derivatives $\partial^4 u/\partial x_i^4$, $i = 1, \cdots, n$, in some subdomain $\Omega' \subset \Omega$, and that also

$$\sum_{i=1}^{n} \frac{\partial^4 u}{\partial x_i^4} \not\equiv 0 \text{ in } \Omega'.$$

It may be assumed without loss of generality that $\overline{R} \subset \Omega'$ and

$$\min_{\mathbf{x} \in \overline{R}} \sum_{i=1}^{n} \frac{\partial^4 u}{\partial x_i^4} = \varepsilon^* > 0. \tag{4.1}$$

Let $h = 1/N$, where $N > 0$ is an integer. Denote by R_h (Γ_h) the set of base points of the cubic mesh in R (on Γ). Obviously $R_h \cup \Gamma_h \subset \Omega_h$.

Let $f = \Delta u$ and let u_h satisfy the difference equations

$$\Delta_h u_h = f \text{ in } R_h, \tag{4.2}$$

where no restrictions are imposed on the values of u_h at boundary points Γ_h.

THEOREM 4.1. *Let u have continuous derivatives $\partial^4 u/\partial x_i^4$, $i = 1, \cdots, n$, in $\bar{R}$, where condition* (4.1) *is satisfied. Then the positive numbers c_0 and N_0 (N_0 an integer) exist such that*

$$\inf_{u_h} \max_{R_h \cup \Gamma_h} |u_h - u| > c_0 h^2, \tag{4.3}$$

where u_h is the solution of system (4.2), *the lower bound is taken over all solutions of* (4.2), *and $h = 1/N$, $N = N_0, N_0 + 1, \cdots$.*

PROOF. The uniform continuity in $\bar{R}$ of the derivatives $\partial^4 u/\partial x_i^4$, $i = 1, \cdots, n$, implies the existence of a $\delta_0 > 0$ such that

$$\max_{1 \leqslant i \leqslant n} \sup_{\substack{\mathbf{x}, \mathbf{y} \in R \\ |\mathbf{x}-\mathbf{y}| < \delta_0}} \left| \frac{\partial^4 u(\mathbf{x})}{\partial x_i^4} - \frac{\partial^4 u(\mathbf{y})}{\partial y_i^4} \right| \leqslant \frac{\varepsilon^*}{2n}.$$

Using this in conjuction with (4.1), it is easily shown by Taylor's formula that

$$f - \Delta_h u \leqslant -\frac{\varepsilon^*}{24} h^2 \text{ in } R_h, \tag{4.4}$$

where $h = 1/N$, $N \geqslant N_0 = [\delta_0^{-1}] + 1$, and [ ] means the integral part.

Let

$$\varepsilon_h = u_h - u = v_h + w_h, \tag{4.5}$$

where u_h is any solution of system (4.2),

$$\Delta_h v_h = 0 \text{ in } R_h, \quad v_h = \varepsilon_h \text{ on } \Gamma_h$$

and from (4.4),

$$\Delta_h w_h \leqslant -\frac{\varepsilon^*}{24} h^2 \text{ in } R_h, \quad w_h = 0 \text{ on } \Gamma_h.$$

Using the maximum principle,

$$\max_{R_h \cup \Gamma_h} |v_h| = \max_{\Gamma_h} |\varepsilon_h| \tag{4.6}$$

and by the comparison theorem ([13], Russian p. 594),

$$w_h \geqslant h^2 \psi \text{ in } R_h \cup \Gamma_h, \tag{4.7}$$

where

$$\begin{gathered} \psi(\mathbf{x}) = \varepsilon^* (1 - |\mathbf{x}|^2)/48n, \\ \Delta_h \psi = -\frac{\varepsilon^*}{24} \text{ in } R_h, \quad \psi \leqslant 0 \text{ on } \Gamma_h. \end{gathered} \tag{4.8}$$

It easily follows from (4.5)–(4.8) that

$$\max_{R_h \cup \Gamma_h} |u_h - u| \geqslant c_0 h^2,$$

where $h = 1/N$, $N = N_0, N_0 + 1, \cdots; N_0 = [\delta_0^{-1}] + 1$, $c_0 = \epsilon^*/96n$, and N_0 and c_0 are independent of the solution of system (4.2). Hence (4.3) holds, and Theorem 4.1 is proved.

From the foregoing and Theorem 4.1, we have

THEOREM 4.2. *If the difference operator* (1.3) *is used for approximating the Laplace operator in a cubic mesh, then, no matter what the individual solution* u^* *of any linear or nonlinear boundary value problem for the Poisson* (*Laplace*) *equation in an arbitrary domain* Ω, *provided only that it has continuous derivatives* $\partial^4 u^*/\partial x_i^4$, $i = 1, \cdots, n$, *in some subdomain* $\Omega' \subset \bar{\Omega}' \subset \Omega$, *and that the equation*

$$\sum_{i=1}^{n} \frac{\partial^4 u^*}{\partial x_i^4} = 0 \tag{4.9}$$

is violated at at least one point of Ω', *the mesh method cannot provide uniform convergence to the individual solution* u^* *at a rate better than* h^2, *regardless of the method of approximating the boundary conditions in* $\gamma_h \cup \Omega_h \backslash \Omega'$.

A particular consequence of this theorem is that the factor h^2 in (2.1) cannot be replaced by $o(h^2)$, no matter what smoothness conditions are imposed on the data of problem (1.1).

An elementary example of a solution which does not satisfy (4.9) is provided by the function $u = |\mathbf{x}|^4$.

This lower bound for the error of the mesh method in a uniform metric, applicable for individual solutions, is similar to the lower bound for the error of quadrature formulas for individual functions obtained by S. M. Nikol'skiĭ in [18], while it extends the result obtained by the present author in [19] for the case of the two-dimensional Laplace equation. The bound reflects the error resulting from approximation of the equation, and is not connected with the boundary conditions.

An approximate method for solving the Dirichlet problem for the two-dimensional Poisson equation is considered in [20]; it gives convergence at a rate $O(h^2)$ in a Hölder weighted space $C_{k,\lambda}^{m,\mu'}$ (in particular, in the space of functions k times continuously differentiable in $\bar{\Omega}$, $0 \leqslant k < \infty$).

BIBLIOGRAPHY

1. E. A. Volkov, *A posteriori and weighted error estimates for solutions of the Poisson equation and their derivatives, computed by the net method*, Dokl. Akad. Nauk SSSR **192** (1970), 717–720 = Soviet Math. Dokl. **11** (1970), 699–703. MR 42 #7085.

2. E. A. Volkov, *The differential properties of the solutions of Laplace's equation, and the errors in the method of nets with boundary values in* C_2 *and* $C_{1,1}$, Ž. Vyčisl. Mat. i Mat. Fiz. **9** (1969), 573–584 = USSR Comput. Math. and Math. Phys. **9** (1969), no. 3, 97–112. MR **40** #8286.

3. N. S. Bahvalov, *Numerical solution of the Dirichlet problem for Laplace's equation*, Vestnik Moskov. Univ. Ser. Mat. Meh. Astr. Fiz. Him. **1959**, no. 5, 171–195. (Russian) MR **22** #6082.

4. E. A. Volkov, *Conditions for the convergence of the method of nets for Laplace's equation at the rate* h^2, Mat. Zametki **6** (1969), 669–679 = Math. Notes 6 (1969), 866–872. MR **41** #6418.

5. ———, *Boundaries of subdomains, Hölder weight classes and solutions in these classes of the Poisson equation*, Trudy Mat. Inst. Steklov. **117** (1972), 75–99 = Proc. Steklov Inst. Math. **117** (1972), 89–117.

6. J. H. Bramble, B. E. Hubbard and V. Thomée, *Convergence estimates for essentially positive type discrete Dirichlet problems*, Math. Comp. **23** (1969), 695–709. MR **42** #1350.

7. E. A. Volkov, *On differential properties of solutions of the Laplace and Poisson equations on a parallelepiped and efficient error estimates of the method of nets*, Trudy Mat. Inst. Steklov. **105** (1969), 46–65 = Proc. Steklov Inst. Math. **105** (1969), 54–78. MR **41** #4836.

8. N. S. Bahvalov, *Approximate computation of multiple integrals*, Vestnik Moskov. Univ. Ser. Mat. Meh. Astr. Fiz. Him. **1959**, no. 4, 3–18. (Russian) MR **22** #6077.

9. E. A. Volkov, *The unimprovability of the error estimate of the method of refinements with higher order differences in a certain class of solutions of Poisson's equation*, Differencial'nye Uravnenija **4** (1968), 136–139 = Differential Equations **4** (1968), 70–71. MR **37** #2468.

10. ———, *The solution by the method of nets of the Dirichlet problem for Laplace's equation on regions with rounded corners*, Differencial'nye Uravnenija **5** (1969), 141–153 = Differential Equations **5** (1969), 122–131. MR **39** #6534.

11. O. A. Ladyženskaja and N. N. Ural'ceva, *Linear and quasilinear equations of elliptic type*, "Nauka", Moscow, 1964; English transl., Academic Press, New York, 1968. MR **35** #1955; **39** #5941.

12. L. Collatz, *Bemerkungen zur Fehlerabschätzung für das Differenzenverfahren bei partiellen Differentialgleichungen*, Z. Angew. Math. Mech. **13** (1933), no. 1, 56.

13. A. N. Tihonov and A. A. Samarskiĭ, *The equations of mathematical physics*, 3rd ed., "Nauka", Moscow, 1966; English transl. of 2nd ed., Pergamon, Press, Oxford; Macmillan, New York, 1963. MR **29** #2498; **35** #1982.

14. E. A. Volkov, *Solving the Dirichlet problem by a method of corrections with higher order differences.* II, Differencial'nye Uravnenija **1** (1965), 1070–1084 = Differential Equations **1** (1965), 835–846. MR **32** #8523.

15. V. I. Lebedev, *Estimation of the error involved in the method of nets for the Dirichlet and Neumann problems*, Dokl. Akad. Nauk SSSR **128** (1959), 665–667. (Russian) MR **22** #1097.

16. J. H. Bramble and B. E. Hubbard, *Approximations of derivatives by finite difference methods in elliptic boundary value problems*, Contributions to Differential Equations **3** (1964), 399–410. MR **29** #4208.

17. E. A. Volkov, *The method of irregular nets for finite and infinite regions with conical points*, Differencial'nye Uravnenija **2** (1966), 1358–1373 = Differential Equations **2** (1966), 702–709. MR **34** #8661.

18. S. M. Nikol'skiĭ, *Quadrature formulas*, Fizmatgiz, Moscow, 1958. (Russian)

19. E. A. Volkov, *The unavoidable errors of the method of nets*, Mat. Zametki **4** (1968), 621–627 = Math. Notes **4** (1968), 865–868. MR **40** #3737.

20. ———, *Approximate solution of the Laplace and Poisson equations in weighted Hölder spaces*, Trudy Mat. Inst. Steklov. **128** (1972), 76–112 = Proc. Steklov. Inst. Math. **128** (1972) (to appear).

Translated by D. E. BROWN

Trudy Mat. Inst. Steklov.
117 (1972)

Proc. Steklov Inst. Math.
117 (1972)

IMBEDDING THEOREMS FOR SPACES OF FUNCTIONS WHOSE MIXED DERIVATIVES SATISFY A MULTIPLE-INTEGRAL HÖLDER CONDITION

A. D. DŽABRAILOV

UDC 517.518.22

Abstract. The differential properties of functions in spaces of type W and B whose multi-index of differential properties is any $n+1$ free vectors $r^0, r^1, \ldots, r^n$ are studied. To study these spaces, a new integral representation for smooth functions is obtained which, in a special case, reduces to a well-known integral representation of V. P. Il'in and O. V. Besov. Corresponding weight function spaces are also considered.

Bibliography: 28 items. Figures: 2.

Introduction

In this paper, we study both the differential and the boundary properties of functions in the function spaces (defined precisely in §1)

$$\bigcap_{i=0}^{n} \mathscr{L}^{\langle \mathbf{r}^i \rangle}_{p_i, \theta_i}(\Omega), \qquad \bigcap_{i=0}^{n} L^{\langle \mathbf{r}^i \rangle}_{p_i}(\Omega), \tag{1}$$

where the points $\mathbf{r}^i = (r^i_1, \cdots, r^i_n)$ $(i = 0, 1, \cdots, n)$, which are the multi-index of the differential properties of this space, do not all lie in the same n-dimensional plane.

The spaces (1) are generalizations of the well-known function spaces

$$B^{\ r_1, \ldots, r_n}_{p_0, p_1, \ldots, p_n}{}_{\theta_1, \ldots, \theta_n}(\Omega), \qquad W^{\ r_1, \ldots, r_n}_{p_0, p_1, \ldots, p_n}(\Omega) \tag{2}$$

(see [1–10]) in the sense that if in (1) we put $\mathbf{r}^0 = (0, \cdots, 0)$ and $\mathbf{r}^i = (0, \cdots, r_i, \cdots, 0)$ $(i = 1, \cdots, n)$, we get the spaces (2).

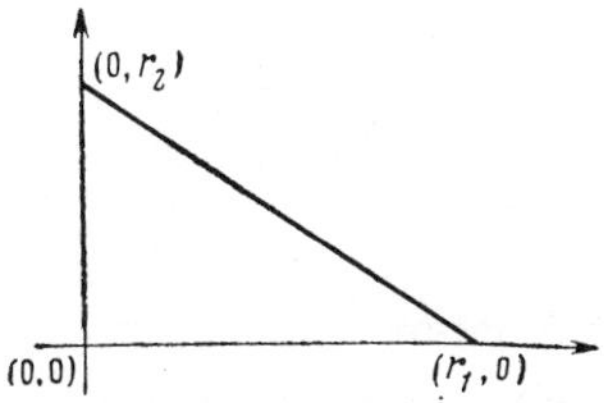

Figure 1

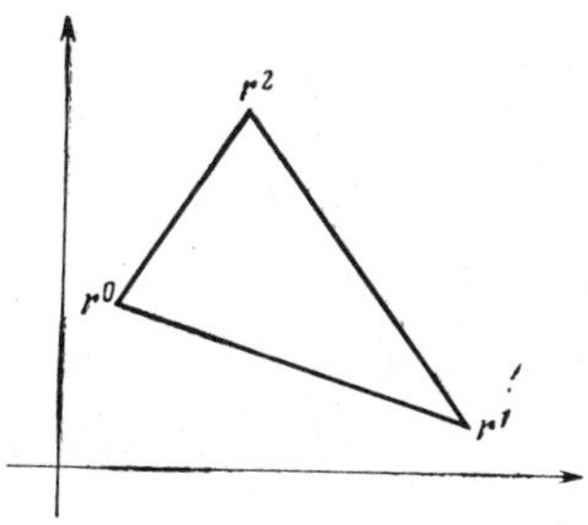

Figure 2

AMS (MOS) subject classifications (1970). Primary 46E35; Secondary 26A69.

The multi-index of the differential properties of the spaces (1) corresponds to the vertices of a certain polyhedron. In the case of the spaces (2), the vertices of this polyhedron are at the points $(0, \cdots, 0)$ and $(0, \cdots, r_i, \cdots, 0)$ $(i = 1, \cdots, n)$ (for $n = 2$, see Figure 1), and in the general case the vertices of the corresponding polyhedron are at the points $\mathbf{r}^i = (r_1^i, \cdots, r_n^i)$ $(i = 0, 1, \cdots, n)$ (for $n = 2$, see Figure 2).

Function spaces whose multi-index of differential properties corresponds to the vertices of a polyhedron (which may lie on the coordinate axes) were first studied by S. M. Nikol'skiĭ [11, 12], and in more generality by V. P. Il'in [13]. The theory of such spaces was further developed in [14–28].

The present work consists of seven sections.

In §1 the fundamental definitions and notation are given.

In §2 the fundamental imbedding theorems are stated, that is, theorems of the type

$$\bigcap_{i=0}^{n} \mathscr{L}_{p_i, \theta_i}^{\langle \mathbf{r}^i \rangle}(\Omega) \to \mathscr{L}_{p, \theta}^{\langle \rho \rangle}(\Omega_m), \qquad \bigcap_{i=0}^{n} L_{p_i}^{\langle \mathbf{r}^i \rangle}(\Omega) \to L_p^{\langle \rho \rangle}(\Omega_m),$$

$$\bigcap_{i=0}^{n} L_{p_i}^{\langle \mathbf{r}^i \rangle}(\Omega) \to \mathscr{L}_{p, \theta}^{\langle \rho \rangle}(\Omega_m).$$

In §3 a certain integral representation of smooth functions is obtained which generalizes the well-known integral representation of Il'in and Besov [10] in the sense that if, in this one, we let $e_{\mathbf{k}0} = \phi, e_{\alpha 0} = \phi, e_{\mathbf{k}i} = \{i\}$ and $e_{\alpha^i} = \{i\}$, then we get the Il'in-Besov representation.

In §4 the results stated in §2 are proved.

In §5 we study the possibility of extending functions beyond the region $\Omega \subset E_n$ to E_n while preserving their differential properties.

In §6 we prove the necessity of some conditions for the validity of inequalities of the type of imbedding theorems.

In §7, weight spaces of functions are considered.

The results of this paper were reported in the seminar "The theory of functions of several real variables and their application to the equations of mathematical physics," directed by S. M. Nikol'skiĭ and L. D. Kudrjavcev at the Steklov Institute for Mathematics.

In conclusion, the author wishes to express his sincere gratitude to L. D. Kudrjavcev for his interest in the present work.

The author wishes to take this opportunity to thank O. V. Besov for reading this paper and for making very valuable suggestions regarding both the results and the exposition; they have been taken into account in the final draft.

§1. Definitions and notation

Let $\mathbf{x} = (x_1, \cdots, x_n)$ be a typical point of n-dimensional Euclidean space E_n. We will denote by e_n the set of the natural numbers $1, \cdots, n$; that is, $e_n = \{1, \cdots, n\}$. For each subset e of the set e_n and for any vector $\mathbf{k} = (k_1, \cdots, k_n)$, the vector with components

$$k_i^e = \begin{cases} k_i, & \text{if} \quad i \in e, \\ 0, & \text{if} \quad i \notin e. \end{cases}$$

will be denoted by $\mathbf{k}^e = (k_1^e, \cdots, k_n^e)$.

The smallest of the subsets e for which the relation $\mathbf{k}^e = \mathbf{k}$ holds is called the support of the vector $\mathbf{k}$, and it will be denoted by $e_{\mathbf{k}}$ (for further detail concerning this, see [11, 12]). For any set e, the number of elements in this set will be denoted by $|e|$; for example, $|e_{\mathbf{r}}|$ is the number of components of the vector $\mathbf{r}$ which are different from zero.

Let $f(\mathbf{x})$ be a sufficiently smooth function defined on E_n. The finite difference of order k_i with respect to x_i with step t_i is denoted by $\Delta_i^{k_i}(t_i)f(\mathbf{x})$, and

$$\Delta^{\mathbf{k}}(\mathbf{t}) f(\mathbf{x}) = \Delta_1^{k_1}(t_1) \ldots \Delta_n^{k_n}(t_n) f(\mathbf{x}).$$

Consequently

$$\Delta^{\mathbf{k}^e}(\mathbf{t}) f(\mathbf{x}) = \Big[\prod_{j \in e} \Delta_j^{k_j}(t_j)\Big] f(\mathbf{x}).$$

Further, let

$$f^{(\mathbf{k})}(\mathbf{x}) \equiv D_1^{k_1} \ldots D_n^{k_n} f(\mathbf{x}) \equiv D^{\mathbf{k}} f(\mathbf{x}) \qquad \left(D_i^{k_i} \equiv \frac{\partial^{k_i}}{\partial x_i^{k_i}}\right).$$

We take

$$\int_{\mathbf{a}}^{\mathbf{b}} f(\mathbf{x})\, d\mathbf{x}^e = \int_{a_1}^{b_1} dx_1^e \ldots \int_{a_n}^{b_n} f(\mathbf{x})\, dx_n^e,$$

where $\mathbf{a} = (a_1, \cdots, a_n)$, $\mathbf{b} = (b_1, \cdots, b_n)$, and

$$\int_{a_i}^{b_i} f(\mathbf{x})\, dx_i^e = \begin{cases} \displaystyle\int_{a_i}^{b_i} f(\mathbf{x})\, dx_i, & \text{if} \quad i \in e, \\ f(\mathbf{x}), & \text{if} \quad i \notin e. \end{cases}$$

Consequently

$$\int_{\mathbf{a}}^{\mathbf{b}} f(\mathbf{x})\, d\mathbf{x}^e = \left(\prod_{j \in e} \int_{a_i}^{b_i} dx_i\right) f(\mathbf{x}) \equiv \int_{\mathbf{a}}^{\mathbf{b}} f(\mathbf{x}) \prod_{j \in e} dx_j.$$

Let $\sigma = (\sigma_1, \cdots, \sigma_n) > 0$; that is, $\sigma_j > 0$ $(j = 1, \cdots, n)$, $h > 0$, $\mathbf{y} = (y_1, \cdots, y_n)$, $\mathbf{y}/h^\sigma = (y_1/h^{\sigma_1}, \cdots, y_n/h^{\sigma_n})$, $h^\sigma = (h^{\sigma_1}, \cdots, h^{\sigma_n})$.

For any two vectors $\alpha = (\alpha_1, \cdots, \alpha_n)$ and $\beta = (\beta_1, \cdots, \beta_n)$, the scalar product of α and β will be denoted by (α, β); that is, $(\alpha, \beta) = \alpha_1\beta_1 + \cdots + \alpha_n\beta_n$. We will always denote the n-dimensional vector all of whose components are 1 by $\omega = (1, \cdots, 1)$; for example, $(\omega, \sigma) = \sigma_1 + \cdots + \sigma_n$, $(\omega^e, \sigma) = \Sigma_{i \in e}\sigma_i$.

Let $\mathbf{r}^i = (r_1^i, \cdots, r_n^i) \geqslant 0$ $(i = 0, 1, \cdots, n)$; that is, $r_j^i \geqslant 0$ $(j = 1, \cdots, n)$. The largest integer less than r_j^i will be denoted by $\bar{r}_j^i$, and for $r_j^i = 0$ we take $\bar{r}_j^i = 0$; consequently $\bar{r}_j^i = (\bar{r}_1^i, \cdots, \bar{r}_n^i)$ $(i = 0, 1, \cdots, n)$ are vectors with integer components.

Let $\mathbf{a} = (a_1, \cdots, a_n) > 0$ (that is, $a_i > 0$), $H > 0$, and $\boldsymbol{\delta} = (\delta_1, \cdots, \delta_n)$, where δ_i $(i = 1, \cdots, n)$ takes on the values $+1$ or -1. There are 2^n different vectors $\boldsymbol{\delta}$.

We will denote by $R(\sigma) \equiv R(\mathbf{a}, H, \epsilon, \boldsymbol{\delta}, \sigma)$ the set of all points $\mathbf{y}$ which satisfy the condition

$$0 < a_i h^{\sigma_i} \leqslant \delta_i y_i \leqslant (a_i + \varepsilon) h^{\sigma_i} \qquad (i = 1, \ldots, n), \; 0 < h \leqslant H.$$

Denoting the operation of vector addition of two sets by the symbol "+", we will call the set $\mathbf{x} + R(\sigma)$ the horn of curvature ω/σ, radius H, and opening $\epsilon > 0$ (for further details, see Besov [9]). Because there are 2^n possible vectors $\boldsymbol{\delta} = (\delta_1, \cdots, \delta_n)$, there are altogether 2^n different horns with fixed parameters $\sigma, H, \mathbf{a}, \epsilon$. We will say that the n-dimensional region Ω belongs to the class $A(R(\sigma))$ if there exists a finite covering by open sets U_k $(k = 1, \cdots, N)$ and a collection $R_k(\sigma)$ such that the following conditions are satisfied:

$$\Omega = \bigcup_{k=1}^{N} U_k, \qquad \operatorname{mes}(U_k \cap U_j) \geqslant q > 0 \qquad (j \neq k),$$

$$U_k + R_k \subset \Omega \qquad (k = 1, \ldots, N).$$

Let $\Delta^{\mathbf{k}}(\mathbf{t}; \Omega) f(\mathbf{x}) = \Delta^{\mathbf{k}}(\mathbf{t}) f(\mathbf{x})$ if the difference is formed from points belonging to Ω together with the polyhedron whose vertices are at these points; otherwise $\Delta^{\mathbf{k}}(\mathbf{t}; \Omega) f(\mathbf{x}) = 0$ (for $\mathbf{k} = (0, \cdots, k_i, \cdots, 0)$, see [10]).

DEFINITION 1. *We will say that* $f(\mathbf{x})$, *defined in* $\Omega \subset E_n$, *belongs to the space* $\mathfrak{L}^{\langle \mathbf{r} \rangle}_{p,\theta}(\Omega)$ *if on* Ω *it has generalized derivatives* $D^{\bar{\mathbf{r}}} f(\mathbf{x})$ *which satisfy the inequality*

$$\| f \|_{\mathscr{L}^{\langle \mathbf{r} \rangle}_{p,\theta}(\Omega)} < \infty,$$

$$\| f \|_{\mathscr{L}^{\langle \mathbf{r} \rangle}_{p,\theta}(\Omega)} \equiv \left(\int \cdots \int \| \Delta^{\omega_r + [\mathbf{r} - \bar{\mathbf{r}}]}(\mathbf{t}; \Omega) D^{\bar{\mathbf{r}}} f \|^{\theta}_{L_p(\Omega)} \prod_{i \in e_{\mathbf{r}}} \frac{dt_i}{|t_i|^{1+\theta(r_i - \bar{r}_i)}} \right)^{1/\theta},$$

where $1 \leqslant p, \theta \leqslant \infty$, *and* $\| \cdot \|_{L_p}$ *denotes the norm of the space* $L_p(\Omega)$; *that is,*

$$\| f \|_{L_p(\Omega)} = \begin{cases} \left(\int_{\Omega} | f(\mathbf{x}) |^p \, d\mathbf{x} \right)^{1/p}, & \text{if } p < \infty, \\ \sup_{\mathbf{x} \in \Omega} | f(\mathbf{x}) |, & \text{if } p = \infty. \end{cases}$$

The set of all finite functions of the space $\mathfrak{L}^{\langle \mathbf{r} \rangle}_{p,\theta}(E_n)$ will be denoted by $\mathring{\mathfrak{L}}^{\langle \mathbf{r} \rangle}_{p,\theta}(E_n)$.

DEFINITION 2. *We will say that* $f(\mathbf{x})$ *belongs to the space*

$$\bigcap_{i=0}^{n} \mathcal{L}_{p_i, \theta_i}^{\langle \mathbf{r}^i \rangle}(\Omega),$$

if it belongs to all of the spaces $\mathfrak{L}_{p_i,\theta_i}^{\langle \mathbf{r}^i \rangle}(\Omega)$ $(i = 0, 1, \cdots, n)$. *We define a seminorm (in the case* $\mathbf{r}^0 = (0, \cdots, 0)$, *a norm) in this space as the sum of the seminorms of the spaces* $\mathfrak{L}_{p_i,\theta_i}^{\langle \mathbf{r}^i \rangle}(\Omega)$ $(i = 0, 1, \cdots, n)$; *that is,*

$$\sum_{i=0}^{n} \| f \|_{\mathcal{L}_{p_i, \theta_i}^{\langle \mathbf{r}^i \rangle}(\Omega)}. \tag{3}$$

It is clear that if $\mathbf{r}^0 = (0, \cdots, 0)$, $\mathbf{r}^i = (0, \cdots, r_i, \cdots, 0)$, then

$$\bigcap_{i=0}^{n} \mathcal{L}_{p_i, \theta_i}^{\langle \mathbf{r}^i \rangle}(\Omega) \equiv B_{p_0, p_1, \ldots, p_n; \theta_1, \ldots, \theta_n}^{r_1, \ldots, r_n}(\Omega).$$

We denote by $\bigcap_{i=0}^{n} \widetilde{\mathfrak{L}}_{p_i,\theta_i}^{\langle \mathbf{r}^i \rangle}(\Omega)$ the closure of the set of functions which are sufficiently smooth in Ω and which satisfy the inequality

$$\sum_{i=0}^{n} \| f \|_{\mathcal{L}_{p_i, \theta_i}^{\langle \mathbf{r}^i \rangle}(\Omega)} < \infty$$

involving the seminorm (3).

DEFINITION 3. *We will say that* $f \in \bigcap_{i=0}^{n} L_{p_i}^{\langle \mathbf{r}^i \rangle}(\Omega)$, *if the generalized derivatives* $D^{[\mathbf{r}^i]} f$ *exist on* $\Omega \subset E_n$ *and*

$$\| f \|_{L_{p_i}^{\langle \mathbf{r}^i \rangle}(\Omega)} = \left(\int \cdots \int \| \Delta^{\omega_*^i}(\mathbf{t};\ \Omega) D^{[\mathbf{r}^i]} f \|_{L_{p_i}(\Omega)}^{p_i} \prod_{j \in e_{\mathbf{r}^i}^*} \frac{dt_j}{| t_j |^{1+p_i(r_j^i - [r_j^i])}} \right)^{1/p_i} < \infty,$$

where $e_{\mathbf{r}^i}^*$ *is the set of indices of the nonintregal components of the vector* $\mathbf{r}^i$; *that is,* $e_{\mathbf{r}^i}^* \equiv e_{\mathbf{r}^i - [\mathbf{r}^i]}$, *where* $[r_j^i]$ *is the integral part of* r_j^i, $[\mathbf{r}^i] = ([r_1^i], \cdots, [r_n^i])$, *and* $\omega_*^i \equiv \omega^{e_{\mathbf{r}^i}^*}$ $(i = 0, 1, \cdots, n)$.

The seminorm in this space is defined as the sum of the seminorms in $L_{p_i}^{\langle \mathbf{r}^i \rangle}(\Omega)$ $(i = 0, 1, \cdots, n)$; *that is,*

$$\sum_{i=0}^{n} \| f \|_{L_{p_i}^{\langle \mathbf{r}^i \rangle}(\Omega)}. \tag{4}$$

For $\mathbf{r}^0 = (0, \cdots, 0)$ and $\mathbf{r}^i = (0, \cdots, r_i, \cdots, 0)$ $(i = 1, \cdots, n)$, the space

$$\bigcap_{i=0}^{n} L_{p_i}^{\langle \mathbf{r}^i \rangle}(\Omega)$$

is identical to the well-known space $W_{p_0, p_1, \ldots, p_n}^{r_1, \ldots, r_n}(\Omega)$ (see [1, 3, 5, 6]).

§2. Fundamental imbedding theorems

The imbedding theorems for the spaces

$$\bigcap_{i=0}^{n} \mathscr{L}^{\langle \mathbf{r}^i \rangle}_{p_i, \theta_i}(\Omega), \qquad \bigcap_{i=0}^{n} \tilde{L}^{\langle \mathbf{r}^i \rangle}_{p_i}(\Omega),$$

are stated in this section; their proofs are given in §4.

THEOREM 1. *Suppose that the points* $\mathbf{r}^i = (r_1^i, \cdots, r_n^i)$ $(i = 0, 1, \cdots, n)$ *do not all belong to the same n-dimensional plane, and that* $r_j^0 \geqslant 0$ $(j = 1, \cdots, n)$, $r_i^i > 0$ *and* $r_j^i \geqslant 0$ $(j \neq i)$ $(i, j = 1, \cdots, n)$.

Let $1 \leqslant p_i \leqslant \theta_i \leqslant \infty$, $p \geqslant p_i$ $(i = 0, 1, \cdots, n)$, *let* $\nu = (\nu_1, \cdots, \nu_n)$ *be a nonnegative integral vector* (*that is,* $\nu_j \geqslant 0, \nu_j$ *integers*), *and let* m *be a natural number,* $m \leqslant n$.

Suppose that

1) $\nu_j \geqslant \bar{r}_j^0 + 1 + [r_j^0 - \bar{r}_j^0] \quad (j \in e_{\mathbf{r}^0}),$

$\nu_j \geqslant \bar{r}_j^i + 1 + [r_j^i - \bar{r}_j^i] \quad (j \neq i) \quad (j \in e_{\mathbf{r}^i};\ i = 1, \ldots, n);$

2) *there exists a positive vector* $\sigma = (\sigma_1, \cdots, \sigma_n)$ (*that is,* $\sigma_j > 0; j = 1, \cdots, n$) *such that*

$$\varepsilon_{m,i} = \left(\mathbf{r}^i - \nu - \frac{1}{p_1}\omega + \frac{1}{p_i}\omega_m, \sigma\right) \geqslant 0 \qquad (i = 1, \ldots, n), \tag{*}$$

where $\omega = (1, \cdots, 1)$, $\omega_m = (\overbrace{1, \cdots, 1}^{m}, 0, \cdots, 0)$;

3) $\Omega \in A(R(\sigma))$;

4) $f \in \bigcap_{i=0}^{n} \mathscr{L}^{\langle \mathbf{r}^i \rangle}_{p_i, \theta_i}(\Omega)$.

Then if $\epsilon_{m,i} > 0$ *or* $\epsilon_{m,i} = 0$, $1 < p_i = \theta_i < p < \infty$, *it follows that the function* $D^\nu f(\mathbf{x})$ *on the intersection of the region with the hyperplane* $x_{m+1} = \text{const}$, $\cdots, x_n = \text{const}$ (*that is, on* Ω_m) *belongs to the space* $L_p(\Omega_m)$. *Moreover,*

$$\| D^\nu f \|_{L_p(\Omega_m)} \leqslant c \sum_{i=0}^{n} h^{\varepsilon_{m,i}} \| f \|_{\mathscr{L}^{\langle \mathbf{r}^i \rangle}_{p_i, \theta_i}(\Omega)}, \tag{5}$$

where $0 < h \leqslant H$, $\epsilon_{m,0} = (\mathbf{r}^0 - \nu - \omega/p_0 + \omega_m/p, \sigma)$, *and* c *is a constant which does not depend on* f *or* h.

THEOREM 2. *Suppose that in Theorem* 1, *condition* 1) *is replaced by the hypothesis*

$$\nu_j \geqslant \begin{cases} [r_j^0] + 1 & \text{for } j \in e^*_{\mathbf{r}^0}, \\ r_j^0 & \text{for } j \in e_{\mathbf{r}^0} \setminus e^*_{\mathbf{r}^0}; \end{cases}$$

$$\nu_j \underset{(j\neq i)}{\geqslant} \begin{cases} [r_j^i]+1, & \text{if } j\in e^*_{\mathbf{r}^i} \\ r_j^i, & \text{if } j\in e_{\mathbf{r}^i}\setminus e^*_{\mathbf{r}^i} \end{cases} \qquad (i=1, \ldots, n),$$

where $e^*_{\mathbf{r}^i}$ *is the set of indices of nonintegral components of the vector* $\mathbf{r}^i$, *and condition* 4) *is replaced by the following*:

$$f\in \bigcap_{i=0}^{n} \tilde{L}_{p_i}^{\langle \mathbf{r}^i\rangle}(\Omega).$$

Then if $\epsilon_{m,i} > 0$ *and* $\epsilon_{m,i} = 0$, $1 < p_i < p < \infty$, *it follows that the function* $D^\nu f(\mathbf{x})$ *on* Ω_m *belongs to the space* $L_p(\Omega_m)$, *and*

$$\| D^\nu f\|_{L_p(\Omega_m)} \leqslant c \sum_{i=0}^{n} h^{\epsilon_{m,i}} \| f\|_{L_{p_i}^{\langle \mathbf{r}^i\rangle}(\Omega)}, \tag{6}$$

where $0 < h \leqslant H$, $\epsilon_{m,0} = (\mathbf{r}^0 - \nu - \omega/p_0 + \omega_m/p, \sigma)$, *and* c *is a constant which does not depend on* f *or* h.

COROLLARY 1. *Let*

$$f\in \bigcap_{i=1}^{n} \mathscr{L}_{p_i, p_i}^{\langle \mathbf{r}^i\rangle}(E_n),$$

where $1 < p_i < p < \infty$, $\mathbf{r}^i = (r_1^i, \cdots, r_n^i)$, $r_i^i > 0$ *and* $r_j^i \geqslant 0$ $(j \neq i)$ $(i, j = 1, \cdots, n)$. *Further, let* $\nu = (\nu_1, \cdots, \nu_n)$ (*where* $\nu_j \geqslant 0$, *and the* ν_j *are integers*), *and let* m *be a natural number*, $m \leqslant n$.

Assume that

1) $\nu_j \geqslant r_j^i + 1 + [r_i^i - r_j^i] \qquad (j\neq i) \quad (i=1, \ldots, n;\ j\in e_{\mathbf{r}^i})$;

2) *there exists a positive vector* $\sigma = (\sigma_1, \cdots, \sigma_n)$ *such that* $\epsilon_{m,i} = 0$ $(i = 1, \cdots, n)$ (*see Theorem* 1, (*)).

Then the function $D^\nu f(\mathbf{x})$ *on* E_m *belongs to the space* $L_p(E_m)$, *and*

$$\| D^\nu f\|_{L_p(E_m)} \leqslant c \sum_{i=1}^{n} \| f\|_{\mathscr{L}_{p_i, p_i}^{\langle \mathbf{r}^i\rangle}(E_n)}, \tag{7}$$

where c *is a constant which does not depend on* f.

COROLLARY 2. *Let*

$$f\in \bigcap_{i=1}^{n} \mathring{L}_{p_i}^{\langle \mathbf{r}^i\rangle}(E_n),$$

where $1 < p_i < p < \infty$, $\mathbf{r}^i = (r_1^i, \cdots, r_n^i)$, $r_i^i > 0$ *and* $r_j^i \geqslant 0$ $(j \neq i)$ $(i, j = 1, \cdots, n)$. *Further, let* $\nu = (\nu_1, \cdots, \nu_n)$ (*where* $\nu_j \geqslant 0$, *and the* ν_j *are integers*), *and let* m *be a natural number*, $m \leqslant n$.

Assume that

1) $$\nu_{j \atop (j\neq i)} \geqslant \begin{cases} [r^i_j]+1 & \text{for } j \in e^*_{\mathbf{r}^i} \\ r^i_j & \text{for } j \in e_{\mathbf{r}^i} \setminus e^*_{\mathbf{r}^i} \end{cases} \qquad (i=1, \ldots, n);$$

2) *there exists a positive vector* $\sigma = (\sigma_1, \cdots, \sigma_n)$ *such that* $\epsilon_{m,i} = 0$ $(i = 1, \cdots, n)$ *(see (*)).*

Then the function $D^\nu f$ *on* E_m *belongs to the space* $L_p(E_m)$, *and*

$$\| D^\nu f \|_{L_p(E_m)} \leqslant c \sum_{i=1}^{n} \| f \|_{L^{\langle \mathbf{r}^i \rangle}_{p_i}(E_n)}, \tag{8}$$

where c *is a constant which does not depend on* f.

THEOREM 3. *Let* $\rho = (\rho_1, \cdots, \rho_n)$ *be a nonnegative vector with support* $e_\rho \subseteq \{1, \cdots, m\}$.

Suppose that the conditions of Theorem 1 *are satisfied, except that condition* 1) *is replaced by the following*:

$$\bar\rho_j + \nu_j \geqslant \bar r^0_j + 1 + [r^0_j - \bar r^0_j] \qquad (j \in e_{\mathbf{r}^0}),$$
$$\bar\rho_j + \nu_j \geqslant \bar r^i_j + 1 + [r^i_j - \bar r^i_j] \qquad (j \neq i) \quad (i=1, \ldots, n;\ j \in e_{\mathbf{r}^i}).$$

Further, let $\theta \geqslant \theta_i$ *and* $\epsilon_{m,i} - (\rho, \sigma) \geqslant 0$ $(i = 1, \cdots, n)$.

Then the function $D^\nu f$ *on* Ω_m *belongs to the space* $\mathcal{L}^{\langle\rho\rangle}_{p,\theta}(\Omega_m)$, *and*

$$\| D^\nu f \|_{\mathscr{L}^{\langle\rho\rangle}_{p,\theta}(\Omega_m)} \leqslant c_1 \sum_{i=0}^{n} \| f \|_{\mathscr{L}^{\langle \mathbf{r}^i \rangle}_{p_i,\theta_i}(\Omega)}. \tag{9}$$

In case $\Omega = E_m$ *and* $\epsilon_{m,i} - (\rho, \sigma) = 0$ $(i = 1, \cdots, n)$, *the inequality*

$$\| D^\nu f \|_{\mathscr{L}^{\langle\rho\rangle}_{p,\theta}(E_m)} \leqslant c_2 \sum_{i=1}^{n} \| f \|_{\mathscr{L}^{\langle \mathbf{r}^i \rangle}_{p_i,\theta_i}(E_n)} \tag{10}$$

also holds, where c_1 *and* c_2 *are constants which do not depend on* f.

COROLLARY 3. *Let* f *belong to the space*

$$\bigcap_{i=1}^{n} \mathring{\mathscr{L}}^{\langle \mathbf{r}^i \rangle}_{p_i,\theta_i}(E_n),$$

where $1 \leqslant p_i \leqslant \theta_i \leqslant \infty$ $(i = 1, \cdots, n)$ *and the coordinates of the vectors* $\mathbf{r}^i = (r^i_1, \cdots, r^i_n)$ $(i = 1, \cdots, n)$ *are subject to the conditions* $r^i_i > 0$ *and* $r^i_j \geqslant 0$ $(j \neq i)$ $(i, j = 1, \cdots, n)$.

Further, let $\nu = (\nu_1, \cdots, \nu_n)$ $(\nu_j \geqslant 0;\ \nu_j$ *integers), and let* m *be a natural number,* $m \leqslant n$.

Assume that

1) $\bar\rho_j + \nu_j \geqslant \bar r^i_j + 1 + [r^i_j - \bar r^i_j] \qquad (j \neq i) \qquad (i, j = 1, \ldots, n);$

2) *there exists a vector* $\sigma = (\sigma_1, \cdots, \sigma_n) > 0$ *such that*

$$\varepsilon_{m,i} = \left(\mathbf{r}^i - \mathbf{v} - \frac{1}{p_i}\boldsymbol{\omega} + \frac{1}{p}\boldsymbol{\omega}_m \cdot \boldsymbol{\sigma}\right) > 0 \quad (i = 1, \ldots, n),$$

$$\varepsilon_{m,i} - (\boldsymbol{\rho} \cdot \boldsymbol{\sigma}) = 0 \quad (i = 1, \ldots, n).$$

Then the function $D^\nu f$ *on the hyperplane* $x^r_{m+1} = \text{const}, \cdots, x_n = \text{const}$ *belongs to the space*

$$\mathscr{L}^{\langle\rho\rangle}_{p,\theta}(E_m),$$

and the inequality

$$\| D^\nu f \|_{\mathscr{L}^{\langle\rho\rangle}_{p,\theta}(E_m)} \leqslant c \sum_{i=1}^{n} \| f \|_{\mathscr{L}^{\langle\mathbf{r}^i\rangle}_{p_i,\theta_i}(E_n)} \tag{11}$$

holds, where c *is a constant which does not depend on* f.

THEOREM 4. *Let* $\rho = (\rho_1, \cdots, \rho_n)$ *be a nonnegative vector* (*that is,* $\rho_j \geqslant 0$, $j = 1, \cdots, n$) *with support* $e_\rho \subseteq \{1, \cdots, m\}$.

Suppose that the conditions of Theorem 1 *are satisfied, except that condition* 1) *is replaced by the following*:

$$[\rho_j] + \nu_j \geqslant \bar{r}^0_j + 1 + [r^0_j - \bar{r}^0_j] \quad (j \in e_{\mathbf{r}^0}),$$

$$[\rho_j] + \nu_j \geqslant \bar{r}^i_j + 1 + [r^i_j - \bar{r}^i_j] \quad (j \neq i) \quad (i = 1, \ldots, n;\ j \in e_{\mathbf{r}^i}),$$

where $[\rho_j]$ *is the integral part of* ρ_j.

Let $\epsilon_{m,i} - (\rho, \sigma) > 0$ $(i = 1, \cdots, n)$ (*see* (*)).

Then the function $D^\nu f$ *on* Ω_m *belongs to the space* $L^{\langle\rho\rangle}_p(\Omega_m)$, *and*

$$\| D^\nu f \|_{L^{\langle\rho\rangle}_p(\Omega_m)} \leqslant c \sum_{i=0}^{n} \| f \|_{\mathscr{L}^{\langle\mathbf{r}^i\rangle}_{p_i,\theta_i}(\Omega)}, \tag{12}$$

where c *is a constant which does not depend on* f.

We get the following statement as a consequence of Theorems 1 and 3.

THEOREM 5. *Under the conditions of Theorem* 3, *the function* $D^\nu f$ *on* Ω_m *belongs to the intersection of the spaces* $L^{\langle\bar{\rho}\rangle}_p(\Omega_m)$ *and* $\mathfrak{L}^{\langle\rho\rangle}_{p,\theta}(\Omega_m)$; *that is,*

$$D^\nu f \in L^{\langle\bar{\rho}\rangle}_p(\Omega_m) \cap L^{\langle\rho\rangle}_{p,\theta}(\Omega_m),$$

and the inequality

$$\| D^\nu f \|_{L^{\langle\bar{\rho}\rangle}_p(\Omega_m)} + \| D^\nu f \|_{\mathscr{L}^{\langle\rho\rangle}_{p,\theta}(\Omega_m)} \leqslant c \sum_{i=0}^{n} \| f \|_{\mathscr{L}^{\langle\mathbf{r}^i\rangle}_{p_i,\theta_i}(\Omega_m)} \tag{13}$$

holds, where c *is a constant which does not depend on* f.

THEOREM 6. *Suppose that* $\mathbf{r}^i = (r_1^i, \cdots, r_n^i)$ $(i = 0, 1, \cdots, n)$ *do not all lie on the same n-dimensional plane and* $r_j^0 \geqslant 0$ $(j = 1, \cdots, n)$, $r_i^i > 0$ *and* $r_j^i \geqslant 0$ $(j \neq i)$ $(i, j = 1, \cdots, n)$.

Let $1 \leqslant p_i \leqslant p \leqslant \infty$ $(i = 0, 1, \cdots, n)$, *let* $\nu = (\nu_1, \cdots, \nu_n)$ *be a nonnegative integral vector (that is,* $\nu_j \geqslant 0$; ν_j *integers), and let* m *be a natural number,* $m \leqslant n$.

We assume that $\rho = (\rho_1, \cdots, \rho_n)$ $(\rho_j \geqslant 0;\ j = 1, \cdots, n)$, $e_\rho \subseteq \{1, \cdots m\}$, *and*

1)
$$[\rho_j] + \nu_j \geqslant \begin{cases} [r_j^0] + 1 & \text{for } j \in e^*_{\mathbf{r}^0}, \\ r_j^0 & \text{for } j \in e_{\mathbf{r}^0} \setminus e^*_{\mathbf{r}^0}; \end{cases}$$

$$[\rho_j] + \underset{(j \neq i)}{\nu_j} \geqslant \begin{cases} [r_j^i] + 1 & \text{for } j \in e^*_{\mathbf{r}^i}, \\ r_j^i & \text{for } j \in e_{\mathbf{r}^i} \setminus e^*_{\mathbf{r}^i} \end{cases} \qquad (i = 1, \ldots, n),$$

where $e^*_{\mathbf{r}^i}$ *is the set of indices of the nonintegral components of the vector* $\mathbf{r}^i$, *and* $[r_j^i]$ *is the integral part of* r_j^i;

2) *there exists a positive vector* $\sigma = (\sigma_1, \cdots, \sigma_n)$ *such that*

$$\varepsilon_{i,m} = \left(\mathbf{r}^i - \boldsymbol{\nu} - \frac{1}{p_i}\,\boldsymbol{\omega} + \frac{1}{p}\,\boldsymbol{\omega}_m, \boldsymbol{\sigma}\right) > 0,$$
$$(\boldsymbol{\rho}, \boldsymbol{\sigma}) < \varepsilon_{m,i} \quad (i = 1, \ldots, n);$$

3) $f \in \bigcap_{i=0}^n \widetilde{L}_{p_i}^{\langle \mathbf{r}^i \rangle}(\Omega)$, *where* $\Omega \in A(R(\sigma))$.

Then the function $D^\nu f$ *on* $\Omega_m \equiv \Omega \cap \{x_{m+1} = \text{const}, \cdots, x_n = \text{const}\}$ *belongs to the space* $L_p^{\langle\rho\rangle}(\Omega_m)$, *and*

$$\| D^\nu f \|_{L_p^{\langle\rho\rangle}(\Omega_m)} \leqslant c \sum_{i=0}^n \| f \|_{L_{p_i}^{\langle \mathbf{r}^i \rangle}(\Omega)}, \tag{14}$$

where c *is a constant which does not depend on* f.

THEOREM 7. *Let the conditions of Theorem* 6 *be satisfied except that, in* 1), $[\rho_j]$ $(j = 1, \cdots, n)$ *are replaced by* $\bar\rho_j$; *that is, the largest integer less than* ρ_j, *but in case* $\rho_j = 0$, *we take* $\bar\rho_j = 0$.

Then the function $D^\nu f$ *on* Ω_m *belongs to the space* $\mathfrak{L}_{p,\theta}^{\langle\rho\rangle}(\Omega_m)$, *where* $\theta \geqslant p$, *and moreover*

$$\| D^\nu f \|_{\mathscr{L}_{p,\theta}^{\langle\rho\rangle}(\Omega_m)} \leqslant c \sum_{i=0}^n \| f \|_{L_{p_i}^{\langle \mathbf{r}^i \rangle}(\Omega)}, \tag{15}$$

where the constant c *is independent of* f.

§3. An integral representation for sufficiently smooth functions

In this section we establish an integral representation for the derivatives of a smooth function in terms of differences of certain given derivatives of the function.

Let $\alpha^i = (\alpha_1^i, \cdots, \alpha_n^i)$ and $\mathbf{k}^i = (k_1^i, \cdots, k_n^i)$ $(i = 0, 1, \cdots, n)$ be nonnegative integral vectors (that is, $\alpha_j^i, k_j^i \geqslant 0$; α_j^i and k_j^i integers).

Further, let $\nu = (\nu_1, \cdots, \nu_n)$ be a given nonnegative integral vector which is related to the vectors α^i and $\mathbf{k}^i$ $(i = 0, 1, \cdots, n)$ in the following way:

$$\begin{aligned} &\nu_j \geqslant \alpha_j^0 + k_j^0 \quad (j = 1, \ldots, n), \\ &\nu_j \geqslant \alpha_j^i + k_j^i \quad (j \neq i)\ (i,\ j = 1, \ldots, n). \end{aligned} \tag{**}$$

The goal of this section is to obtain the following integral representation for sufficiently smooth functions $f(\mathbf{x})$:

$$\begin{aligned} D^\nu f(\mathbf{x}) = (-1)^{(\nu+\alpha^0,\,\omega)} c_0 h^{-\beta_0} \int d\mathbf{z}^0 \int \Delta^{\mathbf{k}^0}\left(\frac{\mathbf{z}^0}{\mathbf{k}^0}\right) D^{\alpha^0} f(\mathbf{x}+\mathbf{y})\, \Phi_{\delta,0}\left(\frac{\mathbf{y}}{h^\sigma};\ \frac{\mathbf{z}^0}{h^\sigma}\right) d\mathbf{y} \\ -\sum_{i=1}^n (-1)^{(\nu+\alpha^i,\,\omega)} c_i \int_0^h \frac{dv}{v^{1+\beta_i}}\, d\mathbf{z}^i \int \Delta^{\mathbf{k}^i}\left(\frac{\mathbf{z}^i}{\mathbf{k}^i}\right) D^{\alpha^i} f(\mathbf{x}+\mathbf{y})\, \Phi_{\delta,i}\left(\frac{\mathbf{y}}{v^\sigma};\ \frac{\mathbf{z}^i}{v^\sigma}\right) d\mathbf{y}, \end{aligned} \tag{16}$$

where $\mathbf{z}^i = \mathbf{z}^{e\mathbf{k}^i}$ $(i = 0, 1, \cdots, n)$, the c_i are positive constants, $\beta_i = (\omega + \omega^i + \nu - \alpha^i, \sigma)$ (here $\omega^i \equiv \omega^{e\mathbf{k}^i}$) $(i = 0, 1, \cdots, n)$, and where the kernels $\Phi_{\delta,i}(\mathbf{y}; \mathbf{z}^i)$ belong to $C_0^\infty(E_\mathbf{y} \times E_{\mathbf{z}^i})$ and are concentrated in an arbitrary given cube in an arbitrary coordinate angle (depending on the vector δ).

To prove the identity (16), we will obtain an explicit form for the kernels $\Phi_{\delta,i}$ and, in addition, we show that any "horn" $\mathbf{x} + R(\sigma)$ can serve as the field of the representation.

We turn to the proof of the identity (16).

Let $\mathcal{K}_\delta(t)$ (where $\delta = +1$ or -1) be a sufficiently smooth finite function defined in E_1 whose support is concentrated on the interval $[\alpha, 1] \subset (0, 1]$ if $\delta = +1$, or on the interval $[-1, -\alpha] \subset [-1, 0)$ if $\delta = -1$; in addition, $\int_{E_1} \mathcal{K}_\delta(t)\,dt = 1$.

If the vector $\delta = (\delta_1, \cdots, \delta_n)$ is given, where $\delta_j = +1$ or -1 $(j = 1, \cdots, n)$, then we take

$$\eta_{\delta_i}(t_i) = \frac{t_i^{s_i-1}}{(s_i-1)!} \int_{-1}^{t_i} \mathcal{K}_{\delta_i}(\xi)\, d\xi \qquad (s_i > 0 \text{ integers}),$$

$$\mathscr{W}_{\delta_i}(t_i) = D^{s_i} \eta_{\delta_i}(t_i) \qquad (i = 1, \ldots, n),$$

$$\mathscr{W}_\delta(\mathbf{t}) = \prod_{i=1}^n \mathscr{W}_{\delta_i}(t_i).$$

We introduce the averages

$$(D^\nu f)_{h^\sigma}(\mathbf{x}) = h^{-(\omega,\,\sigma)} \int_{E_n} D^\nu f(\mathbf{x}+\mathbf{y})\, \mathscr{W}_\delta\left(\frac{\mathbf{y}}{h^\sigma}\right) d\mathbf{y}, \tag{17}$$

$$\widehat{(D^\nu f)}_{h^\sigma}(\mathbf{x}) \equiv ((D^\nu f)_{h^\sigma})_{h^\sigma}(\mathbf{x}) \\ = h^{-2(\omega,\sigma)} \int_{E_n}\int_{E_n} D^\nu f(\mathbf{x}+\mathbf{y}+\mathbf{z}) \mathcal{W}_\delta\left(\frac{\mathbf{y}}{h^\sigma}\right)\mathcal{W}_\delta\left(\frac{\mathbf{z}}{h^\sigma}\right) d\mathbf{y}d\mathbf{z}. \tag{18}$$

It is easy to see that

$$\widehat{(D^\nu f)}_{h^\sigma}(\mathbf{x}) \to D^\nu f(\mathbf{x}) \qquad (h \to 0). \tag{19}$$

It is clear that

$$\frac{\partial}{\partial h}\widehat{(D^\nu f)}_{h^\sigma}(\mathbf{x}) = 2\frac{\partial}{\partial t}((D^\nu f)_{h^\sigma})_{t^\sigma}\Big|_{t=h}. \tag{20}$$

On the other hand,

$$\frac{\partial}{\partial h}\left[\frac{1}{h^{\sigma_i}}\mathcal{W}_{\delta_i}\left(\frac{y_i}{h^{\sigma_i}}\right)\right] = -\sigma_i h^{-\sigma_i-1}\left[\mathcal{W}_{\delta_i}\left(\frac{y_i}{h^{\sigma_i}}\right) + \frac{y_i}{h^{\sigma_i}}\mathcal{W}'_{\delta_i}\left(\frac{y_i}{h^{\sigma_i}}\right)\right],$$

and, in addition,

$$\mathcal{W}_{\delta_i}(t) + t\mathcal{W}'_{\delta_i}(t) = t\mathcal{W}'_{\delta_i}(t) + s_i\mathcal{W}_{\delta_i}(t) - (s_i-1)\mathcal{W}_{\delta_i}(t) \\ = D^{s_i}[t\eta'_{\delta_i}(t) - (s_i-1)\eta_{\delta_i}(t)] = D^{s_i}\left(\frac{t^{s_i}}{(s_i-1)!}\mathcal{K}_{\delta_i}(t)\right) \equiv \zeta_{\delta_i}^{(s_i)}(t).$$

Consequently

$$\frac{\partial}{\partial h}\left[\frac{1}{h^{\sigma_i}}\mathcal{W}_{\delta_i}\left(\frac{y_i}{h^{\sigma_i}}\right)\right] = -\sigma_i h^{-\sigma_i-1}\zeta_{\delta_i}^{(s_i)}\left(\frac{y_i}{h^{\sigma_i}}\right]. \tag{21}$$

With the help of formula (21) we have

$$\mathcal{J} \equiv \int_{E_1}\int_{E_1} D^\nu g(x+y+z)\mathcal{W}_\delta\left(\frac{z}{h^\sigma}\right)\frac{\partial}{\partial h}\left[\frac{1}{h^\sigma}\mathcal{W}_\delta\left(\frac{y}{h^\sigma}\right)\right]dydz \\ = -\sigma h^{\sigma-1}\int_{E_1}\int_{E_1} D^\nu g(x+y+z)\zeta_\delta^{(s)}\left(\frac{y}{h^\sigma}\right)\mathcal{W}_\delta\left(\frac{z}{h^\sigma}\right)dydz \\ = (-1)^{s+1}\sigma h^{-1+(s-1)\sigma}\int_{E_1}\int_{E_1} D^{\alpha+k}g(x+y+z)\zeta_\delta^{(s-\alpha-k+\nu)}\left(\frac{y}{h^\sigma}\right)\mathcal{W}_\delta\left(\frac{z}{h^\sigma}\right)dydz: \tag{22}$$

Now with the help of the formula

$$\frac{\partial}{\partial t}\Delta(t)\psi(x) = \psi'(x+t)$$

and an integration by parts we get

$$\tilde{\mathcal{J}} \equiv \int_{E_1}\int_{E_1} D^k g(x+y+z)\mathcal{W}_\delta\left(\frac{y}{h^\sigma}\right)\mathcal{W}_\delta\left(\frac{z}{h^\sigma}\right)dydz =$$

$$
= \int_{E_1}\int_{E_1} D^k g\left(x + y_1 + \frac{z}{k}\right) Q_{0,\delta}\left(\frac{y_1 - \frac{k-1}{k}z}{h^\sigma};\ \frac{z}{h^\sigma}\right) dy_1 dz
$$

$$
= -\frac{k}{h^\sigma}\int_{E_1}\int_{E_1} \Delta\left(\frac{z}{k}\right) D^{k-1} g(x + y_1) Q_{1,\delta}\left(\frac{y_1 - \frac{k-1}{k}z}{h^\sigma};\ \frac{z}{h^\sigma}\right) dy_1 dz,
$$

whence, repeating this process k times, we have

$$
\tilde{\mathcal{J}} = (-1)^k \frac{k^k}{h^{\sigma k}} \int_{E_1}\int_{E_1} \Delta^k\left(\frac{z}{k}\right) g(x + y_k) Q_{k,\delta}\left(\frac{y_k}{h^\sigma};\ \frac{z}{h^\sigma}\right) dy_k dz, \tag{23}
$$

where

$$
\begin{gathered}
Q_{0,\delta}\left(\frac{y_1 - \frac{k-1}{k}z}{h^\sigma};\ \frac{z}{h^\sigma}\right) = \mathcal{W}_\delta\left(\frac{y_1 - \frac{k-1}{k}z}{h^\sigma}\right)\mathcal{W}_\delta\left(\frac{z}{h^\sigma}\right), \\
Q_{i,\delta}\left(\frac{y_i - \frac{k-i}{k}z}{h^\sigma};\ \frac{z}{h^\sigma}\right) = \frac{\partial}{\partial\left(\frac{z}{h^\sigma}\right)} Q_{i-1,\delta}\left(\frac{y_i - \frac{k-i}{k}z}{h^\sigma};\ \frac{z}{h^\sigma}\right) \quad (i = 1, \ldots, k).
\end{gathered} \tag{23*}
$$

By arguments analogous to those used in the proof of (23), we get from the equality (22)

$$
\mathcal{J} = (-1)^{1+s+k}\sigma k^k h^{-1+(\alpha-\nu)\sigma} \int_{E_1}\int_{E_1} \Delta^k\left(\frac{z}{k}\right) D^\alpha g(x + y_k) \Pi_{k,\delta}\left(\frac{y_k}{h^\sigma};\ \frac{z}{h^\sigma}\right) dy_k dz, \tag{24}
$$

where

$$
\begin{gathered}
\Pi_{0,\delta}\left(\frac{y_1 - \frac{k-1}{k}z}{h^\sigma};\ \frac{z}{h^\sigma}\right) = \zeta_\delta^{(s-\alpha-k+\nu)}\left(\frac{y_1 - \frac{k-1}{k}z}{h^\sigma}\right)\mathcal{W}_\delta\left(\frac{z}{h^\sigma}\right), \\
\Pi_{0,\delta}\left(\frac{y_1 - \frac{k-i}{k}}{h^\sigma};\ \frac{z}{h^\sigma}\right) = \frac{\partial}{\partial\left(\frac{z}{h^\sigma}\right)} \Pi_{i-1,\delta}\left(\frac{y_i - \frac{k-i}{k}z}{h^\sigma};\ \frac{z}{h^\sigma}\right) \quad (i = 1, \ldots, k).
\end{gathered} \tag{24'}
$$

Now, taking (20) into account, we transform

$$
\frac{\partial}{\partial h}\widehat{(D^\nu f)}_{h^\sigma}(\mathbf{x})
$$

in the following way: first transfer the differentiation to the kernel, and then, taking into account the finiteness of the kernels, perform transformations analogous to those in (21)–(24), hence obtaining

$$\frac{\partial}{\partial h}\widehat{(D^\nu f)}_{h^\sigma}(\mathbf{x}) = -2\sum_{i=1}^{n}\sigma_i \prod_{j\in e_{\mathbf{k}^i}} (k_j^i)^{k_j^i}(-1)^{(\alpha^i+\nu,\,\omega)} h^{-1-(\omega+\omega^i+\nu-\alpha^i,\,\sigma)}$$
$$\times \int\int \Delta^{\mathbf{k}^i}\left(\frac{\mathbf{z}^i}{\mathbf{k}^i}\right) D^{\alpha^i} f(\mathbf{x}+\mathbf{y})\, \Phi_{\delta,i}\left(\frac{\mathbf{y}}{h^\sigma};\ \frac{\mathbf{z}^i}{h^\sigma}\right) d\mathbf{y}\,d\mathbf{z}^i. \tag{25}$$

Here $\mathbf{z}^i \equiv \mathbf{z}^{e_{\mathbf{k}^i}}, \omega^i \equiv \omega^{e_{\mathbf{k}^i}}$ $(i = 1, \cdots, n)$, and

$$\Phi_{\delta,i}(\mathbf{y},\ \mathbf{z}^i) = \prod_{j\in e_{\mathbf{k}^i}} \Phi_{\delta_j,i}(y_j;\ z_j) \prod_{j\in e_n\setminus e_{\mathbf{k}^i}} \Phi_{\delta_j,i}(y_j),$$

where

$$\Phi_{\delta_j,i}(y_j) = \begin{cases} \int\limits_{E_1} \mathscr{W}_{\delta_i}(\xi)\, \zeta_{\delta_i}^{(s_i-\alpha_i^i+\nu_i)}(y_i-\xi)\, d\xi & \text{for } j=i, \\ \int\limits_{E_1} \mathscr{W}_{\delta_j}^{(\nu_j-\alpha_j^i)}(y_j-\xi)\, \mathscr{W}_{\delta_j}(\xi)\, d\xi & \text{for } j\neq i, \end{cases}$$

$$\Phi_{\delta_j,i}(y_j;\ z_j) = \begin{cases} Q_{k_j^i,\delta_j}(y_j;\ z_j) & \text{for } j\neq i, \\ \Pi_{k_i^i,\delta_i}(y_j;\ z_i) & \text{for } j=i, \end{cases}$$

where the functions $Q_{k_j^i,\delta_j}(y_j; z_j)$ and $\Pi_{k_i^i,\delta_i}(y_i; z_i)$ are defined by the recursion formulas (23*) and (24*), and where

$$Q_{0,\delta}\left(\frac{y_1-\frac{k-1}{k}z}{h^{\sigma_j}};\ \frac{z}{h^{\sigma_j}}\right) \equiv \mathscr{W}_\delta^{(\nu_j-\alpha_j^i)}\left(\frac{y_1-\frac{k-1}{k}z}{h^{\sigma_j}}\right)\mathscr{W}_\delta\left(\frac{z}{h^{\sigma_j}}\right),$$

$$\Pi_{0,\delta}\left(\frac{y_1-\frac{k-1}{k}z}{h^{\sigma_i}};\ \frac{z}{h^{\sigma_i}}\right) \equiv \zeta_\delta^{(s_i-\alpha_i^i-k_i^i)}\left(\frac{y_1-\frac{k-1}{k}z}{h^{\sigma_i}}\right)\mathscr{W}_\delta\left(\frac{z}{h^{\sigma_i}}\right).$$

The bounds

$$|D^\mu \Phi_{\delta,i}(\mathbf{y};\ \mathbf{z}^i)| \leqslant \text{const} \qquad (\mu = 0,\ 1,\ 2,\ \ldots),$$

hold for the kernels $\Phi_{\delta,i}(\mathbf{y}; \mathbf{z}^i)$ $(i = 1, \cdots, n)$.

By means of similar arguments, we transform the expression

$$\widehat{(D^\nu f)}_{h^\sigma}(\mathbf{x}) = (-1)^{(\nu+\alpha^0,\,\omega)} c_0 h^{-(\omega+\omega^0+\nu-\alpha^0,\,\sigma)}$$
$$\times \int\int \Delta^{\mathbf{k}^0}\left(\frac{\mathbf{z}^0}{\mathbf{k}^0}\right) D^{\alpha^0} f(\mathbf{x}+\mathbf{y})\, \Phi_{\delta,0}\left(\frac{\mathbf{y}}{h^\sigma};\ \frac{\mathbf{z}^0}{h^\sigma}\right) d\mathbf{y}\,d\mathbf{z}^0. \tag{26}$$

Here $\mathbf{z}^0 \equiv \mathbf{z}^{e_{\mathbf{k}^0}}$, $\omega^0 \equiv \omega^{e_{\mathbf{k}^0}}$, $c_0 = \Pi_{j\in e_{\mathbf{k}^0}}(k_j^0)^{k_j^0}$, and

$$\Phi_{\delta,0}(\mathbf{y};\ \mathbf{z}^0) = \prod_{j\in e_{\mathbf{k}^0}} \Phi_{\delta_j,0}(y_j;\ z_j) \prod_{j\in e_n\setminus e_{\mathbf{k}^0}} \Phi_{\delta_j,0}(y_j),$$

$$\Phi_{\delta_j, 0}(y_j) = \int \mathcal{W}_{\delta_j}^{(\nu_j - \alpha_j^0)}(y_j - \xi)\, \mathcal{W}_{\delta_j}(\xi)\, d\xi,$$

$$\Phi_{\delta_j, 0}(y_j;\ z_j) = Q_{k_j^0, \delta_j}(y_i;\ z_j),$$

and, in addition,

$$|D^{\mu}\Phi_{\delta, 0}(\mathbf{y};\ \mathbf{z}^0)| \leqslant \text{const} \qquad (\mu = 0,\ 1,\ 2,\ \ldots).$$

Now with the help of the formula

$$D^{\nu} f(\mathbf{x}) = \widehat{(D^{\nu} f)}_{h^{\sigma}}(\mathbf{x}) - \int_0^h \frac{\partial}{\partial v} \widehat{(D^{\nu} f)}_{v^{\sigma}}(x)\, dv$$

we get the integral representation (16), which was to be proved.

Along with the function $D^{\nu} f(\mathbf{x})$, we will consider the function $\widetilde{f}_{\nu,\delta}(\mathbf{x})$, which is obtained by altering the right-hand side of the representation (16); that is,

$$\begin{aligned} \widetilde{f}_{\nu,\delta}(\mathbf{x}) &= (-1)^{(\nu + \alpha^0, \omega)} c_0 h^{-\beta_0} \\ &\times \int\int \Delta^{\mathbf{k}^0}\left(\frac{\mathbf{z}^0}{\mathbf{k}^0};\ U + R_{\delta}(\sigma)\right) D^{\alpha^0} f(\mathbf{x} + \mathbf{y})\, \Phi_{\delta, 0}\left(\frac{y}{h^{\sigma}};\ \frac{\mathbf{z}^0}{h^{\sigma}}\right) d\mathbf{y} d\mathbf{z}^0 \\ &- 2 \sum_{i=1}^{n} (-1)^{(\nu + \alpha^i, \omega)} c_i \int_0^h \frac{dv}{v^{1+\beta_i}} \\ &\times \int\int \Delta^{\mathbf{k}^i}\left(\frac{\mathbf{z}^i}{\mathbf{k}^i};\ U + R_{\delta}(\sigma)\right) D^{\alpha^i} f(\mathbf{x} + \mathbf{y})\, \Phi_{\delta, i}\left(\frac{\mathbf{y}}{v^{\sigma}};\ \frac{\mathbf{z}^i}{v^{\sigma}}\right) d\mathbf{y} d\mathbf{z}^i. \end{aligned} \tag{26*}$$

In case $e_{\mathbf{k}^i} = \phi$, instead of $\Delta^{\mathbf{k}^i}(\mathbf{z}; G) g$ we have $\chi(G) g$ (where $\chi(G)$ is the characteristic function of the set G).

It is clear that the functions $D^{\nu} f(\mathbf{x})$ and $\widetilde{f}_{\nu,\delta}(\mathbf{x})$ coincide on $U + R_{\delta}$.

§4. Proof of the fundamental results

1. Proof of Theorem 1. In the integral representation (16), we put $\omega^i + [\mathbf{r}^i - \bar{\mathbf{r}}^i]$ $(i = 0, 1, \cdots, n)$, where $\omega^i \equiv \omega^{e_{\mathbf{r}^i}}$, and the $\mathbf{r}^i$ are the vectors given in Theorem 1, in place of $\mathbf{k}^i$ $(i = 0, 1, \cdots, n)$, and we put the vectors $\bar{\mathbf{r}}^i$ $(i = 1, \cdots, n)$ in place of the vectors which indicate the orders of the derivatives of the functions on the right-hand side of the representation (16). We do the same thing in (26*). We denote by $\widetilde{f}_{\nu,\mathbf{k}}(\mathbf{x})$ the function which is obtained from the identity (26*) (after making the change just described) by taking $U_k + R_{\delta k}(\sigma)$ $(R_{\mathbf{k}} \equiv R_{\delta k}(\sigma))$ in place of $U + R_{\delta}(\sigma)$.

It is clear that

$$\| D^{\nu} f \|_{L_p(\mathfrak{Q}_m)} \leqslant c \sum_{k=1}^{N} \| D^{\nu} f \|_{L_p[(U_k + R_k)_m]} \leqslant c \sum_{k=1}^{N} \| \widetilde{f}_{\nu, k} \|_{L_p(E_m)}, \tag{27}$$

where $(U_k + R_k)_m$ is the intersection of the region $U_k + R_k$ and the hyperplane $x_{m+1} = \text{const}, \cdots, x_n = \text{const}$.

We estimate the norm $\|\widetilde{f}_{v,k}\|_{L_p(E_m)}$. To simplify the arguments, we assume that the horn R_k is in the first coordinate angle.

From the identity for $\widetilde{f}_{v,k}(\mathbf{x})$, it follows that

$$\|\widetilde{f}_{v,k}\|_{L_p(E_m)} \leqslant c \sum_{i=0}^{n} \|\mathfrak{F}_{i,k}\|_{L_p(E_m)}. \tag{28}$$

Here

$$|\mathfrak{F}_{0,k}| \leqslant ch^{\alpha_0} \int_0^{ch^\sigma} dy$$

$$\times \int_0^{ch^\sigma} \left| \Delta^{\omega^0 + [\mathbf{r}^0 - \bar{\mathbf{r}}^0]} \left(\frac{\mathbf{z}^0}{\omega^0 + [\mathbf{r}^0 - \bar{\mathbf{r}}^0]};\ U_k + R_k \right) D^{\bar{\mathbf{r}}^0} f(\mathbf{x} + \mathbf{y}) \right| \prod_{j \in e_{\mathbf{r}^0}} \frac{dz_j}{z_j^{\frac{1}{p_0} + r_j^0 - \bar{r}_j^0 - \gamma_j^0}},$$

$$|\mathfrak{F}_{i,k}| \leqslant c \int_0^h \frac{dv}{v^{1-\alpha_i}} \int_0^{cv^\sigma} dy$$

$$\times \int_0^{cv^\sigma} \frac{\left| \Delta^{\omega^i + [\mathbf{r}^i - \bar{\mathbf{r}}^i]} \left(\frac{\mathbf{z}^i}{\omega^i + [\mathbf{r}^i - \bar{\mathbf{r}}^i]};\ U_k + R_k \right) D^{\bar{\mathbf{r}}^i} f(\mathbf{x} + \mathbf{y}) \right|}{\prod_{j \in e_{\mathbf{r}^i}} z_j^{\frac{1}{p_i} + r_j^i - \bar{r}_j^i - \gamma_j^i}} d\mathbf{z}^i,$$

where $\omega^i \equiv \omega^{e_{\mathbf{r}^i}}$, $\mathbf{z}^i \equiv \mathbf{z}^{e_{\mathbf{r}^i}}$, $\alpha_i = (\mathbf{r}^i - \nu - \omega - \omega^i/p_i' - \gamma^i, \sigma)$, and, in addition $0 < \gamma_j^i < 1/p_i + r_j^i - \bar{r}_j^i$ $(j \in e_{\gamma^i})$ and $e_{\gamma^i} = e_{\mathbf{r}^i}$ $(i = 0, 1, \cdots, n)$, with, generally speaking, different constants. For simplicity, we will write h^σ and v^σ in place of the upper limits ch^σ and cv^σ, respectively, in the above integral expressions.

Applying the Hölder inequality with exponents $\lambda_1 = pp_0/(p - p_0)$, $\lambda_2 = p$ and $\lambda_3 = p_0' = p_0/(p_0 - 1)$, we get

$$\|\mathfrak{F}_{0,k}\|_{L_p(E_m)} \leqslant ch^{\left(\mathbf{r}^0 - \nu - \frac{1}{p_0}\omega + \frac{1}{p}\omega_m - \gamma^0,\ \sigma\right)}$$

$$\times \left(\int_0^{h^\sigma} \left\| \Delta^{\omega^0 + [\mathbf{r}^0 - \bar{\mathbf{r}}^0]} \left(\frac{\mathbf{z}^0}{\omega^0 + [\mathbf{r}^0 - \bar{\mathbf{r}}^0]};\ U_k + R_k \right) D^{\bar{\mathbf{r}}^0} f \right\|_{L_{p^0}(E_n)}^{p^0} \prod_{j \in e_{\mathbf{r}^0}} \frac{dz_j}{z_i^{1 + p_0 (r_j^0 - \bar{r}_j^0 - \gamma_j^0)}} \right)^{1/p_0}$$

Now applying the Hölder inequality with exponents $\lambda_1 = \theta_0/p_0$ and $\lambda_2 = \lambda_1(\lambda_1 - 1)$ to the integral expression inside the parentheses on the right-hand side of the last inequality, we arrive at the inequality

$$\|\mathfrak{I}_{0,k}\|_{L_p(E_m)} \leqslant ch^{\varepsilon_{m,0}} \|f\|_{L^{\langle \mathbf{r}^0\rangle}_{p_0,\theta_0}(U_k+R_k)}, \tag{29}$$

where

$$\varepsilon_{m,0} = \left(\mathbf{r}^0 - \boldsymbol{\nu} - \frac{1}{p_0}\boldsymbol{\omega} + \frac{1}{p}\boldsymbol{\omega}_m, \boldsymbol{\sigma}\right), \quad \boldsymbol{\omega}_m = (\overbrace{1, \ldots, 1}^{m}, 0, \ldots, 0), \; 0 < h \leqslant H.$$

The integral expression $\mathfrak{I}_{i,k}$ can be estimated by similar arguments.. In fact, applying the Hölder inequality with exponents $\lambda_1 = pp_i/(p - p_i)$, $\lambda_2 = p$ and $\lambda_3 = p_i' = p_i/(p_i - 1)$, we get

$$\|\mathfrak{I}_{i,k}\|_{L_p(E_m)} \leqslant ch^{\varepsilon_{m,i}-(\gamma,\sigma)+\xi}\left(\int_0^h \frac{dv}{v^{1+\xi p_i}}\right.$$

$$\left.\times \int_0^{v^\sigma} \left\|\Delta^{\omega^i+[\mathbf{r}^i-\bar{\mathbf{r}}^i]}\left(\frac{\mathbf{z}^i}{\omega^i+[\mathbf{r}^i-\bar{\mathbf{r}}^i]}; U_k+R_k\right)D^{\bar{\mathbf{r}}^i}f\right\|^{p_i}_{L_{p_i}(E_n)} \prod_{j\in e_{\mathbf{r}^i}} \frac{dz_j}{z_j^{1+p_i(r^i_j-\bar{r}^i_j-\gamma^i_j)}}\right)^{1/p_i},$$

where $0 < \xi < \epsilon_{m,i} - (\gamma^i, \sigma)$, and the numbers $\epsilon_{m,i}$ $(i = 1, \cdots, n)$ are those mentioned in the theorem.

We interchange the order of integration in the integral expression inside the parentheses in this last inequality. Thus we get an estimate for this expression (which we denote by $(\cdot)$):

$$(\cdot) \leqslant \int_0^{h^\sigma} \left\|\Delta^{\omega^i+[\mathbf{r}^i-\bar{\mathbf{r}}^i]}\left(\frac{\mathbf{z}^i}{\omega^i+[\mathbf{r}^i-\bar{\mathbf{r}}^i]}; U_k+R_k\right)D^{\bar{\mathbf{r}}^i}f\right\|^{p_i}_{L_{p_i}(E_n)} \prod_{j\in e_{\mathbf{r}^i}} \frac{dz_j}{z_j^{1+p_i(r^i_j-\bar{r}^i_j-\gamma^i_j)}} \times \int_{\max_{j\in e_{\mathbf{r}^i}} z_j^{1/\sigma_j}}^{h} \frac{dv}{v^{1+\xi p_i}}.$$

Consequently

$$\|\mathfrak{I}_{i,k}\|_{L_p(E_m)} \leqslant ch^{\varepsilon_{m,i}-(\gamma^i,\sigma)+\xi}$$

$$\times\left(\int_0^{h^\sigma} \frac{\left\|\Delta^{\omega^i+[\mathbf{r}^i-\bar{\mathbf{r}}^i]}\left(\frac{\mathbf{z}^i}{\omega^i+[\mathbf{r}^i-\bar{\mathbf{r}}^i]}; U_k+R_k\right)D^{\bar{\mathbf{r}}^i}f\right\|^{p_i}_{L_{p_i}(E_n)}}{\prod_{j\in e_{\mathbf{r}^i}} z_j^{1+p_i(r^i_j-\bar{r}^i_j)-\left(\gamma^i_j-\frac{\xi}{\sigma_j|e_{\mathbf{r}^i}|}\right)p_i}}\, d\mathbf{z}^i\right)^{1/p_i},$$

where $|e_{\mathbf{r}^i}|$ denotes the number of elements of the set $e_{\mathbf{r}^i}$; that is, the number of components of the vector $\mathbf{r}^i$ which are different from zero.

Now choosing $\gamma^i_j - \xi/\sigma_j|e_{\mathbf{r}^i}| > 0$, we apply the Hölder inequality with exponents $\lambda_1 = \theta_i/p_i$ and $\lambda_2 = \lambda_1/(\lambda_1 - 1)$ to the last expression:

$$\|\mathfrak{I}_{i,k}\|_{L_p(E_m)} \leqslant ch^{\varepsilon_{m,i}} \times \left(\int_0^{h^\sigma} \frac{\left\| \Delta^{\omega^i + [\mathbf{r}^i - \bar{\mathbf{r}}^i]}\left(\frac{\mathbf{z}^i}{\omega^i + [\mathbf{r}^i - \bar{\mathbf{r}}^i]} ; U_k + R_k \right) D^{\bar{\mathbf{r}}^i} f \right\|_{L_{p_i}(E_n)}^{\theta_i}}{\prod_{j \in e_{\mathbf{r}^i}} z_j^{1+\theta_i(r_j^i - \bar{r}_j^i)}} d\mathbf{z}^i \right)^{1/\theta_i} \tag{30}$$

$$\leqslant ch^{\varepsilon_{m,i}} \|f\|_{L_{p_i,\theta_i}^{\langle \mathbf{r}^i \rangle}(U_k+R_k)}$$

Let $\epsilon_{m,i} = 0$ and $1 < p_i = \theta_i < p < \infty$. In this case we apply the Hölder inequality with exponents $\lambda_1 = p_i$ and $\lambda_2 = p_i/(p_i - 1) \equiv p_i'$ to the integral expression

$$\int_0^{v^{\sigma_{m+1}}} dy_{m+1} \cdots \int_0^{v^{\sigma_n}} dy_n \int_0^{v^\sigma} \frac{\left| \Delta^{\omega^i + [\mathbf{r}^i - \bar{\mathbf{r}}^i]}\left(\frac{\mathbf{z}^i}{\omega^i + [\mathbf{r}^i - \bar{\mathbf{r}}^i]} ; U_k + R_k \right) D^{\bar{\mathbf{r}}^i} f(\mathbf{x}+\mathbf{y}) \right|}{\prod_{j \in e_{\mathbf{r}^i}} z_j^{\frac{1}{p_i} + r_j^i - \bar{r}_j^i}} d\mathbf{z}^i$$

$$\leqslant c v^{\frac{1}{p_i'}(\omega_{n-m} + \omega^i, \sigma)} \varphi_{i,k}(\mathbf{x}^m + \mathbf{y}^m),$$

where

$$\omega_{n-m} = (0, \ldots, 0, 1, \ldots, 1), \ \mathbf{x}^m = (x_1, \ldots, x_m), \ \mathbf{y}^m = (y_1, \ldots, y_m),$$

$$\varphi_{i,k}(\mathbf{x}^m + \mathbf{y}^m) = \left(\int_0^{v^{\sigma_{m+1}}} dy_{m+1} \cdots \int_0^{v^{\sigma_n}} dy_n \int_0^{v^\sigma} \frac{\left| \Delta^{\omega^i + [\mathbf{r}^i - \bar{\mathbf{r}}^i]}\left(\frac{\mathbf{z}^i}{\omega^i + [\mathbf{r}^i - \bar{\mathbf{r}}^i]} ; U_k + R_k \right) D^{\bar{\mathbf{r}}^i} f(\mathbf{x}+\mathbf{y}) \right|^{p_i}}{\prod_{j \in e_{\mathbf{r}^i}} z_j^{1+p_i(r_j^i - \bar{r}_j^i)}} d\mathbf{z}^i \right)^{1/p_i}.$$

Now interchanging the order of integration and applying the generalized Minkowski inequality and the Hardy-Littlewood inequality m times, we get

$$\|\mathfrak{I}_{i,k}\|_{L_p(E_m)} \leqslant c \left\| \int_0^h \frac{dv}{v^{1+\left(\frac{1}{p_i'} + \frac{1}{p}\right)(\omega_m, \sigma)}} \int_0^{v^\sigma} \varphi_{i,k}(\mathbf{x}^m + \mathbf{y}^m) \, d\mathbf{y}^m \right\|_{L_p(E_m)}$$

$$\leqslant c \left\| \int_{E_m} \varphi_{i,k}(\mathbf{y}^m) \, d\mathbf{y}^m \int_{\max_{j=1,\ldots,m} |x_j - y_j|^{1/\sigma_j}}^h \frac{dv}{v^{1+\left(\frac{1}{p_i'} + \frac{1}{p_i}\right)(\omega_m, \sigma)}} \right\|_{L_p(E_m)} \tag{31}$$

$$\leqslant c \left\| \int_{E_m} \frac{\varphi_{i,k}(\mathbf{y}^m)}{\prod_{j=1}^m |x_j - y_j|^{\frac{1}{p_i'} + \frac{1}{p_i}}} d\mathbf{y}^m \right\|_{L_p(E_m)}$$

$$\leqslant c \|\varphi_{i,k}\|_{L_p(E_m)} \leqslant c \|f\|_{\mathscr{L}_{p_i,p_i}^{\langle \mathbf{r}^i \rangle}(U_k+R_k)}.$$

From the inequalities (27)–(31) it follows that

$$\| D^{\nu} f \|_{L_p(\mathfrak{Q}_m)} \leqslant c \sum_{i=0}^{n} h^{\varepsilon_{m,i}} \sum_{k=1}^{N} \| f \|_{\mathscr{L}^{\langle \mathbf{r}^i \rangle}_{p_i, \theta_i} (U_k + R_k)}, \tag{32}$$

which proves the theorem, except in the case $\epsilon_{m,i} > 0$, $1 < p_i = p < \infty$, $p_i = \theta_i$ $(i = 1, \cdots, n)$, in which case the theorem is proved similarly, but the arguments are simpler.

2. Proof of Theorem 3. It is clear that

$$\begin{aligned} \| D^{\nu} f \|_{\mathscr{L}^{\langle \rho \rangle}_{p, \theta} (\mathfrak{Q}_m)} &\leqslant c h^{-(\rho - \bar{\rho}, \sigma)} \| D^{\nu + \bar{\rho}} f \|_{L_p(\mathfrak{Q}_m)} \\ &+ \left(\int_0^{h^{\sigma}} \| \Delta^{\omega_{\rho} + [\rho - \bar{\rho}]} (\mathbf{t};\ \mathfrak{Q}_m) D^{\bar{\rho} + \nu} f \|^{\theta}_{L_p(\mathfrak{Q}_m)} \prod_{j \in e_{\rho}} \frac{dt_j}{t_j^{1+\theta(\rho_j - \bar{\rho}_j)}} \right)^{1/\theta} \\ &+ \left(\int_0^{h^{\sigma}} \| \Delta^{\omega_{\rho} + [\rho - \bar{\rho}]} (-\mathbf{t};\ \mathfrak{Q}_m) D^{\bar{\rho} + \nu} f \|^{\theta}_{L_p(\mathfrak{Q}_m)} \prod_{j \in e_{\rho}} \frac{dt_j}{t_j^{1+\theta(\rho_j - \bar{\rho}_j)}} \right)^{1/\theta} \\ &= \mathfrak{R}_1 + \mathfrak{R}_2 + \mathfrak{R}_3 \qquad (\boldsymbol{\omega}_{\bar{\rho}} \equiv \boldsymbol{\omega}^{e_{\rho}}). \end{aligned} \tag{33}$$

Using the estimate (32), we get

$$\mathfrak{R}_1 \leqslant c \sum_{i=1}^{n} h^{\varepsilon_{m,i} - (\rho, \sigma)} \sum_{k=1}^{N} \| f \|_{\mathscr{L}^{\langle \mathbf{r}^i \rangle}_{p_i, \theta_i} (U_k + R_k)} \leqslant c \sum_{i=0}^{h} h^{\varepsilon_{m,i} - (\rho, \sigma)} \| f \|_{\mathscr{L}^{\langle \mathbf{r}^i \rangle}_{p_i, \theta_i} (\mathfrak{Q})}. \tag{34}$$

The integral expressions $\mathfrak{R}_2$ and $\mathfrak{R}_3$ are estimated similarly in terms of the right-hand side of inequality (34), so it is sufficient to estimate only $\mathfrak{R}_2$.

In view of the fact that

$$\mathfrak{R}_2 \leqslant c \sum_{k=1}^{N} \mathfrak{R}_{2,k} \equiv c \sum_{k=1}^{N} \left(\int_0^{h^{\sigma}} \| \Delta^{\omega_{\rho} + [\rho - \bar{\rho}]} (\mathbf{t}) \tilde{f}_{\nu + \bar{\rho}, k} \|^{\theta}_{L_p(E_m)} \prod_{j \in e_{\rho}} \frac{dt_j}{t_j^{1+\theta(\rho_j - \bar{\rho}_j)}} \right)^{1/\theta}, \tag{35}$$

we estimate each $\mathfrak{R}_{2,k}$ separately. Let R_k lie in the first coordinate angle. From the identity (26*) it follows that

$$\begin{aligned} \tilde{f}_{\nu + \bar{\rho}, k} (\mathbf{x}) &= c_0 h^{\alpha_0} \int d\mathbf{y} \int [\cdots]_0 \, d\mathbf{z}_0 \\ &+ c \sum_{i=1}^{n} \left\{ \int_0^{\varphi(\mathbf{t})} \frac{dv}{v^{1-\alpha_i}} \int \int [\cdots]_i \, d\mathbf{y} d\mathbf{z}^i + \int_{\varphi(\mathbf{t})}^{h} \frac{dv}{v^{1-\alpha_i}} \int \int [\cdots]_i \, d\mathbf{y} d\mathbf{z}^i \right\} \\ &\equiv \mathfrak{M}_{0,k} + \sum_{i=1}^{n} \{ \mathfrak{M}^{*}_{i,k} + \mathfrak{M}^{**}_{i,k} \}, \end{aligned} \tag{36}$$

where

$$\alpha_i = (\mathbf{r}^i - \boldsymbol{\nu} - \bar{\boldsymbol{\rho}} - \boldsymbol{\omega} - \boldsymbol{\omega}^i, \boldsymbol{\sigma}) \qquad (i = 0, 1, \ldots, n), \quad \varphi(\mathbf{t}) = \prod_{j \in e_\rho} t_j^{\frac{1}{\sigma_j |e_\rho|}},$$

$$[\cdots]_0 \equiv \Delta^{\omega^0 + [\mathbf{r}^0 - \bar{\mathbf{r}}^0]}\left(\frac{\mathbf{z}^0}{\omega^0 + [\mathbf{r}^0 - \bar{\mathbf{r}}^0]}; U_k + R_k\right) D^{\bar{\mathbf{r}}^0} f(\mathbf{x} + \mathbf{y}) \Phi_{\delta k, 0}\left(\frac{\mathbf{y}}{h^\sigma}; \frac{\mathbf{z}^0}{h^\sigma}\right),$$

$$[\cdots]_i \equiv \Delta^{\omega^i + [\mathbf{r}^i - \bar{\mathbf{r}}^i]}\left(\frac{\mathbf{z}^i}{\omega^i + [\mathbf{r}^i - \bar{\mathbf{r}}^i]}; U_k + R_k\right) D^{\bar{\mathbf{r}}^i} f(\mathbf{x} + \mathbf{y}) \Phi_{\delta k, i}\left(\frac{\mathbf{y}}{v^\sigma}; \frac{\mathbf{z}^i}{v^\sigma}\right).$$

After a few simple transformations we get

$$\begin{aligned} \mathfrak{R}_{2,k} &\leqslant c \prod_{j \in e_\rho} \left(\int_0^{h^{\sigma_j}} \frac{dt_j}{t_j^{1-(1+[\rho_j - \bar{\rho}_j] - \rho_j + \bar{\rho}_j)\theta}} \right)^{1/\theta} \| D^{\omega_\rho + [\rho - \bar{\rho}]} \mathfrak{M}_{0,k} \|_{L_p(E_m)} \\ &+ c \sum_{i=1}^n \left(\int_0^{h^\sigma} \| \mathfrak{M}^*_{i,k} \|^\theta_{L_p(E_m)} \prod_{j \in e_\rho} \frac{dt_j}{t_j^{1+\theta(\rho_j - \bar{\rho}_j)}} \right)^{1/\theta} \\ &+ c \sum_{i=1}^n \left(\int_0^{h^\sigma} \| D^{\omega_\rho + [\rho - \bar{\rho}]} \mathfrak{M}^{**}_{i,k} \|^\theta_{L_p(E_m)} \prod_{j \in e_\rho} \frac{dt_j}{t_j^{1-(1+[\rho_j - \bar{\rho}_j] - \rho_j + \bar{\rho}_j)\theta}} \right)^{1/\theta} \\ &\equiv \mathscr{P}_{0,k} + \sum_{i=1}^n \{ \mathscr{P}^*_{i,k} + \mathscr{P}^{**}_{i,k} \}. \end{aligned} \tag{37}$$

It is clear that

$$\mathscr{P}_{0,k} \leqslant c h^{(\omega_\rho + [\rho - \bar{\rho}] - \rho + \bar{\rho}, \sigma)} \| D^{\omega_\rho + [\rho - \bar{\rho}]} \mathfrak{M}_{0,k} \|_{L_p(E_m)}.$$

On the other hand, the following estimate can be established by a proof similar to that for the inequality (29):

$$\| D^{\omega_\rho + [\rho - \bar{\rho}]} \mathfrak{M}_{0,k} \|_{L_p(E_m)} \leqslant c h^{-(\omega_\rho + [\rho - \bar{\rho}] + \bar{\rho}, \sigma) + \varepsilon_{m,0}} \| f \|_{\mathscr{L}^{\langle \mathbf{r}^0 \rangle}_{p_0, \theta_0}(U_k + R_k)}.$$

Consequently

$$\mathscr{P}_{0,k} \leqslant c h^{\varepsilon_{m,0} - (\rho, \sigma)} \| f \|_{\mathscr{L}^{\langle \mathbf{r}^0 \rangle}_{p_0, \theta_0}(U_k + R_k)}. \tag{38}$$

We now turn to estimates of the integrals $\mathscr{P}^*_{i,k}$ and $\mathscr{P}^{**}_{i,k}$. Applying the Hölder inequality to the integral $\mathfrak{M}^*_{i,k}$, first with exponents $\lambda_1 = pp_i/(p - p_i)$, $\lambda_2 = p$ and $\lambda_3 = p_i' = p_i/(p_i - 1)$, and then with exponents $\lambda_1 = \theta_i/p_i$ and $\lambda_2 = \lambda_1/(\lambda_1 - 1)$, we get

$$\begin{aligned} \| \mathfrak{M}^*_{i,k} \|_{L_p(E_m)} &\leqslant c h^{\varepsilon_{m,i} - (\rho, \sigma)} \varphi^{2\xi} \Bigg(\int_0^{\varphi(\mathbf{t})} \frac{dv}{v^{1+2\xi\theta_i - \left(\rho - \bar{\rho} - \frac{1}{2}\gamma^i, \sigma\right)\theta_i}} \\ &\times \int_0^{v^\sigma} \left\| \Delta^{\omega^i + [\mathbf{r}^i - \bar{\mathbf{r}}^i]}\left(\frac{\mathbf{z}^i}{\omega^i + [\mathbf{r}^i - \bar{\mathbf{r}}^i]}; U_k + R_k\right) D^{\bar{\mathbf{r}}^i} f \right\|^{\theta_i}_{L_{p_i}(E_n)} \prod_{j \in e_{\mathbf{r}^i}} \frac{dz_j}{z_j^{1+\theta_i\left(r^i_j - \bar{r}^i_j - \frac{1}{2}\gamma^i_j\right)}} \Bigg)^{1/\theta_i}, \end{aligned}$$

where $\xi > 0$ is a sufficiently small number.

Applying the generalized Minkowski inequality with exponents $\lambda_1 = \theta/\theta_i$ and $\lambda_2 = \lambda_1/(\lambda_1 - 1)$ to the integral

$$\mathcal{P}^*_{i,k} \leqslant ch^{\varepsilon_{m,i}-(\rho,\sigma)} \left\{ \int_0^{h^\sigma} \prod_{j\in e_\rho} \frac{dt_j}{t_j^{1+\left(\rho_j-\bar\rho_j-\frac{2\xi}{\sigma_j|e_\rho|}\right)\theta}} \left[\int_0^{\varphi(t)} \frac{A_k dv}{v^{1+2\theta_i\xi-\left(\rho-\bar\rho-\frac{1}{2}\gamma^i,\sigma\right)\theta_i}} \right]^{\theta/\theta_i} \right\}^{1/\theta},$$

we get the estimate

$$\mathcal{P}^*_{i,k} \leqslant ch^{\varepsilon_{m,i}-(\rho,\sigma)} \left\{ \int_0^h \frac{A_k dv}{v^{1+\frac{1}{2}(\gamma^i,\sigma)\theta_i}} \right\}^{1/\theta_i}, \qquad (39)$$

where

$$A_k = \int_0^{v^\sigma} \frac{\left\| \Delta^{\omega^i+[\mathbf{r}^i-\bar{\mathbf{r}}^i]}\left(\frac{\mathbf{z}^i}{\omega^i+[\mathbf{r}^i-\bar{\mathbf{r}}^i]};\ U_k+R_k \right) D^{\bar{\mathbf{r}}^i} f \right\|^{\theta_i}_{L_p(E_m)}}{\prod_{j\in e_{\mathbf{r}^i}} z_j^{1+\theta_i\left(r^i_j-\bar r^i_j-\frac{1}{2}\gamma^i_j\right)}} d\mathbf{z}^i.$$

The integral on the right-hand side of (39), after interchanging the order of integration, is bounded by $\|f\|_{\mathfrak{L}^{\langle \mathbf{r}^i\rangle}_{p_i,\theta_i}(U_k+R_k)}$, from which it follows that

$$\mathcal{P}^*_{i,k} \leqslant ch^{\varepsilon_{m,i}-(\rho,\sigma)} \|f\|_{\mathcal{L}^{\langle \mathbf{r}^i\rangle}_{p_i,\theta_i}(U_k+R_k)}. \qquad (40)$$

After transferring the differentiation to the kernel and applying the Hölder inequality with exponents $\lambda_1 = pp_i/(p - p_i)$, $\lambda_2 = p$ and $\lambda_3 = p_i' = p_i/(p_i - 1)$, and then with exponents $\lambda_1 = \theta_i/p_i$ and $\lambda_2 = \lambda_1/(\lambda_1 - 1)$ to the integral $D^{\omega_\rho+[\rho-\bar\rho]}\mathfrak{M}^{**}_{i,k}$, we get

$$\| D^{\omega_\rho+[\rho-\bar\rho]}\mathfrak{M}^{**}_{i,k} \|_{L_p(E_m)} \leqslant ch^{\varepsilon_{m,i}-(\rho,\sigma)} \varphi^{-2\xi} \left[\int_{\varphi(t)}^h \frac{dv}{v^{1+\theta_i\left(\beta+\frac{1}{2}\gamma^i,\sigma\right)-2\xi\theta_i}} \times \int_0^{v^\sigma} \|\cdot\|^{\theta_i}_{L_{p_i}(E_n)} \prod_{j\in e_{\mathbf{r}^i}} \frac{dz_j}{z_j^{1+\theta_i\left(r^i_j-\bar r^i_j-\frac{1}{2}\gamma^i_j\right)}} \right]^{1/\theta_i}, \qquad (41)$$

where $\beta = (\beta_1, \cdots, \beta_n)$, $e_\beta = e_\rho$, $\beta_j = 1 + [\rho_j - \bar\rho_j] - \rho_j + \bar\rho_j$ $(j \in e_\beta)$, and

$$\|\cdot\| = \left\| \Delta^{\omega^i+[\mathbf{r}^i-\bar{\mathbf{r}}^i]}\left(\frac{\mathbf{z}^i}{\omega^i+[\mathbf{r}^i-\bar{\mathbf{r}}^i]};\ U_k+R_k \right) D^{\bar{\mathbf{r}}^i} f \right\|.$$

We estimate the integral in the square brackets on the right-hand side of (41), after changing the order of integration, in the following way:

$$[\cdot]\leqslant\left(\int_0^{h^\sigma}\|\cdot\|^{\theta_i}_{L_{p_i}(E_n)}\prod_{j\in e_{\mathbf{r}^i}}\frac{dz_j}{z_j^{1+\theta_i\left(r^i_j-\bar r^i_j-\frac{1}{2}\gamma^i_j\right)}}\right.$$
$$\left.\times\int_{\max\left(\max_{j\in e_{\mathbf{r}^i}} z_j^{1/\sigma_j};\ \varphi(\mathbf{t})\right)}^{h}\frac{dv}{v^{1+\theta_i\left(\beta+\frac{1}{2}\gamma^i,\ \sigma\right)-2\xi\theta_i}}\right)^{1/\theta_i}\tag{42}$$
$$\leqslant c\left\{\int_0^{h^\sigma}\frac{\prod_{j\in e_{\mathbf{r}^i}} z_j^{-1-\left(r^i_j-\bar r^i_j-\frac{1}{2}\gamma^i_j\right)\theta_i}\|\cdot\|^{\theta_i}_{L_{p_i}(E_n)}d\mathbf{z}^i}{\left(\sqrt{\left(\max_{j\in e_{\mathbf{r}^i}} z_j^{1/\sigma_j}\right)^2+\varphi^2}\right)^{\theta_i\left(\beta+\frac{1}{2}\gamma^i,\ \sigma\right)-2\xi\theta_i}}\right\}^{1/\theta_i}.$$

Consequently

$$\mathscr{P}^{**}_{i,\,k}\leqslant ch^{\varepsilon_{m,\,i}-(\rho,\,\sigma)}\left[\int_0^{h^\sigma}\{\ldots\}^{\theta/\theta_i}\prod_{j\in e_{\mathbf{r}^i}}\frac{dz_j}{z_j^{1-\beta_j\theta-\frac{2\xi\theta}{\sigma_j|e_\rho|}}}\right]^{1/\theta},$$

where $\{\cdots\}$ is the integral in the braces on the right-hand side of (42).

After applying the generalized Minkowski inequality with exponent θ/θ_i to the right-hand side of the last inequality and making the substitution

$$t_k=\left(\max_{j\in e_{\mathbf{r}^i}} z_j^{1/\sigma_j}\right)\tau_k\qquad(k\in e_\rho),$$

we arrive at the estimate

$$\mathscr{P}^{**}_{i,\,k}\leqslant ch^{\varepsilon_{m,\,i}-(\rho,\,\sigma)}\left\{\int_0^{h^\sigma}\|\cdot\|^{\theta_i}_{L_{p_i}(E_n)}\prod_{j\in e_{\mathbf{r}^i}}\frac{dz_j}{z_j^{1+\theta_i(r^i_j-\bar r^i_j)}}\right\}^{1/\theta_i}\tag{43}$$
$$\leqslant ch^{\varepsilon_{m,\,i}-(\rho,\,\sigma)}\|f\|_{\mathscr{L}^{\langle\mathbf{r}^i\rangle}_{p_i,\,\theta_i}(U_k+R_k)}.$$

From the inequalities (37)–(40) and (43) it follows that

$$\mathfrak{N}_{2,\,k}\leqslant c\sum_{i=0}^{n}h^{\varepsilon_{m,\,i}-(\rho,\,\sigma)}\|f\|_{\mathscr{L}^{\langle\mathbf{r}^i\rangle}_{p_i,\,\theta_i}(U_k+R_k)}.\tag{44}$$

The inequalities (33)–(35) and (44) prove the following statement:

$$\|D^{\nu}f\|_{\mathscr{L}^{\langle\rho\rangle}_{p,\theta}(\mathfrak{L}_m)} \leqslant c\sum_{i=0}^{n} h^{\varepsilon_{m,i}-(\rho,\sigma)} \sum_{k=1}^{N} \|f\|_{\mathscr{L}^{\langle \mathbf{r}^i\rangle}_{p_i,\theta_i}(U_k+R_k)} \leqslant c\sum_{i=0}^{n} h^{\varepsilon_{m,i}-(\rho,\sigma)} \|f\|_{\mathscr{L}^{\langle \mathbf{r}^i\rangle}_{p_i,\theta_i}(\mathfrak{L})}, \tag{45}$$

from which the inequality (9) of Theorem 3 follows.

If we assume that $\Omega \subseteq E_n$ is an unbounded region and that $\epsilon_{m,i} - (\rho, \sigma) = 0$ $(i = 1, \cdots, n)$, then from (45) for $h \to \infty$, the inequality (10) of Theorem 3 follows, which was to be proved.

3. Theorem 2 is proved by arguments which are similar to those used in the proof of Theorem 1.

Theorems 4, 6, and 7 are proved with the help of the representations (16) and (26*), with an appropriate choice for the vectors α^i and $\mathbf{k}^i$ $(i = 0, 1, \cdots, n)$, and the proof uses arguments similar to those in the proof of Theorem 3.

§5. Theorems concerning the extension of functions beyond the boundary of the region $\Omega \subset E_n$, to E_n, while preserving their differential properties

In the conditions of Theorem 3, §2 (for $m = n$), from the fact that the function $f(\mathbf{x})$ belongs to the space

$$\bigcap_{i=0}^{n} \mathscr{L}^{\langle \mathbf{r}^i\rangle}_{p_i,\theta_i}(\mathfrak{L})$$

it follows that the function $D^{\nu}f$ on Ω belongs to the space $\mathfrak{L}^{\langle\rho\rangle}_{p,\theta}(\Omega)$, and, in addition,

$$\|D^{\nu}f\|_{\mathscr{L}^{\langle\rho\rangle}_{p,\theta}(\mathfrak{L})} \leqslant c\sum_{i=0}^{n} \|f\|_{\mathscr{L}^{\langle \mathbf{r}^i\rangle}_{p_i,\theta_i}(\mathfrak{L})},$$

where c is a constant which does not depend on f.

The goal of this section is the extension of the function $D^{\nu}f(\mathbf{x})$ beyond the boundary of the region $\Omega \subset E_n$ to E_n (under additional conditions on the region Ω), keeping the differential properties invariant; that is, it consists in the construction of a function $f_{\nu}(\mathbf{x})$ which is defined on E_n, which belongs to the space $\mathfrak{L}^{\langle\rho\rangle}_{p,\theta}(E_n)$, which coincides with the function $D^{\nu}f(\mathbf{x})$ on Ω, and which satisfies the inequality

$$\|f_{\nu}\|_{\mathscr{L}^{\langle\rho\rangle}_{p,\theta}(E_n)} \leqslant c\sum_{i=0}^{n} \|f\|_{\mathscr{L}^{\langle \mathbf{r}^i\rangle}_{p_i,\theta_i}(\mathfrak{L})},$$

where c is a constant which does not depend on f.

DEFINITION. *We will say that the region Ω belongs to the class $A_{\epsilon}(R(\sigma))$ of regions if $\Omega \in A(R(\sigma))$ and for some covering $\{U_k\}$ and some $\epsilon > 0$ the condition*

$$\Omega \subset \bigcup_{k=1}^{N} U_k^{(\varepsilon)}$$

is satisfied, where

$$U_k^{(\varepsilon)} = \{\mathbf{x};\ \mathbf{x} \in U_k;\ \rho(\mathbf{x},\ \Omega \setminus U_k) > \varepsilon\}.$$

Let $\mathbf{k} = (k_1, \cdots, k_n)$ and $\alpha = (\alpha_1, \cdots, \alpha_n)$ be nonnegative integral vectors (that is, k_i and α_i are nonnegative integers), let $\beta = (\beta_1, \cdots, \beta_n)$ be a vector such that $0 < \beta_j < k_j$ $(j \in e_\beta)$, and let $e_{\mathbf{k}} = e_\beta$ and $e_\alpha \subseteq e_\beta$.

We introduce the seminorm

$$\|f\|_{\mathscr{L}^{\langle\alpha+\beta;\ \mathbf{k}\rangle}_{p,\theta}(\Omega)} = \left(\int \cdots \int \left\|\Delta^{\mathbf{k}}\left(\frac{\mathbf{t}}{\mathbf{k}};\ \Omega\right) D^\alpha f\right\|^\theta_{L_p(\Omega)} \prod_{j \in e_{\mathbf{k}}} \frac{dt_j}{|t_j|^{1+\beta_j\theta}}\right)^{1/\theta}.$$

LEMMA. *Let α^i and $\mathbf{k}^i$ $(i = 0, 1, \cdots, n)$ be nonnegative integral vectors, let β^i be nonnegative vectors such that $\beta^i < \mathbf{k}^i$ (that is, $\beta^i_j < k^i_j$ $(j \in e_{\beta^i})$), and let $e_{\mathbf{k}^i} = e_{\beta^i}$ and $e_{\alpha^i} \subseteq e_{\beta^i}$.*

Assume that the points $\alpha^i + \beta^i$ $(i = 0, 1, \cdots, n)$ do not all lie on the same n-dimensional plane, and that $\alpha^0_j + \beta^0_j \geqslant 0$ $(j = 1, \cdots, n)$, $\alpha^i_j + \beta^i_j \geqslant 0$ $(j \neq i)$ and $\alpha^i_i + \beta^i_i > 0$ $(i, j = 1, \cdots, n)$.

Further, let $1 \leqslant p_i \leqslant \theta_i \leqslant \infty$, $p \geqslant p_i$, $\theta \geqslant \theta_i$ $(i = 0, 1, \cdots, n)$, and let $\nu = (\nu_1, \cdots, \nu_n)$ and $\rho = (\rho_1, \cdots, \rho_n)$ be nonnegative integral vectors (that is, ρ_j, $\nu_j \geqslant 0$ are integers).

Assume that

1) $\nu_j + \rho_j \geqslant \alpha^0_j + k^0_j$ *for* $j \in e_{\alpha^0+\beta^0}$,
$\nu_j + \rho_j \geqslant \alpha^i_j + k^i_j$ $(j \neq i)$ $(i = 1, \ldots, n;\ j \in e_{\alpha^i+\beta^i})$;

2) *there exists a positive vector $\sigma = (\sigma_1, \cdots, \sigma_n)$ (that is, $\sigma_j > 0; j = 1, \cdots, n$) such that*

$$\left(\alpha^i + \beta^i - (\rho + \gamma) - \left(\frac{1}{p_i} - \frac{1}{p}\right)\omega,\ \sigma\right) \geqslant 0 \quad (i = 1, \ldots, n).$$

Let the functions $\tilde{f}_{\nu,\delta}(\mathbf{x})$ be defined by the formula (see (26))

$$\begin{aligned} \tilde{f}_{\nu,\delta}(\mathbf{x}) = (-1)^{(\alpha^0+\nu,\ \omega)} c_0 h^{-\mu_0} \int d\mathbf{z}_0 \int \Delta^{\mathbf{k}_0}\left(\frac{\mathbf{z}^0}{\mathbf{k}^0};\ U + R_\delta(\sigma)\right) D^{\alpha^0} f(\mathbf{x} + \mathbf{y}) \\ \times \Phi_{\delta,0}\left(\frac{\mathbf{y}}{h^\sigma};\ \frac{\mathbf{z}^0}{h^\sigma}\right) d\mathbf{y} - \sum_{i=1}^n (-1)^{(\alpha^i+\nu,\ \omega)} c_i \int_0^h \frac{dv}{v^{1+\mu_i}} \int d\mathbf{z}^i \\ \times \int \Delta^{\mathbf{k}^i}\left(\frac{\mathbf{z}^i}{\mathbf{k}^i};\ U + R^i_\delta(\sigma)\right) D^{\alpha^i} f(\mathbf{x} + \mathbf{y})\, \Phi_{\delta,i}\left(\frac{\mathbf{y}}{v^\sigma};\ \frac{\mathbf{z}^i}{v^\sigma}\right) d\mathbf{y}, \end{aligned} \qquad (46)$$

where $\mathbf{z}^i \equiv \mathbf{z}^{e}\mathbf{k}^i$, $\omega^i \equiv \omega^{e}\mathbf{k}^i$, $\mu_i = (\omega + \omega^i + \nu - \alpha^i, \sigma)$ $(i = 0, 1, \cdots, n)$.

The inequality

$$\|\tilde{f}_{\nu,\delta}\|_{\mathscr{L}^{\langle\rho+\gamma;\,\mathbf{s}\rangle}_{p,\theta}(E_n)} \leqslant c \sum_{i=0}^{n} \|f\|_{\mathscr{L}^{\langle\alpha^i+\beta^i;\,\mathbf{k}^i\rangle}_{p_i,\theta_i}(U+R_\delta)} \tag{47}$$

holds for the functions $\tilde{f}_{\nu,\delta}(\mathbf{x})$, *where* $\mathbf{s} = (s_1, \cdots, s_n)$, $\gamma = (\gamma_1, \cdots, \gamma_n)$ (s_j *being nonnegative integers*), *and* $0 \leqslant \gamma_j < s_j$, $e_\gamma = e_{\mathbf{s}}$, *and* c *is a constant which does not depend on* f.

This lemma is proved using arguments like those in §4 (see the proofs of Theorems 1 and 3).

THEOREM 8. *Let* $f \in \bigcap_{i=0}^{n} \tilde{\mathfrak{L}}^{\langle\mathbf{r}^i\rangle}_{p_i,\theta_i}(\Omega)$, *where* $\mathbf{r}^i = (r_1^i, \cdots, r_n^i)$ $(i = 0, 1, \cdots, n)$, $r_j^0 \geqslant 0$ $(j = 1, \cdots, n)$, $r_i^i > 0$, $r_j^i \geqslant 0$ $(j \neq i)$ $(i, j = 1, \cdots, n)$, $1 < p_i \leqslant \theta_i < \infty$ *and* $\Omega \in A_\epsilon(R(\sigma))$.

Further, let $p \geqslant p_i$, $\theta \geqslant \theta_i$ $(i = 0, 1, \cdots, n)$, $\nu = (\nu_1, \cdots, \nu_n)$ ($\nu_j \geqslant 0$ *are integers*), $\rho = (\rho_1, \cdots, \rho_n)$, $\rho_j \geqslant 0$ *and* $e_\rho \neq \phi$.

Assume that

1) $\rho_j + \nu_j \geqslant r_j^0 + [r_j^0 - r_j^0] + 1 \qquad (j \in e_{\mathbf{r}^0})$,

$\rho_j + \nu_j \geqslant r_j^i + 1 + [r_j^i - r_j^i] \qquad (j \neq i) \quad (i = 1, \ldots, n;\ j \in e_{\mathbf{r}^i})$;

2) $\left(\mathbf{r}^i - \nu - \rho - \frac{1}{p_i}\omega + \frac{1}{p}\omega, \sigma\right) \geqslant 0 \qquad (i = 1, \ldots, n)$.

Then the function $D^\nu f(\mathbf{x})$ (*which, by the corollary to Theorem* 3, *belongs to the space* $\mathfrak{L}^{\langle\rho\rangle}_{p,\theta}(\Omega)$) *can be extended beyond the boundary of the region* $\Omega \subset E_n$ *to* E_n, *the differential properties remaining invariant*; *that is, it is possible to construct a function* $\tilde{f}_\nu(\mathbf{x})$, *defined on* E_n, *belonging to the space* $\mathfrak{L}^{\langle\rho\rangle}_{p,\theta}(E_n)$, *coinciding with the function* $D^\nu f(\mathbf{x})$ *on* Ω, *and satisfying the inequality*

$$\|\tilde{f}_\nu\|_{\mathscr{L}^{\langle\rho\rangle}_{p,\theta}(E_n)} \leqslant c \sum_{i=0}^{n} \|f\|_{\mathscr{L}^{\langle\mathbf{r}^i\rangle}_{p_i,\theta_i}(\Omega)}. \tag{48}$$

PROOF. This theorem, for $\mathbf{r}^0 = (0, \cdots, 0)$ and $\mathbf{r}^i = (0, \cdots, r_i, \cdots, 0)$, is proved by O. V. Besov and V. P. Il'in in [10] for a more general class of regions. The proof of our theorem uses the same arguments as those in [10], except that, in certain places, we use the lemma of this section.

Let the infinitely differentiable functions $\eta_k(\mathbf{x})$ $(k = 1, \cdots, N)$ constitute a partition of unity corresponding to the covering $\{U_k\}$ of the region Ω; that is,

$$0 \leqslant \eta_k(\mathbf{x}) \leqslant 1 \quad \text{for} \quad \mathbf{x} \in E_n,$$

$$\sum_{k=1}^{N} \eta_k(\mathbf{x}) = 1 \quad \text{for} \quad \mathbf{x} \in \Omega,$$

$$\eta_k(\mathbf{x}) = 0 \quad \text{outside of the } \epsilon\text{-neighborhood} \quad U_k^{(\epsilon)}$$

In addition, we require that

$$\sup_{\mathbf{x}\in E_n} |D^{\alpha}\eta_k(\mathbf{x})| \leqslant c, \qquad (\boldsymbol{\alpha},\ \boldsymbol{\omega}) \leqslant \max_{i=0,\,1,\,\ldots,\,n} (\mathbf{r}^i,\ \boldsymbol{\omega})$$

(for further details, see [10]).

We denote by $\widetilde{f}_{\nu,k}(\mathbf{x})$ the function defined by formula (46) in the case where $\alpha^i = \overline{\mathbf{r}}^i$, $\mathbf{k}^i = \omega^i + [\mathbf{r}^i - \overline{\mathbf{r}}^i]$ $(i = 0, 1, \cdots, n)$, and $U + R_\delta(\sigma) \equiv U_k + R_k$, $\Phi_{\delta,i} \equiv \Phi_{\delta k,i}$.

It is clear that the function $\widetilde{f}_{\nu,k}$ coincides with the function $D^\nu f$ on $U_k + R_k$.

Using the lemma of this section, we can show that the function defined by the formula

$$\tilde{f}_\nu(\mathbf{x}) = \sum_{k=1}^{N} \eta_k(\mathbf{x}) \tilde{f}_{\nu,k}(\mathbf{x}),$$

satisfies the conditions in the theorem.

§6. Some necessary conditions

In this section we will follow the ideas and the notation of V. P. Il′in [13].

Let $\mathfrak{S}$ be a given finite set of nonnegative vectors (points) $\mathbf{r} = (r_1, \cdots, r_n)$ (that is, $r_j \geqslant 0$) and let $\nu = (\nu_1, \cdots, \nu_n)$ be a nonnegative integral vector (that is, $\nu_j \geqslant 0$; ν_j integers).

We denote by $\hat{\mathfrak{S}}$ the set of vectors $\mathbf{l} \equiv \overline{\mathbf{r}} + [\mathbf{r} - \overline{\mathbf{r}}] + \omega_{\mathbf{r}}$, where $\mathbf{r} \in \mathfrak{S}$, $\overline{\mathbf{r}} = (\overline{r}_1, \cdots, \overline{r}_n)$, $\overline{r}_j$ is the largest integer less that r_j, $\omega_r \equiv \omega^{e_r}$ ($e_{\mathbf{r}}$ is the support of the vector $\mathbf{r}$), and $\omega = (1, \cdots, 1)$.

The goal of this section is to prove some conditions which are necessary for the inequality

$$\|D^\nu f\|_{L_q(\Omega)} \leqslant c \sum_{\mathbf{r}\in\mathfrak{S}} \|f\|_{\mathscr{L}^{\langle\mathbf{r}\rangle}_{p,\,\theta}(\Omega)} \qquad (1 \leqslant p \leqslant \theta \leqslant \infty,\ q \geqslant p), \tag{49}$$

to hold, where c is a constant which does not depend on f.

We denote by $\hat{\mathfrak{S}}^e(\nu)$ the set of all vectors $\mathbf{l} \in \hat{\mathfrak{S}}$ for which $l_j \leqslant \nu_j$ for all $j \in e' \equiv e_n \backslash e$, and by $P\hat{\mathfrak{S}}^e(\nu)$ the projection of the set $\hat{\mathfrak{S}}^e(\nu)$ on the hyperplane $\{l_j = 0,\ \forall\ j \in e'\}$.

We denote by $\hat{\Pi}^e(\nu)$ the closed convex polyhedron spanning the points of the set $P\hat{\mathfrak{S}}^e(\nu)$.

The set of points $\mathbf{x} = (x_1, \cdots, x_n)$ satisfying the inequality

$$0 \leqslant x_i \leqslant M \quad (i = 1, \ldots, n),$$

we denote by $\square(M)$; that is,

$$\square(M) = \{\mathbf{x};\ 0 \leqslant x_i \leqslant M, \quad i = 1, \ldots, n\}.$$

THEOREM 9. *Let* $\mathfrak{S}$ *be a given finite set of nonnegative vectors* $\mathbf{r}=(r_1,\cdots,r_n)$ (*that is,* $r_j\geqslant 0$), *and let* $\nu=(\nu_1,\cdots,\nu_n)$ *be some nonnegative integral vector* (*that is,* $\nu_j\geqslant 0$; ν_j *integers*). *In order that the inequality* (49) *hold for* $\Omega=\square(M)$ *and for any* $q\geqslant p$, *it is necessary that the points* $\nu^e\in\hat{\Pi}^e(\nu)$ *for all* $e\subseteq e_n$.

PROOF. From the obvious inequality

$$\|f\|_{\mathscr{L}^{\langle\mathbf{r}\rangle}_{p,\theta}(\square)}\leqslant c\|D^{\bar{\mathbf{r}}+[\mathbf{r}-\bar{\mathbf{r}}]+\omega_{\mathbf{r}}}f\|_{L_p(\square)}$$

it follows that if $f\in\hat{\mathfrak{S}}L_p(\square)$; that is, if

$$\sum_{\mathbf{l}\in\hat{\mathscr{E}}}\|D^{\mathbf{l}}f\|_{L_p(\square)}<\infty,$$

then $f(\mathbf{x})$ on $\square$ also belongs to $\bigcap_{r\in\mathfrak{S}}\mathfrak{L}^{\langle\rho\rangle}_{p,\theta}(\square)$, and, in addition,

$$\sum_{\mathbf{r}\in\mathscr{E}}\|f\|_{\mathscr{L}^{\langle\mathbf{r}\rangle}_{p,\theta}(\square)}\leqslant c\sum_{\mathbf{l}\in\hat{\mathscr{E}}}\|D^{\mathbf{l}}f\|_{L_p(\square)},$$

from which we get, as a consequence of the inequality (49), that for some set of functions contained in $\bigcap_{\mathbf{r}\in\mathfrak{S}}\mathfrak{L}^{\langle\rho\rangle}_{p,\theta}(\square)$ the inequality

$$\|D^{\nu}f\|_{L_q(\square)}\leqslant c\sum_{\mathbf{l}\in\hat{\mathscr{E}}}\|D^{\mathbf{l}}f\|_{L_p(\square)}$$

holds.

Then, as a consequence of Theorem 1.1 of V. P. Il'in (see [13], p. 149 = 166), we get that the condition $\nu^e\in\hat{\Pi}^e(\nu)$ is necessary, which was to be proved.

COROLLARY 4. *In order that the inequality of Theorem* 1, §2, *hold*; *that is, the inequality*

$$\|D^{\nu}f\|_{L_p(\Omega)}\leqslant c\sum_{i=0}^{*}\|f\|_{\mathscr{L}^{\langle\mathbf{r}^i\rangle}_{p_i,\theta_i}(\Omega)}$$

for $\Omega=\square(M)$, *it is necessary that condition* 1) *of this theorem hold*; *that is,*

$$\nu_j\geqslant\bar{r}^0_j+1+[r^0_j-\bar{r}^0_j]\qquad(j\in e_{\mathbf{r}^0}),$$

$$\nu_i<r^i_i+1+[r^i_i-\bar{r}^i_i],\qquad\nu_j\geqslant\bar{r}^i_j+1+[r^i_j-\bar{r}^i_j]\quad(j\neq i)$$

$$(i=1,\ \ldots,\ n;\ j\in e_{\mathbf{r}^i}).$$

Now let $\hat{\hat{\mathfrak{S}}}$ be the set of vectors $\mathbf{l}=[\mathbf{r}]+\omega_{\mathbf{r}}$, where $\mathbf{r}\in\mathfrak{S}$, $[\mathbf{r}]=([r_1],\cdots,[r_n])$, $[r_j]$ being the integral part of r_j. The sets

$$\hat{\hat{\mathscr{E}}}^e(\nu),\qquad P\hat{\hat{\mathscr{E}}}^e(\nu),\qquad\hat{\hat{\Pi}}^e(\nu)$$

are defined analogously.

THEOREM 10. *Let $\mathfrak{E}$ be a given finite set of nonnegative vectors* $\mathbf{r}=(r_1, \cdots, r_n)$, *and let* $\nu=(\nu_1, \cdots, \nu_n)$ *be a nonnegative integral vector.*

In order that the inequality

$$\| D^\nu f \|_{L_q(\Omega)} \leqslant c \sum_{\mathbf{r}\in\mathfrak{E}} \| f \|_{L_p^{\langle \mathbf{r} \rangle}(\Omega)} \tag{50}$$

hold for $\Omega=\square(M)$ *and for any* $q \geqslant p$, *it is necessary that the points* $\nu^e \in \hat{\hat{\Pi}}^e(\nu)$ *for all* $e \subseteq e_n$.

The proof of this theorem is similar to that of Theorem 9.

COROLLARY 5. *In order that the inequality*

$$\| D^\nu f \|_{L_q(\square)} \leqslant c \sum_{i=0}^{n} \| f \|_{L_{p_i}^{\langle \mathbf{r}^i \rangle}(\square)} \tag{51}$$

hold, it is necessary that

$$\nu_j \geqslant \begin{cases} [r_j^0]+1 & \text{for } j \in e^*_{\mathbf{r}^0}, \\ r_j^0 & \text{for } j \in e_{\mathbf{r}^0} \setminus e^*_{\mathbf{r}^0}; \end{cases}$$

$$\nu_i < \begin{cases} [r_i^i]+1 & \text{for } i \in e^*_{\mathbf{r}^i}, \\ r_i^i & \text{for } i \in e_{\mathbf{r}^i} \setminus e^*_{\mathbf{r}^i}; \end{cases}$$

$$\underset{(j\neq i)}{\nu_j} \geqslant \begin{cases} [r_j^i]+1 & \text{for } j \in e^*_{\mathbf{r}^i} \\ r_j^i & \text{for } j \in e_{\mathbf{r}^i} \setminus e^*_{\mathbf{r}^i} \end{cases} \qquad (i=1, \ldots, n).$$

§7. The case of weight spaces

If in the definition of the spaces (1) (see the Introduction), in place of the norm $\| \cdot \|_{L_{p_i}(\Omega)}$, we take the norm

$$\| \cdot \|_{L_{p_i}, \alpha(\Omega)} = \left(\int_\Omega \rho^{\alpha p_i} | \cdot |^{p_i} d\mathbf{x} \right)^{1/p_i},$$

where $\rho=\rho(\mathbf{x})$ is the distance from the point $\mathbf{x}$ to the boundary of the region Ω, this defines new weight classes of functions, denoted by

$$\bigcap_{i=0}^{n} \mathscr{L}^{\langle \mathbf{r}^i \rangle}_{\substack{p_i, \theta_i \\ \alpha}}(\Omega), \qquad \bigcap_{i=0}^{n} L^{\langle \mathbf{r}^i \rangle}_{p_i, \alpha}(\Omega) \tag{52}$$

respectively.

A complete theory of the weight spaces (Ω) has been constructed.

In the special case where $\alpha \geqslant 0$, $m=n$ (that is, $\Omega_m=\Omega$), Theorems 1–7 and

Corollaries 1–3 for the spaces (52) are exactly like Theorems 1–7 and Corollaries 1–3 above, except that, in the appropriate places, one must replace the norms in the spaces

$$L_p(\Omega), \quad \mathscr{L}^{\langle\rho\rangle}_{p,\theta}(\Omega), \quad L^{\langle\rho\rangle}_p(\Omega), \quad \mathscr{L}^{\langle r^i\rangle}_{p_i,\theta_i}(\Omega), \quad L^{\langle r^i\rangle}_{p_i}(\Omega)$$

by the corresponding norms in the weight spaces

$$L_{p,\alpha}(\Omega), \quad \mathscr{L}^{\langle\rho\rangle}_{\substack{p,\theta\\ \alpha}}(\Omega), \quad L^{\langle\rho\rangle}_{p,\alpha}(\Omega), \quad \mathscr{L}^{\langle r^i\rangle}_{\substack{p_i,\theta_i\\ \alpha}}(\Omega), \quad L^{\langle r^i\rangle}_{p_i,\alpha}(\Omega).$$

The corresponding results for the special case of the spaces $W^{\mathfrak{E}}_{p,\alpha}(\Omega)$ are given on [28].

BIBLIOGRAPHY

1. S. L. Sobolev, *Applications of functional analysis in mathematical physics,* Izdat. Leningrad. Gos. Univ., Leningrad, 1950; English transl., Transl. Math. Monographs, vol. 7, Amer. Math. Soc., Providence, R. I., 1963. MR **14**, 565; **29** #2624.

2. S. M. Nikol'skiĭ, *Inequalities for entire functions of exponential type and their application to the theory of differentiable functions of several variables,* Trudy Mat. Inst. Steklov. **38** (1951), 244–278; English transl., Amer. Math. Soc. Transl. (2) **80** (1969), 1–38. MR **14**, 32.

3. L. N. Slobodeckiĭ, *Generalized Sobolev spaces and their application to boundary problems for partial differential equations,* Leningrad. Gos. Ped. Inst. Učen. Zap. **197** (1958), 54–112; English transl., Amer. Math. Soc. Transl. (2) **57** (1966), 207–276. MR **34** #3075.

4. L. D. Kudrjavcev, *Direct and inverse imbedding theorems. Applications to solving elliptic equations by the variational method,* Trudy Mat. Inst. Steklov. **55** (1959); English transl., Transl. Math. Monographs, Amer. Math. Soc., Providence, R. I. (to appear). MR **33** #7838.

5. O. V. Besov, *Investigation of a family of function spaces in connection with imbedding and extension theorems,* Trudy Mat. Inst. Steklov. **60** (1961), 42–81; English transl., Amer. Math. Soc. Transl. (2) **40** (1964), 85–126. MR **24** #A3501.

6. V. P. Il'in, *Properties of some classes of differentiable functions of several variables defined in an n-dimensional region,* Trudy Mat. Inst. Steklov. **66** (1962), 227–363; English transl., Amer. Math. Soc. Transl. (2) **81** (1969), 91–256. MR **27** #3750.

7. S. V. Uspenskiĭ, *Imbedding theorems for weighted classes,* Trudy Mat. Inst. Steklov. **60** (1961), 282–303; English transl., Amer. Math. Soc. Transl. (2) **87** (1970), 121–145. MR **25** #440; **41** #3202.

8. V. P. Il'in and V. A. Solonnikov, *On some properties of differentiable functions of several variables,* Trudy Mat. Inst. Steklov. **6** (1962), 205–226; English transl., Amer. Math. Soc. Transl. (2) **81** (1969), 67–90. MR **27** #2768.

9. O. V. Besov, *On the density of functions with compact support in $\mathfrak{L}_{p,\theta}$ and the extension of functions,* Trudy Mat. Inst. Steklov. **89** (1967), 18–30 = Proc. Steklov Inst. Math. **89** (1967), 16–31. MR **35** #7118.

10. O. V. Besov and V. P. Il'in, *Natural extension of the class of regions in imbedding theorems,* Mat. Sb. **75** (**117**) (1968), 483–495 = Math. USSR Sb. **4** (1968), 445–456. MR **37** #1987.

11. S. M. Nikol'skiĭ, *Functions with dominant mixed derivative satisfying a multiple Hölder condition,* Sibirsk. Mat. Ž. **4** (1963), no. 6, 1342–1364; English transl., Amer. Math. Soc. Transl. (2) **102** (1973), (2) **102** (1973), 27–51. MR **28** #3322.

12. S. M. Nilol'skiĭ, *A variational problem*, Mat. Sb. **62** (**104**) (1963), 53–75; English transl., Amer. Math. Soc. Transl. (2) **51** (1966), 132–154. MR **28** #1391.

13. V. P. Il'in, *On inequalities between the norms of partial derivatives of functions of several variables*, Trudy Mat. Inst. Steklov. **84** (1965), 144–173 = Proc. Steklov Inst. Math. **84** (1965), 161–190. MR **33** #448.

14. N. S. Bahvalov, *Imbedding theorems for classes of functions with a number of bounded derivatives*, Vestnik Moskov. Univ. Ser. I Mat. Meh. **1963**, no. 3, 7–16. (Russian) MR **28** #4067.

15. A. D. Džabrailov, *On some function spaces. Direct and inverse imbedding theorems*, Dokl. Akad. Nauk SSSR **159** (1964), 254–257 = Soviet Math. Dokl. **5** (1964), 1457–1462. MR **31** #3867.

16. P. Grisvard, *Commutativité de deux foncteurs d'interpolation et applications*, J. Math. Pures Appl. (9) **45** (1966), 143–206. MR **36** #4361.

17. P. I. Lizorkin and S. M. Nikol'skiĭ, *A classification of differentiable functions in some fundamental spaces with dominant mixed derivatives*, Trudy Mat. Inst. Steklov. **77** (1965), 143–167 = Proc. Steklov Inst. Math. **77** (1965), 160–187. MR **33** #450.

18. T. I. Amanov, *Representation and imbedding theorems for the function spaces* $S^{\mathbf{r}}_{p,\theta}B(R_n)$ *and* $S^{\mathbf{r}}_{p^*,\theta}B(0 \leqslant x_j \leqslant 2\pi;\ j = 1, 2, \dots, n)$, Trudy Mat. Inst. Steklov. **77** (1965), 3–34 = Proc. Steklov Inst. Math. **77** (1965), 3–36. MR **33** #1718.

19. A. D. Džabrailov, *On the theory of imbedding theorems*, Trudy Mat. Inst. Steklov. **89** (1967), 80–118 = Proc. Steklov Inst. Math. **89** (1967), 92–135. MR **36** #676.

20. ———, *Properties of functions in certain weight spaces*, Trudy Mat. Inst. Steklov. **89** (1967), 70–79 = Proc. Steklov Inst. Math. **89** (1967), 80–91. MR **36** #677.

21. A. D. Džabrailov and V. P. Il'in, *Inequalities between norms of partial derivatives of functions of several variables*, Imbedding Theorems and Their Applications, Proc. All-Union Sympos. Imbedding Theorems (Baku, 1966), "Nauka", Moscow, 1970, pp. 73–75. (Russian)

22. V. P. Il'in, *Conditions of validity of inequalities between* L_p*-norms of partial derivatives of functions of several variables*, Trudy Mat. Inst. Steklov. **96** (1968), 205–242 = Proc. Steklov Inst. Math. **96** (1968), 259–306. MR **42** #6605.

23. O. V. Besov and A. D. Džabrailov, *Interpolation theorems for some spaces of differentiable functions*, Trudy Mat. Inst. Steklov. **105** (1969), 15–20 = Proc. Steklov Inst. Math. **105** (1969), 16–23.

24. A. D. Džabrailov, *Interpolation spaces of differentiable functions defined in an n-dimensional domain*, Dokl. Akad. Nauk SSSR **187** (1969), 259–262 = Soviet Math. Dokl. **10** (1969), 883–837.

25. Ja. S. Bugrov, *Approximation of a class of functions with dominant mixed derivatives*, Mat. Sb. **64** (**106**) (1964), 410–418. (Russian) MR **29** #2587.

26. A. D. Džabrailov, *Weight spaces of functions with dominant mixed derivatives and properties of functions from these spaces*, Dokl. Akad. Nauk SSSR **175** (1967), 276–279 = Soviet Math. Dokl. **8** (1967), 835–839, MR **35** #5921.

27. G. G. Kazarjan, *Bounds for derivatives of functions in terms of differential operators*, Candidate's Dissertation, Steklov Institute of Mathematics, Moscow, 1970. (Russian)

28. A. D. Džabrailov, *Boundary properties of functions in weight spaces*, Imbedding Theorems and Their Applications, Proc. All-Union Sympos. Imbedding Theorems (Baku, 1966), "Nauka", Moscow, 1970, pp. 64–67. (Russian)

Translated by H. T. JONES

Trudy Mat. Inst. Steklov.
117 (1972)

Proc. Steklov Inst. Math.
117 (1972)

FAMILIES OF SPACES OF FUNCTIONS WHOSE MIXED DERIVATIVES SATISFY A MULTIPLE-INTEGRAL HÖLDER CONDITION

A. D. DŽABRAILOV

UDC 517.518.22

Abstract. The differential properties of functions in spaces whose multi-index of differential properties is 2^n free vectors are studied. For a suitable choice of vectors, these spaces coincide with the well-known spaces of types SW and SB.

A new integral representation for smooth functions is obtained to study these spaces. For $n = 1$, this representation is a generalization of the well-known representation of V. P. Il'in and O. V. Besov. The corresponding weight spaces of functions are also considered.

Bibliography: 21 items. Figures: 2.

Introduction

Let $e_n = \{1, \cdots, n\}$ be the set consisting of the natural numbers $1, \cdots, n$. We denote by e any fixed subset of the set e_n (including the sets ϕ and e_n). The number of these subsets is 2^n. We assume that these subsets are indexed; that is, the sets e^i $(i = 1, \cdots, 2^n)$ are all of the subsets of the set e_n (including the set e_n itself and the empty set ϕ).

With each set e^i we associate some fixed vector $\mathbf{r}^i = (r_1^i, \cdots, r_n^i)$ with components $r_j^i \geqslant 0$ and with support $e_{\mathbf{r}^i} \supseteq e^i$; that is, $r_j^i > 0$ if $j \in e^i$ and $r_j^i \geqslant 0$ if $j \in e_n \backslash e^i$ (the set of indices of the nonzero components of a vector $\mathbf{r} = (r_1, \cdots, r_n)$ is called the support of $\mathbf{r}$ and is denoted by $e_\mathbf{r}$ (see [1])).

The purpose of this paper is to study the differential properties of functions in the function spaces

$$\bigcap_{i=1}^{2^n} L_{p_i}^{\langle \mathbf{r}^i \rangle}(\Omega), \tag{1}$$

$$\bigcap_{i=1}^{2^n} \mathscr{L}_{p_i,\theta_i}^{\langle \mathbf{r}^i \rangle}(\Omega). \tag{2}$$

Let $\mathbf{r} = (r_1, \cdots, r_n)$ $(r_j > 0; j = 1, \cdots, n)$ be any vector. In the case where

$$r_j^i = \begin{cases} r_j & \text{for } j \in e^i, \\ 0 & \text{for } j \notin e^i \end{cases}$$

AMS (MOS) subject classifications (1970). Primary 46E35; Secondary 26A69.

$p_i = p$, and $\theta_i = \theta$ $(i = 1, \cdots, 2^n)$, the spaces (1) and (2) are the well-known spaces $S_p^{\mathbf{r}} W(\Omega)$ and $S_{p,\theta}^{\mathbf{r}} B(\Omega)$ which have been studied by S. M. Nikol'skiĭ [1,2], N. S. Bahvalov [3], V. P. Il'in [4], A. D. Džabrailov [5, 6], T. I. Amanov [7], P. I. Lizorkin [8], O. V. Besov [9], O. V. Besov and A. D. Džabrailov [10], S. Ja. Bugrov [11], and others.

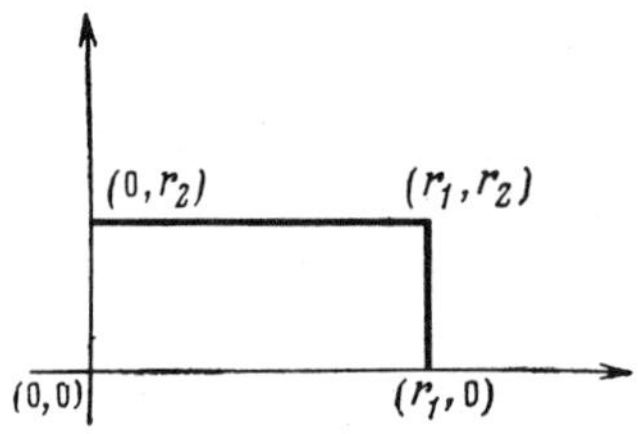

Figure 1.

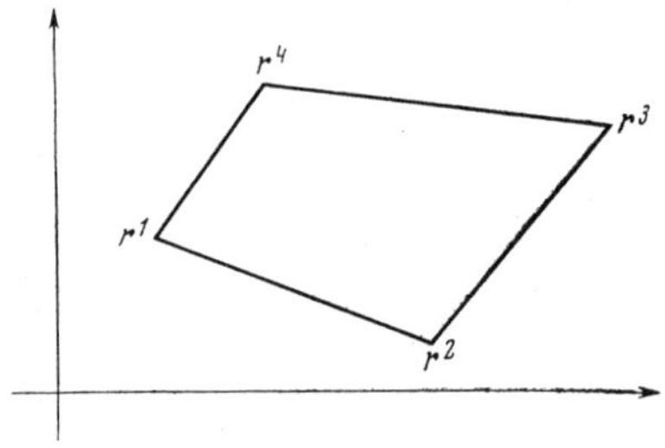

Figure 2.

In case $n = 2$, the "multi-index" of the differential properties of the spaces $S_p^{\mathbf{r}} W$ and $S_{p,\theta}^{\mathbf{r}} B$ is the vertices of the polygon in Figure 1 (the rectangle with vertices at the points $(0, 0)$, $(r_1, 0)$, $(0, r_2)$, (r_1, r_2)).

The "multi-index" of the differential properties of the space (1) and (2) is the polyhedron with vertices at the points $\mathbf{r}^i$ $(i = 1, \cdots, 2^n)$ (for $n = 2$, see Figure 2).

The spaces (1) for the case of vectors $\mathbf{r}^i$ $(i = 1, \cdots, 2^n)$ with *integer* components have been studied by V. P. Il'in [12–14], A. D. Džabrailov [15], Džabrailov and Il'in [16], and others.

This paper consists of six sections.

In §1 the notation and basic definitions are given.

In §2 the basic imbedding theorems and their corollaries are stated.

In §3 a certain integral representation of the derivatives of a smooth function in terms of multiple differences of specified derivatives of the function is established. This integral representation contains new results in the one-dimensional case. Certain ideas from the works of Besov and Il'in [17] and Džabrailov [6] are used in the proof of this integral representation.

In §4 the integral representation obtained in §3 is used to prove the fundamental theorems of §2.

In §5 certain theorems are proved concerning the extension of a function beyond the boundary of the region $\Omega \subset E_n$ to E_n, certain differential properties being preserved.

In §6 the case of weight spaces is considered.

The results of this paper were communicated to the Mathematical Institute of the Academy of Sciences of the USSR in December, 1969, in the seminar "The theory of functions of several variables and its applications to the equations of mathematical physics", directed by L. D. Kudrjavcev and S. M. Nikol'skiĭ.

The author wishes to express his gratitude to L. D. Kudrjavcev for his interest in this work. The author also wishes to take this opportunity to thank O. V. Besov for reading this paper and for making very valuable suggestions concerning both the results and their exposition; these suggestions have been taken into account in the final draft.

§1. Fundamental definitions and notation

The present paper is (to a certain extent) a continuation of the work in [18]; therefore in some places we will follow the notation of that paper (see §1).

Let the vectors $\mathbf{r}^i = (r_1^i, \cdots, r_n^i)$ $(i = 1, \cdots, 2^n)$ be associated with the subsets e^i $(i = 1, \cdots, 2^n)$ of the set $e_n \equiv \{1, \cdots, n\}$ (see the Introduction). We assume that $\bar{r}_j^i$ is the largest integer which is less than r_j^i; that is, $\bar{r}_j^i \geqslant 0$ is an integer and $0 < r_j^i - \bar{r}_j^i \leqslant 1$. In case $r_j^i = 0$, we let $\bar{r}_j^i = 0$; consequently to each vector $\mathbf{r}^i = (r_1^i, \cdots, r_n^i)$ there corresponds a unique vector $\bar{\mathbf{r}}^i = (\bar{r}_1^i, \cdots, \bar{r}_n^i)$.

Each vector $\mathbf{r} = (r_1, \cdots, r_n)$ (where $r_j \geqslant 0$) defines a unique vector $[\mathbf{r}] = ([r_1], \cdots, [r_n])$, where $[r_j]$ is the integral part of r_j.

Let $\mathbf{k} = (k_1, \cdots, k_n)$ (where $k_j \geqslant 0$ are integers), and let $\Delta^{\mathbf{k}}(\mathbf{t})f = \Delta_1^{k_1}(t_1) \cdots \Delta_n^{k_n}(t_n)f$, where $\Delta_j^{k_j}(t_j)f$ is the finite difference of order k_j of the function $f(\mathbf{x})$ in the direction of x_j, with step t_j.

Let $\Delta^{\mathbf{k}}(\mathbf{t}; \Omega)f(\mathbf{x}) = \Delta^{\mathbf{k}}(\mathbf{t})f(\mathbf{x})$, if the multiple differences are formed from points belonging to the region Ω together with the parallelepiped whose vertices are at these points; otherwise $\Delta^{\mathbf{k}}(\mathbf{t}; \Omega)f(\mathbf{x}) = 0$ (for $\mathbf{k} = (0, \cdots, k_i, \cdots, 0)$, see [17]).

Further, let

$$\mathbf{h} = (h_1, \ldots, h_n),\ h_j > 0\ (j = 1, \ldots, n),\ \mathbf{y} = (y_1, \ldots, y_n),$$

$$\frac{\mathbf{y}}{\mathbf{h}} = \left(\frac{y_1}{h_1}, \ldots, \frac{y_n}{h_n}\right),\ D^{\mathbf{k}} f(\mathbf{x}) \equiv D_1^{k_1} \ldots D_n^{k_n} f(\mathbf{x}) \equiv f^{(\mathbf{k})}(\mathbf{x}),$$

$$\int_{\mathbf{a}}^{\mathbf{b}} f(\mathbf{x})\, d\mathbf{x}^e = \int_{a_1}^{b_1} dx_1^e \ldots \int_{a_n}^{b_n} f(\mathbf{x})\, dx_n^e \equiv \left(\prod_{j \in e} \int_{a_j}^{b_j} dx_j\right) f(\mathbf{x}) \equiv \int_{\mathbf{a}}^{\mathbf{b}} f(\mathbf{x}) \prod_{j \in e} dx_j;$$

that is, the integration takes place only with respect to those variables whose indices belong to the set e.

DEFINITION 1. *We will say that the function* $f(\mathbf{x})$ *belongs to the space*

$$\bigcap_{i=1}^{2^n} \mathscr{L}_{p_i, \theta_i}^{\langle \mathbf{r}^i \rangle}(\Omega),$$

if the generalized derivatives $D^{\bar{\mathbf{r}}^i} f$ *(in the sense of S. L. Sobolev* [19]) *exist on* $\Omega \subseteq E_n$ *and satisfy the conditions*

$$\|f\|_{\mathscr{L}^{\langle \mathbf{r}^i\rangle}_{p_i,\theta_i}(\Omega)} \equiv \left(\int\cdots\int \|\Delta^{\omega^i+[\mathbf{r}^i-\bar{\mathbf{r}}^i]}(\mathbf{t};\Omega) D^{\bar{\mathbf{r}}^i} f\|^{\theta_i}_{L_{p_i}(\Omega)} \prod_{j\in e_{\mathbf{r}^i}} \frac{dt_j}{|t_j|^{1+\theta_i(r^i_j-\bar{r}^i_j)}}\right)^{1/\theta_i} < \infty,$$

where $\omega=(1,\cdots,1)$ *is the n-dimensional vector with unit components,* $\omega^i\equiv\omega^{e_{\mathbf{r}^i}}$, $1\leqslant p_i,\theta_i\leqslant\infty$.

The seminorm in this space is defined as

$$\sum_{i=1}^{2^n}\|f\|_{\mathscr{L}^{\langle \mathbf{r}^i\rangle}_{p_i,\theta_i}(\Omega)}. \tag{1.1}$$

DEFINITION 2. *We will say that the function* $f(\mathbf{x})$ *belongs to the space*

$$\bigcap_{i=1}^{2^n} L^{\langle \mathbf{r}^i\rangle}_{p_i}(\Omega),$$

if the generalized derivatives $D^{[\mathbf{r}^i]}f(\mathbf{x})$ *(in the sense of S. L. Sobolev) exist on the region* $\Omega\subseteq E_n$ *and satisfy the conditions*

$$\|f\|_{L^{\langle \mathbf{r}^i\rangle}_{p_i}(\Omega)} \equiv \left(\int\cdots\int \|\Delta^{\tilde{\omega}^i}(\mathbf{t};\Omega) D^{[\mathbf{r}^i]}f\|^{p_i}_{L_{p_i}(\Omega)} \prod_{j\in e^*_{\mathbf{r}^i}} \frac{dt_j}{|t_j|^{1+p_i(r^i_j-[r^i_j])}}\right)^{1/p_i} < \infty \quad (i=1,\dots,2^n),$$

where $e^*_{\mathbf{r}^i}$ *is the set of indices of the nonintegral components of the vector* $\mathbf{r}^i$, $\tilde{\omega}^i\equiv\omega^{e^*_{\mathbf{r}^i}}$.

We define the seminorm in this space as

$$\sum_{i=1}^{2^n}\|f\|_{L^{\langle \mathbf{r}^i\rangle}_{p_i}(\Omega)}.$$

We denote by

$$\bigcap_{i=1}^{2^n}\tilde{\mathscr{L}}^{\langle \mathbf{r}^i\rangle}_{p_i,\theta_i}(\Omega) \qquad \left(\bigcap_{i=1}^{2^n}\tilde{L}^{\langle \mathbf{r}^i\rangle}_{p_i}(\Omega)\right)$$

the closed set of sufficiently smooth functions on Ω which satisfy the inequality

$$\sum_{i=1}^{2^n}\|f\|_{\mathscr{L}^{\langle \mathbf{r}^i\rangle}_{p_i,\theta_i}(\Omega)}<\infty \quad \left(\sum_{i=1}^{2^n}\|f\|_{L^{\langle \mathbf{r}^i\rangle}_{p_i}(\Omega)}<\infty\right)$$

involving the seminorm (1.1) (involving the seminorm (1.2)).

Let $\mathbf{H}=(H_1,\cdots,H_n)$, $H_j>0$ $(j=1,\cdots,n)$, and $\delta=(\delta_1,\cdots,\delta_n)$, where $\delta_j=\pm1$ $(j=1,\cdots,n)$. There are 2^n possible different vectors δ.

For each pair of fixed vectors $\mathbf{H}$ and δ, the rectangular parallelepiped in E_n defined by the inequalities $0\leqslant\delta_i x_i\leqslant H_i$ $(i=1,\cdots,n)$ will be denoted by $\square=\square(\mathbf{H};\delta)$. For a fixed $\mathbf{H}$ there are 2^n different parallelepipeds defined by the vectors δ^s $(s=1,\cdots,2^n)$.

We say that the region $\Omega\subseteq E_n$ belongs to the class $C(\mathbf{H})$ if there exists a finite covering by open sets U_k $(k = 1, \cdots, N)$ and a collection $\square_k = \square(\mathbf{H}; \delta^k)$ such that the following conditions are fulfilled:

$$1)\ \Omega = \bigcup_{k=1}^{N} U_k, \quad \operatorname{mes}(U_k \cap U_j) \geqslant q > 0;$$
$$2)\ U_k + \square_k \subset \Omega \quad (k = 1, \ldots, N).$$

For further details concerning this class of regions, see V. P. Il'in [14].

§2. Fundamental results

In this section the fundamental results on imbedding theorems are stated. The proofs of these theorems are given in §4.

THEOREM 1. *Let $f(\mathbf{x})$ belong to the space*

$$\bigcap_{i=1}^{2^n} \mathscr{L}_{p_i, \theta_i}^{\langle \mathbf{r}^i \rangle}(\Omega),$$

where $\Omega \in C(H_1, \cdots, H_n)$ and where $\mathbf{r}^i = (r_1^i, \cdots, r_n^i)$ are the vectors corresponding to the subsets e^i $(i = 1, \cdots, 2^n)$ of $e_n = \{1, \cdots, n\}$; that is, $r_j^i > 0$ for $j \in e^i$ and $r_j^i \geqslant 0$ for $j \in e_n \backslash e^i$, $1 \leqslant p_i \leqslant \theta_i \leqslant \infty$ $(i = 1, \cdots, 2^n)$.

Further, let $\nu = (\nu_1, \cdots, \nu_n)$ be a nonnegative integral vector; that is, $\nu_j \geqslant 0$, ν_j integers, m a natural number, $m \leqslant n, p \geqslant p_i$ $(i = 1, \cdots, 2^n)$.

Assume that

1) $\nu_j \geqslant \bar{r}_j^i + 1 + [r_j^i - \bar{r}_j^i]$ $(i = 1, \cdots, 2^n; j \in e_{\mathbf{r}'} \backslash e^i)$ *and* $\epsilon^i = (\epsilon_1^i, \cdots, \epsilon_n^i)$ $(i = 1, \cdots, 2^n)$, *where*

$$\varepsilon_j^i = \begin{cases} r_j^i - \nu_i - \dfrac{1}{p_i} + \dfrac{1}{p} & \textit{for } j \in e_m \equiv \{1, \ldots, m\}, \\ r_j^i - \nu_j - \dfrac{1}{p_i} & \textit{for } j \in e_n \setminus e_m \equiv \{m+1, \ldots, n\}; \end{cases}$$

2) *for each fixed $i = 1, \cdots, 2^n$*

$$\varepsilon_j^i \geqslant 0 \quad \textit{for } j \in e^i \cdot e_m,$$
$$\varepsilon_j^i > 0 \quad \textit{for } j \in e^i \setminus e_m,$$

where it is assumed that $\epsilon_j^i = 0$ for $j \in e^i \cdot e_m$ only in case $1 < p_i = \theta_i < p < \infty$.

Then the function $D^\nu f(\mathbf{x})$ on the intersection of the region Ω with the hyperplane $x_{m+1} = \text{const}, \cdots, x_n = \text{const}$ (denoted by Ω_m) belongs to the space $L_p(\Omega_m)$, and the following inequality holds:

$$\| D^\nu f \|_{L_p(\Omega_m)} \leqslant c \sum_{i=1}^{2^n} \prod_{j=1}^{n} h_j^{\varepsilon_j^i} \| f \|_{\mathscr{L}_{p, \theta_i}^{\langle \mathbf{r}^i \rangle}(\Omega)}, \tag{2.1}$$

where $0 < h_j \leqslant H_j \leqslant \infty$ $(j = 1, \cdots, n)$, *and* c *is a constant which depends neither on the function* f *nor on the vector* $\mathbf{h} = (h_1, \cdots, h_n)$.

COROLLARY 1. *Let* $\mathbf{r} = (r_1, \cdots, r_n)$, $r_j > 0$ $(j = 1, \cdots, n)$, *let* $1 < p < q < \infty$, *and let* $\nu = (\nu_1, \cdots, \nu_n)$ *be a nonnegative integral vector, and let*

$$r_j - \nu_j - \frac{1}{p} + \frac{1}{q} = 0 \qquad (j = 1, \ldots, n).$$

Then the inequality

$$\| D^\nu f \|_{L_q(E_n)} \leqslant c \| f \|_{\mathscr{L}^{\langle \mathbf{r} \rangle}_{p,\, p}(E_n)}, \tag{2.2}$$

where c *is a constant which does not depend on* f, *holds for all sufficiently smooth functions* $f(\mathbf{x})$.

This result is well known (see O. V. Besov [9]).

THEOREM 2. *Let* $f(\mathbf{x})$ *belong to the space*

$$\bigcap_{i=1}^{2^n} \tilde{L}_{p_i}^{\langle \mathbf{r}^i \rangle}(\Omega),$$

where $\Omega \in C(H_1, \cdots, H_n)$, *where* $\mathbf{r}^i = (r_1^i, \cdots, r_n^i)$ $(i = 1, \cdots, 2^n)$ *are vectors which correspond to the subsets* e^i $(i = 1, \cdots, 2^n)$ *of the set* $e_n = \{1, \cdots, n\}$, *and where* $1 \leqslant p_i \leqslant \infty$ $(i = 1, \cdots, 2^n)$.

Further, let $\nu = (\nu_1, \cdots, \nu_n)$ *be a nonnegative integral vector, let* m *be a natural number such that* $m \leqslant n$, *and let* $p \geqslant p_i$ $(i = 1, \cdots, 2^n)$. *Assume that*

1)

$$\nu_j \geqslant \begin{cases} [r_j^i] + 1 & \text{for } j \in e^*_{\mathbf{r}^i} \setminus e^i \\ r_j^i & \text{for } j \in e_{\mathbf{r}^i} \setminus (e^*_{\mathbf{r}^i} + e^i) \end{cases} \qquad (i = 1, \ldots, 2^n)$$

and $\epsilon^i = (\epsilon_1^i, \cdots, \epsilon_n^i)$ $(i = 1, \cdots, 2^n)$ *(for the meaning of* ϵ_j^i, *see* (*), *Theorem* 1);

2) *for each fixed* $i = 1, \cdots, 2^n$

$$\varepsilon_j^i \geqslant 0 \quad \text{for} \quad j \in e^i \cdot e_m,$$
$$\varepsilon_j^i > 0 \quad \text{for} \quad j \in e^i \setminus e_m,$$

where it is assumed that the equality $\epsilon_j^i = 0$ *for* $j \in e^i \cdot e_m$ *holds only in case* $1 < p_i < p < \infty$.

Then the function $D^\nu f(\mathbf{x})$ *on* Ω_m *belongs to the space* $L_p(\Omega_m)$, *and the inequality*

$$\| D^\nu f \|_{L_p(\Omega_m)} \leqslant c \sum_{i=1}^{2^n} \prod_{j=1}^{n} h^{\varepsilon_j^i} \| f \|_{L_{p_i}^{\langle \mathbf{r}^i \rangle}(\Omega)} \tag{2.3}$$

holds, where $0 < h_j \leqslant H_j \leqslant \infty$ $(j = 1, \cdots, n)$, *and where* c *is a constant which depends neither on the function* $f(\mathbf{x})$ *nor on the vector* $\mathbf{h} = (h_1, \cdots, h_n)$.

THEOREM 3. *Let* $f(\mathbf{x})$ *belong to the space*

$$\bigcap_{i=1}^{2^n} \widetilde{\mathscr{L}}^{\langle \mathbf{r}^i \rangle}_{p_i, \theta_i}(\Omega),$$

where $\Omega \in C(H_1, \cdots, H_n)$, $1 \leqslant p_i \leqslant \theta_i \leqslant \infty$, *and* $\mathbf{r}^i = (r_1^i, \cdots, r_n^i)$ $(i = 1, \cdots, 2^n)$ *are vectors corresponding to the subsets* e^i $(i = 1, \cdots, 2^n)$ *of the set* $e_n \doteq \{1, \cdots, n\}$.

Further, let $\nu = (\nu_1, \cdots, \nu_n)$ *be a nonnegative integral vector, let* m *be a natural number such that* $m \leqslant n$, *and let* $p \geqslant p_i$ *and* $\theta \geqslant \theta_i$ $(i = 1, \cdots, 2^n)$.

Assume that $\rho = (\rho_1, \cdots, \rho_n)$, $\rho_j \geqslant 0$, $e_\rho \subseteq e_m = \{1, \cdots, m\}$, $\bar{\rho}_j + \nu_j \geqslant \bar{r}_j^i + 1 + [r_j^i - \bar{r}_j^i]$ $(i = 1, \cdots, 2^n; j \in e_{\mathbf{r}^i} \backslash e^i)$, $\epsilon^i = (\epsilon_1^i, \cdots, \epsilon_n^i)$ (ϵ_j^i *are defined in* (*), *Theorem* 1), *and we let* $\rho_j \leqslant \epsilon_j^i$ $(j \in e^i; i = 1, \cdots, 2^n)$.

Then the function $D^\nu f(\mathbf{x})$ *on* Ω_m *belongs to the space* $\mathfrak{L}^{\langle \rho \rangle}_{p,\theta}(\Omega_m)$, *and the inequality*

$$\| D^\nu f \|_{\mathscr{L}^{\langle \rho \rangle}_{p,\theta}(\Omega_m)} \leqslant c \sum_{i=1}^{2^n} \prod_{j=1}^{n} h_j^{\epsilon_j^i - \rho_j} \| f \|_{\mathscr{L}^{\langle \mathbf{r}^i \rangle}_{p_i, \theta_i}(\Omega)} \tag{2.4}$$

holds, where $0 < h_j \leqslant H_j \leqslant \infty$ $(j = 1, \cdots, n)$, *and where* c *is a constant which depends neither on the function* $f(\mathbf{x})$ *nor on the vector* $\mathbf{h} = (h_1, \cdots, h_n)$.

A well-known result follows from this theorem (see [9]).

COROLLARY 2 (O. V. BESOV [9]). *Let* $\mathbf{r} = (r_1, \cdots, r_n)$, $r_j > 0$ $(j = 1, \cdots, n)$, $1 \leqslant p \leqslant \theta \leqslant \infty$, $q \geqslant p$, $\sigma \geqslant \theta$, $\nu = (\nu_1, \cdots, \nu_n)$ ($\nu_j \geqslant 0$ *are integers*), *and* $\rho = (\rho_1, \cdots, \rho_n)$, $\rho_j = r_j - \nu_j - 1/p + 1/q > 0$ $(j = 1, \cdots, n)$.

Then for sufficiently smooth finite functions $f(\mathbf{x})$ *the inequality*

$$\| D^\nu f \|_{\mathscr{L}^{\langle \rho \rangle}_{q,\sigma}(E_n)} \leqslant c \| f \|_{\mathscr{L}^{\langle \mathbf{r} \rangle}_{p,\theta}(E_n)} \tag{2.5}$$

holds, where c *is a constant which does not depend on* f.

The following interesting result also follows from Theorem 3.

COROLLARY 3. *Let* $f(\mathbf{x})$ *be a sufficiently smooth function defined on the region* Ω *belonging to* $C(\infty, \cdots, \infty, H_{k+1}, \cdots, H_n)$, *such that the function* $f(\mathbf{x})$ *itself and its derivatives vanish for large* $|\mathbf{x}| = \sqrt{x_1^2 + \cdots + x_n^2}$.

Further, let $\mathbf{r} = (r_1, \cdots, r_n)$, $r_j > 0$ $(j = 1, \cdots, n)$, $1 \leqslant p \leqslant \theta \leqslant \infty$, $q \geqslant p$, $\sigma \geqslant \theta$, $\nu = (\nu_1, \cdots, \nu_n)$ ($\nu_j \geqslant 0$ *are integers*), $\epsilon_j = r_j - \nu_j - 1/p + 1/q > 0$ $(j = 1, \cdots, m)$, $\epsilon_j = r_j - \nu_j - 1/p_j > 0$ $(j = m + 1, \cdots, n)$, $\rho = (\rho_1, \cdots, \rho_m, 0, \cdots, 0)$ $(1 \leqslant k \leqslant m \leqslant n)$, $\rho_j = \epsilon_j$ $(j = 1, \cdots, k)$, $0 \leqslant \rho_j \leqslant \epsilon_j$ $(j = k + 1, \cdots, m)$.

Then the inequality

$$\|D^\nu f\|_{\mathscr{L}^{\langle\rho\rangle}_{q,\sigma}(\Omega_m)} \leqslant c \sum_{e_k\subseteq e\subseteq e_n} \|f\|_{\mathscr{L}^{\langle \mathbf{r}^e\rangle}_{p,\theta}(\Omega)} \tag{2.5*}$$

holds, where the sum is taken over all possible subsets $e \subseteq e_n$ *containing the set* $e_k = \{1, \cdots, k\}$, *and* c *is a constant which does not depend on the function* $f(\mathbf{x})$.

THEOREM 4. *Let the conditions of Theorem* 2 *be satisfied, except that in place of condition* 1), *the following holds. Let* $\rho = (\rho_1, \cdots, \rho_n)$ *be a nonnegative vector with support* $e_\rho \subseteq e_m \equiv \{1, \cdots, m\}$, *and*

$$\nu_j + [\rho_j] \geqslant \begin{cases} [r^i_j]+1 & \text{for } j \in e^*_{\mathbf{r}^i} \setminus e^i \\ r^i_j & \text{for } j \in e_{\mathbf{r}^i} \setminus (e^i + e^*_{\mathbf{r}^i}) \end{cases} \quad (i = 1, \ldots, 2^n).$$

Let $\epsilon^i_j > 0$ $(j \in e^i; i = 1, \cdots, 2^n)$, *where*

$$\begin{aligned} \rho_j &\leqslant \varepsilon^i_j \quad \text{for } j \in e^*_\rho \cdot e^i \cdot e^*_{\mathbf{r}^i} \\ \rho_j &< \varepsilon^i_j \quad \text{for } j \in e^i \setminus e^*_\rho \cdot e^*_{\mathbf{r}^i} \end{aligned} \quad (1 = 1, \ldots, 2^n).$$

Then the function $D^\nu f$ *on* $\Omega_m \equiv \Omega \cap \{x_{m+1} = \text{const}, \cdots, x_n = \text{const}\}$ *belongs to the space* $L^{\langle\rho\rangle}_p(\Omega_m)$, *and the inequality*

$$\|D^\nu f\|_{L^{\langle\rho\rangle}_p(\Omega_m)} \leqslant c \sum_{i=1}^{2^n} \prod_{j=1}^{n} h_j^{\varepsilon^i_j - \rho_j} \|f\|_{L^{\langle \mathbf{r}^i\rangle}_{p_i}(\Omega)} \tag{2.6}$$

holds, where $0 < h_j \leqslant H_j \leqslant \infty$ $(j = 1, \cdots, n)$ *and where* c *is a constant which depends neither on the function* f *nor on the vector* $\mathbf{h} = (h_1, \cdots, h_n)$.

THEOREM 5. *Let the conditions of Theorem* 4 *hold, except that, in* 1), $\nu_j + [\rho_j]$ *is replaced by* $\nu_j + \bar{\rho}_j$. *Let* $\epsilon^i_j > 0$ $(j \in e^i; i = 1, \cdots, 2^n)$, *where*

$$\begin{aligned} \rho_j &\leqslant \varepsilon^i_j \quad \text{for } j \in e_\rho \cdot e^i \cdot e^*_{\mathbf{r}^i} \\ \rho_i &< \varepsilon^i_j \quad \text{for } j \in e^i \setminus e_\rho \cdot e^*_{\mathbf{r}^i} \end{aligned} \quad (i = 1, \ldots, 2^n).$$

Then the function $D^\nu f$ *on* Ω_m *belongs to the space* $\mathfrak{L}^{\langle\rho\rangle}_{p,\theta}(\Omega_m)$ (*where* $0 \geqslant p_i$; $i = 1, \cdots, 2^n$), *and the inequality*

$$\|D^\nu f\|_{\mathscr{L}^{\langle\rho\rangle}_{p,\theta}(\Omega_m)} \leqslant c \sum_{i=1}^{2^n} \prod_{j=1}^{h} h_j^{\varepsilon^i_j - \rho_j} \|f\|_{L^{\langle \mathbf{r}^i\rangle}_{p_i}(\Omega)} \tag{2.7}$$

holds, where $0 < h_j \leqslant H_j \leqslant \infty$ $(j = 1, \cdots, n)$ *and where* c *is a constant which depends neither on the function* $f(\mathbf{x})$ *nor on the vector* $\mathbf{h} = (h_1, \cdots, h_n)$.

§3. An integral representation for smooth functions

In this section the proof is given for an integral representation for the derivatives $D^\nu f(\mathbf{x})$ of a smooth function in terms of multiple differences of certain given derivatives of the function. (Similar proofs were given in §3 of [18].)

Let $\mathcal{K}_\delta(t)$ (where $\delta = \pm 1$) be a sufficiently smooth finite function defined on E_1, whose support is contained in the interval $[\alpha, 1] \subset (0, 1]$, if $\delta = +1$, and in the interval $[-1, -\alpha] \subset [-1, 0)$, if $\delta = -1$, and, in addition, let

$$\int_{E_1} \mathcal{K}_\delta(t)\, dt = 1.$$

If $\delta = (\delta_1, \cdots, \delta_n)$ is a vector with components $\delta_j = +1$ $(j = 1, \cdots, n)$, then for sufficiently large integers s_j we put

$$\eta_{\delta_j}(t_j) = \frac{t_j^{s_j - 1}}{(s_j - 1)!} \int_{-1}^{t_j} \mathcal{K}_{\delta_j}(\xi)\, d\xi \qquad (j = 1, \ldots, n),$$

$$\mathcal{W}_{\delta_j}(t_j) = D^{s_j} \eta_{\delta_j}(t_j) \equiv \eta_{\delta_j}^{(s_j)}(t_j) \qquad (j = 1, \ldots, n),$$

$$\mathcal{W}_\delta(\mathbf{t}) = \prod_{j=1}^{n} \mathcal{W}_{\delta_j}(t_j).$$

We introduce averages of the function $D^\nu f(\mathbf{x})$ (where $\nu = (\nu_1, \cdots, \nu_n)$ is a nonnegative integral vector; that is, $\nu_j \geqslant 0$, and the ν_j are integers):

$$(D^\nu f)_{\mathbf{h}}(\mathbf{x}) = \int_{E_n} D^\nu f(\mathbf{x} + \mathbf{y}) \frac{1}{h_1} \cdots \frac{1}{h_n} \mathcal{W}_\delta\left(\frac{\mathbf{y}}{\mathbf{h}}\right) d\mathbf{y}, \tag{3.1}$$

$$\begin{aligned} \widehat{(D^\nu f)}_{\mathbf{h}}(\mathbf{x}) &\equiv ((D^\nu f)_{\mathbf{h}})_{\mathbf{h}}(\mathbf{x}) \\ &= \int_{E_n}\int_{E_n} D^\nu f(\mathbf{x} + \mathbf{y} + \mathbf{z}) \frac{1}{h_1^2} \cdots \frac{1}{h_n^2} \mathcal{W}_\delta\left(\frac{\mathbf{y}}{\mathbf{h}}\right) \mathcal{W}_\delta\left(\frac{\mathbf{z}}{\mathbf{h}}\right) d\mathbf{y}\, d\mathbf{z}, \end{aligned} \tag{3.2}$$

where $\mathbf{h} = (h_1, \cdots, h_n)$, $h_j > 0$ $(j = 1, \cdots, n)$, $\mathbf{y}/\mathbf{h} = (y_1/h_1, \cdots, y_n/h_n)$ and $\mathbf{z}/\mathbf{h} = (z_1/h_1, \cdots, z_n/h_n)$. It is clear that

$$\widehat{(D^\nu f)}_{\mathbf{h}}(\mathbf{x}) \to D^\nu f(\mathbf{x}) \qquad (h_1^2 + \ldots + h_n^2 \to \infty). \tag{3.3}$$

It is not difficult to show that, for any fixed $e^i \subseteq e_n$,

$$\begin{aligned} D_{h_1}^{\omega_1^{e^i}} \ldots D_{h_n}^{\omega_n^{e^i}} \widehat{(D^\nu f)}_{\mathbf{h}}(\mathbf{x}) &\equiv \left(\prod_{j \in e^i} \frac{\partial}{\partial h_j}\right) \widehat{(D^\nu f)}_{\mathbf{h}}(\mathbf{x}) \\ &\equiv D_{\mathbf{h}}^{\omega^{e^i}} \widehat{(D^\nu f)}_{\mathbf{h}}(\mathbf{x}) = 2^{|\omega^{e^i}|} D_{\mathbf{t}}^{\omega^{e^i}} ((D^\nu f)_{\mathbf{h}})_{\mathbf{t}}^{(\mathbf{x})} \big|_{\mathbf{t}=\mathbf{h}}, \end{aligned} \tag{3.4}$$

where $\omega \equiv (\omega_1, \cdots, \omega_n) = (1, \cdots, 1)$, $\omega^{e^i} = (\omega_1^{e^i}, \cdots, \omega_n^{e^i})$ and $|\omega^{e^i}| = \Sigma_{j \in e^i} \omega_j$.

Now we prove some auxiliary equalities.

Let $j \in e^i$; then, after certain transformations, one can see that

$$\frac{\partial}{\partial h_j}\left[\frac{1}{h_j} \mathcal{W}_{\delta_j}\left(\frac{y_j}{h_j}\right)\right] = -\frac{1}{h_j^2} \zeta_{\delta_j}^{(s_j)}\left(\frac{y_j}{h_j}\right), \tag{3.5}$$

where

$$\zeta_{\delta_j}(t_j) = \frac{t_j^{s_j}}{(s_j-1)!}\mathscr{K}_{\delta_j}(t_j). \tag{3.5'}$$

Making use of formula (3.5), we get some integral equalities (we consider the one-dimensional case, so we omit the index j on the symbols $x_j, y_j, z_j, \nu_j, s_j, \delta_j$ and h_j)

$$\mathfrak{I} \equiv \int_{E_1}\int_{E_1} D^\nu g(x+y+z)\mathscr{W}_\delta\left(\frac{z}{h}\right)\frac{\partial}{\partial n}\left[\frac{1}{h}\mathscr{W}_\delta\left(\frac{y}{h}\right)\right]dydz$$
$$= \int_{E_1}\int_{E_1} D^\nu g(x+y+z)\mathscr{W}_\delta\left(\frac{z}{h}\right)\zeta_\delta^{(s)}\left(\frac{y}{h}\right)dydz,$$

whence, integrating by parts and making the substitution $y_1 = y + (k-1)z/k$, we have

$$\mathfrak{I} = \frac{(-1)^{\alpha+k-\nu}}{h^{\nu-\alpha-k}}\int_{E_1}\int_{E_1} D^{\alpha+k} g(x+y+z)\mathscr{W}_\delta\left(\frac{z}{h}\right)\zeta_\delta^{(s-\alpha-k+\nu)}\left(\frac{y}{h}\right)dydz$$
$$= \frac{(-1)^{\alpha+k-\nu}}{h^{\nu-\alpha-k}}\int_{E_1}\int_{E_1} D^{\alpha+k} g\left(x+y_1+\frac{z}{k}\right)\mathscr{W}_\delta\left(\frac{z}{h}\right)\zeta_\delta^{(s-\alpha-k+\nu)}\left(\frac{J_1-\frac{k-1}{k}z}{h}\right)dy_1dz. \tag{3.6}$$

From this last equality, taking into account the formula

$$\frac{\partial}{\partial t}\Delta(t)\psi(x) = \psi'(x+t) \tag{**}$$

and integrating by parts, we get

$$\mathfrak{I} = \frac{(-1)^{\alpha+k-\nu-1}k}{h^{\nu-\alpha-k+1}}\int_{E_1}\int_{E_1}\Delta\left(\frac{z}{k}\right)D^{\alpha+k-1}g(x+y_1)$$
$$\times\frac{\partial}{\partial\left(\frac{z}{h}\right)}\left\{\mathscr{W}_\delta\left(\frac{z}{h}\right)\zeta_\delta^{(s-\alpha-k+\nu)}\left(\frac{y_1-\frac{k-1}{k}z}{h}\right)\right\}dy_1dz.$$

In the last integral we make the substitution $y_2 = y_1 - z/k$ and apply the formula (**). After an integration by parts we arrive at the equality

$$\mathfrak{I} = \frac{(-1)^{\alpha+k-\nu-2}k^2}{h^{\nu-\alpha-k+2}}\int_{E_1}\int_{E_1}\Delta^2\left(\frac{z}{k}\right)D^{\alpha+k-2}g(x+y_2)\,Q_{2,\delta}\left(\frac{y_2-\frac{k-2}{k}z}{h};\ \frac{z}{h}\right)dy_2dz,$$

where

$$Q_{2,\delta}\left(\frac{y_2 - \frac{k-2}{k}z}{h};\ \frac{z}{h}\right) \equiv \frac{\partial}{\partial\left(\frac{z}{h}\right)} Q_{1,\delta}\left(\frac{y_2 - \frac{k-2}{k}z}{h};\ \frac{z}{h}\right),$$

$$Q_{1,\delta}\left(\frac{y_1 - \frac{k-1}{k}z}{h};\ \frac{z}{h}\right) \equiv \frac{\partial}{\partial\left(\frac{z}{h}\right)} Q_{0,\delta}\left(\frac{y_1 - \frac{k-1}{k}}{h};\ \frac{z}{h}\right)$$

$$\equiv \frac{\partial}{\partial\left(\frac{z}{h}\right)}\left\{\mathscr{W}_\delta\left(\frac{z}{h}\right)\zeta_\delta^{(s-\alpha-k+\nu)}\left(\frac{y_1 - \frac{k-1}{k}z}{h}\right)\right\}.$$

Now, applying the process described above $k - 2$ times more; that is, making the substitution $y_i = y_{i-1} - z/k$, applying the formula (**), and integrating by parts, we get

$$\mathfrak{J} = \frac{(-1)^{\alpha-\nu}k^k}{h^{\nu-\alpha}} \int\limits_{E_1}\int\limits_{E_1} \Delta^k\left(\frac{z}{k}\right) D^\alpha g(x + y_k)\, Q_{k,\delta}\left(\frac{y_k}{h};\ \frac{z}{h}\right) dy_k dz, \tag{3.7}$$

where $Q_{k,\delta}(y_k/h, z/h)$ is defined by the recursion formulas

$$Q_{0,\delta}\left(\frac{y_1 - \frac{k-1}{k}z}{h};\ \frac{z}{h}\right) = \mathscr{W}_\delta\left(\frac{z}{h}\right)\zeta_\delta^{(s-\alpha-k+\nu)}\left(\frac{y_1 - \frac{k-1}{k}z}{h}\right),$$

$$Q_{i,\delta}\left(\frac{y_i - \frac{k-i}{k}z}{h};\ \frac{z}{h}\right) \equiv \frac{\partial}{\partial\left(\frac{z}{h}\right)} Q_{i-1,\delta}\left(\frac{y_i - \frac{k-i}{k}z}{h};\ \frac{z}{h}\right) \qquad (i = 1, \ldots, k). \tag{3.8}$$

We transform the following integral:

$$\mathscr{F} \equiv \int\limits_{E_1}\int\limits_{E_1} D^\nu g(x + y + z)\, \mathscr{W}_\delta\left(\frac{y}{h}\right)\mathscr{W}_\delta\left(\frac{z}{h}\right) dy dz.$$

If we assume that $\nu \geqslant \alpha + k$, then, after an integration by parts, we get

$$\mathscr{F} = (-1)^{\nu-\alpha-k} h^{-\nu+\alpha+k} \int\limits_{E_1}\int\limits_{E_1} D^{\alpha+k} g(x + y + z)\, \mathscr{W}_\delta^{(\nu-\alpha-k)}\left(\frac{y}{h}\right)\mathscr{W}_\delta\left(\frac{z}{h}\right) dy dz. \tag{3.9}$$

Using arguments similar to those given in the proof of (3.7), we get

$$\mathscr{F} = (-1)^{-\alpha+\nu} k^k h^{\alpha-\nu} \int\limits_{E_1}\int\limits_{E_1} \Delta^k\left(\frac{z}{k}\right) D^\alpha g(x + y_k)\, \Pi_{k,\delta}\left(\frac{y_k}{h};\ \frac{z}{h}\right) dy_k dz, \tag{3.10}$$

where $\Pi_{k,\delta}(y_k/h; z/h)$ is defined by the following recursion formula:

$$\Pi_{0,\delta}\left(\frac{y_1-\frac{k-1}{k}z}{h};\ \frac{z}{h}\right)\equiv\mathscr{W}_\delta^{(\nu-\alpha-k)}\left(\frac{y_1-\frac{k-1}{k}z}{h}\right)\mathscr{W}_\delta\left(\frac{z}{h}\right),$$

$$\Pi_{i,\delta}\left(\frac{y_i-\frac{k-i}{k}z}{h};\ \frac{z}{h}\right)=\frac{\partial}{\partial\left(\frac{z}{h}\right)}\Pi_{i-1,\delta}\left(\frac{y_i-\frac{k-i}{k}z}{h};\ \frac{z}{h}\right)\quad (i=1,\ldots,k).\tag{3.11}$$

The equalities (3.7) and (3.10) were obtained for the case $k\neq 0$. Now let $k=0$; then in the first integral in (3.6), making the substitution $y+z=y^*$, we get

$$\mathfrak{J}=(-1)^{\alpha+\nu}h^{-\nu+\alpha-1}\int_{E_1}D^\alpha g(x+y^*)\,\mathscr{M}_\delta\left(\frac{y^*}{h}\right)dy^*,\tag{3.12}$$

where

$$\mathscr{M}_\delta(t)=\int_{E_1}\mathscr{W}_\delta(\xi)\,\zeta_\delta^{(s-\alpha+\nu)}(t-\xi)\,d\xi.\tag{3.12*}$$

Reasoning similarly, from (3.9) in the case $k=0$ we get the equality

$$\mathscr{F}=(-1)^{\nu+\alpha}h^{-\nu+\alpha-1}\int_{E_1}D^\alpha g(x+y^*)\,\mathscr{N}_\delta\left(\frac{y^*}{h}\right)dy^*,\tag{3.13}$$

where

$$\mathscr{N}_\delta(t)=\int_{E_1}\mathscr{W}_\delta(\xi)\,\mathscr{W}_\delta^{(\nu-\alpha)}(t-\xi)\,d\xi.\tag{3.13*}$$

Now we pass to the proof of the fundamental identity; that is, to the proof of the required integral representation for smooth functions.

In [6] (page 83 = 95), we found the formula

$$F(0,\ \ldots,\ 0)=\sum_{e\subseteq e_n}(-1)^{|\omega^e|}\int_0^{\mathbf{h}}D_{\mathbf{v}}^{\omega^e}F(\mathbf{v}^e+\mathbf{h}^{e_n\setminus e})\,d\mathbf{v}^e,\tag{3.14}$$

which was proved with the help of a fundamental formula of integral calculus; that is, if

$$A_iF(h_1,\ \ldots,\ h_n)\equiv F(h_1,\ \ldots,\ h_n)-\int_0^{h_i}\frac{\partial}{\partial v_i}F(h_1,\ \ldots,\ h_{i-1},\ v_i,\ h_{i+1},\ \ldots,\ h_n)\,dv_i,$$

then formula (3.14) is just

$$F(0,\ \ldots,\ 0)=A_1\ldots A_nF(h_1,\ \ldots,\ h_n).\tag{3.14*}$$

From (3.3) and (3.14) it follows that

$$D^{\nu}f(\mathbf{x})=\sum_{i=1}^{2^n}(-1)^{|\omega^{e^i}|}\int_{0}^{\mathbf{h}} D_{\mathbf{v}}^{\omega^{e^i}}\widehat{(D^{\nu}f)}_{\mathbf{v}^{e^i}+\mathbf{h}^{e_n\setminus e^i}}(\mathbf{x})\,d\mathbf{v}^{e^i}. \tag{3.15}$$

On the other hand, if we assume that the vector $\nu=(\nu_1,\cdots,\nu_n)$ (where $\nu_j\geqslant 0$ and the ν_j are integers) is related to the vectors $\alpha^i=(\alpha_1^i,\cdots,\alpha_n^i)$ and $\mathbf{k}^i=(k_1^i,\cdots,k_n^i)$ $(i=1,\cdots,2^n)$ (where the $\alpha_j^i, k_j^i\geqslant 0$ are integers) in the following way:

$$\nu_j\geqslant\alpha_j^i+k_j^i\quad (j\in e_n\setminus e^i;\ i=1,\ \ldots,\ 2^n),$$

then, with the help of (3.4), (3.7), (3.10), (3.12) and (3.13), we get that

$$\begin{aligned}D_{\mathbf{h}}^{\omega^{e^i}}\widehat{(D^{\nu}f)}_{\mathbf{h}}(\mathbf{x})&=(-1)^{|\nu|+|\alpha^i|}2^{|\omega^{e^i}|}c_i\prod_{j\in e^i}h_j^{-2-\nu_j+\alpha_j^i-\omega_j^i}\prod_{j\in e_n\setminus e^i}h^{-1-\nu_j+\alpha_j^i-\omega_j^i}\\&\times\int\int\Delta^{\mathbf{k}^i}\left(\frac{\mathbf{z}^i}{\mathbf{k}^i}\right)D^{\alpha^i}f(\mathbf{x}+\mathbf{y})\,\Phi_{\delta,i}\left(\frac{\mathbf{y}}{\mathbf{h}};\ \frac{\mathbf{z}^i}{\mathbf{h}}\right)d\mathbf{y}d\mathbf{z}^i.\end{aligned} \tag{3.16}$$

Here

$$|\nu|=\nu_1+\ldots+\nu_n,\quad |\alpha^i|=\alpha_1^i+\ldots+\alpha_n^i,$$

$$c_i=\prod_{j\in e_{\mathbf{k}^i}}(k_j^i)^{k_j^i},\qquad \mathbf{z}^i\equiv z^{e_{\mathbf{k}^i}},\qquad \omega^i\equiv(\omega_1^i,\ \ldots,\ \omega_n^i)\equiv\omega^{e_{\mathbf{k}^i}},\qquad \omega=(1,\ \ldots,\ 1),$$

$$\Phi_{\delta,i}(\mathbf{y};\ \mathbf{z}^i)=\prod_{j\in e_{\mathbf{k}^i}}\Phi_{\delta_j}(y_j;\ z_j)\prod_{j\in e_n\setminus e_{\mathbf{k}^i}}\Phi_{\delta_j}(y_j),$$

where

$$\Phi_{\delta_j}(y_j)=\mathcal{M}_{\delta_j}(y_j)\qquad\text{for } j\in e^i,$$

$$\Phi_{\delta_j}(y_j)=\mathcal{N}_{\delta_j}(y_j)\qquad\text{for } j\notin e^i,$$

$$\Phi_{\delta_j}(y_j;\ z_j)=\begin{cases}Q_{k_j^i,\delta_j}(y_j;\ z_j) & \text{for } j\in e^i,\\ \Pi_{k_j^i,\delta_j}(y_j;\ z_j) & \text{for } j\notin e^i.\end{cases}$$

From formulas (3.15) and (3.16) there follows the integral representation for smooth functions which we need; that is,

$$\begin{aligned}D^{\nu}f(\mathbf{x})&=\sum_{i=1}^{2^n}(-1)^{|\omega^{e^i}|+|\nu|+|\alpha^i|}2^{|\omega^{e^i}|}c_i\prod_{j\in e_n\setminus e^i}h_j^{-1-\nu_j+\alpha_j^i-\omega_j^i}\\&\times\int_0^{\mathbf{h}}\frac{d\mathbf{v}^{e^i}}{\prod_{j\in e^i}v_j^{2+\nu_j-\alpha_j^i+\omega_j^i}}\int\int\Delta^{\mathbf{k}^i}\left(\frac{\mathbf{z}^i}{\mathbf{k}^i}\right)D^{\alpha^i}f(\mathbf{x}+\mathbf{y})\\&\times\Phi_{\delta,i}\left(\frac{\mathbf{y}}{\mathbf{v}^{e^i}+\mathbf{h}^{e_n\setminus e^i}};\ \frac{\mathbf{z}^i}{\mathbf{v}^{e^i}+\mathbf{h}^{e_n\setminus e^i}}\right)d\mathbf{y}d\mathbf{z}^i.\end{aligned} \tag{3.17}$$

We note that the integral representation (3.17) contains new results for $n = 1$. In what follows we will need the function $\widetilde{f}_{\nu,\delta}(\mathbf{x})$, which is obtained from the representation (3.17) in the following way:

$$\tilde{f}_{\nu,\delta}(\mathbf{x}) = \sum_{i=1}^{2^n} (-1)^{|\omega^{e^i}|+|\nu|+|\alpha^i|} 2^{|\omega^{e^i}|} c_i \prod_{j \in e_n \setminus e^i} h_j^{-1-\nu_j+\alpha_j^i-\omega_j^i}$$
$$\times \int_0^{\mathbf{h}} \frac{d\mathbf{v}^{e^i}}{\prod_{j\in e^i} v_j^{2+\nu_j-\alpha_j^i+\omega_j^i}} \int\int \Delta^{\mathbf{k}^i}\left(\frac{\mathbf{z}^i}{\mathbf{k}^i}; U + \square_\delta\right) D^{\alpha^i} f(\mathbf{x}+\mathbf{y})$$
$$\times \Phi_{\delta,i}\left(\frac{\mathbf{y}}{\mathbf{v}^{e^i}+\mathbf{h}^{e_n\setminus e^i}}; \frac{\mathbf{z}^i}{\mathbf{v}^{e^i}+\mathbf{h}^{e_n\setminus e^i}}\right) d\mathbf{y}\, d\mathbf{z}^i, \tag{3.18}$$

where $\square_\delta \equiv \square(\delta; H)$ is the rectangle defined previously.

In case $e_{k^i} = \phi$ (that is, when $\mathbf{k}^i$ is the zero vector), instead of

$$\Delta^{\mathbf{k}^i}\left(\frac{\mathbf{z}^i}{\mathbf{k}^i}; U + \square_\delta\right) D^{\alpha^i} f(\mathbf{x}+\mathbf{y})$$

we take

$$\chi(U + \square_\delta) D^{\alpha^i} f(\mathbf{x}+\mathbf{y}),$$

where $\chi(G)$ is the characteristic function of the set G, and it is taken at the points $\mathbf{x} + \mathbf{y}$ (the "+" sign denotes the direct sum).

REMARK. In the case $n = 1$ the representation (3.17) contains new results. In fact, for $n = 1$ we have the following representations as consequences of (3.17):

$$D^\nu f(x) = (-1)^{\nu+\alpha_1} k_1^{k_1} h^{-2-\nu+\alpha_1} \int_{E_1}\int_{E_1} \Delta^{k_1}\left(\frac{z}{k_1}\right) D^{\alpha_1} f(x+y)\, \Phi_{\delta,1}\left(\frac{y}{h}; \frac{z}{h}\right) dy\, dz$$
$$+ (-1)^{1+\nu+\alpha_2} 2 k_2^{k_2} \int_0^h \frac{dv}{v^{3+\nu-\alpha_2}} \int_{E_1}\int_{E_1} \Delta^{k_2}\left(\frac{z}{k_2}\right) D^{\alpha_2} f(x+y)\, \Phi_{\delta,2}\left(\frac{y}{v}; \frac{z}{v}\right) dy\, dz, \tag{3.19}$$

which corresponds to the case $k_1 \neq 0$, $k_2 \neq 0$;

$$D^\nu f(x) = (-1)^{\nu+\alpha_1} k_1^{k_1} h^{-2-\nu+\alpha_1} \int_{E_1}\int_{E_1} \Delta^{k_1}\left(\frac{z}{k_1}\right) D^{\alpha_1} f(x+y)\, \Phi_{\delta\,1}\left(\frac{y}{h}; \frac{z}{h}\right) dy\, dz$$
$$+ (-1)^{1+\nu+\alpha_2} 2 \int_0^h \frac{dv}{v^{2+\nu-\alpha_2}} \int_{E_2} D^{\alpha_2} f(x+y)\, \mathcal{M}_\delta\left(\frac{y}{v}\right) dy, \tag{3.20}$$

which corresponds to the case $k_1 \neq 0$, $k_2 = 0$;

$$D^{\nu}f(x)=(-1)^{\nu+\alpha_1}h^{-1-\nu+\alpha_1}\int\limits_{E_1} D^{\alpha_1}f(x+y)\mathcal{N}_\delta\left(\frac{y}{h}\right)dy$$

$$+(-1)^{1+\nu+\alpha_2}2k_2^{k_2}\int\limits_0^h \frac{dv}{v^{3+\nu-\alpha_2}}\int\limits_{E_1}\int\limits_{E_1}\Delta^{k_2}\left(\frac{z}{k_2}\right)D^{\alpha_2}f(x+y)\Phi_{\delta,2}\left(\frac{y}{v};\ \frac{z}{v}\right)dydz, \quad (3.21)$$

which corresponds to the case $k_1 = 0, k_2 \neq 0$;

$$D^{\nu}f(x)=(-1)^{\nu+\alpha_1}h^{-1-\nu+\alpha_1}\int\limits_{E_1} D^{\alpha_1}f(x+y)\mathcal{N}_\delta\left(\frac{y}{h}\right)dy$$

$$+(-1)^{1+\nu+\alpha_2}2\int\limits_0^h \frac{dv}{v^{2+\nu-\alpha_2}}\int\limits_{E_1} D^{\alpha_2}f(x+y)\mathcal{M}_\delta\left(\frac{y}{v}\right)dy, \quad (3.22)$$

which corresponds to the case $k_1 = 0, k_2 = 0$. The integral representations (3.19) and (3.20) are new results, but the representations (3.21) and (3.22) are the well-known representations of Il′in and Besov (see [17]).

§4. Proof of the fundamental results of §2

PROOF OF THEOREM 1. We prove the theorem for $1 < p_i < p < \infty$; it is proved in the other cases by similar, but simpler, arguments.

In view of the fact that $\Omega \in C(\mathrm{H})$, there exist a covering $\{U_k\}$ and a collection $\square_k$ satisfying the conditions 1) and 2) of §1. Then from the identities (3.17) and (3.18), for $\alpha^i = \bar{r}^i$, $\mathbf{k}^i = \omega^i + [\mathbf{r}^i - \bar{r}^i]$ (where $\omega^i \equiv \omega^{e_{\mathbf{r}^i}}$) $(i = 1, \cdots, 2^n)$, it follows that

$$\|D^{\nu}f\|_{L_p(\Omega_m)} \leqslant \sum_{k=1}^{N}\|D^{\nu}f\|_{L_p[(U_k+\square_k)m]} \leqslant \sum_{k=1}^{N}\|\tilde{f}_{\nu,k}\|_{L_p(E_m)}$$

$$\leqslant c\sum_{k=1}^{N}\sum_{i=1}^{2^n}\prod_{j\in e_n\setminus e^i} h_j^{-1-\nu_j+\bar{r}^i_j-\left(1-\frac{1}{p_i}\right)\omega^i_j-\gamma^i_j}|\mathfrak{J}_{i,k}|. \quad (4.1)$$

Here

$$|\mathfrak{J}_{i,k}| \leqslant \int\limits_0^{\mathbf{h}}\mathcal{P}_{i,k}\left(\prod_{j\in e^i} v_j^{-1+\varepsilon^i_j-\lambda^i_j-\gamma^i_j}\right)d\mathbf{v}^{e^i}, \quad (4.2)$$

$$\mathcal{P}_{i,k}=\int\limits_0^{(\mathbf{v}+\mathbf{h})_i}dy\int\limits_0^{(\mathbf{v}+\mathbf{h})_i}|F_{i,k}(\mathbf{x}+\mathbf{y};\ \mathbf{t}^i)|\left(\prod_{j\in e_{\mathbf{r}^i}} t_j^{\gamma^i_j}\right)d\mathbf{t}^i, \quad (4.2^*)$$

where

$$(\mathbf{v}+\mathbf{h})_i \equiv \mathbf{v}^{e^i}+\mathbf{h}^{e_n\setminus e^i},\quad 0<\gamma^i_j\leqslant\frac{1}{p_i}+(r^i_j-\bar{r}^i_j),\quad \mathbf{t}^i\equiv\mathbf{t}^{e_{\mathbf{r}^i}},$$

$$\lambda_j^i = \begin{cases} \dfrac{2}{p_i'} + \dfrac{1}{p} & \text{for } j \in e^i \cdot e_m \\ \dfrac{2}{p_i'} & \text{for } j \in e^i \setminus e_m \end{cases} \qquad \left(\frac{1}{p_i} + \frac{1}{p_i'} = 1\right),$$

$$F_{i,k}(\mathbf{y};\ \mathbf{t}^i) = \frac{\Delta^{\omega^i + [\mathbf{r}^i - \bar{\mathbf{r}}^i]}\left(\dfrac{t^i}{\omega^i + [\mathbf{r}^i - \bar{\mathbf{r}}^i]};\ U_k + \square_k\right) D^{\bar{\mathbf{r}}^i} f(\mathbf{y})}{\prod\limits_{j \in e_{\mathbf{r}^i}} t_j^{\frac{1}{p_i} + (r_j^i - \bar{r}_j^i)}},$$

the ϵ_j^i being defined in the hypotheses of the theorem.

For some fixed i let $1 < p_i < p < \infty$ and $\epsilon_j^i > 0$ $(j \in e^i)$. In this case we apply the Hölder inequality with exponents $\lambda_1 = pp_i/(p - p_i)$, $\lambda_2 = p$ and $\lambda_3 = p_i' = p_i/(p_i - 1)$ to the integral on the right-hand side of (4.2), integrate the resulting inequality over E_m with respect to the variables $x_1, \cdots, x_m$, and change the order of integration; after some simple transformations we get

$$\|\mathfrak{I}_{i,k}\|_{L_p(E_m)} \leqslant c(\mathbf{h}^{e_n \setminus e^i}) \prod_{j \in e^i} h_j^{\epsilon_j^i - \frac{1}{2}\gamma_j^i} \left[\int_0^{\mathbf{h}} \prod_{j \in e_{\mathbf{r}^i}} t_j^{\frac{1}{2}\gamma_j^i p_i} \|F_{i,k}\|_{L_{p_i}(E_n)}^{p_i} d\mathbf{t}^i\right]^{1/p_i}. \tag{4.3}$$

Consequently, for $\theta_i = p_i$,

$$\|\mathfrak{I}_{i,k}\|_{L_p(E_m)} \leqslant c(\mathbf{h}^{e_n \setminus e^i}) \prod_{j \in e^i} h_j^{\epsilon_j^i} \|f\|_{\mathscr{L}_{p_i, p_i}^{\langle \mathbf{r}^i \rangle}(U_k + \square_k)}. \tag{4.4}$$

Now let $\theta_i > p_i$; then, applying the Hölder inequality with exponents $\lambda_1 = \theta_i/p_i$, $\lambda_2 = \lambda_1/(\lambda_1 - 1)$, to the integral on the right-hand side of (4.3), we arrive at the inequality

$$\|\mathfrak{I}_{i,k}\|_{L_p(E_m)} \leqslant c(\mathbf{h}^{e_n \setminus e^i}) \prod_{j \in e^i} h_j^{\epsilon_j^i - \frac{1}{2}\gamma_j^i} \|f\|_{\mathscr{L}_{p_i, \theta_i}^{\langle \mathbf{r}^i \rangle}(U_k + \square_k)}$$
$$\times \left[\int_0^{\mathbf{h}} \frac{d\mathbf{t}^i}{\prod\limits_{j \in e_{\mathbf{r}^i}} t_j^{1 - \frac{1}{2}\gamma_j^i \lambda_2 p_i}}\right]^{1/\lambda_2 p_i} \leqslant c(\mathbf{h}^{e_n \setminus e^i}) \prod_{j \in e^i} h_j^{\epsilon_j^i} \|f\|_{\mathscr{L}_{p_i, \theta_i}^{\langle \mathbf{r}^i \rangle}(U_k + \square_k)}. \tag{4.5}$$

Now we assume that for some fixed i $(i = 1, \cdots, 2^n)$ the condition $p_i = \theta_i$ is satisfied, and, in addition, there exists some set e from $e^i \cdot e_m$ (where $e_m \equiv \{1, \cdots, m\}$) such that $\epsilon_j^i = 0$ for $j \in e$.

In this case, from (4.2) it follows that

$$|\mathfrak{I}_{i,k}| \leqslant \int_0^{\mathbf{h}} \frac{d\mathbf{v}^e}{\prod_{j\in e} v_j^{1+\frac{2}{p'_i}+\frac{1}{p}}} \int_0^{\mathbf{v}} [\cdot]\, d\mathbf{y}^e, \tag{4.6}$$

where

$$[\cdot] = \int_0^{\mathbf{h}} \left(\prod_{j\in e^i\setminus e} v_j^{-1+\varepsilon_j^i - \lambda_j^i - \gamma_j^i} \right) d\mathbf{v}^{e^i\setminus e}$$

$$\times \int_0^{(\mathbf{v}+\mathbf{h})_i} d\mathbf{y}^{e_n\setminus e} \int_0^{(\mathbf{v}+\mathbf{h})_i} \left(\prod_{j\in e^i\setminus e} t_j^{\gamma_j^i} \right) |F_{i,k}(\mathbf{x}+\mathbf{y};\, t^i)|\, dt^i. \tag{4.6*}$$

We apply the Hölder inequality with exponents $\lambda_1 = pp_i/(p - p_i)$, $\lambda_2 = p$ and $\lambda_3 = p'_i = p_i/(p_i - 1)$ to the integral $[\cdot]$ (see (4.6*)), raise the resulting inequality to the power p, and integrate with respect to the variables x_j (for all $j \in e_m\setminus e$) over $E_{|\omega e_m\setminus e|}$. After changing the order of integration and making some simple transformations, we get

$$\|[\cdot]\|_{L_p(E_{|\omega^{e_m}\setminus e|})} \leqslant c(\mathbf{h}^{e_n\setminus e^i}) \prod_{j\in e^i\setminus e} h_j^{\varepsilon_j^i} \prod_{j\in e} v_j^{1/p_i}$$

$$\times \left(\int_0^{\mathbf{h}} \|F_{i,k}(\mathbf{x}+\mathbf{y};\, \mathbf{t}^i)\|_{L_{p_i}(E_{|\omega^{e_m}\setminus e|})}^{p_i} dt^i \right)^{1/p_i} \equiv c(\mathbf{h}^{e_n\setminus e^i}) \prod_{j\in e^i\setminus e} h_j^{\varepsilon_j^i} \prod_{j\in e} v_j^{1/p'_i} \varphi(\mathbf{x}^e + \mathbf{y}^e). \tag{4.7}$$

From the inequalities (4.6) and (4.7) it follows that if we apply the generalized Minkowski inequality to the integral $\|\mathfrak{I}_{i,k}\|_{L_p(E_m)}$, we get

$$\|\mathfrak{I}_{i,k}\|_{L_p(E_m)} \leqslant c(h^{e_n\setminus e^i}) \prod_{j\in e^i\setminus e} h_j^{\varepsilon_j^i} \left\| \int_0^{\mathbf{h}} \left(\prod_{j\in e} v_j^{1-\frac{1}{p'_i}-\frac{1}{p}} \right) d\mathbf{v}^e \int_0^{\mathbf{v}} \varphi(\mathbf{x}^e + \mathbf{y}^e)\, d\mathbf{y}^e \right\|_{L_p(E_{|\omega^e|})}$$

$$\leqslant c(\mathbf{h}^{e_n\setminus e^i}) \prod_{j\in e^i\setminus e} h_j^{\varepsilon_j^i} \left\| \int_0^{\mathbf{h}} \frac{\varphi(\mathbf{y}^e)\, d\mathbf{y}^e}{\prod_{j\in e} |x_j - y_j|^{\frac{1}{p'_i}+\frac{1}{p}}} \right\|_{L_p(E_{|\omega^e|})} \tag{4.8}$$

Now applying the generalized Minkowski inequality and the Hardy-Littlewood inequality $|\omega^e|$ times in succession, we get

$$\|\mathfrak{I}_{i,k}\|_{L_p(E_m)} \leqslant c(\mathbf{h}^{e_n\setminus e^i}) \prod_{j\in e^i\setminus e} h_j^{\varepsilon_j^i} \|\varphi\|_{L_{p_i}(E_{|\omega^e|})}$$

$$\leqslant c(h^{e_n\setminus e^i}) \prod_{j\in e^i\setminus e} h_j^{\varepsilon_j^i} \|f\|_{\mathscr{L}^{\langle \mathbf{r}^i\rangle}_{p_i,p_i}(U_k+\square_k)}. \tag{4.9}$$

From the inequalities (4.5) and (4.9) it follows that

$$\|\mathfrak{J}_{i,k}\|_{L_p(E_m^\circ)} \leqslant c(\mathbf{h}^{e_n \setminus e^i}) \prod_{j \in e^i} h_j^{\varepsilon_j^i} \|f\|_{\mathscr{L}^{\langle \mathbf{r}^i \rangle}_{p_i, \theta_i}(U_k + \square_k)}. \tag{4.10}$$

The inequality (4.10) is proved by similar arguments if $\square_k$ is not in the first coordinate angle.

The inequalities (4.1) and (4.10) prove the statement

$$\|D^\nu f\|_{L_p(\Omega_m)} \leqslant c \sum_{k=1}^{N} \sum_{i=1}^{2^n} \prod_{j=1}^{n} h_j^{\varepsilon_j^i} \|f\|_{\mathscr{L}^{\langle \mathbf{r}^i \rangle}_{p_i, \theta_i}(U_k + \square_k)} \leqslant c \sum_{i=1}^{2^n} \prod_{j=1}^{h} h_j^{\varepsilon_j^i} \|f\|_{\mathscr{L}^{\langle \mathbf{r}^i \rangle}_{p_i, \theta_i}(\Omega)}, \tag{4.11}$$

which was to be proved.

PROOF OF THEOREM 3. Let $\mathscr{P}_{i,k} = \mathscr{P}_{i,k}(\mathbf{x}; v^{e^i})$ $(i = 1, \cdots, 2^n)$ be functions defined by the formula (4.2*). Further, let

$$\mathscr{M}_{i,k} = \int_{G_{\mathbf{t}}} \mathscr{P}_{i,k} \prod_{j \in e^i} v_j^{-m_j} d\mathbf{v}^{e^i}, \tag{4.12}$$

$$\mathscr{M}^*_{i,k} = \int_{G^*_{\mathbf{t}}} \mathscr{P}_{i,k} \prod_{j \in e^i} v_j^{-m^*_j} d\mathbf{v}^{e^i}. \tag{4.13}$$

Here

$$G_{\mathbf{t}} = \{\mathbf{v}^{e^i};\ 0 < v_j \leqslant t_j\ (j \in e^i \cdot e_\rho),\ \ 0 < v_j \leqslant h_j\ (j \in e^i \setminus e_\rho)\},$$

$$G^*_{\mathbf{t}} = \{\mathbf{v}^{e^i};\ t_j \leqslant v_j \leqslant h_j\ (j \in e^i \cdot e_\rho),\ \ 0 < v_j \leqslant h_j\ (j \in e^i \setminus e_\rho)\},$$

$$m_j = \begin{cases} 1 + \left(\frac{2}{p_i'} + \frac{1}{p}\right) - (\varepsilon_j^i - \rho_j) - (\rho_j - \bar\rho_j - \gamma_j^i) & \text{for } j \in e^i \cdot e_\rho, \\ 1 + \left(\frac{2}{p_i'} + \frac{1}{p}\right) - (\varepsilon_j^i - \gamma_j^i) & \text{for } j \in e^i \cdot e_m \setminus e_\rho, \\ 1 + \frac{2}{p_i'} - (\varepsilon_j^i - \gamma_j^i) & \text{for } j \in e^i \setminus e_m, \end{cases}$$

$$m_j^* = \begin{cases} 1 + \left(\frac{2}{p_i'} + \frac{1}{p}\right) - (\varepsilon_j^i - \rho_j) + (1 + [\rho_j - \bar\rho_j] - \rho_j + \bar\rho_j) + \gamma_j^i & \text{for } j \in e^i \setminus e_\rho, \\ m_j & \text{for } j \in e^i \cdot e_m \setminus e_\rho, \end{cases}$$

where $1/p_i + 6/p_i' = 1$, ϵ_j^i and ρ_j are given in Theorem 3, $[\rho_j - \bar\rho_j]$ is the integral part of $\rho_j - \bar\rho_j$, and γ_j^i are sufficiently small positive numbers.

LEMMA. *Under the conditions of Theorem* 3, *the following inequalities hold*:

$$\left(\int_0^{\mathbf{h}} \|\mathscr{M}_{i,k}\|^\theta_{L_p(E_m)} \prod_{j \in e^i \cdot e_\rho} t_j^{-1-\theta(\rho_j - \bar\rho_j)} dt_j\right)^{1/\theta} \leqslant c(\mathbf{h}^{e_n \setminus e^i}) \prod_{j \in e^i} h_j^{\varepsilon_j^i - \rho_j} \|f\|_{\mathscr{L}^{\langle \mathbf{r}^i \rangle}_{p_i, \theta_i}(U_k + \square_k)}, \tag{4.14}$$

$$\left(\int_0^h \|\mathcal{M}^*_{i,k}\|^\theta_{L_p(E_m)} \prod_{j\in e^i\cdot e_\rho} t_j^{-1+\theta(1+[\rho_j-\bar\rho_j]-\rho_j+\bar\rho_j)}\,dt_j\right)^{1/\theta}$$
$$\leqslant c\,(h^{e_n\setminus e^i})\prod_{j\in e^i} h_j^{\varepsilon^i_j-\rho_j}\|f\|_{\mathcal{L}^{\langle \mathbf{r}^i\rangle}_{p_i,\theta_i}(U_k+\square_k)}. \quad (4.15)$$

For $p_i = \theta_i = p$, this lemma is contained in [6] (see Theorems 2 and 3, §3, pp. 90 and 95 = 102 and 108), and for the case $\theta \geqslant \theta_i \geqslant p_i$ it is proved by means of arguments which are similar to those in the proof of those same Theorems 2 and 3, but the generalized Minkowski inequality, with suitable exponents, must be used several more times.

In view of the fact that $\Omega \in C(\mathbf{H})$, there exists a covering $\{U_k\}$ of the region Ω and a collection $\square_k = \square(\delta^k; \mathbf{H})$ satisfying the conditions 1) and 2) of §1. Let $\widetilde{f}_{\nu,k}(\mathbf{x})$ be a function defined by the inequality (3.18) for $\alpha^i = \bar{\mathbf{r}}^i$, $\mathbf{k}^i = \omega^i + [\mathbf{r}^i - \bar{\mathbf{r}}^i]$, $\square(\delta; \mathbf{H}) = \square_k$,

$$\Phi_{i,k} = \Phi_{\delta^k,i} \qquad (i = 1, \ldots, 2^n).$$

It is clear that

$$\|D^\nu f\|_{\mathcal{L}^{\langle\rho\rangle}_{p,\theta}(\mathfrak{L}_m)} \leqslant c\prod_{j\in e_\rho} h_j^{-\rho_j+\bar\rho_j}\|D^{\nu+\bar\rho}f\|_{L_p(\mathfrak{L}_m)}$$
$$+\left(\int_0^h \|\Delta^{\omega_\rho+[\rho-\bar\rho]}(\mathbf{t};\mathfrak{L}_m)D^{\nu+\bar\rho}f\|^\theta_{L_p(\mathfrak{L}_m)}\prod_{j\in e_\rho} t_j^{-1-\theta(\rho_j-\bar\rho_j)}\,dt_j\right)^{1/\theta}$$
$$+\left(\int_0^h \|\Delta^{\omega_\rho+[\rho-\bar\rho]}(-\mathbf{t};\mathfrak{L}_m)D^{\nu+\bar\rho}f\|^\theta_{L_p(\mathfrak{L}_m)}\prod_{j\in e_\rho} t_j^{-1-\theta(\rho_j-\bar\rho_j)}\,dt_j\right)^{1/\theta} = Q_1+Q_2+Q_3, \quad (4.16)$$

where $\omega_\rho \equiv \omega^{e_\rho}$.

With the aid of inequality (4.11), we get

$$Q_1 \leqslant c\sum_{i=1}^{2^n}\prod_{j=1}^{n} h_j^{\varepsilon^i_j-\rho_j}\sum_{k=1}^{N}\|f\|_{\mathcal{L}^{\langle\mathbf{r}^i\rangle}_{p_i,\theta_i}(U_k+\square_k)}. \quad (4.17)$$

The integrals Q_2 and Q_3 are both estimated by the same argument in terms of the right-hand side of the inequality (4.17), so it is sufficient to estimate Q_2.

From the integral representation (3.18), for $\alpha^i = \bar{\mathbf{r}}^i$, $\mathbf{k}^i = \omega^i + [\mathbf{r}^i - \bar{\mathbf{r}}^i]$ $(i = 1, \cdots, 2^n)$ it follows that

$$\tilde f_{\nu+\bar\rho,k}(\mathbf{x}) = \sum_{i=1}^{2^n}(-1)^{|\omega^{e^i}|+|\nu|+|\bar\rho|+|\bar{\mathbf{r}}^i|}2^{|\omega^{e^i}|}\prod_{j\in e_n\setminus e^i} h_j^{-1-\nu_j+\bar r^i_j-\bar\rho_j-\omega^i_j}\{\mathfrak{N}_{i,k}+\mathfrak{N}^*_{i,k}\}. \quad (4.18)$$

Here

$$\mathfrak{N}_{i,k}+\mathfrak{N}^*_{i,k}=\left(\int_0^t+\int_t^h\right)d\mathbf{v}^{e_\rho\cdot e^i}\int_0^h\prod_{j\in e^i}v_j^{-3-\nu_j-\bar\rho_j+\bar r^i_j}d\mathbf{v}^{e^i\setminus e_\rho}$$

$$\times\int\int\Delta^{\omega^i+[\mathbf{r}^i-\bar{\mathbf{r}}^i]}\left(\frac{\mathbf{z}^i}{\omega^i+[\mathbf{r}^i-\bar{\mathbf{r}}^i]};\ U_k+\square_k\right)D^{\bar{\mathbf{r}}^i}f(\mathbf{x}+\mathbf{y})\,\Phi_{i,k}d\mathbf{y}d\mathbf{z}^i. \tag{4.19}$$

From (4.18) and (4.19) it follows that

$$Q_2\leqslant\sum_{i=1}^{2^n}c_i\left(h^{e_n\setminus e^i}\right)\sum_{k=1}^{N}\left\{\left(\int_0^h\left\|\Delta^{\omega_\rho+[\rho+\bar\rho]}(\mathbf{t})\,\mathfrak{N}_{i,k}\right\|^\theta_{L_p(E_m)}\prod_{j\in e_\rho}t^{-1-\theta(\rho_j-\bar\rho_j)}dt_j\right)^{1/\theta}\right.$$

$$\left.+\left(\int_0^h\left\|\Delta^{\omega_\rho+[\rho-\bar\rho]}(\mathbf{t})\,\mathfrak{N}^*_{i,k}\right\|^\theta_{L_p(E_m)}\prod_{j\in e_\rho}\frac{dt_j}{t_j^{1+\theta(\rho_j-\bar\rho_j)}}\right)^{1/\theta}\right\}$$

$$=\sum_{i=1}^{2^n}c_i\left(\mathbf{h}^{e_n\setminus e^i}\right)\sum_{k=1}^{N}\{\mathfrak{M}_{i,k}+\mathfrak{M}^*_{i,k}\}. \tag{4.20}$$

From the inequality

$$\left\|\Delta^{\omega_\rho+[\rho+\bar\rho]}(\mathbf{t})\,\mathfrak{N}_{i,k}\right\|_{L_p(E_m)}\leqslant c\prod_{j\in e_\rho\setminus e^i}t_j^{1+[\rho_j-\bar\rho_j]}\left\|\left(\prod_{\ell\in e_\rho\setminus e^i}D_{x_j}^{1+[\rho_j-\bar\rho_j]}\right)\mathfrak{N}_{i,k}\right\|_{L_p(E_m)}$$

after some transformations, it follows that

$$\mathfrak{M}_{i,k}\leqslant c\left(\int_0^h\prod_{j\in e_\rho\setminus e^i}t_j^{-1+\theta(1+[\rho_j-\bar\rho_j]-\rho_j+\bar\rho_j)}dt_j\right)^{1/\theta}$$

$$\times\left(\int_0^h\left\|\mathscr{M}_{i,k}\right\|^\theta_{L_p(E_m)}\prod_{j\in e_\rho e^i}t_j^{-1-\theta(\rho_j-\bar\rho_j)}dt_j\right)^{1/\theta}, \tag{4.21}$$

where $\mathscr{M}_{i,k}$ is defined by formula (4.12).

Applying (4.14) of the lemma to the right-hand side of the last inequality, we arrive at the estimate

$$\mathfrak{M}_{i,k}\leqslant c_i\left(h^{e_n\setminus e^i}\right)\prod_{j\in e^i}h_j^{\varepsilon^i_j-\rho_j}\|f\|_{\mathscr{L}^{\langle\mathbf{r}^i\rangle}_{p_i,\theta_i}(U_k+\square_k)}. \tag{4.22}$$

Now we estimate the integral $\mathfrak{M}^*_{i,k}$. Using the inequality

$$\left\|\Delta^{\omega_\rho+[\rho-\bar\rho]}(\mathbf{t})\,\mathfrak{N}^*_{i,k}\right\|_{L_p(E_m)}\leqslant c\prod_{j\in e_\rho}t_j^{1+[\rho_j-\bar\rho_j]}\left\|D^{\omega_\rho+[\rho-\bar\rho]}\mathfrak{N}^*_{i,k}\right\|_{L_p(E_m)},$$

differentiating the integral $\mathfrak{N}^*_{i,k}$, taking into account the properties of the kernel $\Phi_{i,k}$, and making some transformations, we get

$$\mathfrak{M}^*_{i,k} \leqslant c \prod_{j \in e_\rho \setminus e^i} \left(\int_0^{h_j} t_j^{-1+\theta(1+[\rho_j-\bar\rho_j]-\rho_j+\bar\rho_j)}\, dt_j \right)^{1/\theta}$$
$$\times \left[\int_0^{\mathbf{h}} \| \mathcal{M}^*_{i,k} \|^\theta_{L_p(E_m)} \prod_{j \in e^i \cdot e_\rho} t_j^{-1+\theta(1+[\rho_j+\bar\rho_j]-\rho_j+\bar\rho_j)} dt_j \right]^{1/\theta}, \quad (4.23)$$

where $\mathcal{M}^*_{i,k}$ is defined by the formula (4.13).

Applying the inequality (4.15) to the integral in the square brackets on the right-hand side of (4.23), we arrive at the inequality

$$\mathfrak{M}^*_{i,k} \leqslant c_1 (\mathbf{h}^{e_n \setminus e^i}) \prod_{j \in e^i} h_j^{\varepsilon^i_j - \rho_j} \| f \|_{\mathscr{L}^{\langle \mathbf{r}^i \rangle}_{p_i,\theta_i}(U_k + \square_k)}. \quad (4.24)$$

From the inequalities (4.20), (4.22) and (4.24) it follows that

$$Q_2 \leqslant c \prod_{j=1}^n h_j^{\varepsilon^i_j - \rho_j} \| f \|_{\mathscr{L}^{\langle \mathbf{r}^i \rangle}_{p_i,\theta_i}(U_k + \square_k)}. \quad (4.25)$$

Similar arguments show that

$$Q_3 \leqslant c \prod_{j=1}^n h_j^{\varepsilon^i_j - \rho_j} \| f \|_{\mathscr{L}^{\langle \mathbf{r}^i \rangle}_{p_i,\theta_i}(U_k + \square_k)}. \quad (4.26)$$

The inequalities (4.16), (4.17), (4.25) and (4.26) prove the statement

$$\| D^\nu f \|_{\mathscr{L}^{\langle \rho \rangle}_{p,\theta}(\Omega_m)} \leqslant c \sum_{i=1}^{2^n} \prod_{j=1}^n h_j^{\varepsilon^i_j - \rho_j} \sum_{k=1}^N \| f \|_{\mathscr{L}^{\langle \mathbf{r}^i \rangle}_{p_i,\theta_i}(U_k + \square_k)},$$

from which the inequality (2.4) of Theorem 3 follows.

PROOF OF THE REMAINING THEOREMS OF §2. The proof of Theorem 2 follows the same arguments as those in the proof of Theorem 1, but in this case one must use the identities (3.17) and (3.18) for $\alpha^i = [\mathbf{r}^i]$ and $\mathbf{k}^i = \omega^{e^*_{\mathbf{r}^i}}$ ($i = 1, \cdots, 2^n$) (where $\omega = (1, \cdots, 1)$ and $e^*_{\mathbf{r}^i}$ is the set of nonintegral components of the vector $\mathbf{r}^i$).

Theorems 4 and 5 are proved by the same methods as those in the proof of Theorem 3; that is, with the aid of the integral representation for smooth functions stated in §3, for $\alpha^i = [\mathbf{r}^i]$ and $\mathbf{k}^i = \omega^{e^*_{\mathbf{r}^i}}$ ($i = 1, \cdots, 2^n$).

§5. Theorems on the extension of functions beyond the boundary of the region $\Omega \subset E_n$ to E_n, keeping the differential properties invariant

DEFINITION. *We will say that the region Ω belongs to the class $C_\epsilon(\mathbf{H})$ if $\Omega \in C(\mathbf{H})$, and, in addition, for the covering $\{U_k\}$, for some $\epsilon > 0$ the additional condition*

$$\Omega \subset \bigcup_{i=1}^{N} U_k^{(\varepsilon)}$$

is satisfied, where

$$U_k^{(\varepsilon)} = \{\mathbf{x};\ \mathbf{x} \in U_k,\ \rho(\mathbf{x},\ \Omega \setminus U_k) > \varepsilon\}.$$

THEOREM 6. *Let $f(\mathbf{x})$ belong to the space*

$$\bigcap_{i=1}^{2^n} \mathscr{L}_{p_i, \theta_i}^{\langle \mathbf{r}^i \rangle}(\Omega),$$

where $\Omega \in C_\epsilon(\mathbf{H})$, and let all of the conditions of Theorem 3 (*for the case $m = n$*) *be satisfied.*

Then the function $D^\nu f(\mathbf{x})$ (which, as a consequence of Theorem 3, *belongs to the space $\mathscr{L}_{p,\theta}^{\langle \rho \rangle}(\Omega)$) can be extended beyond the boundary of the region $\Omega \subset E_n$ to E_n, keeping the differential properties invariant; that is, it is possible to construct a function $\tilde{f}_\nu(\mathbf{x})$, defined on E_n and belonging to the space $\mathscr{L}_{p,\theta}^{\langle \rho \rangle}(E_n)$, which coincides on Ω with the function $D^\nu f$ and which satisfies the inequality*

$$\|\tilde{f}_\nu\|_{\mathscr{L}_{p,\theta}^{\langle \rho \rangle}(E_n)} \leqslant c \sum_{i=1}^{2^n} \|f\|_{\mathscr{L}_{p_i, \theta_i}^{\langle \mathbf{r}^i \rangle}(\Omega)},$$

where c is a constant which does not depend on f.

The proof of this theorem is carried out by means of arguments used by the author in [18] (see Theorem 8), but the integral representation (3.18), given in §3, is used in the proof of this theorem. The idea of the proof of this theorem was borrowed from [17].

§6. Weight spaces

If in the definition of the space (1) and (2) (see the Introduction) instead of the norm $\|\cdot\|_{L_{p_i}(\Omega)}$ we take the norm

$$\|\cdot\|_{L_{p_i, \alpha}(\Omega)} = \left(\int_\Omega [\rho(\mathbf{x})]^{\alpha p_i} |\cdot|^{p_i} d\mathbf{x} \right)^{1/p_i}, \tag{6.1}$$

where $\rho(\mathbf{x})$ is the distance from the point $\mathbf{x}$ to the boundary of the region Ω, then this defines new weight spaces of functions, which we denote by

$$\bigcap_{i=1}^{2^n} L_{p_i, \alpha}^{\langle \mathbf{r}^i \rangle}(\Omega), \qquad \bigcap_{i=1}^{2^n} \mathscr{L}_{p_i, \theta_i, \alpha}^{\langle \mathbf{r}^i \rangle}(\Omega), \tag{6.2}$$

respectively.

A complete theory of the weight spaces (6.2) can be constructed.

In particular, if we take $\alpha \geqslant 0$ and $m = n$ (that is, $\Omega_m = \Omega$), then Theorems 1–5 and Corollaries 1–2 for the spaces (6.2) are stated in the same way as the corresponding

Theorems 1–5 and Corollaries 1–2, § 2, except that, in the appropriate places, instead of the norms in the spaces

$$L_p(\Omega), \qquad \mathcal{L}^{\langle\rho\rangle}_{p,\theta}(\Omega), \qquad L^{\langle\rho\rangle}_p(\Omega), \qquad \mathcal{L}^{\langle r^i\rangle}_{p_i,\theta_i}(\Omega), \qquad L^{\langle r^i\rangle}_{p_i}(\Omega)$$

one must take the corresponding norms in the weight spaces

$$L_{p,\alpha}(\Omega), \qquad \mathcal{L}^{\langle\rho\rangle}_{\substack{p,\theta\\ \alpha}}(\Omega), \qquad L^{\langle\rho\rangle}_{p,\alpha}(\Omega), \qquad \mathcal{L}^{\langle r^i\rangle}_{\substack{p_i,\theta_i\\ \alpha}}(\Omega), \qquad L^{\langle r^i\rangle}_{p_i,\alpha}(\Omega).$$

For the case of the spaces $S^{\mathbf{r}}_{p,\alpha}W(\overset{+}{E}_n)$ and $S^{\mathbf{r}}_{p_\alpha,\theta}B(\overset{+}{E}_n)$, see [20, 21].

BIBLIOGRAPHY

1. S. M. Nikol′skiĭ, *Functions with dominant mixed derivatives satisfying a multiple Hölder condition*, Sibirsk. Mat. Ž. **4** (1963), 1342–1346; English transl., Amer. Math. Soc. Transl. (2) **102** (1973), 27–51. MR **28** #3322.

2. ———, *A variational problem*, Mat. Sb. **62 (104)** (1963), 53–75; English transl., Amer. Math. Soc. Transl. (2) **51** (1966), 132–154. MR **28** #1391.

3. N. S. Bahvalov, *Imbedding theorems for classes of functions with a number of bounded derivatives*, Vestnik Moskov. Univ. Ser. I Mat. Meh. **1963**, no. 3, 7–16. (Russian) MR **28** #4067.

4. V. P. Il′in, *On inequalities between the norms of partial derivatives of functions of several variables*, Trudy Mat. Inst. Steklov. **84** (1965), 144–173 = Proc. Steklov Inst. Math. **84** (1965), 161–190. MR **33** #448.

5. A. D. Džabrailov, *On some function spaces. Direct and inverse imbedding theorems*, Dokl. Akad. Nauk SSSR **159** (1964), 254–257 = Soviet Math. Dokl. **5** (1964), 1457–1462. MR **31** #3867.

6. ———, *On the theory of imbedding theorems*, Trudy Mat. Inst. Steklov. **89** (1967), 80–118 = Proc. Steklov Inst. Math. **89** (1967), 92–135. MR **36** #676.

7. T. I. Amanov, *Representation and imbedding theorems for the functions spaces $S^{\mathbf{r}}_{p,\theta}B(R_n)$ and $S^{\mathbf{r}}_{p^*,\theta}B$ $(0 \leq x_j \leq 2\pi; j = 1, 2, \ldots, n)$*, Trudy Mat. Inst. Steklov. **77** (1965), 3–34 = Proc. Steklov Inst. Math. **77** (1965), 3–36. MR **33** #1718.

8. P. I. Lizorkin and S. M. Nikol′skiĭ, *A classification of differentiable functions in some fundamental spaces with dominant mixed derivative*, Trudy Mat. Inst. Steklov. **77** (1965), 143–167 = Proc. Steklov Inst. Math. **77** (1965), 160–187. MR **33** #450.

9. O. V. Besov, *Classes of functions with a generalized mixed Hölder condition*, Trudy Mat. Inst. Steklov. **105** (1969), 21–29 = Proc. Steklov Inst. Math. **105** (1969), 24–34.

10. O. V. Besov and A. D. Džabrailov, *Interpolation theorems for some spaces of differentiable functions*, Trudy Mat. Inst. Steklov. **105** (1969), 15–20 = Proc. Steklov Inst. Math. **105** (1969), 16–23.

11. Ja. S. Bugrov, *Approximation of a class of functions with a dominant mixed derivative*, Mat. Sb. **64 (106)** (1964), 410–418. (Russian) MR **29** #2587.

12. V. P. Il′in, *On inequalities between norms of partial derivatives of functions of several variables*, Dokl. Akad. Nauk SSSR **150** (1963), 975–977 = Soviet Math. Dokl. **4** (1963), 784–787. MR **28** #2177.

13. ———, *Conditions for the validity of inequalities between L_p-norms of partial derivatives of functions of several variables*, Imbedding Theorems and Their Applications, Proc. All-Union Sympos. Imbedding Theorems (Baku, 1966), "Nauka", Moscow, 1970, pp. 95–98. (Russian)

14. V. P. Il'in, *Conditions of validity of an inequality between L_p-norms of partial derivatives of functions of several variables*, Trudy Mat. Inst. Steklov. **96** (1968), 205–242 = Proc. Steklov Inst. Math. **96** (1968), 259–306. MR **42** #6605.

15. A. D. Džabrailov, *Interpolation spaces of differentiable functions defined on an n-dimensional domain*, Dokl. Akad. Nauk SSSR **187** (1969), 259–262 = Soviet Math. Dokl. **10** (1969), 833–837.

16. A. D. Džabrailov and V. P. Il'in, *Inequalities between norms of partial derivatives of functions of several variables*, Imbedding Theorems and Their Applications, Proc. All-Union Sympos. Imbedding Theorems (Baku, 1966), "Nauka", Moscow, 1970, pp. 73–75. (Russian)

17. O. V. Besov and V. P. Il'in, *Natural extension of the class of regions in imbedding theorems*, Mat. Sb. **75** (**117**) (1968), 483–495 = Math. USSR Sb. **4** (1968), 445–456. MR **37** #1987.

18. A. D. Džabrailov, *Imbedding theorems for spaces of functions whose mixed derivatives satisfy a multiple-integral Hölder condition*, Trudy Mat. Inst. Steklov. **117** (1972), 113–138 = Proc. Steklov Inst. Math. **117** (1972), 135–164.

19. S. L. Sobolev, *Applications of functional analysis in mathematical physics*, Izdat. Leningrad. Gos. Univ., Leningrad, 1950; English transl., Transl. Math. Monographs, vol. 7, Amer. Math. Soc., Providence, R. I., 1963. MR **14**, 565; **29** #2624.

20. A. D. Džabrailov, *Weight spaces of functions with dominant mixed derivatives and properties of functions from these spaces*, Dokl. Akad. Nauk SSSR **175** (1967), 276–279 = Soviet Math. Dokl. **8** (1967), 835–839. MR **35** #5921.

21. ———, *Imbedding theorem inequalities for weight classes of functions*, Imbedding Theorems and Their Applications, Proc. All-Union Sympos. Imbedding Theorems (Baku, 1966), "Nauka", Moscow, 1970, pp. 67–73. (Russian)

Translated by H. T. JONES

Trudy Mat. Inst. Steklov.
117 (1972)

Proc. Steklov. Inst. Math.
117 (1972)

ASYMPTOTIC DISTRIBUTION OF THE EIGENVALUES AND EIGENFUNCTIONS OF A CLASS OF SINGULAR ELLIPTIC OPERATORS

I. A. KIPRIJANOV

UDC 518.333

Abstract. In a domain situated in the half-space $y = 0$, a linear differential operator $\mathfrak{L}$ containing a differential Bessel operator is considered. Formulas are given showing the asymptotic distribution of the eigenvalues and eigenfunctions of the operator $\mathfrak{L}$. These formulas are generalizations to degenerate equations of results of Gårding for regular elliptic operators.

Bibliography: 7 items.

The asymptotic distribution of the eigenvalues and eigenfunctions of regular elliptic operators has by now been studied in sufficient detail (see, for example, the survey by Clark [1]). The analogous problem for singular elliptic operators has been little studied. In this paper we study the distribution of eigenvalues and eigenfunctions of elliptic operators $\mathfrak{L}$ containing a singular differential Bessel operator. The asymptotic formulas obtained are generalizations of the corresponding asymptotic formulas obtained by L. Gårding [2] for regular elliptic operators. The paper consists of three sections.

In the first section the estimates for the fundamental solutions of B-elliptic operators are proved [3]. For the case of constant coefficients the proof uses the Fourier-Bessel transform, and in the case of variable coefficients it uses the Levi method. In the latter case, as a preliminary, a real parameter t is introduced in an appropriate manner into the operator.

In the second and third sections the Green's function G_t corresponding to the operator $\mathfrak{L}_t = \mathfrak{L} + t$ is constructed by the parametrix method, and the asymptotic behavior of G_t as $t \to +\infty$ is investigated. The asymptotic formulas found for G_t in conjunction with Tauberian theorems allow one to give asymptotic formulas for the distribution of eigenvalues and eigenfunctions of B-elliptic operators.

1. Let R_{n+1} denote the $\overline{n+1}$-dimensional Euclidean space of points (x, y), where $x = (x_1, \cdots, x_n)$, $x_{n+1} = y$. We denote by R^+_{n+1} the open half-space

AMS (MOS) subject classifications (1970). Primary 35P20, 47F05; Secondary 35B45.

$\{(x, y); y > 0\}$. Let $C_0^\infty(\bar{R}_{n+1}^+)$ denote the collection of infinitely differentiable functions, even with respect to y, and having compact support contained in $\bar{R}_{n+1}^+$. For any function $f \in C_0^\infty(\bar{R}_{n+1}^+)$ the Fourier-Bessel transform is defined by the formula

$$\varphi(\xi, \eta) = \int_{R_{n+1}^+} f(x, y) e^{ix\xi} j_\nu(y\eta) y^{2\nu+1} dx dy, \tag{1.1}$$

where $x\xi = x_1\xi_1 + \cdots + x_n\xi_n$, and the number ν is fixed and larger than $-\frac{1}{2}$. The inverse transform has the following form:

$$f(x, y) = C_\nu \int_{R_{n+1}^+} \varphi(\xi, \eta) e^{ix\xi} j_\nu(y\eta) \eta^{2\nu+1} d\xi d\eta, \tag{1.2}$$

where $C_\nu = [(2\pi)^n 2^{2\nu}\Gamma^2(\nu + 1)]^{-1}$ For each N we have $\varphi(\xi, \eta) = O((1 + \sqrt{\xi^2 + \eta^2})^{-N})$, where $\xi = (\xi_1, \cdots, \xi_n)$.

Now let $\psi(\xi, \eta)$ be a locally integrable function such that $\psi(\xi, \eta) = O(1 + \sqrt{\xi^2 + \eta^2})^N$ for some N. Put

$$l(f) = C_\nu \int_{R_{n+1}^+} \psi(\xi, \eta) \varphi(\xi, \eta) \eta^{2\nu+1} d\xi d\eta; \tag{1.3}$$

this defines a linear functional whose domain of definition is $C_0^\infty(\bar{R}_{n+1}^+)$. This functional is called the generalized inverse Fourier-Bessel transform of the function $\psi(\xi, \eta)$. And if

$$C_\nu \int_{R_{n+1}^+} |\psi(\xi, \eta)| \eta^{2\nu+1} d\xi d\eta = C < \infty,$$

then the function

$$l(x, y) = C_\nu \int_{R_{n+1}^+} \psi(\xi, \eta) e^{-ix\xi} j_\nu(\eta y) \eta^{2\nu+1} d\xi d\eta$$

is continuous and $|l| \leqslant C$. In this case

$$l(f) = \int_{R_{n+1}^+} l(x, y) f(x, y) y^{2\nu+1} dx dy. \tag{1.4}$$

Let Ω^+ be a domain in R_{n+1}^+ contiguous with the hyperplane $y = 0$. We will say that l is a function $l(x, y)$ in Ω^+ if there exists a locally integrable function $l(x, y)$ in Ω^+ for which (1.4) is satisfied for any $f \in C_0^\infty(\Omega^+)$. By $C_0^\infty(\Omega^+)$ we mean the collection of infinitely differentiable functions, even with respect to y, having compact support contained in Ω^+ and "outgoing" in the appropriate section of the hyperplane $y = 0$.

Put

$$\mathcal{D}^\alpha = \mathcal{D}_x^\alpha = i^{|\alpha|} \frac{\partial^{|\alpha|}}{\partial x_1^{\alpha_1} \dots \partial x_n^{\alpha_n}} \qquad (|\alpha| = \alpha_1 + \dots + \alpha_n)$$

$$\mathcal{B}_y = i^2 \left(\frac{\partial^4}{\partial y^2} + \frac{2\nu+1}{y} \frac{\partial}{\partial y} \right) \qquad (2\nu + 1 > 0,\ y > 0).$$

By $\mathfrak{D}_x^\alpha \mathfrak{B}_y^\beta l$ for l we mean the expression

$$\mathcal{D}_x^\alpha \mathcal{B}_y^\beta l(f) = C_\nu \int_{R_{n+1}^+} \xi^\alpha \eta^{2\beta} \psi(\xi, \eta) \varphi(\xi, \eta) \eta^{2\nu+1} d\xi d\eta. \tag{1.5}$$

The product λl of a polynomial λ and l is defined by the formula

$$\lambda l(f) = C_\nu \int_{R_{n+1}^+} \psi(\xi, \eta) \lambda(\mathcal{D}_\xi, \mathcal{B}_\eta) \varphi(\xi, \eta) \eta^{2\nu+1} d\xi d\eta, \tag{1.6}$$

where $\lambda(\mathfrak{D}_\xi, \mathfrak{B}_\eta) = \lambda(i\partial/\partial\xi_1, \cdots, i\partial/\partial\xi_n, i^2\mathfrak{B}_\eta)$. The usual method (see [4]) proves that if the product of l and the polynomial λ does not become zero as a function of (x, y), then $l = b/\lambda$ is also a function. If all the derivatives of l are functions, then l is infinitely differentiable and its usual derivatives are defined by

$$\mathcal{D}_x^\alpha \mathcal{B}_y^\beta l(f) = \int_{R_{n+1}^+} \mathcal{D}_x^\alpha \mathcal{B}_y^\beta l(x, y) f(x, y) y^{2\nu+1} dx dy. \tag{1.7}$$

The following assertion is valid.

LEMMA. *Let $p(\xi, \eta^2)$ be a polynomial of degree μ whose coefficients are majorized by a constant c_1, and let*

$$|p(\xi, \eta^2)| \geqslant c_2 (1 + \sqrt{\xi^2 + \eta^2})^\mu.$$

Then the generalized inverse Fourier-Bessel transform of $1/p(\xi, \eta^2)$ is a function $P(x, y)$ infinitely differentiable everywhere except at the origin, and we have the following estimates:

$$|\mathcal{D}^\alpha \mathcal{B}^\beta P(x, y)| \leqslant C l_{|\alpha|+2\beta}(x, y)(1 + r^N)^{-1}, \tag{1.8}$$

where

$$l_{|\alpha|+2\beta}(x, y) = 1, \quad \textit{if} \quad \mu - |\alpha| - 2\beta - (n + 1 + 2\nu + 1) > 0;$$
$$l_{|\alpha|+2\beta} = r^{\mu - |\alpha| - 2\beta - (n+1+2\nu+1) - \varepsilon}, \quad \textit{if} \quad \mu - |\alpha| - 2\beta - (n + 1 + 2\nu + 1) \leqslant 0.$$

REMARK. Here $r^2 = \sum_{i=1}^n x_i^2 + y^2$, the numbers N and $0 < \epsilon < 1$ are arbitrary, and the number C depends on $c_1, c_2, |\alpha|, \beta, N$ and ϵ but does not depend on the polynomial p.

PROOF. Let λ be a polynomial. Then

$$\lambda \mathcal{D}^\alpha \mathcal{B}^\beta P(l) = C_\nu \int_{R_{n+1}^+} \xi^\alpha \eta^{2\beta} p^{-1}(\xi, \eta^2) \lambda(\mathcal{D}_\xi, \mathcal{B}_\eta) \varphi(\xi, \eta) \eta^{2\nu+1} d\xi d\eta.$$

Because of the restrictions imposed on $p(\xi, \eta^2)$ we can integrate by parts in this formula. After integration by parts we obtain

$$\lambda\mathcal{D}^\alpha\mathcal{B}^\beta P(l) = C_\nu \int_{R_{n+1}^+} \varphi\lambda(-\mathcal{D}_\xi, \mathcal{B}_\eta)(\xi^\alpha\eta^{2\beta}p^{-1}(\xi, \eta^2))\eta^{2\nu+1}d\xi d\eta.$$

Let λ be a homogeneous polynomial of degree $\tau \leqslant N$ and let the absolute value of its coefficients not exceed unity. Then we have

$$|\lambda(-\mathcal{D}_\xi, \mathcal{B}_\eta)(\xi^\alpha\eta^{2\beta}p^{-1}(\xi, \eta^2))| \leqslant C(1+r)^{|\alpha|+2\beta-\mu-\tau},$$

where C is a constant and $r^2 = \xi^2 + \eta^2$. If $\mu - |\alpha| - 2\beta > n + 1 + 2\nu + 1$ and λ is the above suitably chosen polynomial, then

$$|\lambda\mathcal{D}^\alpha\mathcal{B}^\beta| \leqslant C.$$

This proves the first part of the lemma.

Now let $\mu - |\alpha| - 2\beta \leqslant n + 1 + 2\nu + 1$. Let λ be a homogeneous polynomial of degree $\tau + n + 1 + 2\nu + 1 + |\alpha| + 2\beta - \mu$, where $\tau + 2\nu$ is an integer. Then we have

$$|\lambda(-\mathcal{D}_\xi, \mathcal{B}_\eta)(\xi^\alpha\eta^{2\beta}p^{-1}(\xi, \eta^2))| \leqslant C(1+r)^{-(n+1+2\nu+1)-\tau},$$

where $r^2 = \xi^2 + \eta^2$ and N is chosen so that $[\tau] = N + 1$. If $\lambda\mathfrak{D}^\alpha\mathfrak{B}^\beta P$ is a function, then

$$\lambda\mathcal{D}^\alpha\mathcal{B}^\beta P = C_\nu \int_{R_{n+1}^+} e^{-ix\xi}j_\nu(\eta y)\lambda(-\mathcal{D}_\xi, \mathcal{B}_\eta)(\xi^\alpha\eta^{2\beta}p^{-1}(\xi, \eta^2))\eta^{2\nu+1}d\xi d\eta.$$

The left-hand side of the latter formula becomes zero for $x = y = 0$. Therefore on the right, in the expression under the integral sign, we can replace $e^{-ix\xi}j_\nu(\eta y)$ by $e^{-ix\xi}j^\nu(\eta y) - 1$. If we take into account that $\nu > -½$, then, using the properties of the Bessel function j_ν, we will have

$$|e^{-ix\xi}j_\nu(\eta y) - 1| \leqslant C(\sqrt{x^2+y^2})^{1-\varepsilon}(\sqrt{\xi^2+\eta^2})^{1-\varepsilon}.$$

So for $\lambda\mathfrak{D}^\alpha\mathfrak{B}^\beta P$ we have the estimate

$$|\lambda\mathcal{D}^\alpha\mathcal{B}^\beta P| \leqslant C(\sqrt{x^2+y^2})^{1-\varepsilon}.$$

Therefore the second part of the lemma holds.

Now consider the linear differential operator $p(-\mathfrak{D}_x, \mathfrak{B}_y)$, where p is the polynomial occurring in the condition of the lemma. A fundamental solution of the differential operator $p(-\mathfrak{D}_x, \mathfrak{B}_y)$ is a distribution $\mathfrak{E}(x, y) \in [C_0^\infty(\overline{R}_{n+1}^+)]'$ satisfying in $\overline{R}_{n+1}^+$ the equation

$$p(-\mathcal{D}_x, \mathcal{B}_y)\mathcal{E}(x, y) = \delta(x, y), \tag{1.9}$$

where $\delta(x, y)$ is the Dirac function in the half-space $\bar{R}^+_{n+1}$. The definition of the Dirac δ-function occurring here may be found, for example, in [3]. In the same place the behavior of the fundamental solution $\mathcal{E}$ is studied when $\mathcal{E} \in S'_B \subset [C_0^\infty(\bar{R}^+_{n+1})]'$ (for the definition and properties of S_B and S'_B see the same place).

Equation (1.9) means that for any $f \in C_0^\infty(\bar{R}^+_{n+1})$

$$\int\limits_{R^+_{n+1}} \mathcal{E}(x, y)\, p(\mathcal{D}_x, \mathcal{B}_y) f(x, y)\, y^{2\nu+1}\, dxdy = \int\limits_{R^+_{n+1}} \delta(x, y) f(x, y)\, y^{2\nu+1} dxdy = f(0, 0). \tag{1.10}$$

Thus the fundamental solution $\mathcal{E}(x, y)$ defined here has a singularity at the origin. To obtain a fundamental solution $\mathcal{E}$ with a singularity at $(x, 0)$ $(x > 0)$ we must translate along the hyperplane $y = 0$. The relation (1.10) now take the following form:

$$\int\limits_{R^+_{n+1}} \mathcal{E}(x-u, y)\, p(\mathcal{D}_u, \mathcal{B}_y) f(u, y)\, y^{2\nu+1} dudy = f(x, 0). \tag{1.11}$$

A fundamental solution $\mathcal{E}$ with a singularity at (x, y) $(y > 0)$ is obtained from $\mathcal{E}(x-u, y)$ by applying the operation of generalized translation to the variable y. In this case we have

$$\int\limits_{R^+_{n+1}} T^v_y \mathcal{E}(x-u, y)\, p(\mathcal{D}_u, \mathcal{B}_v) f(u, v)\, v^{2\nu+1} dudv = f(x, y), \tag{1.12}$$

where T^v_y is a generalized translation operator defined by the formula

$$T^v_y \varphi(y) = \frac{\Gamma(\nu+1)}{\Gamma\left(\frac{1}{2}\right)\Gamma\left(\nu+\frac{1}{2}\right)} \int\limits_0^\pi \varphi\left(\sqrt{y^2+v^2-2yv\cos\alpha}\right) \sin^{2\nu}\alpha\, d\alpha.$$

Taking into account that $\mathcal{B}_v T^v_y = \mathcal{B}_y T^v_y$ and that the operators $\mathcal{B}_y$ and T^v_y commute (see [3, 5]), we see that (1.12) can also be written in the form

$$\int\limits_{R^+_{n+1}} T^v_y \delta(x-u, y) f(u, v)\, v^{2\nu+1} dudv = f(x, y). \tag{1.13}$$

The connection between $\mathcal{E}$ and p may now be symbolically expressed in the following way with the help of the inverse Fourier-Bessel transform:

$$\mathcal{E}(x, y) \sim C_\nu \int\limits_{R^+_{n+1}} e^{-ix\xi} j_\nu(\eta y)\, p^{-1}(\xi, \eta^2)\, \eta^{2\nu+1} d\xi d\eta. \tag{1.14}$$

For the corresponding translations we have

$$T^v_y \mathcal{E}(x-u, y) \sim C_\nu \int\limits_{R^+_{n+1}} e^{-i(x-u)\xi} T^v_y j_\nu(\eta y)\, p^{-1}(\xi, \eta^2)\, \eta^{2\nu+1} d\xi d\eta, \tag{1.15}$$

where $T_y^v j_\nu(\eta y) = j_\nu(\eta y) j_\nu(\eta v)$.

With the help of the lemma proved above we obtain the following estimates for the fundamental solution. Let the singularity of $\mathfrak{E}$ be at the origin.

THEOREM. *Let the conditions of the lemma be satisfied. The the following estimates are valid:*

$$|\mathscr{D}_x^\alpha \mathscr{B}_y^\beta \mathscr{E}(x, y)| \leqslant C r^{1-\varepsilon-(n+1+2\nu+1)} (1 + r^N)^{-1}, \tag{1.16}$$

if $|\alpha| + 2\beta < \mu$;

$$|\mathscr{D}_x^\alpha \mathscr{B}_y^\beta \mathscr{E}(x, y)| \leqslant C r^{-\varepsilon-(n+1+2\nu+1)} (1 + r^N)^{-1},$$

if $|\alpha| + 2\beta \leqslant \mu$, *where the constant* C *does not depend on the polynomial* p,

$$r^2 = x^2 + y^2.$$

PROOF. First let $|\alpha| + 2\beta < \mu$. Let λ be a homogeneous polynomial of degree $\tau + n + 1 + 2\nu + 1$, where $\tau + 2\nu$ is an integer. Choose $N = [\tau]$. Then we have

$$|\lambda(-\mathscr{D}_\xi, \mathscr{B}_\eta)(\xi^\alpha \eta^{2\beta} p^{-1}(\xi, \eta^2))| \leqslant C(1 + r_1)^{|\alpha|+2\beta-\mu-(\tau+n+1+2\nu+1)}.$$

Consequently for the corresponding derivatives we obtain

$$|\mathscr{D}_x^\alpha \mathscr{B}_y^\beta \mathscr{E}(x, y)| \leqslant \frac{C}{(1 + r^N)} r^{1-\varepsilon-(n+1+2\nu+1)} \int_{R_{n+1}^+} \frac{\eta^{2\nu+1} d\xi d\eta}{(1 + r_1)^{n+1+2\nu+1+\mu-|\alpha|-2\beta+\varepsilon-1}},$$

where $r_1^2 = \xi^2 + \eta^2$ and $r^2 = x^2 + y^2$. For the proof of the second part it is necessary to take $N + 1 = [\tau]$.

We note the following circumstances. Under the translation of the singularity of the fundamental solution $\mathfrak{E}$ along the hyperplane $y = 0$ the inequality (1.16) is preserved. But if the singularity is translated to the interior of the domain R_{n+1}^+, the latter being achieved using the generalized translation operator, then the singularity in the estimates (1.16) is removed. In this case $2\nu + 1 = 0$ and the operator $p(-\mathfrak{D}_x, \mathfrak{B}_y)$ behaves like a usual elliptic operator.

Let τ be a real parameter. Consider the differential operator of the series

$$b = b(\tau; x, y; \mathscr{D}_x, \mathscr{B}_y) = \sum_{|\alpha|+2\beta\leqslant\mu} b_{\alpha,\beta}(\tau; x, y)\tau^{-|\alpha|-2\beta} \mathscr{D}_x^\alpha \mathscr{B}_y^\beta,$$

where $b_{\alpha,\beta}(\tau; x, y)$ is a polynomial with respect to τ^{-1} with infinitely differentiable coefficients in Ω^+. We assume that the polynomial with respect to (ξ, η)

$$b_0(\tau; x, y; \xi, \eta^2) = b(\tau; x, y; \tau\xi, (\tau\eta)^2) = \sum_{|\alpha|+2\beta=\mu} b_{\alpha,\beta}(\tau; x, y)\xi^\alpha \eta^{2\beta}$$

has the property that

$$b_0^{-1}(\infty;\ x,\ y;\ \xi,\ \eta^2) = O(1)(1 + (\sqrt{\xi^2 + \eta^2})^{\mu})^{-1} \tag{1.17}$$

uniformly in suitable compact subsets of Ω^+ up to the hyperplane $y = 0$. A concrete form of the polynomial will be mentioned later.

Formally the adjoint operator is defined by the integral identity

$$\int_{R_{n+1}^+} b^* f(x,\ y)\, g(x,\ y)\, y^{2\nu+1} dx dy = \int_{R_{n+1}^+} f(x,\ y)\, bg(x,\ y)\, y^{2\nu+1} dx dy$$

for any $f, g \in C_0^\infty$.

The operator formally adjoint to b will have the form

$$b^*(\tau;\ x,\ y;\ \mathscr{D}_x,\ \mathscr{B}_y) = \sum_{|\alpha|+2\beta \leqslant \mu} b^*_{\alpha,\beta}(\tau;\ x,\ y)\, \tau^{-|\alpha|-2\beta} \mathscr{D}_x^{\alpha} \mathscr{B}_y^{\beta},$$

where b^* is a polynomial in τ^{-1}. The polynomial $b_0^*(\infty; x, y; \xi, \eta^2)$ satisfies (1.17); thus

$$b_0^*(\infty;\ x,\ y;\ \xi,\ \eta^2) = b_0(\infty;\ x,\ y;\ -\xi,\ \eta^2).$$

Let U^+ be any open subset of Ω^+ contiguous to the hyperplane $y = 0$ and entirely contained in Ω^+. For sufficiently many values of the parameter τ it is possible to construct a fundamental solution Γ of the differential operator b; that is, a function $\Gamma(\tau; x, y; u, v)$, defined in $U^+ \times U^+$ and satisfying the equation

$$\int_{R_{n+1}^+} \Gamma(\tau;\ u,\ v;\ x,\ y)\, b^*(\tau;\ x,\ y;\ \mathscr{D}_x,\ \mathscr{B}_y)\, f(x,\ y)\, y^{2\nu+1} dx dy = f(u,\ v)$$

for any suitable infinitely differentiable finite function f. The point (u, v) is called the pole of Γ.

If the pole occurs in the interior of the domain; that is, $v > 0$, then the following estimates for Γ is valid:

$$\begin{aligned} \Gamma(\tau;\ u,\ v;\ x,\ y) &= O(1)\, l_0(\tau(x-u),\ \tau(y-v)) \\ &\times (1 + (\sqrt{|\tau(x-u)|^2 + |\tau(y-v)|^2})^N)^{-1}\, \tau^{n+1}, \end{aligned} \tag{1.18}$$

where $l_0(z) = |z|^{\mu-n-1-\epsilon}$ if $\mu - n - 1 \leqslant 0$, and $l_0(z) = 1$ otherwise.

For the derivatives we have estimates of the form

$$\begin{aligned} &\mathscr{D}_x^{\alpha} \mathscr{B}_y^{\beta} \Gamma(\tau;\ u,\ v;\ x,\ y) = O(1)\, \tau^{|\alpha|+2\beta+1-\varepsilon} \\ &\times (\sqrt{|x-u|^2 + |y-v|^2})^{1-\varepsilon-n-1} (1 + (\sqrt{|\tau(x-u)|^2 + |\tau(y-v)|^2})^N)^{-1} \end{aligned} \tag{1.19}$$

for $|\alpha| + 2\beta < \mu$ and $v > 0$. These estimates coincide with the corresponding estimates obtained by Gårding [2] for usual regular elliptic operators. But if the pole (u, v) occurs on the hyperplane $v = 0$, then the following estimates hold for the fundamental solution Γ:

$$\Gamma(\tau;\ u,\ 0;\ x,\ y)=O(1)\,\tau^{n+1+2\nu+1}\times l_0(\tau(x-u),\ \tau y)\,(1+(\sqrt{|\tau(x-u)|^2+\tau^2y^2})^N)^{-1}, \tag{1.20}$$

where $l_0(z)=|z|^{\mu-(n+1+2\nu+1)-\epsilon}$, if $\mu-(n+1+2\nu+1)\leqslant 0$; $l_0(z)=1$ otherwise. With regard to the derivatives we have for $v=0$

$$\mathscr{D}_x^\alpha\mathscr{B}_y^\beta\Gamma(\tau;\ u,\ 0;\ x,\ y)=O(1)\,\tau^{|\alpha|+2\beta+1-\varepsilon}\times(\sqrt{|x-u|^2+y^2})^{1-\varepsilon-(n+1+2\nu+1)}\,(1+(\sqrt{|\tau(x-u)|^2+\tau^2y^2})^N)^{-1} \tag{1.21}$$

for $|\alpha|+2\beta<\mu$. In (1.18)–(1.21) the numbers $N\geqslant 0$ and $0<\epsilon<1$ are arbitrary and fixed. The estimates hold uniformly in $U^+\times U^+$ for sufficiently large τ.

It will also be shown that if $\mu>n+1+2\nu+1$ and (x,y) is interior to the domain U^+, then

$$\lim_{\tau\to\infty}\tau^{-(n+1+2\nu+1)}\Gamma(\tau;\ x,\ y;\ x,\ y)=C_\nu\int_{R_{n+1}^+}\frac{[j_\nu(\eta y\tau)]^2\,\eta^{2\nu+1}}{b_0(\infty;\ x,\ y;\ \xi,\ \eta^2)}\,d\xi d\eta. \tag{1.22}$$

But if (x,y) lies in the hyperplane $y=0$, then (1.22) takes the form

$$\lim_{\tau\to\infty}\tau^{-(n+1+2\nu+1)}\Gamma(\tau;\ x,\ 0;\ x,\ 0)=\int_{R_{n+1}^+}\frac{\eta^{2\nu+1}}{b_0(\infty;\ x,\ 0;\ \xi,\ \eta^2)}\,d\xi d\eta. \tag{1.23}$$

Exactly the same estimate holds for the fundamental solution $\Gamma^*(\tau;x,y;u,v)$ with a pole at (u,v), defined in $U^+\times U^+$ for sufficiently large τ. The fundamental solution Γ^* satisfies the relation

$$\int_{R_{n+1}^+}\Gamma^*(\tau;\ x,\ y;\ u,\ v)\,b(\tau;\ x,\ y;\ \mathscr{D}_x,\ \mathscr{B}_y)\,f(x,\ y)\,y^{2\nu+1}\,dxdy=f(u,\ v).$$

For the construction of Γ and Γ^* we use the parametrix method. The parametrix is the fundamental solution $\mathscr{E}(\tau;u,v;x,y)$ of the differential operator $b(\tau;u,v;-\mathscr{D}_x,\mathscr{B}_y)$ with a pole at (u,v).

We assume

$$\mathscr{E}(\tau;\ u,\ v;\ x,\ y)\sim C_\nu\int_{R_{n+1}^+}e^{-i(x-u)\xi}T_y^v j_\nu(\eta y)\,b^{-1}(\tau;\ u,\ v;\ \xi,\ \eta^2)\,\eta^{2\nu+1}d\xi d\eta$$

$$\sim C_\nu\int_{R_{n+1}^+}\tau^{n+1+2\nu+1}e^{-ix'\xi}j_\nu(\tau\eta v)\,j_\nu(\tau\eta v)\,b^{-1}(\tau;\ u,\ v;\ \tau\xi,\ \tau^2\eta^2)\,\eta^{2\nu+1}d\xi d\eta. \tag{1.24}$$

Let the pole (u,v) be in the hyperplane $v=0$. With respect to the condition (1.17) the polynomial $b(\tau;u,0;\tau\xi,\tau^2\eta^2)$ satisfies the conditions of the lemma in a suitable compact subset of Ω^+. Therefore by the theorem we have

$$\mathscr{D}_{x'}^\alpha\mathscr{B}_{y'}^\beta\mathscr{E}(\tau;\ u,\ 0;\ x,\ y)=O(1)\,\tau^{n+2+2\nu+1}\times|x'|^{1-\varepsilon-(n+1+2\nu+1)}\,(1+(\sqrt{|x'|^2+|y'|^2})^N)^{-1}\quad(|\alpha|+2\beta<\mu), \tag{1.25}$$

where $x' = \tau(x - u)$ and $y' = \tau y$. Therefore we finally obtain

$$\mathcal{D}_x^\alpha \mathcal{B}_y^\beta \mathcal{E}(\tau;\ u,\ 0;\ x,\ y) = O(1)\,\tau^{|\alpha|+2\beta+1-\varepsilon} \times \left(\sqrt{|x-u|^2+y^2}\right)^{1-\varepsilon-(n+1+2\nu+1)} \left(1+\left(\sqrt{|\tau(x-u)|^2+\tau^2 y^2}\right)^N\right)^{-1}. \tag{1.26}$$

Similarly we find that

$$\mathcal{D}_x^\alpha \mathcal{B}_y^\beta \mathcal{E}(\tau;\ u,\ 0;\ x,\ y) = O(1)\,\tau^{|\alpha|+2\beta-\varepsilon} \times \left(\sqrt{|x-u|^2+y^2}\right)^{-\varepsilon-(n+1+2\nu+1)} \left(1+\left(\sqrt{|x'|^2+|y'|^2}\right)^N\right)^{-1} \tag{1.27}$$

for $|\alpha| + 2\beta \leqslant \mu$.

But if the pole (u, v) is interior to the domain; that is, if $v > 0$, then after applying the generalized translation operator (see formula (1.24)) the singularities in the estimates (1.26) and (1.27) are removed. In this case we must put $2\nu + 1 = 0$ on the right in (1.26) and (1.27).

Put

$$r(\tau;\ u,\ v;\ x,\ y) = (b(\tau;\ u,\ v;\ \mathcal{D}_x,\ \mathcal{B}_y) - b(\tau;\ x,\ y;\ \mathcal{D}_x,\ \mathcal{B}_y))\,\mathcal{E},$$

and let $q(\tau; u, v; x, y)$ be a solution of the integral equation

$$q(\tau;\ u,\ v;\ x,\ y) - \int_{U+} q(\tau;\ u,\ v;\ s,\ t)\, r(\tau;\ s,\ t;\ x,\ y)\, t^{2\nu+1}\,ds\,dt = r(\tau;\ u,\ v;\ x,\ y). \tag{1.28}$$

Then

$$\Gamma(\tau;\ u,\ v;\ x,\ y) = \mathcal{E}(\tau;\ u,\ v;\ x,\ y) + \int_{U+} q(\tau;\ u,\ v;\ s,\ t)\,\mathcal{E}(\tau;\ s,\ t;\ x,\ y)\, t^{2\nu+1}\,ds\,dt \tag{1.29}$$

will be a fundamental solution in $U^+ \times U^+$ (see [6]).

We will study the conditions under which (1.28) has a solution. For this it is sufficient to consider the case when the pole lies on the corresponding hyperplane, because the case when the pole is interior to the domain reduces to the usual elliptic operator with the calculation of (1.26) and (1.27) when $2\nu + 1 = 0$ on the right-hand sides, and this may be found in [2].

We will write out $r(\tau; u, v; x, y)$ in more detail:

$$\sum (b_{\alpha,\beta}(\tau;\ u,\ v) - b_{\alpha,\beta}(\tau;\ x,\ y))\,\tau^{-|\alpha|-2\beta}\,\mathcal{D}_x^\alpha \mathcal{B}_y^\beta \mathcal{E}(\tau;\ u,\ v;\ x,\ y).$$

Using (1.26) and (1.27), we find

$$r(\tau;\ u,\ 0;\ x,\ y) = O(1)\,\tau^{-\varepsilon}\left(\sqrt{|x-u|^2+y^2}\right)^{1-\varepsilon-(n+1+2\nu+1)}\left(1+\left(\sqrt{|\tau(x-u)|^2+\tau^2y^2}\right)^N\right)^{-1}$$

for arbitrary fixed N and $0 < \epsilon < 1$ and sufficiently large τ uniformly in a suitable compact set of $\Omega^+ \times \Omega^+$. Therefore the Neumann series for (1.28)

$$r(\tau;\ u,\ 0;\ x,\ y)+\int_{U+} r(\tau;\ u,\ 0;\ x',\ y')\, r(\tau;\ x',\ y';\ x,\ y)\, y'^{2\nu+1}dx'dy'+\dots$$

is majorized by the series

$$\left(1+\left(\sqrt{|\tau(x-u)|^2+(\tau y)^2}\right)^N\right)^{-1}\Big\{c\tau^{-\varepsilon}\left(\sqrt{|x-u|^2+y^2}\right)^{1-\varepsilon-(n+1+2\nu+1)}$$

$$+c(\tau^{-\varepsilon})^2\int_{U+}\left(\sqrt{|u-s|^2+t^2}\right)^{1-\varepsilon-(n+1+2\nu+1)}\left(\sqrt{|x-s|^2+|y-t|^2}\right)^{1-\varepsilon-(n+1)}$$

$$\times\, t^{2\nu+1}dsdt+\dots\Big\}.$$

For the construction of the majorant we use the inequality

$$\left(1+\left(\sqrt{|\tau(s-u)|^2+(\tau t)^2}\right)^N\right)^{-1}\left(1+\left(\sqrt{|\tau(x-u)|^2+|\tau(y-t)|^2}\right)^N\right)^{-1}$$

$$\leqslant\left(1+\left(\sqrt{|\tau(x-u)|^2+(\tau y)^2}\right)^N\right)^{-1}.$$

Consequently for sufficiently large τ a solution of (1.28) may be found in the form of a Neumann series, and for the solution we have the estimate

$$q(\tau;\ u,\ 0;\ x,\ y)$$

$$=O(1)\,\tau^{-\varepsilon}\left(\sqrt{|x-u|^2+y^2}\right)^{1-\varepsilon-(n+1+2\nu+1)}\left(1+\left(\sqrt{|\tau(x-u)|^2+(\tau y)^2}\right)^N\right)^{-1}. \qquad (1.30)$$

Estimates (1.30) and (1.29) imply the necessary estimates (1.21) for the derivatives of Γ. With regard to estimates for Γ itself, we find these by similar means. Namely, from (1.24) and the lemma we find that

$$\mathcal{E}(\tau;\ u,\ 0;\ x,\ y)$$

$$=O(1)\,\tau^{n+1+2\nu+1}\, l_0(\tau(x-u),\ \tau y)\left(1+\left(\sqrt{|\tau(x-u)|^2+(\tau y)^2}\right)^N\right)^{-1}, \qquad (1.31)$$

which together with (1.21) and (1.30) will give an estimate of the form of (1.20).

For the proof of (1.22) and (1.23) we note the following circumstances. From (1.24) we find that

$$\lim_{\tau\to\infty}\tau^{-(n+1+2\nu+1)}\mathcal{E}(\tau;\ x,\ y;\ x,\ y)=C_\nu\int_{R^+_{n+1}}\frac{[j_\nu(\tau\eta y)]^2\,\eta^{2\nu+1}}{b_0(\infty;\ x,\ y;\ \xi,\ \eta^2)}\,d\xi d\eta$$

taking into account the properties of the polynomial b uniformly in a suitable compact subset of $\Omega^+ \times \Omega^+$. If $\mu > n+1+2\nu+1$, then from the properties of the fundamental solution $\mathcal{E}$ and the estimate (1.30) we obtain the necessary estimate for Γ. The construction of Γ^* and the corresponding proof is no different from the construction of Γ and the corresponding proofs.

For convenience we formulate the final result for Γ^*. For sufficiently large values of τ we may construct fundamental solutions Γ^* of the differential operator b with the following properties. If $v > 0$, then (1.18) and (1.19) are valid for $\Gamma^*(\tau; u, v; x, y)$. But if $v = 0$, then (1.20) and (1.21) are valid for Γ^*. For

$\mu > n + 1 + 2\nu + 1$ the limiting relations (1.22) and (1.23) are satisfied. In accordance with the known results for fundamental solutions (see [6]), $\Gamma(\tau; u, v; x, y)$ is infinitely differentiable in $U^+ \times U^+$; if $u \neq x, v \neq y$ and if $f \in C_0^\infty$, then

$$\Gamma f(x, y) = \int_{R_{n+1}^+} \Gamma(\tau; u, v; x, y) f(u, v) v^{2\nu+1} du dv \tag{1.32}$$

is also infinitely differentiable in U^+. In addition

$$b(\tau; x, y; \mathscr{D}_x, \mathscr{B}_y) \Gamma(\tau; u, v; x, y) = 0, \tag{1.33}$$

if $(x, y) \neq (u, v)$;

$$b(\tau; x, y; \mathscr{D}_x, \mathscr{B}_y) \Gamma f(x, y) = f(x, y).$$

Similar results hold for Γ^*.

2. Let $\mathfrak{L}_{2,\nu}(\Omega^+)$ be the set of functions which are square summable with weight $y^{2\nu+1}$. Let there be given a bilinear form on $\mathfrak{L}_{2,\nu} \times \mathfrak{L}_{2,\nu}$

$$C(f, g) = C(\tau; f, g),$$

which is uniformly bounded for sufficiently large τ and satisfies the identity

$$C(bf, g) = C(f, b^* g) = \int_{R_{n+1}^+} f(x, y) g(x, y) y^{2\nu+1} dx dy \tag{2.1}$$

in the set $C_0^\infty(\Omega^+) \times C_0^\infty(\Omega^+)$.

We will show that C has a kernel $c(\tau; x, y; u, v)$,

$$C(f, g) = \int_{\Omega^+ \times \Omega^+} c(\tau; x, y; u, v) f(x, y) g(u, v) y^{2\nu+1} v^{2\nu+1} dx dy du dv, \tag{2.2}$$

which is infinitely differentiable if $(x, y) \neq (u, v)$ and satisfies the following condition: if $v > 0$, then

$$\begin{aligned} &c(\tau; x, y; u, v) \\ &= O(1) \tau^{n+1} l_0(\tau(x-u), \tau(y-v)) \left(1 + \left(\left(\sqrt{|\tau(x-u)|^2 + |\tau(y-v)|^2}\right)^N\right)\right)^{-1}, \end{aligned} \tag{2.3}$$

where $l_0(z) = |z|^{\mu-n-1-\epsilon}$ if $\mu - n - 1 \leqslant 0$; $l_0(z) = 1$ if $\mu - n - 1 > 0$. And if $v = 0$, then we have the estimate

$$c(\tau; x, y; u, 0) = O(1) \tau^{n+1+2\nu+1} l_0(\tau(x-u), \tau y) \left(1 + \left(\sqrt{|\tau(x-u)|^2 + \tau^2 y^2}\right)^N\right)^{-1},$$

where

$$\begin{aligned} l_0(z) &= |z|^{\mu-n-1-2\nu-1-\varepsilon}, &&\text{if } \mu - (n+1+2\nu+1) \leqslant 0; \\ l_0(z) &= 1, &&\text{if } \mu - (n+1+2\nu+1) > 0. \end{aligned} \tag{2.4}$$

The estimate $O(1)$ is uniform for sufficiently large τ in a compact subset of $\Omega^+ \times \Omega^+$ contiguous with the corresponding hyperplane. If $\mu - (n + 1 + 2\nu + 1) > 0$, then

$$\lim_{\tau\to\infty} \frac{c(\tau;\ x,\ y;\ u,\ v)}{\tau^{n+1+2\nu+1}} = C_\nu \delta_{x,\,y}^{(u,\,v)} \int\limits_{R_{n+1}^+} \frac{[j_\nu(\eta y\tau)]^2\,\eta^{2\nu+1}}{b_0(\infty;\ x,\ y;\ \xi,\ \eta^2)}\,d\xi d\eta \tag{2.5}$$

uniformly in a compact subset of $\Omega^+ \times \Omega^+$ contiguous with the corresponding hyperplane. The symbol $\delta_{x,y}^{(u,v)}$ is equal to one if $x = u$ and $y = v$, and is zero otherwise.

Let V^+ be an arbitrary open subset of Ω^+ entirely contained in Ω^+ and contiguous to the corresponding hyperplane $v = 0$. Choose neighborhoods V_1^+, V_2^+ and V_3^+ contiguous to the corresponding hyperplane $v = 0$ and such that

$$V^+ \subset \bar V^+ \subset V_1^+ \subset \bar V_1^+ \subset V_2^+ \subset \bar V_2^+ \subset V_3^+ \subset \bar V_3^+ \subset \Omega^+.$$

We will construct fundamental solutions Γ and Γ^* in $U^+ = V_3^+$ satisfying the conditions mentioned in the preceding section (conditions (1.18)–(1.23)). Let $\psi \in C_0^\infty(V_3^+)$ and let it be equal to one in V_2^+. We will show that

$$\begin{aligned}\sigma(\tau;\ x,\ y;\ x^*,\ y^*) &= \Gamma^*(\tau;\ x,\ y;\ x^*,\ y^*) - \Gamma(\tau;\ x,\ y;\ x^*,\ y^*)\\ &= \int\limits_{V_3^+ - V_2^+} \Gamma(\tau;\ x,\ y;\ u,\ v)\, b^*(\tau;\ u,\ v;\ \mathcal{D}_u,\ \mathcal{B}_v)\\ &\quad\times (\psi(u,\ v)\,\Gamma^*(\tau;\ u,\ v;\ x^*,\ y^*))\, v^{2\nu+1} du dv,\end{aligned} \tag{2.6}$$

if (x, y) and (x^*, y^*) belong to V_2^+. The points may lie in the corresponding hyperplane. Firstly let $(x, y) \neq (x^*, y^*)$. Choose neighborhoods W^+, W_1^+ and W_2^+ of (x^*, y^*) such that

$$\bar W^+ \subset W_1^+ \subset \bar W_1^+ \subset \bar W_2^+ \subset V_2^+$$

and are contiguous to the corresponding hyperplane. The neighborhood $\bar W_2^+$ does not contain (x, y). Let $\varphi \in C_0^\infty(W_1^+)$ and be equal to one in W^+; let $\varphi_1 \in C_0^\infty(W_2^+)$ and be equal to one in W_1^+.

Put $f(u, v) = \Gamma(\tau; x, y; u, v)$ and $f^*(u, v) = \Gamma^*(\tau; u, v; x^*, y^*)$. By the properties of Γ and Γ^* we have

$$\int\limits_{R_{n+1}^+} f(u,\ v)\, b^*(\tau;\ u,\ v;\ \mathcal{D}_u,\ \mathcal{B}_v)\,(\psi(u,\ v) - \varphi(u,\ v))\, f^*(u,\ v)\, v^{2\nu+1}\, du dv = f^*(x,\ y).$$

Since $b^*(\tau; u, v; \mathcal{D}_u, \mathcal{B}_v) f^*(u, v) = 0$ except at the point $(x, y) = (u, v)$, the left-hand side of the latter equality becomes the following:

$$\int\limits_{R_{n+1}^+} f b^*(\psi f^*)\, v^{2\nu+1} du dv + \int\limits_{R_{n+1}^+} f b^*(1 - \varphi)\, f^* v^{2\nu+1} du dv.$$

On the other hand,

$$\int_{R_{n+1}^+} fb^*(1-\varphi)f^*v^{2\nu+1}dudv = \int_{R_{n+1}^+} \varphi_1 fb^*(1-\varphi)f^*v^{2\nu+1}dudv$$

$$= \int_{R_{n+1}^+} b\varphi_1 f(1-\varphi)f^*v^{2\nu+1}dudv = \int_{R_{n+1}^+} b\varphi_1 ff^*v^{2\nu+1}dudv = f(x^*, y^*).$$

Consequently

$$f^*(x, y) - f(x^*, y^*) = \int_{R_{n+1}^+} fb^*(\psi f^*)v^{2\nu+1}dudv.$$

The last formula proves (2.6). By the properties of the chosen sets and by continuity, formula (2.6) also holds in the case $(x, y) = (x^*, y^*)$. Therefore σ is infinitely differentiable in $V_2^+ \times V_3^+$.

Put

$$\Gamma^*(\tau; f, g) = \int_{R^+_{+1}} \Gamma^*(\tau; x, y; x^*, y^*)f(x, y)g(x^*, y^*)y^{2\nu+1}y^{*2\nu+1}dxdydx^*dy^*$$

and

$$C'(f, g) = C(f, g) - \Gamma^*(\tau, f, g).$$

Owing to (1.19) and (1.21) with $\alpha = \beta = 0$, the norm $|C'| = \sup|C'(f, g)|/\|f\|_{\mathfrak{L}_{2,\nu}}\|g\|_{\mathfrak{L}_{2,\nu}}$ of the bilinear form C' is $O(1)\tau^{1-\epsilon}$.

By the properties of Γ^* we have

$$C'(bf, g) = \int_{R_{n+1}^+} f(x, y)g(x, y)y^{2\nu+1}dxdy - \int_{R_{n+1}^+} f(x, y)g(x, y)y^{2\nu+1}dxdy = 0;$$

and if $\Gamma(\tau; f, g)$ and $\sigma(\tau; f, g)$ are defined the same as for Γ^*, then we have

$$C'(f, b^*g) = C(f, b^*g) - \Gamma(\tau; f, b^*g) - \sigma(\tau; f; b^*g) = -\sigma(\tau; f, b^*g)$$

in $C_0^\infty(V_2) \times C_0^\infty(V_2)$.

Now put

$$\begin{aligned} p(x, y; u, v) &= b(\tau; u, v; \mathcal{D}_u, \mathcal{B}_v)(\Gamma(\tau; x, y; u, v)(1-\varphi)),\\ p^*(u, v; x^*, y^*) &= b^*(\tau; u, v; \mathcal{D}_u, \mathcal{B}_v)(\Gamma^*(\tau; u, v; x^*, y^*)(1-\varphi)), \end{aligned} \tag{2.7}$$

where the function $\varphi \in C_0^\infty(V_2^+)$ and is equal to one on V_1^+. For given (x, y) and (x^*, y^*) from V_1 the functions $p(x, y; u, v)$ and $p^*(u, v; x^*, y^*)$ become zero except in the half-ring $V_2^+ - V_1^+$.

Further, put

$$c'(\tau; x, y; x^*, y^*) = C'(p(x, y; \ldots), p^*(\ldots, x^*, y^*)). \tag{2.8}$$

By the continuity of p, p^* and C' we have

$$c'(\tau;\ f,\ g) = \int_{R^+_{n+1}} c'(\tau;\ x,\ y;\ x^*,\ y^*) f(x,\ y) g(x^*,\ y^*) y^{2\nu+1} y^{*2\nu+1} dx dy dx^* dy^*$$

$$= C'(f - b\varphi\Gamma f,\ g - b^*\varphi\Gamma^* g),$$

where $f, g \in C_0^\infty(V)$, and, in addition,

$$\Gamma f(u,\ v) = \int_{R^+_{n+1}} \Gamma(\tau;\ x,\ y;\ u,\ v) f(u,\ v) v^{2\nu+1} du dv,$$

$$\Gamma^* g(u,\ v) = \int_{R^+_{n+1}} \Gamma^*(\tau;\ u,\ v;\ x^*,\ y^*) g(x^*,\ y^*) y^{*2\nu+1} dx^* dy^*.$$

By the properties of the form C',

$$c'(\tau;\ f,\ g) = C'(f,\ g) + \sigma(\tau;\ f,\ b^*\varphi\Gamma^* g).$$

Now we have

$$b^*\varphi\Gamma^* g(u, v) = b^*(\tau;\ u,\ v,\ \mathscr{D}_u,\ \mathscr{B}_v)\varphi \int_{R^+_{n+1}} \Gamma^*(\tau;\ u,\ v;\ x^*,\ y^*) g\, y^{*2\nu+1} dx^* dy^*$$

$$= g(u,\ v) + \int_{R^+_{n+1}} b^*\varphi(u, v) \Gamma^*(\tau;\ u,\ v,\ x^*,\ y^*) g(x^*, y^*) y^{*2\nu+1} dx^* dy^* \qquad (2.9)$$

with the kernel $b^*(\varphi\Gamma^*) = \sigma_1(\tau; u, v; x^*, y^*)$, which is not zero only if $(u, v) \in V_2^+ - V_1^+$. Therefore

$$C(f,\ g) = \Gamma^*(\tau;\ f,\ g) + c'(\tau;\ f,\ g) + \sigma(\tau;\ f,\ g) + \lambda(\tau;\ f,\ g), \qquad (2.10)$$

where

$$\lambda = \int_{V_2^+ - V_1^+} \sigma(\tau;\ x,\ y;\ u,\ v) \sigma_1(\tau;\ u,\ v;\ x^*,\ y^*) v^{2\nu+1} du dv.$$

Now we consider the functions $p(x, y; u, v)$ and $p^*(u, v; x^*, y^*)$. Since

$$b(\tau;\ u,\ v,\ \mathscr{D}_u,\ \mathscr{B}_v)\Gamma(\tau;\ x,\ y;\ u,\ v) = 0 \quad \text{for} \quad (x,\ y) \neq (u,\ v),$$

$$b^*(\tau;\ u,\ v;\ \mathscr{D}_u,\ \mathscr{B}_v)\Gamma^*(\tau;\ u,\ v;\ x^*,\ y^*) = 0 \quad \text{for} \quad (x^*,\ y^*) \neq (u,\ v),$$

only the derivatives of order $< \mu$ of Γ and Γ^* will actually occur in p and p^*. On the basis of the corresponding properties of Γ and Γ^* we have

$$p(x,\ y;\ u,\ v) = O(1)\, \tau^{1-\varepsilon} r^{1-\varepsilon-(n+1+2\nu+1)} (1 + |r\tau|^N)^{-1},$$

where r is the corresponding distance of (x, y) to V_1^+. The estimate is valid up to the corresponding hyperplane. For this we must put $2\nu + 1$ equal to zero on the right-hand side.

Similarly, we have

$$p^*(u, v; x^*, y^*) = O(1)\,\tau^{1-\varepsilon} r^{*(1-\varepsilon-(n+1+2\nu+1))} \left(1 + |r^*\tau|^N\right)^{-1}$$

up to the corresponding hyperplane on replacement of $2\nu + 1$ by zero in the right-hand side. Therefore

$$p = O(1)\,\tau^{1-\varepsilon-N}, \qquad p^* = O(1)\,\tau^{1-\varepsilon-N} \tag{2.11}$$

uniformly in $V^+ \times (V_2^+ - V_1^+)$ and $(V_2^+ - V_1^+) \times V^+$ respectively. In a similar way we find from (2.6) that

$$\sigma(\tau; x, y; u, v) = O(1)\,\tau^{2(1-\varepsilon)-N}$$

uniformly in $V_1^+ \times V_2^+$, and

$$\sigma(\tau; x, y; x^*, y^*) = O(1)\,\tau^{2(1-\varepsilon)-2N} \tag{2.12}$$

uniformly in $V_1^+ \times V_1^+$. In addition, $\sigma_1 = O(1)\tau^{1-\epsilon-N}$ uniformly in $V_2^+ \times V_1^+$. Therefore by the formula for λ we have

$$\lambda(\tau; x, y; x^*, y^*) = O(1)\,\tau^{3(1-\varepsilon)-2N}$$

uniformly in $V_1^+ \times V_1^+$.

From the preceding estimates (2.11) and the estimates for $|C'|$ we obtain that

$$c(\tau; x, y; x^*, y^*) = \Gamma^*(\tau; x, y; x^*, y^*) + O(1)\,\tau^{3(1-\varepsilon)-2N},$$

and, moreover, uniformly in $V^+ \times V^+$. Therefore (2.3) and (2.4) follow from the corresponding estimates of the previous section for Γ^*.

3. Let $\mathfrak{L}(x, y; \mathfrak{D}_x, \mathfrak{B}_y)$ be a linear differential operator of the form

$$\mathcal{L}(x, y; \mathcal{D}_x, \mathcal{B}_y) = \sum_{|\alpha|+2\beta\leqslant 2m} a_{\alpha,\beta}(x, y)\,\mathcal{D}_x^\alpha \mathcal{B}_y^\beta \quad (m \geqslant 1),$$

where $x = (x_1, \cdots, x_{n+1})$, $x_{n+1} = y$ and $|\alpha| = \alpha_1 + \cdots + \alpha_n$. The coefficients $a_{\alpha,\beta}$ are assumed to be infinitely differentiable in Ω^+ and the operator is assumed to be $\mathfrak{B}$-elliptic; that is, the polynomial

$$\mathcal{L}_0(x, y; \xi, \eta^2) = \sum_{|\alpha|+2\beta=2m} a_{\alpha,\beta}(x, y)\,\xi^\alpha \eta^{2\beta} \tag{3.1}$$

is positive definite with respect to ξ and η for all $y \geqslant 0$ in Ω^+. Let Ω_1^+ be a subdomain of Ω, entirely contained in it, and contiguous to the hyperplane $y = 0$. In the set $C_0^\infty(\Omega_1^+)$ we put

$$(f, g)_\nu^{(m)} = \int_{\Omega_1^+} \sum_{|\alpha|+2\beta\leqslant m} \mathcal{D}_x^\alpha \mathcal{B}_y^\beta f(x, y)\,\mathcal{D}_x^\alpha \mathcal{B}_y^\beta g(x, y)\, y^{2\nu+1}\,dx\,dy, \tag{3.2}$$

$$\left.\begin{aligned}(f, g)_\nu &= \int_{\Omega_1^+} f(x, y)\, g(x, y)\, y^{2\nu+1} dx dy,\\ \|f\|^2_{\mathring{H}^m_\nu} &= (f, f)^{(m)}_\nu, \qquad \|f\|^2_{\mathscr{L}_{2,\nu}} = (f, f)_\nu.\end{aligned}\right\} \tag{3.3}$$

Completing C_0^∞ with respect to the indicated norm, we obtain the Hilbert space of functions $\mathring{H}^m_\nu$.

Let f and $g \in C_0^\infty(\Omega_1^+)$. Integrating by parts, we obtain

$$(\mathscr{L}f, g)_\nu = \mathscr{L}(f, g) = \int_{\Omega_1^+} \sum_{\alpha, \beta, \alpha', \beta'} a^{(\alpha, \beta)}_{\alpha', \beta'}(x, y)\, \mathscr{D}^\alpha_x \mathscr{B}^\beta_y f\, \mathscr{D}^{\alpha'}_x \mathscr{B}^{\beta'}_y g\, y^{2\nu+1} dx dy \tag{3.4}$$

$$\left(\frac{|\alpha|}{2} + \beta \leqslant m, \quad \frac{|\alpha'|}{2} + \beta' \leqslant m\right).$$

Obviously the indicated form of f and g is bounded above in $\mathring{H}^m_\nu$.

We note that the results in [7] on the boundedness from below of the form considered there do carry over to the case of the form (3.4). The fact is that the proof given there of Gårding's inequalities is based on Fourier-Bessel transforms, on certain relations between differential operators generated by these transforms, and on the corresponding weighted Parseval equalities. All of this remains valid for this case. So the proof is repeated almost verbatim.

Let t be a sufficiently large positive number. Put

$$\mathscr{L}_t(f, g) = \mathscr{L}(f, g) + t(f, g)_\nu,$$
$$(f, g)_{\nu, t} = (f, g)^{(m)}_\nu + t(f, g)_\nu.$$

Then there exists a positive number t and $c > 0$ so that

$$c^{-1}(f, f)_{\nu, t} \leqslant |\mathscr{L}_t(f, f)| \leqslant c(f, f)_{\nu, t} \quad (t > t_0) \tag{3.5}$$

for all $f \in \mathring{H}^m_\nu$. This follows immediately from [7], if it is noted that even powers of the corresponding operators which occur in the definition of the norms of the spaces considered there may be replaced by powers of the operator $\mathscr{B}_y$. Therefore the relation

$$\mathscr{L}_t(f, f') = (N_t f, f')_{\nu, t} \qquad f, f', N_t f \in \mathring{H}^m_\nu,$$

defines a bounded linear operator N_t acting from $\mathring{H}^m_\nu$ to $\mathring{H}^m_\nu$, and this operator has a bounded inverse.

As above, let $\mathfrak{L}_{2,\nu}(\Omega_1^+)$ be the set of functions square summable with weight $y^{2\nu+1}$. The equation

$$(f, f')_\nu = (M_t f, f')_{\nu, t} \qquad f \in \mathscr{L}_{2, \nu};\ M_t f, f' \in \mathring{H}^m_\nu$$

defines a completely continuous operator acting from $\mathfrak{L}_{2,\nu}$ to $\mathring{H}_\nu^m$. Therefore

$$(f, f')_\nu = \mathcal{L}_t(G_t f, f'), \qquad f \in \mathcal{L}_{2,\nu};\ f' \in \mathring{H}_\nu^m,$$

where $G_t = N_t^{-1} M_t$ defines a completely continuous operator G_t acting from $\mathfrak{L}_{2,\nu}$ to $\mathring{H}_\nu^m$, which, following Gårding [2], we will call a Green's transform corresponding to the differential operator $\mathfrak{L}_t = \mathfrak{L} + t$. The operator G_t maps $\mathfrak{L}_{2,\nu}$ to $\mathring{H}_\nu^m$, and the operator G_t^{-1} is an extension of the differential operator $\mathfrak{L}_t$, whose graph is the set of pairs $\{f, \mathfrak{L}_t f\}$, where $f \in \mathring{H}_\nu^m$ and is $2m$ times differentiable (with respect to y it has the Bessel operator applied to it the corresponding number of times) and $\mathfrak{L}_t f \in \mathfrak{L}_{2,\nu}$. In fact, if $h \in C_0^\infty$, then we have

$$\mathcal{L}_t(G_t \mathcal{L}_t f, h) = (\mathcal{L}_t f, h)_\nu = \mathcal{L}_t(f, h);$$

in other words $G_t \mathfrak{L}_t f = f$.

We now obtain an estimate for the Green's function. For sufficiently large t the Green's function was defined by the formula

$$(f, f')_\nu = \mathcal{L}_t(G_t f, f'), \tag{3.6}$$

where $f \in \mathfrak{L}_{2,\nu}$ and $G_t f, f' \in \mathring{H}_\nu^m$.

Since

$$\|f\|_{\nu,t}^2 = (f, f)_\nu^{(m)} + t(f, f)_\nu \geqslant t\|f\|_{\mathcal{L}_{2,\nu}}^2,$$

on taking (3.5) into account we have

$$tc^{-1}\|G_t f\|_{\mathcal{L}_{2,\nu}}^2 \leqslant c^{-1}\|G_t f\|_{\nu,t}^2 \leqslant |\mathcal{L}_t(G_t f, G_t f)| \leqslant \|f\|_{\mathcal{L}_{2,\nu}} \|G_t f\|_{\mathcal{L}_{2,\nu}}. \tag{3.7}$$

Therefore

$$\|G_t\|_{\mathcal{L}_{2,\nu}} \leqslant \frac{c}{t}. \tag{3.8}$$

It was proved above that if $f \in C_0^\infty(\Omega_1^+)$, then

$$G_t \mathcal{L}_t f = f. \tag{3.9}$$

Let $\mathfrak{L}_t^*$ be the operator adjoint to $\mathfrak{L}_t$:

$$(\mathcal{L}_t f, f')_\nu = (f, \mathcal{L}_t^* f')_\nu, \quad f, f' \in C_0^\infty.$$

If $f \in \mathring{H}_\nu^m$ and $f' \in C_0^\infty$, then

$$(G_t f, \mathcal{L}_t^* f')_\nu = \mathcal{L}_t(G_t f, f') = (f, f')_\nu. \tag{3.10}$$

Let k be a positive integer. Put

$$C(f, f') = (G_t^k f, f')_\nu t^k$$

and in addition, $b = \mathfrak{L}_t^k/t^k$. The bilinear form $C(f, f')$ is bounded; that is,

$$|C(f, f')| \leqslant \|f\|_{\mathscr{L}_{2,\nu}} \|f'\|_{\mathscr{L}_{2,\nu}},$$

and according to (3.6)

$$C(bf, f') = (G_t^k \mathscr{L}_t^k f, f')_\nu = (f, f')_\nu,$$

where $f, f' \in C_0^\infty(\Omega_1^+)$.

If b^* is the operator adjoint to b, then we find from (3.10) that

$$C(f, b^* f') = (G_t^k f, \mathscr{L}_t^{*k} f')_\nu = (f, f')_\nu$$

for any $f, f' \in C_0^\infty$.

Putting $t = \tau^{2m}$, we get that

$$b = b(\tau; x, y; \mathscr{D}_x, \mathscr{B}_y) = \sum b_{\alpha,\beta}(\tau, x, y)\, \tau^{-|\alpha|-2\beta} \mathscr{D}_x^\alpha \mathscr{B}_y^\beta$$

possesses property (1.17) described above. Therefore applying the results of the preceding part we see that G_t^k has a continuous kernel $g_t^{(k)}(x, y; u, v)$ with the following property:

$$\lim_{t\to\infty} t^{k-\gamma} g_t^{(k)}(x, y; u, v) = \delta_{x,y}^{(u,v)} C_\nu \int_{R_{n+1}^+} \frac{[j_\nu(t\eta y)]^2 \eta^{2\nu+1}}{[\mathscr{L}_0(x, y; \xi, \eta^2) + 1]^k} d\xi d\eta, \tag{3.11}$$

if $2mk > n + 1 + 2\nu + 1$, where $\gamma = (n + 1 + 2\nu + 1)/2m$. The symbol $\delta_{x,y}^{(u,v)}$ has the same meaning as before. Namely, it is equal to one when $x = u$ and $y = v$, and is equal to zero otherwise. For $x = u$ and $y = v$ the limit equality (3.11) remains valid up to the hyperplane $y = 0$. In this case the presence of the $2\nu + 1$ in the power of t is essential, and formula (3.11) takes the form

$$\lim_{t\to\infty} t^{k-\gamma} g_t^{(k)}(x, 0; x, 0) = C_\nu \int_{R_{n+1}^+} \frac{\eta^{2\nu+1}}{[\mathscr{L}_0(x, 0; \xi, \eta^2) + 1]^k} d\xi d\eta \tag{3.12''}$$

for $\gamma = (n + 1 + 2\nu + 1)/2m$ and $2mk > n + 1 + 2\nu + 1$.

If $\mathfrak{L}_t$ is selfadjoint, then $\mathfrak{L}_t^* = \mathfrak{L}_t$, and then $\mathfrak{L}_t(f, f) = (\mathfrak{L}_t f, f)_\nu = (f, \mathfrak{L}_t^* f)_\nu$ is a real number for $f \in C_0^\infty$. The fact is that the case of complex coefficients is not excluded. Therefore $\mathfrak{L}_t(f, f)$ is real for $f \in \mathring{H}_\nu^m$, and consequently $(G_t f, f)_\nu = \mathfrak{L}_t(G_t f, G_t f)$ is real for $f \in \mathfrak{L}_{2,\nu}$. Thanks to property (3.5), the form $(G_t f, f)_\nu \geqslant 0$ if t is sufficiently large, and that is why it is a selfadjoint operator. But then $g_t^{(k)}(x, y; x, y) \geqslant 0$, and by Fatou's Lemma

$$\begin{aligned} \lim_{t\to\infty} {}^{k-\gamma} \int_{\Omega_1^+} & g_t^{(k)}(x, y; x, y)\, y^{2\nu+1} dx dy \\ & \geqslant C_\nu \int_{\Omega_1^+} y^{2\nu+1} dx dy \int_{R_{n+1}^+} \frac{[j_\nu(t\eta y)]^2 \eta^{2\nu+1}}{[\mathscr{L}_0(x, y; \xi, \eta^2) + 1]^k} d\xi d\eta. \end{aligned} \tag{3.12}$$

Let $\widetilde{\Omega}^+$ be an open set in Ω^+ contiguous with the hyperplane $y = 0$ and contained in the closure of Ω_1^+ (we note that Ω_1^+ may also be an open bounded set contiguous to $y = 0$), and whose closure is contained in Ω^+. Let $\widetilde{G}_t$ be a Green's function, corresponding to $\widetilde{\Omega}^+$. Taking (3.6) into account, we have

$$(G_t f,\ f)_\nu = \mathscr{L}_t (G_t f,\ G_t f) = \sup \frac{|(f,\ g)_\nu|}{\mathscr{L}_t (g, g)},$$

where $g \in \overset{\circ}{H}{}_\nu^m(\Omega_1^+)$.

Since $\overset{\circ}{H}{}_\nu^m(\widetilde{\Omega}^+)$ contains $\overset{\circ}{H}{}_\nu^m(\Omega_1^+)$, we get

$$(G_t f,\ f)_\nu \leqslant (\widetilde{G}_t f,\ f)_\nu = (E\widetilde{G}_t f,\ f)_\nu,$$

where E is the projection of $\mathfrak{L}_{2,\nu}(\widetilde{\Omega}^+)$ onto $\mathfrak{L}_{2,\nu}(\Omega_1^+)$. Therefore all the eigenvalues of G_t, taken in decreasing order, do not exceed the corresponding eigenvalues of the restriction $\Gamma_t = E\widetilde{G}_t E$ of the operator $\widetilde{G}_t$ to $\mathfrak{L}_{2,\nu}$. Consequently

$$tr G_t^k \leqslant tr \Gamma_t^k$$

for all k.

The bilinear forms $(\widetilde{G}_t^k f, f')_\nu t^k$ and $(\Gamma_t^k f, f')_\nu t^k$ both satisfy the required conditions of the preceding lemma relative to the operator $b = \mathfrak{L}_t^k/t^k$ in the domains $\widetilde{\Omega}^+$ and Ω_1^+ respectively. In particular, Γ_t^k has a kernel $\gamma_t^{(k)}(x, y; u, v)$ which is continuous for $2mk > n + 1 + 2\nu + 1$ and satisfies the relation

$$\lim_{t\to\infty} t^{k-\tilde{\gamma}} \gamma_t^{(k)} (x,\ y;\ x,\ y) = C_\nu \int\limits_{R_{n+1}^+} \frac{[j_\nu (\eta t y)]^2 \eta^{2\nu+1}}{[\mathscr{L}_0 (x,\ y;\ \xi,\ \eta^2) + 1]^k} d\xi d\eta \tag{3.13}$$

with $\tilde{\gamma} = (2\nu + 1 + n + 1)/2m$. This relation remains valid up to the hyperplane $y = 0$.

The kernel $\gamma_t(x, y; u, v)$ of the operator Γ_t is the restriction of the kernel $\tilde{g}(x, y; u, v)$ of the operator G_t to the corresponding domain $\widetilde{\Omega}^+$. Therefore we have the following estimate for γ when $v > 0$:

$$\gamma_t (x,\ y;\ u,\ v) = t^{\tilde{\gamma}-1-l} O(1)\, e(\tau (x-u),\ \tau (y-v))\big(1 + (\sqrt{|\tau (x-u)|^2 + |\tau (y-v)|^2})^N\big)^{-1} \tag{3.14}$$

for any N and $0 < \epsilon < 1$, uniformly in $\Omega_1^+ \times \Omega_1^+$ for sufficiently large t ($\tau^{2m} = t$), $\tilde{\gamma} = (n + 1)/2m$. Here $e(z) = 1$ if $2m - (n + 1) > 0$, and $e(z) = |z|^{2m-(n+1)-\epsilon}$ otherwise. The number l is chosen equal to $((2\nu + 1)/2m)k$. But if $v = 0$, then from the estimates obtained earlier we find that

$$\gamma_t (x,\ y;\ u,\ 0) = t^{\frac{n+1+2\nu+1}{2m}-1} O(1)\, e(\tau (x-u),\ \tau y)\big(1 + (\sqrt{|\tau (x-u)|^2 + \tau^2 y^2})^N\big)^{-1},$$

where

$$e(z) = |z|^{2m-(n+1+2\nu+1)-\varepsilon} \quad \text{for} \quad 2m-(n+1+2\nu+1) \leqslant 0,$$
$$e(z) = 1, \qquad\qquad \text{if} \quad 2m-(n+1+2\nu+1) > 0. \tag{3.15}$$

Thanks to the indicated estimates and Fubini's Theorem the kernel of Γ_t^k has a number of properties, and may be written in the form

$$\gamma_t^{(k)}(x, y; u, v) = \int\limits_{\Omega_1^+ \times \dots \times \Omega_1^+} \gamma_t(x, y; u_1, v_1)\gamma_t(u_1, v_1; u_2, v_2) \dots \gamma_t(u_{k-1}, v_{k-1}; u, v) v_1^{2\nu+1} du_1 dv_1 \dots v_{k-1}^{2\nu+1} du_{k-1} dv_{k-1}.$$

Putting $x = u$, $y = v$, $u_i = x + \tau^{-1} z_i$ and $v_i = y + \tau^{-1} s_i$, we get that

$$\begin{aligned} 0 &\leqslant t^{k-\tilde{\gamma}} \gamma_t^{(k)}(x, y; x, y) \\ &\leqslant O(1) \int\limits_{R_{n+1}^+} e(z_1, s_1) e(z_1 - z_2, s_1 - s_2) \dots e(z_{k-1}, s_{k-1}) (1 + (\sqrt{z_1^2 + s_1^2})^N)^{-1} \\ &\times (1 + (\sqrt{|z_1 - z_2|^2 + |s_1 - s_2|^2})^N)^{-1} \dots (1 + (\sqrt{|z_{k-1}|^2 + |s_{k-1}|^2})^N)^{-1} \\ &\times s_1^{2\nu+1} dz_1 ds_1 \dots s_{k-1}^{2\nu+1} dz_{k-1} ds_{k-1}. \end{aligned} \tag{3.16}$$

If $y > 0$ and $2mk > n + 1$, then we may choose N so large and ϵ so small that the right-hand side of (3.16) is finite for $2mk - \epsilon k > n + 1$. And if $y = 0$ and $2mk > n + 1 + 2\nu + 1$, then N may be chosen so large and ϵ so small that the right-hand side of (3.16) is finite for $2mk - \epsilon k > n + 1 + 2\nu + 1$.

Applying Lebesgue's theorem, we find that

$$\begin{aligned} &\lim_{t\to\infty} t^{k-\tilde{\gamma}} \int\limits_{\Omega_1^+} \gamma_t^{(k)}(x, y; x, y) y^{2\nu+1} dx dy \\ &\qquad = c_\nu \int\limits_{\Omega_1^+} y^{2\nu+1} dx dy \int\limits_{R_{n+1}^+} \frac{[j_\nu(\eta t y)]^2 \eta^{2\nu+1}}{[\mathscr{L}_0(x, y; \xi, \eta^2) + 1]^k} d\xi d\eta, \end{aligned} \tag{3.17}$$

where $\tilde{\gamma} = (2\nu + 1 + n + 1)/2m$. Since the kernel $\gamma_t^{(k)}$ is continuous and positive, the integral on the left-hand side is the trace of Γ_t^k, and therefore

$$\overline{\lim}\, t^{k-\tilde{\gamma}} tr G_t^k \leqslant C_\nu \int\limits_{\Omega_1^+} y^{2\nu+1} dx dy \int\limits_{R_{n+1}^+} \frac{[j_\nu(\eta t y)]^2 \eta^{2\nu+1}}{[\mathscr{L}_0(x, y; \xi, \eta^2) + 1]^k} d\xi d\eta. \tag{3.18}$$

This inequality together with (3.12) gives that

$$\begin{aligned} &\lim_{t\to\infty} t^{k-\tilde{\gamma}} \int\limits_{\Omega_1^+} g_t^{(k)}(x, y; x, y) y^{2\nu+1} dx dy \\ &= C_\nu \int\limits_{\Omega_1^+} y^{2\nu+1} dx dy \int\limits_{R_{n+1}^+} \frac{[j_\nu(\eta t y)]^2 \eta^{2\nu+1}}{[\mathscr{L}_0(x, y; \xi, \eta^2) + 1]^k} d\xi d\eta = \lim_{t\to\infty} t^{k-\tilde{\gamma}} tr G_t^k, \\ &\qquad \tilde{\gamma} = \frac{n+1+2\nu+1}{2m}, \qquad 2mk > n+1+2\nu+1. \end{aligned} \tag{3.19}$$

Since G_t is selfadjoint and positive, there exists a complete system $\varphi_1, \varphi_2, \cdots$ of eigenfunctions of the operator G_t (t fixed) with positive eigenvalues $(\lambda_1 + t)^{-1} \geqslant (\lambda_2 + t)^{-1} \geqslant \cdots$. By the properties of G_t a necessary and sufficient condition that $\varphi \in \mathfrak{L}_{2,\nu}$ and satisfies $G_t\varphi = (\lambda + t)^{-1}\varphi$ is that $\varphi \in \overset{\circ}{H}{}^m_\nu$ and satisfies $(\lambda + t)(\varphi, f)_\nu = \mathfrak{L}_t(\varphi, f)$ for all $f \in \overset{\circ}{H}{}^m_\nu$. Therefore $G_t\varphi = (\lambda + t)^{-1}\varphi$ means that $G_s\varphi = (\lambda + s)^{-1}\varphi$, and conversely, if s and t are sufficiently large. Consequently $G_t\varphi_j = (\lambda_j + t)^{-1}\varphi_j$ for all t. In addition, $(\lambda_j + t)(\varphi_j, f)_\nu = \mathfrak{L}_t(\varphi_j, f)$ means that $(\varphi_j, (\lambda_j - \mathfrak{L})f) = 0$ if $f \in C_0^\infty(\Omega_1^+)$; and by a property of weak solutions φ_j is infinitely differentiable and $\mathfrak{L}\varphi_j = \lambda_j\varphi_j$.

If we prove that

$$g_t^{(k)}(x, y; u, v) = \sum_j (\lambda_j + t)^{-k} \varphi_j(x, y)\varphi_j(u, v) \tag{3.20}$$

for $2mk > n + 1 + 2\nu + 1$, then we will have

$$\int_{\Omega_1^+} g_t^{(k)}(x, y; x, y)\, y^{2\nu+1} dx dy = \sum_j (\lambda_j + t)^{-k} = tr G_t^k. \tag{3.21}$$

For the proof of (3.20) we consider the kernel $g_t^{(k)}(x, y; u, v)$. This kernel, as was shown above, has the following properties. If $v > 0$, then it is bounded on $\Omega_1^+ \times \Omega_1^+$ for $2mk > n + 1$ and has the estimate

$$O(1) \times (\sqrt{|x - u|^2 + |y - v|^2})^{2mk-(n+1)-\varepsilon},$$

if $2mk \leqslant n + 1$; if $v = 0$ then either it is bounded in the corresponding domain for $2mk > n + 1 + 2\nu + 1$ or it has an estimate of the form

$$O(1)(\sqrt{|x - u|^2 + y^2})^{2mk-(n+1+2\nu+1)-\varepsilon} \text{ for } 2mk \leqslant n + 1 + 2\nu + 1.$$

Therefore if $2mk - (n + 1 + 2\nu + 1) - \epsilon > (n + 1 + 2\nu + 1)/2$; that is, if $2mk > (n + 1 + 2\nu + 1)/2 + \epsilon$, then the integral

$$\int_{V+} |g_t^{(k)}(x, y; u, v)|^2 v^{2\nu+1} du dv$$

is finite, and V^+ is contiguous to the hyperplane $v = 0$ and is entirely contained in Ω_1^+. We will also prove that

$$\int_{\Omega_1^+} |g_t^{(k)}(x, y; u, v)|^2 v^{2\nu+1} du dv < \infty. \tag{3.22}$$

We suppose $C(f, g) = t^k(G_t^k f, g)_\nu$ and $b = \mathfrak{L}_t^k/t^k$, and use the methods of the preceding part. Let U^+ be an open subset of Ω_1^+ contiguous to the corresponding hyperplane and entirely contained in Ω_1^+; we choose another open set U_1^+ that adjoins the same hyperplane and is entirely contained in Ω_1^+ and entirely contains U^+. Let $\eta \in C_0(U_1^+)$ be equal to one on U^+ and let $\tau = t^{1/2m}$ and in addition

$$q(x, y; u, v) = b(\tau; u, v, \mathscr{D}_u, \mathscr{B}_v)(\Gamma(\tau, x, y; u, v)(1-\eta))$$

for $(x, y) \in U^+$, and

$$r(x, y) = C(q(x, y, \ldots), g) \qquad \text{for} \qquad (x, y) \in U^+,$$

where $g \in C_0^\infty(\Omega_1^+ - U_1^+)$. If $f \in C_0^\infty(U^+)$, then it follows from the properties of C and Γ that

$$\int_{R_{n+1}^+} r(x, y) f(x, y) y^{2\nu+1} dx dy = C(q(f, \ldots), g)$$

$$= C(b(\Gamma(\tau, f, \ldots))(1-\eta), g) = C(f, g) - (b\Gamma(\tau, f, \ldots), g) = C(f, g).$$

Now the bilinear form C has the kernel $c(\tau; x, y; u, v)$, so that the latter result may be written in the form

$$\int_{R_{n+1}^+} r(x, y) f(x, y) y^{2\nu+1} dx dy = \int_{R_{n+1}^+} cf(x, y) g(u, v) y^{2\nu+1} dx dy v^{2\nu+1} du dv.$$

Since f is arbitrary in $C_0^\infty(U^+)$, since r is continuous and since c' is continuous if $x \neq u$ and $y \neq v$, we obtain

$$C(q(x, y; \ldots), g) = \int_{R_{n+1}^+} c(\tau; x, y; u, v) g(u, v) v^{2\nu+1} du dv,$$

if $(x, y) \in U^+$ and $g \in C_0^\infty(\Omega_1^+ - U_1^+)$. This shows that

$$\int_{\Omega_1^+ - U_1^+} |c(\tau; x, y; u, v)|^2 v^{2\nu+1} du dv \leqslant |C|^2 q|(x, y, \ldots)|^2,$$

if $(x, y) \in U^+$. Since $g_t^{(k)} = c(\tau; x, y; u, v)/t^k$, we obtain (3.22).

The kernel $g_t^{(k)}(x, y; \cdots)$ considered as an element of $\mathfrak{L}_{2,\nu}$ is uniformly continuous with respect to (x, y) in compact subsets of Ω_1^+ contiguous to the hyperplane $y = 0$.

By Fubini's Theorem and by the properties of the operator G_t we have

$$(\lambda_j + t)^{-k}(f, \varphi_j)_\nu = (G_t^k f, \varphi_j)$$

$$= \int_{R_{n+1}^+} \varphi_j(u, v) v^{2\nu+1} du dv \int_{R_{n+1}^+} g_t^{(k)}(x, y; u, v) f(x, y) y^{2\nu+1} dx dy$$

$$= \int_{R_{n+1}^+} \left\{ \int_{R_{n+1}^+} g_t^{(k)}(x, y; u, v) \varphi_j(u, v) v^{2\nu+1} du dv \right\} f y^{2\nu+1} dx dy,$$

if $f \in C_0^\infty(\Omega_1^+)$; this means that

$$\int_{R_{n+1}^+} g_t^{(k)}(x, y; u, v)\varphi_j v^{2\nu+1} du dv = (\lambda_j + t)^{-k}\varphi_j(x, y),$$

where both sides are continuous with respect to (x, y). Therefore by Parseval's formula we have

$$(g_t^{(k)}(x, y), \ldots), g_t^{(k)}(u, v, \ldots)) = \sum (\lambda_j + t)^{-2k}\varphi_j(x, y)\varphi_j(u, v). \tag{3.23}$$

By Fubini's Theorem we have

$$(G_t^k f, G_t^k f)_\nu = (G_t^{2k} f, f)_\nu = \int_{R_{n+1}^+} g_t^{(2k)}(x, y; u, v) f(x, y) f(u, v) y^{2\nu+1} dx dy v^{2\nu+1} du dv$$

for $f \in C_0^\infty(\Omega_1^+)$. Consequently the left-hand side of (3.23) is equal to $g_t^{(2k)}(x, y; u, v)$, and, since ϵ is arbitrary, this also proves (3.20) for $2mk > n + 1 + 2\nu + 1$. We note that under this condition the kernel $g_t^{(k)}(x, y; u, v)$ is continuous up to the corresponding hyperplane.

Thus we have shown above that for $2mk > n + 1 + 2\nu + 1$

$$g_t^{(k)} = \sum_j (\lambda_j + t)^{-k}\varphi_j(x, y)\varphi_j(u, v) \tag{3.24}$$

and that

$$\int_{\Omega_1^+} g_t^{(k)} y^{2\nu+1} dx dy = \sum_j (\lambda_j + t)^{-k} = tr G_t^k\,. \tag{3.25}$$

In addition we have the limiting equality

$$\lim_{t\to\infty} t^{k-\gamma} g_t^{(k)} = \delta_{x, y}^{(u, v)} C_\nu \int_{R_{n+1}^+} \frac{[j_\nu(t\eta y)]^2\,\eta^{2\nu+1}}{[\mathscr{L}_0(x, y; \xi, \eta^2) + 1]^k} d\xi d\eta, \tag{3.26}$$

where also $2mk > n + 1 + 2\nu + 1$ and $\gamma = (n + 1 + 2\nu + 1)/2m$.

Put

$$\sigma(\lambda) = \sum_{\lambda_j \leqslant \lambda} |\varphi_j(x, y)|^2.$$

Then by formula (3.24) we have for $k = 1$

$$g_t = g_t^{(1)}(x, y; x, y) = \sum_j \frac{|\varphi_j(x, y)|^2}{\lambda_j + t} = \int_0^\infty \frac{d\sigma(\lambda)}{\lambda + t}\,.$$

By the Tauberian theorem it follows from (3.26) that

$$\sigma(\lambda) \sim C_\nu \frac{2m}{\pi(n+1+2\nu+1)} \sin \frac{(n+1+2\nu+1)\pi}{2m} \lambda^{\frac{n+1+2\nu+1}{2m}} \times \int\limits_{R^+_{n+1}} \frac{[j_\nu(\lambda\eta y)]^2 \eta^{2\nu+1}}{\mathscr{L}_0(x;\ y;\ \xi,\ \eta^2)+1} d\xi d\eta. \tag{3.27}$$

For the application of the Tauberian theorem we must consider the cases $y > 0$ and $y = 0$ separately.

Since

$$S(x,\ y;\ x,\ y;\ \lambda) \sum_{\lambda_j \leqslant \lambda} |\varphi_j(x,\ y)|^2 = \sigma(\lambda),$$

we obtain an asymptotic formula for $S(x, y; x, y; \lambda)$. For $(x, y) \neq (u, v)$ consider the function

$$\sigma_1(\lambda) = \sum_{\lambda_j \leqslant \lambda} |\varphi_j(x,\ y) + \theta\varphi_j(u,\ v)|^2,$$

where θ is an arbitrary complex parameter $|\theta| = 1$. From the Tauberian theorem and the results obtained above it follows that

$$\lambda^{\frac{n+1+2\nu+1}{2m}} S(x,\ y;\ u,\ v;\ \lambda) \to 0 \tag{3.28}$$

for $\lambda \to \infty$.

Finally we can also obtain an asymptotic law for $N(\lambda)$. With regard to (3.25), for $k = 1$ it is necessary to apply the Tauberian theorem to the function $\sigma_2(\lambda) = N(\lambda)$. Finally we obtain the following asymptotic formula. If $2m > n + 1 + 2\nu + 1$, then for $\lambda \to \infty$

$$S(x,\ y;\ u,\ v;\ \lambda) = \sum_{\lambda_j \leqslant \lambda} \varphi_j(x,\ y)\varphi_j(u,\ v) \sim \delta^{(u,\ v)}_{x,\ y} C_\nu h^{(\nu)}_{m,\ n} \lambda^{\frac{n+1+2\nu+1}{2m}} \int\limits_{R^+_{n+1}} \frac{[j_\nu(\lambda\eta y)]^2 \eta^{2\nu+1}}{\mathscr{L}_0(x,\ y;\ \xi,\ \eta^2)+1} d\xi d\eta, \tag{3.29}$$

where

$$C_\nu = \frac{1}{(2\pi)^n 2^{2\nu}\Gamma^2(\nu+1)}, \quad h^{(\nu)}_{m,\ n} = \frac{2m}{(n+1+2\nu+1)\pi} \sin \frac{n+1+2\nu+1}{2m}\pi,$$

$$N(\lambda) = \sum_{\lambda_j \leqslant \lambda} 1 \sim h^{(\nu)}_{m,\ n} C_\nu \lambda^{\frac{n+1+2\nu+1}{2m}} \int\limits_{\Omega^+_1} y^{2\nu+1} dx dy \int\limits_{R^+_{n+1}} \frac{[j_\nu(\lambda\eta y)]^2 \eta^{2\nu+1}}{\mathscr{L}_0(x,\ y;\ \xi,\ \eta^2)+1} d\xi d\eta. \tag{3.30}$$

If (x, y) is in the interior of the domain Ω^+, then the operator $\mathfrak{L}$ becomes a usual regular elliptic operator and our asymptotic formulas coincide with the corresponding formulas obtained by Gårding [2] for regular elliptic operators.

BIBLIOGRAPHY

1. C. Clark, *The asymptotic distribution of eigenvalues and eigenfunctions for elliptic boundary value problems*, SIAM Rev. 9 (1967), 627–646.

2. L. Gårding, *On the asymptotic distribution of the eigenvalues and eigenfunctions of elliptic differential operators*, Math. Scand. **1** (1953), 237–255. MR **16**, 366.

3. I. A. Kiprijanov and V. I. Kononenko, *Fundamental solutions of B-elliptic equations*, Differencial'nye Uravnenija **3** (1967), 114–129 = Differential Equations **3** (1967), 57–64. MR **35** #1945.

4. L. Schwartz, *Théorie des distributions.* Tomes I, II, Actualités Sci. Indust., nos. 1091, 1122, Hermann, Paris, 1950, 1951. MR **12**, 31; 833.

5. B. M. Levitan, *Expansion in Fourier series and integrals with Bessel functions*, Uspehi Mat. Nauk **6** (1951), no. 2 (42), 102–143. (Russian) MR **14**, 163.

6. I. A. Kiprijanov and V. I. Kononenko, *The fundamental solutions of certain singular partial differential equations*, Differencial'nye Uravnenija **5** (1969), 1470–1483 = Differential Equations **5** (1969), 1082–1091. MR **40** #6054.

7. I. A. Kiprijanov, *On an inequality of Gårding for degenerate elliptic operators, and its applications*, Trudy Mat. Inst. Steklov. **105** (1969), 77–88 = Proc. Steklov Inst. Math. **105** (1969), 92–104. MR **41** #4302.

Translated by D. NEWTON

Trudy Mat. Inst. Steklov.
117 (1972)

Proc. Steklov Inst. Math.
117 (1972)

ON POLYNOMIAL TRACES AND MODULI OF SMOOTHNESS OF FUNCTIONS OF SEVERAL VARIALBES

L. D. KUDRJAVCEV

UDC 517.518.22

Abstract. Let $\varphi_j(t)$ $(1 \leqslant j \leqslant n; j$ integral) be nonnegative measurable functions defined for $t > 0$, let the function $f(x)$, $x = (x_1, \cdots, x_n)$, be defined on the "strip" $E_b^n = \{0 < x_n < b \leqslant +\infty\}$, $E^{n-1} = \{x_n = 0\}$, $p \geqslant 1$, and let

$$\left| f, L^{r_i r_n}_{p, \varphi_i, \varphi_n} \right| = \left| \varphi_i \frac{\partial^{r_i} f}{\partial x_i^{r_i}}, \quad L_p\left(\hat{E}_b^n\right) \right| + \left| \varphi_n \frac{\partial^{r_n} f}{\partial x_n^{r_n}}, \ L_p\left(\bar{E}_b^n\right) \right|.$$

The inequality

$$\left| \Delta^k_{h_i} f, L_p\left(E^{n-1}\right) \right| \leqslant c \left| f, L^{r_i r_n}_{p, \varphi_i, \varphi_n} \right| \left[| h_i |^{r_i} G_i(t) + t^{r_n} G_n(t) \right]$$

is proved, where

$$G_j(t) = \frac{1}{t} \left[\int_0^t \varphi_j^{-q}(t)\, dt \right]^{\frac{1}{q}}, \quad j = i, n; \quad \frac{1}{p} + \frac{1}{q} = 1,$$

$t > 0$ is an arbitrary parameter and the constant $c > 0$ does not depend on the function f. With the aid of this inequality the differential properties of the traces of functions from anisotropic weight spaces are studied, and, in particular, a criterion is given as to when the trace of the function on the hyperplane E^{n-1} is a polynomial.

Bibliography: 15 items.

The study of the properties of the traces of functions is related to the theory of imbedding of function spaces. Here on the one hand there are theorems establishing smoothness properties of the traces of functions of various classes on the boundary of a region (cf. the monographs of S. L. Sobolev [1] and S. M. Nikol'skiĭ [2]), and on the other hand there are theorems on the transition of functions at infinity into polynomials (cf., for example, S. L. Sobolev [3], S. V. Uspenskiĭ [4]). Up to now theorems of these two types have usually been proved independently of each other. In the present paper a method is presented for obtaining simultaneously the various properties of traces. It enables us to prove estimates from which, in every case, follow theorems on the transition of a function to a constant, as well as estimates of the Hölder integral norms of the traces of functions. This method is based on

---

AMS (MOS) subject classifications (1970). Primary 46E35; Secondary 26A69.

examination of anisotropic weight spaces and on obtaining in them estimates, depending on a parameter, for higher order differences.

The paper consists of two sections. The first section is supportive; the fundamental theorems are proved in the second section. First of all a criterion for the trace of a function to be a polynomial is established. This criterion contains, as a special case, earlier known criteria for the transition of a function to a constant (in addition to those indicated above, see also the papers of V. R. Portnov [5], V. N. Sedov [6], G. N. Jakovlev [7]). Then the smoothness modulus of the trace of a function belonging to the weight space is evaluated. In this connection note should be taken of the paper of G. N. Jakovlev [7], who studies the smoothness properties of the trace of functions possessing first derivatives and belonging to a weight space with nondecreasing weight. In the present paper conditions of the type of monotonicity are not imposed on the weight; differentiability of the weight, of different orders with respect to different variables, is assumed (i.e. what is usually called "anisotropism" of the function spaces).

§1. Definitions and notation

1.1. Notation for sets and functions. We shall denote by E^n the n-dimensional euclidean space of points $x = (x_i) = (x_1, \cdots, x_n)$, where x_i is the ith coordinate of the point x in some fixed coordinate system $i = 1, 2, \cdots, n$, $|x| = \sqrt{\Sigma_1^n x_i^2}$, $x^{(n-1)} = (x_1, \cdots, x_{n-1})$, and thus $x = (x^{(n-1)}, x_n)$.

Let $-\infty \leqslant a < b \leqslant +\infty$, $0 < d \leqslant +\infty$. Let us set

$$E^n(a, b; d) = \{x : a < x_n < b, \ |x_i| < d, \ i = 1, 2, \ldots, n-1\}, \tag{1.1}$$

$$E^n(b; d) = E^n(0, b; d), \tag{1.2}$$

$$\overset{+}{E}{}^n(a; d) = E^n(a, +\infty; d), \tag{1.3}$$

$$E^{n-1}(a; d) = \{x : x_n = a, \ |x_i| < d, \ i = 1, 2, \ldots, n-1\}, \tag{1.4}$$

$$E^{n-1}(d) = E^{n-1}(0; d), \tag{1.5}$$

$$\overset{+}{E}{}^n_a = \overset{+}{E}{}^n(a; +\infty), \qquad \overset{+}{E}{}^n = \overset{+}{E}{}^n_0 \text{ (a subspace)}, \tag{1.6}$$

$$E^n_b = E^n(b; +\infty) \text{ (a "strip")}, \tag{1.7}$$

$$E^{n-1}_a = E^{n-1}(a; +\infty), \qquad E^{n-1} = E^{n-1}_0 \text{ (a hyperplane)}. \tag{1.8}$$

Let $r = (r_1, \cdots, r_n)$, the r_i being nonnegative integers, $i = 1, \cdots, n$. We shall always understand derivatives to be generalized derivatives (cf. [1, 2]), and denote them by the symbols

$$D^{r_i} = \frac{\partial^{r_i}}{\partial x_i^{r_i}}, \qquad D^r = D^{r_1} D^{r_2} \ldots D^{r_n}.$$

1.2. Function spaces. Let B be a Banach space. We shall denote the norm of an element x in the space B by $|x, B|$.

Let G be an open set, $G \subset E^n$, and let $\varphi = \varphi(x)$ be a nonnegative measurable function defined on the set G, and let $1 \leqslant p \leqslant +\infty$.

For any nonnegative integer r_i we shall denote by $L^{r_i}_{p,\varphi} = L^{r_i}_{p,\varphi}(G)$ (if $r_i = 0$, simply $L_{p,\varphi}$ is written instead of $L^0_{p,\varphi}$) the normed linear space of functions f having in G derivatives of order r_i with respect to x_i, in which the norm (more precisely, the seminorm) is defined by the formula

$$|f, L^{r_i}_{p,\varphi}| = |\varphi D^{r_i} f, L_p(G)| \quad (i = 1, 2, \ldots, n). \tag{1.9}$$

If $r = (r_{i_1}, \cdots, r_{i_k})$ and $\varphi = (\varphi_{i_1}, \cdots, \varphi_{i_k})$, $1 \leqslant i_j \leqslant n, j = 1, 2, \cdots, k$, then we set

$$\left.\begin{aligned} L^r_{p,\varphi} &= \bigcap_{j=1}^{k} L^{r_{i_j}}_{p,\varphi_{i_j}}, \\ |f, L^r_{p,\varphi}| &= \sum_{j=1}^{k} \left| f, L^{r_{i_j}}_{p,\varphi_{i_j}} \right| \quad (1 \leqslant k \leqslant n). \end{aligned}\right\}. \tag{1.10}$$

The functions $\varphi_i(t)$ are called weights or weight functions, and the spaces $L^{r_i}_{p,\varphi_i}$, $L^r_{p,\varphi}$, weight spaces.

If $\varphi_i(x) \equiv 1$, then $L^{r_i}_{p,1} = L^{r_i}_p$ and $L^r_{p,1} = L^r_p$ are the usual Sobolev spaces [1].

1.3. Differences. Let the function $f(x)$ of the real variable x be defined on the entire real axis, and let h be a real number. Let us set

$$\left.\begin{aligned} \Delta_h f(x) &= \Delta^1_h f(x) = f(x+h) - f(x) \\ \Delta^{k+1}_h f(x) &= \Delta_h \left[\Delta^k_h f(x)\right], \quad k = 1, 2, \ldots \end{aligned}\right\}. \tag{1.11}$$

From this we get

$$\Delta^k_h f(x) = \sum_{j=0}^{k} (-1)^j C^j_k f[x + (k-j)h] = \sum_{j=0}^{k} (-1)^{j+k} C^j_k f(x + jh). \tag{1.12}$$

If the function f has a summable kth derivative, then by induction we get

$$\Delta^k_h f(x) = \int_0^h du_1 \int_0^h du_2 \ldots \int_0^h f^{(k)}\left(x + \sum_{i=1}^{k} u_i\right) du_k. \tag{1.13}$$

We shall obtain an estimate for the differences. From (1.12) we have

$$\begin{aligned} \Delta^{k+l}_h f(x) &= \Delta^l_h \int_0^h du_1 \ldots \int_0^h f^{(k)}\left(x + \sum_{i=1}^{k} u_i\right) du_k \\ &= \sum_{j=0}^{l} (-1)^{j+k} C^j_l \int_0^h du_1 \ldots \int_0^h f^{(k)}\left(x + jh + \sum_{i=1}^{k} u_i\right) du_k. \end{aligned}$$

Effecting in each integral the change of variable $v_1 = jh + \Sigma_{i=1}^k u_i$, $v_\nu = u_\nu$, $\nu = 2, 3, \cdots, k$ and noting that $|v_1| \leqslant (k+l)|h|$ and $|v_\nu| \leqslant |h|$, $\nu = 2, \cdots, k$, we obtain

$$|\Delta_h^{k+l} f(x)| \leqslant c\,|h|^{k-1} \int_0^{(k+l)|h|} |f^{(k)}(x + v \operatorname{sign} h)|\, dv, \qquad l \geqslant 0, \tag{1.14}$$

where the constant c does not depend on the function f.

If the function $f(x)$ belongs locally to the space L_p and $\Delta_h^r f(x) = 0$ for almost all h and x, then the function f is equivalent to a polynomial of degree no greater than $r - 1$.

Indeed, for $r = 1$ this is obvious. For $r > 1$ it follows from the condition $\Delta_h^r f(x) = 0$ that $\Delta_h^k f(x) = 0$ for all $k \geqslant r$; consequently the function f belongs, for example, to the class H_p^{r+1} (cf. [2], p. 190), and hence it automatically has derivatives of orders up to and including r.

From (1.13) we have

$$\int_0^h du_1 \int_0^h du_2 \ldots \int_0^h f^{(r)}\left(x + \sum_{i=1}^r u_i\right) du_k = \Delta_h^r f(x) = 0.$$

Differentiating this equality r times with respect to h, we get $f^{(r)}(x + rh) = 0$ for almost all h, i.e. the derivative $f^{(r)}$ is zero almost everywhere. Hence, taking, for example, Taylor's formula with integral form of remainder, we get that the function $f(x)$ is a polynomial of degree not greater than $r - 1$.

Now let the function $f(x)$ have a summable rth derivative on the interval $(0, b)$, $0 < b \leqslant +\infty$. We shall derive a formula giving the sth derivative $(0 \leqslant s \leqslant r - 1)$ at the origin in terms of the values of the function itself at certain nodes and the next higher derivative; that is, the derivative of order r.

Specifically we shall prove that

$$f^{(s)}(0) = \frac{1}{x^s} \sum_{\nu=2}^{r+1} a_{\nu s} f(\nu x) + b_{\nu s} \int_x^{\nu x} (\nu x - u)^{r-1} f^{(r)}(u)\, du \\ + c_s \int_0^x u^{r-s-1} f^{(r)}(u)\, du, \qquad 0 \leqslant s \leqslant r - 1. \tag{1.15}$$

The representation (1.15) is analogous to the representation of S. V. Uspenskiĭ ([8], formula (5)), and differs from it only in the choice of nodes. Both these representations are obtained by S. M. Nikol'skiĭ's method, by means of which the intermediate derivatives are expressed in terms of a higher order derivative and the function itself (cf. [9], [2]). This method consists in the following: we fix $x \in (0, b/(r+1))$ and substitute into the Taylor formula

$$f(x+h)=\sum_{j=0}^{r-1}\frac{f^{(j)}(x)}{j!}h^j+\frac{1}{(r-1)!}\int_x^{x+h}(x+h-u)^{r-1}f^{(r)}(u)\,du$$

the values $h = kx$, $k = 1, 2, \cdots, r$. We obtain

$$f[(k+1)x]=\sum_{j=0}^{r-1}k^j\frac{f^{(j)}(x)}{j!}x^j+\frac{1}{(r-1)!}\int_x^{(k+1)x}[(k+1)x-u]^{r-1}f^{(r)}(u)\,du. \quad (1.16)$$

We solve this system of equations for the $f^{(j)}(x)x^j/j!$, $j = 0, 1, \cdots, r-1$. This is possible, since with respect to these unknowns the system (1.16) is a linear system with the Vandermonde determinant $W(1, 2, \cdots, r) \neq 0$. We substitute the resulting values

$$\frac{f^{(j)}(x)}{j!}x^j=\sum_{k=2}^{r+1}a_kf(kx)+b_k\int_x^{kx}(kx-u)^{r-1}f^{(r)}(u)\,du, \qquad j=0,\ 1,\ \ldots,\ r-1,$$

(a_k and b_k, $k = 2, \cdots, r+1$, are constants) into the Taylor formula for the derivative $f^{(k)}$ at the point x with increment $-x$, i.e. into the formula

$$f^{(k)}(0)=\sum_{\nu=0}^{r-k-1}(-1)^\nu f^{(\nu+k)}(x)\frac{x^\nu}{\nu!}+\frac{(-1)^{r-k}}{(r-k+1)!}\int_0^x u^{r-k+1}f^{(r)}(u)\,du,$$

and as a result we obtain (1.15).

We make note of an estimate of the remainder term in this formula which we shall need later on:

$$\left|\int_x^{kx}(kx-u)^{r-1}f^{(r)}(u)\,du\right|\leqslant(k-1)\int_0^{kx}u^{r-1}\,|f^{(r)}(u)|\,du. \quad (1.17)$$

This follows directly from the fact that if $x \leqslant u \leqslant kx$, then $0 \leqslant kx - u \leqslant (k-1)u$.

Now let $f(x)$ be a function of several variables: $x = (x_1, \cdots, x_n)$. Let e_i be the unit vector in the direction of the ith coordinate axis.

Let us set

$$\Delta_{h_i}f(x)=\Delta^1_{h_i}f(x)=f(x+h_ie_i)-f(x),$$
$$\Delta^{k+1}_{h_i}f(x)=\Delta_{h_i}[\Delta^k_{h_i}f(x)], \qquad k=1,\ 2,\ \ldots.$$

LEMMA 1. *Let the function* $f(x_1, x_2)$ *possess on* E^2_b $(0 < b < +\infty)$ *the partial derivatives* $\partial^{r_1}f/\partial x_1^{r_1}$ *and* $\partial^{r_2}f/\partial x_2^{r_2}$, $r_1 \geqslant 1$, $r_2 \geqslant 1$. *Let the derivative* $\partial^{r_2}f/\partial x_2^{r_2}$ *be summable on the interval* $[0, b]$ *for almost all* x_1, *and let* $0 \leqslant s \leqslant r_2$ *and* $k \geqslant r_1$. *Then*

$$\left|\Delta_{h_1}^k \frac{\partial^s f(x_1, 0)}{\partial x_2^s}\right| \leqslant \frac{a_s}{x_2^s}|h_1|^{r_1-1}\sum_{\nu=2}^{r_2+1}\int_0^{k|h_1|}\left|\frac{\partial^{r_1} f(x_1+v\,\mathrm{sign}\,h_1, \nu x_2)}{\partial x_1^{r_1}}\right|dv$$

$$+\frac{b_s}{x_2^s}\sum_{j=0}^{k}\int_0^{(r_2+1)x_2} u^{r_2-1}\left|\frac{\partial^{r_2} f(x_1+jh_1, u)}{\partial x_2^{r_2}}\right|du$$

$$+c_s\sum_{j=0}^{k}\int_0^{x_2} u^{r_2-s-1}\left|\frac{\partial^{r_2} f(x_1+jh_1, u}{\partial x_2^{r_2}}\right|du, \tag{1.18}$$

where a_s, b_s *and* c_s $(s = 0, 1, \cdots, r_2 - 1)$ *are constants not depending on the function* f.

PROOF. Using formulas (1.15) and (1.12), we get

$$\Delta_{h_1}^k \frac{\partial^s f(x_1, 0)}{\partial x_2^s} = \frac{1}{x_2^s}\sum_{\nu=2}^{r_2+1} a_{\nu s}\Delta_{h_1}^k f(x_1, \nu x_2) + b_{\nu s}\Delta_{h_1}^k\int_{x_2}^{\nu x_2}(\nu x_2 - u)^{r_2-1}\frac{\partial^{r_2} f(x_1, u)}{\partial x_2^{r_2}}du$$

$$+ c_s\Delta_{h_1}^k\int_0^{x_2} u^{r_2-s-1}\frac{\partial^{r_2} f(x_1, u)}{\partial x_2^{r_2}}du = \frac{1}{x_2^s}\sum_{\nu=2}^{r_2+1}\Big[a_{\nu s}\Delta_{h_1}^{k-r_1}\Delta_{h_1}^{r_1} f(x_1, \nu x_2)$$

$$+ b_{\nu s}\sum_{j=0}^{k}(-1)^{j+k}C_k^j\int_{x_2}^{\nu x_2}(\nu x_2 - u)^{r_2-1}\frac{\partial^{r_2} f(x_1+jh_1, u)}{\partial x_2^{r_2}}du\Big]$$

$$+ c_s\sum_{j=0}^{k}(-1)^{j+k}C_k^j\int_0^{x_2} u^{r_2-s-1}\frac{\partial^{r_2} f(x_1+jh_1, u)}{\partial x_2^{r_2}}du.$$

We apply the inequalities (1.14) and (1.17) and increase the upper limits of integration from νx_2 to $(r_2 + 1)x_2$. Then, introducing new constants a_s, b_s and c_s, we obtain (1.18). This estimate is analogous to the corresponding estimate by S. V. Uspenskiĭ [8].

1.4. Traces.

DEFINITION 1. *Let the function* $f(x)$ *be defined on a "parallelepiped"* $E^n(a, b; d)$ *(see (1.1)),* $-\infty \leqslant a < b \leqslant +\infty$, $0 < d \leqslant +\infty$, *and let* $a \leqslant \eta \leqslant b$. *If it is possible to modify the function* f *on a set of* n*-dimensional measure zero so that after the modification*(1) *the finite limit* $\lim_{x_n\to\eta, x\in E^n(a,b;d)} f(x)$, *denoted by* $f(x^{(n-1)}, \eta)$, *will exist for almost all (in the sense of* $(n-1)$*-dimensional measure) points* $x^{(n-1)} \in E_d^{n-1}$ *(see (1.4)), then this limit is called the trace, in the sense of convergence almost everywhere, on the function on the* $(n-1)$*-dimensional "parallelepiped"* $E^{n-1}(\eta; d)$.

Since the case $\eta = b = +\infty$, $\eta = a = -\infty$, is not excluded, the concept of the "trace at infinity" is, in particular, defined, i.e. for $x_n = +\infty$ and $x_n = -\infty$.

---

(1) We shall denote the modified function by the same symbol f.

It is not difficult to check that the trace of a function in the sense of convergence almost everywhere (if it exists) is uniquely determined up to a set of $(n-1)$-dimensional measure zero.

The trace $f(x^{(n-1)}, \eta)$ of a function f on $E^{n-1}(\eta; d)$ is also denoted by $f|_{E^{n-1}(\eta;d)}$.

DEFINITION 2. *Let the function $f(x)$ be defined on the parallelepiped $E^n(a, b; d)$ and let $-\infty \leqslant a \leqslant \eta \leqslant b \leqslant +\infty$, $0 < d \leqslant +\infty$. If there exists a function $\varphi(x^{(n-1)})$ defined on E_d^{n-1}, and if the function f can be so modified on a set of n-dimensional measure zero that after the modification*

$$\lim_{h\to 0} \int |f(x^{(n-1)}, \eta+h) - \varphi(x^{(n-1)})|^p \, dE_d^{n-1} = 0, \quad (x^{(n-1)}, \eta+h) \in E^n(a, b; d),$$

then the function $\varphi(x^{(n-1)})$ is called the trace, in the sense of convergence in the mean (in the L_p metric), of the function f on the n-dimensional parallelepiped $E^{n-1}(\eta; d)$ (2)

It is not difficult to check that the trace of a function in the sense of convergence in the mean is determined uniquely up to a set of $(n-1)$-dimensional measure zero.

For the cases to be considered here it will be proved that the trace of a function in the sense of convergence almost everywhere coincides with the trace in the sense of convergence in the mean. Hence for these two concepts of trace the same notation will be used; and, where no ambiguity will arise, simply the term "trace" will be used.

We remark, for example, that if $f \in L_p^1(E^n(a, b; d))$, $p \geqslant 1$, then for any finite η there exist on $E^{n-1}(\eta; d)$ traces of the function f in the sense of convergence almost everywhere as well as in the sense of convergence in the mean in the L_p metric; and, further, they coincide and locally belong to L_p. In the case $p = 1$, this is true also for $\eta = +\infty$.

1.5. Functional inequalities. In what follows use will frequently be made of the so-called generalized Minkowski inequality

$$\left\| \int F(u, v)\, du, L_p \right\| \leqslant \int \| F(u, v), L_p \| \, dv \tag{1.19}$$

as well as of two generalized inequalities of Hardy, on which we shall dwell in somewhat more detail.

Let $1 < p < +\infty$, $1/p + 1/q = 1$. Let the nonnegative function $\varphi(x)$ be defined on the interval (a, b), $-\infty \leqslant a < b \leqslant +\infty$, and for any $c \in (a, b)$ let

$$\int_c^b \varphi^{-q}(x)\, dx < +\infty \tag{1.20}$$

(2) Sometimes in the case of weight spaces the trace in the sense of convergence in the mean is understood in another sense, namely in the sense of convergence in the mean with weight; see, for example, G. N. Jakovlev [7].

(it is essential to observe that finiteness of this integral for $c = a$ is not assumed). From the condition (1.20) it follows, in particular, that the function $\varphi(x)$ is different from zero almost everywhere on (a, b).

If

$$\psi(x) = \varphi^{-\frac{q}{p}}(x)\left[\int_x^b \varphi^{-q}(t)\,dt\right]^{-1}, \quad a < x < b, \tag{1.21}$$

then for any function $f(x)$ measurable on (a, b) the inequality

$$\left\{\int_a^b \left[\psi(x)\int_x^b |f(t)|\,dt\right]^p dx\right\}^{\frac{1}{p}} \leqslant \frac{p}{p-1}\left[\int_a^b |\varphi(x) f(x)|^p\,dx\right]^{\frac{1}{p}} \tag{1.22}$$

is valid (a so-called generalized inequality of Hardy). This inequality, as well as inequality (1.34) derived below, was earlier treated in papers of F. A. Sysoeva [10], V. R. Portnov [5], V. N. Sedov [6], and G. N. Jakovlev [7].

The function $\psi(x)$ determined by formula (1.21) can also be written in the form

$$\psi^p(x) = \frac{1}{p-1}\frac{d}{dx}\left[\int_x^b \varphi^{-q}(t)\,dt\right]^{-p+1}. \tag{1.23}$$

LEMMA 2. *If* $c \in (a, b)$ *and condition* (1.20) *is satisfied, then the function* $\psi(x)$ *belongs to* $L_p(a, c)$:

$$\int_a^c \psi^p(x)\,dx < +\infty. \tag{1.24}$$

We emphasize that the inequality (1.24) is satisfied independently of whether the integral

$$\int_a^c \varphi^{-q}(x)\,dx \tag{1.25}$$

is finite or infinite.

PROOF OF THE LEMMA. Employing formula (1.23) for an arbitrary $\eta \in (a, c)$, we obtain

$$\int_\eta^c \psi^p(x)\,dx = \frac{1}{p-1}\left\{\left[\int_c^b \varphi^{-q}(t)\,dt\right]^{-p+1} - \left[\int_\eta^b \varphi^{-q}(t)\,dt\right]^{-p+1}\right\}. \tag{1.26}$$

Passing now to the limit as $\eta \to a + 0$, both in the case where the integral (1.25) converges and in the case where it diverges (since it occurs in the denominator), we get a finite limit in the right member of the equality. The lemma is proved.

We note that even in the simplest case where $\varphi(x) \asymp 1$ as $x \to b-0$,(³) we have $\psi(x) \asymp (b-x)^{-1}$ as $x \to b-0$, and consequently $\psi(x)$ has a nonintegrable singularity for $x=b$. Thus in Lemma 2 the interval (a, c) cannot be replaced by the whole interval (a, b).

We note two estimates for the function $\psi(x)$ in the case where it is monotonic.

1. If the function $\varphi(x)$ is nonincreasing on the interval (a, b), then

$$(b-x)\psi(x) \leqslant \varphi(x), \qquad a<x<b. \tag{1.27}$$

In fact, noting that $q/p=q-1$ and that $\varphi(x)/\varphi(t) \geqslant 1$ for $x<t<b$, we have (see (1.21))

$$\psi(x) = \frac{\varphi(x)}{\int\limits_x^b \frac{\varphi^q(x)}{\varphi^q(t)}\,dt} \leqslant \frac{\varphi(x)}{b-x}. \tag{1.28}$$

2. If the function $\varphi(x)$ is nondecreasing on the interval (a, b), then

$$(b-x)\psi(x) \geqslant \varphi(x). \tag{1.29}$$

This is proved in a way analogous to the preceding.

There are also analogs of the inequality (1.22) for the cases $p=1$ and $p=\infty$.

If the function φ is nonnegative and possesses a generalized derivative on the interval (a, b), then the inequalities

$$\int\limits_a^b \varphi'(x)\,dx \int\limits_x^b |f(t)|\,dt \leqslant \int\limits_a^b \varphi(x)|f(x)|\,dx \qquad (p=1), \tag{1.30}$$

$$\varphi'(x)\int\limits_x^b |f(t)|\,dt \leqslant \int\limits_a^b \varphi(x)|f(x)|\,dx + \operatorname*{ess\,sup}_{a<x<b} \varphi(x)|f(x)| \quad (p=+\infty) \tag{1.31}$$

hold.

EXAMPLES. 1. Let $\varphi(x)^\alpha = x$ and $a=0$. Then as $x \to +0$

$$\psi(x) \asymp \begin{cases} x^{\alpha-1}, & \text{if } \alpha > \frac{1}{q}, \\ x^{-\frac{1}{p}} \ln^{-1}\frac{1}{x}, & \text{if } \alpha = \frac{1}{q}, \\ x^{-\frac{\alpha q}{p}}, & \text{if } \alpha < \frac{1}{q}. \end{cases}$$

(³) $f(x) \asymp g(x)$ as $x \to x_0$ signifies that $f(x) = O(g(x))$ and $g(x) = O(f(x))$ as $x \to x_0$.

2. Let $\varphi(x) = xe^{-1/x}$, $p = q = 2$ and $a = 0$. Then $\psi(x) \asymp x^{-2}\varphi(x)$ as $x \to +0$.

We proceed now to the second generalized Hardy inequality.

Again let $1 < p < +\infty$ and let the nonnegative function $\varphi(x)$ be defined on the interval (a, b), $-\infty \leqslant a < b \leqslant +\infty$; but now for any $c \in (a, b)$ let the condition

$$\int_a^c \varphi^{-q}(x)\,dx < +\infty \tag{1.32}$$

be satisfied. If

$$\omega(x) = \varphi^{-\frac{q}{p}}(x)\left[\int_a^x \varphi^{-q}(t)\,dt\right]^{-1}, \qquad a < x < b, \tag{1.33}$$

then for any function f measurable on (a, b) there holds the inequality

$$\left\{\int_a^b \left[\omega(x)\int_a^x |f(t)|\,dt\right]^p dx\right\}^{\frac{1}{p}} \leqslant \frac{p}{p-1}\left[\int_a^b |\varphi(x) f(x)|^p\,dx\right]^{\frac{1}{p}}. \tag{1.34}$$

By analogy with Lemma 2, starting from the formula

$$\omega^p(x) = -\frac{1}{p-1}\frac{d}{dx}\left[\int_a^x \varphi^{-q}(t)\,dt\right]^{-p+1}, \tag{1.35}$$

we prove

LEMMA 3. *If $c \in (a, b)$ and condition* (1.32) *is satisfied, then*

$$\omega(x) \in L_p(c, b).$$

In the case corresponding to monotonicity of the function $\varphi(x)$ the following estimates for the function $\omega(x)$ are valid.

1. If $\varphi(x)$ is nonincreasing on (a, b), then

$$(x-a)\,\omega(x) \geqslant \varphi(x). \tag{1.36}$$

2. If $\varphi(x)$ is nondecreasing on (a, b), then

$$(x-a)\,\omega(x) \leqslant \varphi(x). \tag{1.37}$$

If $\varphi(x)$ possesses a generalized derivative on (a, b), then

$$\int_a^b \varphi'(x)\,dx \int_a^x |f(t)|\,dt \leqslant \int_a^b \varphi(x)|f(x)|\,dx \qquad (p=1), \tag{1.38}$$

$$|\varphi'(x)|\int_a^x |f(t)|\,dt \leqslant \int_a^b \varphi(x)|f(x)|\,dx + \operatorname*{ess\,sup}_{a<x<b} \varphi(x)|f(x)| \qquad (p=+\infty). \tag{1.39}$$

EXAMPLE. If $\varphi(x) = x^\alpha$, $b = +\infty$, then as $x \to +\infty$

$$\omega(x) \asymp \begin{cases} x^{\alpha-1}, & \text{if } \alpha < \frac{1}{q}, \\ x^{-\frac{1}{p}} \ln^{-1} \frac{1}{x}, & \text{if } \alpha = \frac{1}{q}, \\ x^{-\frac{\alpha q}{p}}, & \text{if } \alpha > \frac{1}{q}. \end{cases}$$

REMARK 1. Thus if the integrals $\int_c^b \varphi^{-q}(x)dx$ converge for all $c \in (a, b)$, then inequality (1.22) holds; but if the integrals $\int_a^c \varphi^{-q}(x)dx$ converge, then (1.34) is valid. Finally, if the integral $\int_a^b \varphi^{-q}(x)dx$ converges, then both (1.22) and (1.34) are valid. Moreover, in the neighborhood of the point $x = a$ the inequality (1.34) becomes "sharper" in the sense that $\psi(x) = o(\omega(x))$ as $x \to a + 0$, and in the neighborhood of the point $x = b$ (1.22) becomes "sharper" in the sense that $\omega(x) = o(\psi(x))$ as $x \to b - 0$. This follows from formulas (1.21) and (1.33).

REMARK 2. We shall now prove that in the cases of importance from the point of view of the preceding note the functions $\psi(x)$ and $\omega(x)$ in their behavior as $x \to a + 0$ are close, in a well-defined sense, to the derivative $\varphi'(x)$ of the function $\varphi(x)$ (if, of course, it exists); concerning this behavior see also [15].

Let us first consider the case where condition (1.32) is satisfied. In this case in the neighborhood of the point a the inequality (1.34) is "sharper". Let us consider the case where the finite or infinite limits

$$\lim_{x\to a+0} \varphi(x), \qquad \lim_{x\to a+0} \varphi'(x)$$

exist.

Case 1. If

$$\lim_{x\to a+0} \varphi(x) = +\infty \tag{1.40}$$

(then in some neighborhood of the point a the condition (1.32) holds), and if $\lim_{x\to a+0}\varphi'(x)$ exists, then

$$\lim_{x\to a+0} \omega(x) = -\lim_{x\to a+0} \varphi'(x) = +\infty. \tag{1.41}$$

PROOF. Applying l'Hospital's rule for determination of the limit of the function $\omega(x)$ defined by formula (1.33), we obtain

$$\lim_{x\to a+0} \omega(x) = \lim_{x\to a+0} \frac{\varphi^{-\frac{q}{p}}(x)}{\int_a^x \varphi^{-q}(t)\,dt} = -\frac{q}{p} \lim_{x\to a+0} \varphi'(x), \tag{1.42}$$

since $q - q/p - 1 = 0$. Let us further choose any sequences $\{x_n\}$ and $\{y_n\}$, $a < x_n < y_n < c < b$, such that

$$\lim_{n\to\infty} x_n = \lim_{n\to\infty} y_n = a, \quad \lim_{n\to\infty} [\varphi(x_n) - \varphi(y_n)] = +\infty \tag{1.43}$$

(this is always possible). By Lagrange's formula

$$\varphi(y_n) - \varphi(x_n) = \varphi'(\xi_n)(y_n - x_n), \quad x_n < \xi_n < y_n, \quad n = 1, 2, \ldots .$$

From (1.43) it follows that

$$\lim_{n\to\infty} \varphi'(\xi_n) = -\infty, \quad \lim_{n\to\infty} \xi_n = a,$$

and since $\lim_{x\to a+0}\varphi'(x)$ exists, we have $\lim_{x\to a+0}\varphi'(x) = -\infty$, and hence, on the basis of (1.42),

$$\lim_{x\to a+0} \omega(x) = +\infty.$$

Formula (1.41) is proved.

Case 2. If

$$\lim_{x\to a+0} \varphi(x) = 0, \tag{1.44}$$

condition (1.32) is satisfied and a summable derivative $\varphi'(x)$ exists, with $\lim_{x\to a+0}\varphi'(x)$ existing, then

$$\lim_{x\to a+0} \omega(x) = \lim_{x\to a+0} \varphi'(x) = +\infty. \tag{1.45}$$

PROOF. Under the assumptions made the equality $\lim_{x\to a+0}\omega(x) = +\infty$ follows directly from (1.33). Further, condition (1.44) implies

$$\varphi(x) = \int_a^x \varphi'(t)\,dt. \tag{1.46}$$

Substituting this expression into (1.32), we conclude that the convergence of the integral

$$\int_a^c \left[\int_a^x \varphi'(t)\,dt\right]^{-q} dx$$

implies the existence of a sequence

$$\{x_n\}, \quad \lim_{n\to\infty} x_n = a, \quad x_n > a,$$

such that

$$\left[\int_a^{x_n} \varphi'(t)\,dt\right]^{-q} < \frac{1}{x_n - a},$$

whence (on noting that (1.46) implies that $\int_a^x\varphi'(t)dt > 0$ for almost all $x > a$) follows

$$(x_n - a)^{\frac{1}{q}} \leqslant \int_a^{x_n} \varphi'(t)\,dt. \tag{1.47}$$

If a constant $M > 0$ were found such that in some neighborhood of the point a the inequality $\varphi'(t) \leqslant M$ were satisfied, then from (1.47) it would follow that

$$(x_n - a)^{\frac{1}{q}} \leqslant M(x_n - a),$$

and hence $M(x_n - a)^{1/p} \geqslant 1$. However, this is not fulfilled for sufficiently large indices n. Hence the indicated $M > 0$ does not exist, i.e. there is a sequence $\{\xi_n\}$, $\lim_{n\to\infty}\xi_n = a$, $a < \xi_n < b$, $n = 1, 2, \cdots$, such that $\lim_{n\to\infty}\varphi'(\xi_n) = +\infty$. Consequently on the basis of the existence of $\lim_{x\to a+0}\varphi'(x)$ we have $\lim_{x\to a+0}\varphi'(x) = +\infty$. The equality (1.45) is proved.

Case 3. If the finite limit

$$\lim_{x\to a+0} \varphi(x) \neq 0 \tag{1.48}$$

exists (and consequently condition (1.32) is satisfied in some neighborhood of the point a), then formula (1.33) implies that

$$\lim_{x\to a+0} \omega(x) = +\infty.$$

Along with this, even for $\varphi(x) \equiv 1$ we have $\lim_{x\to a+0}\varphi'(x) = 0$. Thus in the case of (1.48) the limits

$$\lim_{x\to a+0} \omega(x) \text{ and } \lim_{x\to a+0} \varphi'(x),$$

generally speaking, do not coincide even up to a constant.

Case 4. If for all $c \in (a, b)$

$$\int_a^c \varphi^{-q}(x)\,dx = +\infty, \tag{1.49}$$

condition (1.20) is satisfied and the finite or infinite limits

$$\lim_{x\to a+0} \varphi(x), \quad \lim_{x\to a+0} \varphi'(x),$$

exist, then the finite or infinite limit $\lim_{x\to a+0}\psi(x)$ exists, and

$$\lim_{x\to a+0} \psi(x) = \frac{q}{p} \lim_{x\to a+0} \varphi'(x).$$

PROOF. By virtue of the existence of the limit $\lim_{x\to a+0}\varphi(x)$, condition (1.49) implies that $\lim_{x\to a+0}\varphi(x) = 0$. Further, applying l'Hospital's rule, we find that from (1.21) that

$$\lim_{x\to a+0}\psi(x)=\lim_{x\to a+0}\frac{\varphi^{-\frac{q}{p}}(x)}{\int\limits_x^b\varphi^{-q}(t)\,dt}=\frac{q}{p}\lim_{x\to a+0}\varphi'(x).$$

On the basis of what has been proved the natural question arises as to whether the functions $\varphi'(x)$ and $\omega(x)$ in Cases 1 and 2, and the functions $\varphi'(x)$ and $\psi(x)$ in Case 4, may not be functions of the same order, i.e. $\varphi'(x)\asymp\omega(x)$ and $\varphi'(x)\asymp\psi(x)$, respectively, as $x\to a+0$, at least under supplementary constraints on the monotonicity of the functions $\varphi(x)$ and $\varphi'(x)$.

Everything which has been said carries over in a natural way also to the case $x\to b-0$.

§2. Properties of the traces of functions from the spaces $L^r_{p,\varphi}(E^n(b;d))$ and $L^r_{p,\varphi}(\overset{+}{E}{}^n(a;d))$

2.1. Weight functions. We shall always assume that the nonnegative function $\varphi(x)$ is defined on a "parallelepiped" $E^n(a,b;d)\subset E^n$, with $0<d\leqslant+\infty$ (see (1.1)). Two fundamental cases will be considered: $a=0$ and $b=+\infty$, i.e. the cases $E^n(b;d)$ (see (1.2)) and $\overset{+}{E}{}^n(a;d)$ (see (1.3)). However, everything said below will carry over also to any parallelepiped $E^n(a,b;d)$, only the exposition becomes more cumbersome.

DEFINITION 1. *The condition*

$$\int\varphi^{-q}(x)\,dE^n(c;\ d)<+\infty \tag{2.1}$$

for all $c\in(0,b)$ *is called a condition of weak degeneracy of the function* $\varphi(x)$ *at the origin.*

DEFINITION 2. *The condition*

$$\int\varphi^{-q}(x)\,dE^n(c;\ d)=+\infty \tag{2.2}$$

for all $c\in(0,b)$ *is called a condition of strong degeneracy of the function* $\varphi(x)$ *at the origin.*

DEFINITION 3. *The condition*

$$\int\varphi^{-q}(x)\,d\overset{+}{E}{}^n(c;\ d)<+\infty \tag{2.3}$$

for all $c\in(a,+\infty)$ *is called a condition of weak degeneracy of the function* $\varphi(x)$ *at infinity.*

DEFINITION 4. *The condition*

$$\int\varphi^{-q}(x)\,d\overset{+}{E}{}^n(c;\ d)=+\infty \tag{2.4}$$

for all $c\in(a,+\infty)$ *is called a condition of strong degeneracy at infinity.*

Let $\varphi(x) = x_n^\alpha$ and $x = (x^{(n-1)}, x_n)$; then for $b \leqslant +\infty$ the condition of weak degeneracy at the origin is equivalent to the condition $\alpha < 1/q$, and the condition of strong degeneracy at the origin is equivalent to the condition $\alpha \geqslant 1/q$. For $a \geqslant 0$ the condition of weak degeneracy at infinity is equivalent to the condition $\alpha > 1/q$, and the condition of strong degeneracy is equivalent to the condition $\alpha \leqslant 1/q$.

If the function $\varphi(x)$ is defined on $E^n(b; d)$ and is such that for all $c \in (0, b)$

$$\int \varphi^{-q}(x)\, dE^n(c, b; d) < +\infty, \tag{2.5}$$

then we shall set

$$\psi(x) = \psi(x^{(n-1)}, x_n) = \varphi^{-\frac{q}{p}}(x^{(n-1)}, x_n)\left[\int \varphi^{-q}(x^{(n-1)}, t)\, dE^n(x_n, b; d)\right]^{-1}. \tag{2.6}$$

But if the function $\varphi(x)$ is such that for all $c \in (0, b)$

$$\int \varphi^{-q}(x)\, dE^n(c; d) < +\infty, \tag{2.7}$$

then we set

$$\omega(x) = \omega(x^{(n-1)}, x_n) = \varphi^{-\frac{q}{p}}(x^{(n-1)}, x_n)\left[\int \varphi^{-q}(x^{(n-1)}, t)\, dE^n(a, x_n; d)\right]^{-1}. \tag{2.8}$$

The functions $\psi(x)$ and $\omega(x)$ are defined analogously also in the case where $\varphi(x)$ is given on $\overset{+}{E}{}^n(a; d)$.

Below, when the function $\varphi(x)$ may have some index, for example $\varphi_i(x)$, then we shall ascribe that same index to the functions corresponding to it on the basis of formulas (2.6) and (2.8); for example, in the case indicated we shall write $\psi_i(x)$ and $\omega_i(x)$.

We shall call the functions $\varphi_i(x)$, $\psi_i(x)$ and $\omega_i(x)$ weight functions.

2.2. The case of strong degeneracy. Let us first examine degeneracy around the origin. Let the nonnegative function $\varphi_n(x)$ and the function $f(x)$ together with its generalized derivative $\partial f(x)/\partial x_n$ be defined on $E^n(b; d)$. Then the following lemma holds.

LEMMA 1. *For any $\eta \in (0, b)$ let there exist constants $c_1 = c_1(\eta) > 0$ and $c_2 = c_2(\eta) > 0$ such that for any $x_n \in (\eta, b)$ the inequality*

$$c_1 \leqslant \varphi_n(x^{(n-1)}, x_n) \leqslant c_2 < +\infty \tag{2.9}$$

is satisfied. Then for any $\xi \in (0, b)$ the inequality

$$|f, L_{p,\psi_n}(E^n(\xi; d))| \leqslant c[|f, L_p(E^{n-1}(\xi; d))| + |f, L^{1_n}_{p,\varphi_n}(E^n(\xi; d))|] \tag{2.10}$$

holds, where $c = c(\xi)$ does not depend on the function f, and $d \leqslant +\infty$.

PROOF. If $0 < x_n < \xi < b$, then from the definition of generalized derivative it follows that

$$f(x^{(n-1)}, x_n) = f(x^{(n-1)}, \xi) - \int_{x_n}^{\xi} \frac{\partial f(x^{(n-1)}, t)}{\partial x_n} dt$$

almost everywhere on $E^n(b; d)$. Hence

$$\begin{aligned} |f, L_{p,\psi_n}(E^n(\xi; d))| &= \left[\int |\psi_n(x) f(x)|^p dE^n(\xi; d)\right]^{\frac{1}{p}} \\ &\leqslant \left\{\int \left|\psi_n(x)\left[f(x^{(n-1)}, \xi) - \int_{x_n}^{\xi} \frac{\partial f(x^{(n-1)}, t)}{\partial x_n} dt\right]\right|^p dE^n(\xi; d)\right\}^{\frac{1}{p}} \\ &\leqslant \left\{\int |f(x^{(n-1)}, \xi)|^p dE^{n-1}(d) \int_0^{\xi} \psi_n^p(x^{(n-1)}, x_n)\, dx_n\right\}^{\frac{1}{p}} \\ &+ \left\{\int dE^{n-1}(d) \int_0^{\xi} \left[\psi_n(x^{(n-1)}, x_n) \int_{x_n}^{\xi} \left|\frac{\partial f(x^{(n-1)}, t)}{\partial x_n}\right| dt\right]^p\right\}^{\frac{1}{p}}. \end{aligned} \tag{2.11}$$

On the basis of formulas (1.23) and the property (2.9) of the function φ_n we obtain

$$\int_0^{\xi} \psi_n^p(x^{(n-1)}, x_n)\, dx_n \leqslant \frac{1}{p-1}\left[\int_{\xi}^{b} \varphi_n^{-q}(x^{(n-1)}, t)\, dt\right]^{-p+1} \leqslant c(\xi), \tag{2.12}$$

where $c(\xi) > 0$ is some constant. Now from (2.11), applying the generalized Hardy inequality (1.22) to the first term of the right member of (2.12) and to the inner integral of the second term, we get (2.10). The lemma is proved.

REMARK 1. If $f \in L^{1n}_{p,\varphi_n}(E^n(b; d)) \cap L_p(E^n(a, b; d))$ for some $a \in (0, b)$, then from the property (2.9) of the function φ it follows that for any $\xi \in (0, b)$ the function f belongs to the space $L^{1n}_p(E^n(\xi, b; d))$, and consequently, according to the imbedding theorem of S. L. Sobolev (cf. [1]), has on $E^{n-1}(\xi; d)$ a locally summable trace. Hence for $f \in L^{1n}_{p,\varphi_n}(E^n(b; d)) \cap L_p(E^n(a, b; d))$, $a > 0$ and $d < +\infty$ both terms of the right member of the inequality (2.10) are finite.

REMARK 2. If the trace of the function f on $E^{n-1}(\xi; d)$ is equal to zero, $f(x^{(n-1)}, \xi) = 0$, then inequality (2.10) is valid without appeal to property (2.9) of the function $\varphi_n(x)$. In fact, this property is used only for evaluation of the first term of the right member of inequality (2.11), which in this case is equal to zero. For this case the inequality (2.10) was obtained earlier in papers of V. N. Sedov [13] and A. Kufner [14].

REMARK 3. For establishment of the validity of (2.10) neither an assumption on the weak degeneracy of the weight function $\varphi_n(x)$ nor an assumption on its strong degeneracy was used. However, in a well-defined sense this inequality is characteristic expressly for the case of strong degeneracy, since in the sequel (see Lemma 6) it will be proved that in the case of weak degeneracy and null traces this inequality can be strengthened (see also Remark 1 in §1.5).

Let us now consider the case of degeneracy at infinity. Let the nonnegative function $\varphi_n(x)$ and the function $f(x)$ together with its generalized derivative $\partial f/\partial x_n$ be defined on $\overset{+}{E}{}^n(a; d)$. Then the following lemma is valid.

LEMMA 2. *For any* $\eta \in (a, +\infty)$ *let there exist constants* $c_1 = c_1(\eta) > 0$ *and* $c_2 = c_2(\eta) > 0$ *such that for all* $x_n \in (a, \eta)$ *the inequality*

$$c_1 \leqslant \varphi(x^{(n-1)}, x_n) \leqslant c_2 < +\infty \tag{2.13}$$

is satisfied. Then for any $\xi \in (a, +\infty)$ *the inequality*

$$|f, L_{p,\omega_n}(\overset{+}{E}{}^n(\xi; d))| \leqslant c[|f, L_p(E^{n-1}(\xi; d))| + |f, L^{1n}_{p,\varphi_n}(\overset{+}{E}{}^n(\xi; d))|] \tag{2.14}$$

holds, where the constant c *does not depend on the function* f, *and* $d \leqslant +\infty$.

Lemma 2 is proved in a manner analogous to Lemma 1 with the use of the second generalized Hardy inequality (1.34), and remarks to it can be made which are analogous to Remarks 1, 2 and 3 to Lemma 1. We shall not dwell on this further.

REMARK 4. If the function $\varphi(x)$ possesses a derivative, then with the aid of inequalities (1.30), (1.31), (1.38) and (1.39) it is possible to obtain analogs of Lemmas 1 and 2 for the case $p = 1$ and $p = +\infty$.

REMARK 5. Let the function f be such that

$$D_{\alpha,p}(f) = \int |x_n|^{\alpha p} \sum_{i=1}^{n} \left|\frac{\partial f}{\partial x_i}\right|^p d\overset{+}{Q}{}^n < +\infty, \tag{2.15}$$

where $\overset{+}{Q}{}^n = E^n/Q^n$ is the interior of the unit ball $Q^n = \{x: |x| \leqslant 1\}$. In the integral (2.15) we convert to the spherical coordinates (ρ, s), where ρ is the radius vector of the point and $s = (\varphi_1, \cdots, \varphi_{n-1}) \in S^{n-1} = \{x: |x| = 1\}$. Denoting a volume element of the unit sphere S^{n-1} by dS^{n-1}, we obtain

$$\int dS^{n-1} \int_1^{+\infty} \left[\rho^{\left(\frac{n-1}{p}+\alpha\right)p} \left|\frac{\partial f}{\partial \rho}\right|^p + \rho^{\left(\frac{n-1}{p}+\alpha-1\right)p} \sum_{i=1}^{n} \left|\frac{\partial f}{\partial \varphi_i}\right|^p\right] d\rho \leqslant D_{\alpha,p}(f).$$

We shall have here, with respect to the variable ρ, the case of strong degeneracy at infinity, if $(n-1)/p + \alpha \leqslant 1/q$, i.e. if $\alpha \leqslant 1 - n/p$.

Spaces of functions satisfying (2.15) (and also the analogous conditions for derivatives of higher order), were studied earlier by the author in [11]. In particular, there under the condition $\alpha < 1 - n/p$ an inequality of the type of (2.14) was obtained.

2.3. The case of weak degeneracy.

LEMMA 3. *Let*

$$f(x) \in L^1_{p,\varphi_n}(E^n(b;\ d)),\ 0 < b \leqslant +\infty,\ 0 < d \leqslant +\infty \tag{2.16}$$

and for the function $\varphi_n(x)$ *let the condition* (2.1) *of weak degeneracy be satisfied at the origin. Then for the function* f *the trace* $f(x^{(n-1)}, 0)$, $x^{(n-1)} \in E^{n-1}(d)$, *exists in the sense of convergence almost everywhere.*

PROOF. From the existence of the generalized derivative $\partial f/\partial x_n$ for $0 < x_n < c$ we have

$$f(x^{(n-1)},\ x_n) = f(x^{(n-1)},\ c) - \int_{x_n}^{c} \frac{\partial f(x^{(n-1)},\ t)}{\partial x_n}\, dt. \tag{2.17}$$

From the condition (2.1) of weak degeneracy of the function φ_n it follows by Fubini's theorem that for almost all $x^{(n-1)} \in E^{n-1}(d)$ the integral

$$\int_0^c \varphi^{-q}(x^{(n-1)},\ t)\, dt$$

is finite. On the basis of that same Fubini theorem it follows from (2.16) that for almost all $x^{(n-1)} \in E^{n-1}(d)$ the integral

$$\int_0^b \varphi^p(x^{(n-1)},\ t) \left| \frac{\partial f(x^{(n-1)},\ t)}{\partial x_n} \right|^p dt$$

is finite. Hence, applying the Hölder inequality, we get for almost all $x^{(n-1)} \in E^{n-1}(d)$

$$\int_0^c \left| \frac{\partial f(x^{(n-1)},\ t)}{\partial x_n} \right| dt$$

$$\leqslant \left[\int_0^c \varphi^{-q}(x^{(n-1)},\ t)\, dt \right]^{\frac{1}{q}} \left[\int_0^c \varphi^p(x^{(n-1)},\ t) \left| \frac{\partial f(x^{(n-1)},\ t)}{\partial x_n} \right|^p dt \right]^{\frac{1}{p}} < +\infty,$$

i.e. the derivative $\partial f/\partial x_n$ is summable on the segment $[0, c]$. Hence in the equality (2.17) the right member has a limit as $x_n \to +0$; consequently for almost all $x^{(n-1)} \in E^{n-1}(d)$ the limit

$$\lim_{x_n \to +0} f(x^{(n-1)},\ x_n) = f(x^{(n-1)},\ 0)$$

exists. The lemma is proved.

For the case $\varphi_n(x) = x_n^\alpha$ the existence of the trace for $\alpha < 1/q$ was proved by the author in [12], and for the weight function $\varphi(x_n)$ of the single variable x_n it was proved by G. N. Jakovlev [7].

REMARK 6. We observe that in the proof of Lemma 3 it was shown, in particular, that if for some function $g(x)$ defined on $E^n(c; d)$ the inequality

$$\int \varphi^p(x)\,|g(x)|^p\,dE^n(c;\ d) < +\infty$$

is satisfied, where $\int\varphi^{-q}(x)dE^n(c; d) < +\infty$, then for almost all $x^{(n-1)} \in E^{n-1}(d)$ the integral $\int_0^c g(x^{(n-1)}, x_n)dx_n$ is finite (in our case $g(x)$ was $\partial f(x)/\partial x_n$).

If the weight function depends on the single variable x_n, then for $d < +\infty$ the condition (2.1) of weak degeneracy at the origin is equivalent to the following condition: for any $c \in (0, b)$

$$\int_0^c \varphi^{-q}(x_n)\,dx_n < +\infty. \tag{2.18}$$

LEMMA 4. *Let*

$$f(x) \in L^{1_n}_{p,\,\varphi_n}(E^n(b;\ d)) \cap L_p(E^n(a,\ b;\ d)),\quad 0<a<b\leqslant+\infty,\quad 0<d\leqslant+\infty,$$

and let the weight function $\varphi_n = \varphi_n(x_n)$ *satisfy the condition* (2.18) *of weak degeneracy at the origin and condition* (2.9). *Then the trace* $f(x^{(n-1)}, 0)$ *of the function* f *in the sense of convergence almost everywhere belongs locally to the space* L_p.

PROOF. From (2.17) for $x_n \to +0$ we get

$$f(x^{(n-1)},\ 0) = f(x^{(n-1)},\ c) - \int_0^c \frac{\partial f(x^{(n-1)},\ t)}{\partial x_n}\,dt,\qquad 0<c<b. \tag{2.19}$$

Taking the norm in $L_p(E^{n-1}(\delta))$ of both members of this equality, $0 < \delta < d$, and applying the Hölder inequality, we get

$$\Big[\int |f(x^{(n-1)},\ 0)|^p\,dE^{n-1}(\delta)\Big]^{\frac{1}{p}}$$

$$\leqslant\Big[\int |f(x^{(n-1)},\ c)|^p\,dE^{n-1}(\delta)\Big]^{\frac{1}{p}} + \Bigg[\int dE^{n-1}(\delta)\left|\int_0^c \frac{\partial f(x^{(n-1)},\ t}{\partial x_n}\,dt\right|^p\Bigg]^{\frac{1}{p}}$$

$$\leqslant |f,\ L_p(E^{n-1}(c;\ d))| + \left(\int_0^c \varphi_n^{-q}(t)\,dt\right)^{\frac{1}{q}} |f,\ L^{1_n}_{p,\,\varphi_n}(E^n(c;\ d))|.$$

The finiteness of the first term of the right member of this inequality follows from the fact that the trace of a function f such that $f \in L_p$ and $\partial f/\partial x_n \in L_p$ belongs locally to L_p. In our case condition (2.9) implies that for any $\eta \in (0, c)$ we have

$$\frac{\partial f(x)}{\partial x_n} \in L_p(E^n(\eta, c; d)),$$

and since $f \in L_p(E^n(a, b; d))$, it follows that $|f, L_p(E^{n-1}(c; \delta))| < +\infty$, because $\delta < +\infty$. The second term is finite by the condition of the lemma. The lemma is proved.

Even the simplest example of a function which is equal to a nonzero constant shows that membership of the function in the class $L^{1\,n}_{p,\varphi u}(E^n(b; +\infty))$, $b > 0$, does not imply membership of its trace in L_p on the entire boundary.

For the weight function $\varphi(x_n)$ let the condition (2.18) of weak degeneracy at the origin be satisfied. Let us set

$$\chi_\varphi(h) = \left[\int_0^h \varphi^{-q}(t)\, dt\right]^{\frac{1}{q}}, \qquad 0 < h < b. \tag{2.20}$$

It is evident that $\chi_\varphi(h) \downarrow 0$ as $h \to 0$.(4)

If $\varphi_n(x) = x_n^\alpha$, $\alpha < 1/q$, then $\chi_{x_n^\alpha}(h) \asymp h^{1/q-\alpha}$ as $h \to 0$.

LEMMA 5. *Let $f(x) \in L^{1\,n}_{p,\varphi_n}(E^n(b; d))$, $0 < b \leqslant +\infty$, $0 < d < +\infty$, and let the weight function φ_n depend on the single variable x_n and satisfy conditions* (2.9) *and* (2.18). *Then*

$$\left[\int |f(x^{(n-1)}, x_n) - f(x^{(n-1)}, 0)|^p\, dE^{n-1}(d)\right]^{\frac{1}{p}} = o\,(\chi_{\varphi_n}(x_n)) \tag{2.21}$$

as $x_n \to 0$. In particular, the trace of the function f on $E^{n-1}(d)$ in the sense of convergence almost everywhere is also the trace in the sense of convergence in the mean in L_p.

PROOF. Using (2.19) and applying the Hölder inequality, we get

$$\left[\int |f(x^{(n-1)}, x_n) - f(x^{(n-1)}, 0)|^p\, dE^{n-1}(d)\right]^{\frac{1}{p}}$$

$$= \left[\int \left|\int_0^{x_n} \frac{\partial f(x^{(n-1)}, t)}{\partial x_n}\, dt\right|^p dE^{n-1}(d)\right]^{\frac{1}{p}} \leqslant \left[\int_0^{x_n} \varphi_n^{-q}(t)\, dt\right]^{\frac{1}{q}}$$

$$\times \left[\int \left|\varphi(x_n) \frac{\partial f(x^{(n-1)}, t)}{\partial x_n}\right|^p dE^n(x_n; d)\right]^{\frac{1}{p}} = o\,(\chi_{\varphi_n}(x_n)).$$

The lemma is proved.

We observe that if in the conditions of Lemma 5 the function $g(x^{(n-1)})$ defined on $E^{n-1}(d)$ is the trace of the function f in the sense of convergence in the mean (such a trace exists on the strength of that same Lemma 5), then it is also the trace in the sense of convergence almost everywhere. This follows from the uniqueness of the trace of a function up to a set of $(n-1)$-dimensional measure zero.

(4) The symbolism $\downarrow 0$ signifies monotonic decrease and convergence to zero.

We shall now prove a certain strengthened version of Lemma 1 of §2.2 for the case of weak degeneracy with zero trace. Specifically, we shall prove that under the assumption of weak degeneracy at the origin the function $f \in L^{1n}_{p,\varphi_n}$ is summable not only with weight ψ_n, but also with weight ω_n (see Remark 1 in §1.5).

LEMMA 6. *Let* $f(x) \in L^{1n}_{p,\varphi_n}(E^n(b; d))$, $0 < b \leqslant +\infty$, $0 < d \leqslant +\infty$; *let the condition* (2.1) *of weak degeneracy at the origin be satisfied for the function* $\varphi_n(x)$, *and let* $f(x^{(n-1)}, 0)$ *for* $x^{(n-1)} \in E^{n-1}(d)$. *Then*

$$|f,\ L_{p,\omega_n}(E^n(b;\ d))| \leqslant c\,|f,\ L^{1n}_{p,\varphi_n}(E^n(b;\ d))|. \tag{2.22}$$

PROOF. Since under the assumptions of the lemma

$$f(x^{(n-1)},\ x_n) = \int_0^{x_n} \frac{\partial f(x^{(n-1)},\ t)}{\partial x_n}\,dt,$$

inequality (2.22) is simply the generalized Hardy inequality (1.34) integrated over $E^{n-1}(d)$:

$$|f,\ L_{p,\omega_n}(E^n(b;\ d))| = \left[\int dE^{n-1}(d) \int_0^b |\omega_n(x) f(x)|^p\,dx_n\right]^{\frac{1}{p}}$$

$$\leqslant \left[\int dE^{n-1}(d) \int_0^b \left|\omega_n(x) \int_0^{x_n} \frac{\partial f(x^{(n-1)},\ t)}{\partial x_n}\,dt\right|^p dx_n\right]^{\frac{1}{p}} \leqslant |f,\ L^{1n}_{p,\varphi_n}(E^n(b;\ d))|.$$

The lemma is proved.

LEMMA 7. *Let* $f(x) \in L^{1n}_{p,\varphi_n}(E^n(b; d))$, $0 < b < +\infty$, $0 < d < +\infty$, *and let the weight function* $\varphi_n(x_n)$ *satisfy conditions* (2.9) *and* (2.18). *Then the following inequalities are valid*:

$$|f,\ L_p(E^n(b;\ d))| \leqslant c\,[|f,\ L_p(E^{n-1}(0;\ d))| + |f,\ L^{1n}_{d,\varphi_n}(E^n(b;\ d))|], \tag{2.23}$$

$$|f,\ L_p(E^n(b;\ d))| \leqslant c\,[|f,\ L_p(E^{n-1}(b;\ d))| + |f,\ L^{1n}_{p,\varphi_n}(E^n(b;\ d))|], \tag{2.24}$$

where the constant c *does not depend on* f.

PROOF. Employing (2.19), increasing the upper limit of integration from x_n to b and applying the Hölder inequality, we obtain

$$f,\ L_p(E^n(b;\ d))| \leqslant \left[\int |f(x^{(n-1)},\ 0)|^p\,dE^{n-1}(d) \int_0^b dx_n\right]^{\frac{1}{p}} +$$

$$+\left[\int dE^n(b;\ d)\left(\int_0^b\left|\frac{\partial f(x^{(n-1)},\ t)}{\partial x_n}\right|dt\right)^p\right]^{\frac{1}{p}}$$

$$\leqslant b^{\frac{1}{p}}\left[|f,\ L_p(E^{n-1}(0;\ d))|+\left(\int_0^b \varphi_n^{-q}(t)\,dt\right)^{\frac{1}{q}}|f,\ L^{1n}_{p,\varphi_n}(E^n(b;\ d))|\right].$$

Inequality (2.23) is proved. In an analogous manner in equality (2.24) is also proved. Lemma 7 is proved.

We observe that it was proved above that if in addition to the conditions of the lemma it is assumed that $f\in L_p(E^n(a,b;d))$ for some $a\in(0,b)$ and $d<+\infty$, then the first terms of the right members of the inequalities (2.23) and (2.24) are finite.

LEMMA 8. *Let* $f(x)\in L^{1n}_{p,\varphi_n}(E^n(b;d))$, $0<b<+\infty$, $0<d<+\infty$, *and let the weight function* $\varphi_n(x_n)$ *satisfy the conditions* (2.9) *and* (2.18). *Then*

$$|f,\ L_p(E^{n-1}(0;\ d))|\leqslant c\,[|f,\ L_p(E^n(b;\ d))|+|f,\ L^{1n}_{p,\varphi_n}(E^n(b;\ d))|],\quad (2.25)$$

$$|f,\ L_p(E^{n-1}(b;\ d))|\leqslant c\,[|f,\ L_p(E^n(b;\ d))|+|f,\ L^{1n}_{p,\varphi_n}(E^n(b;\ d))|],\quad (2.26)$$

where the constant c *does not depend on* f.

PROOF. Employing (2.19), applying the Hölder inequality and increasing from x_n to b the upper limit of integration in the last of the integrals obtained, we get

$$|f,\ L_p(E^{n-1}(0;\ d))|^p\leqslant 2^{p-1}\left[\int|f(x^{(n-)},\ x_n)|^p\,dE^{n-1}(d)\right.$$

$$\left.+\int dE^{n-1}(d)\left|\int_0^{x_n}\frac{\partial f(x^{(n-1)},\ t)}{\partial x_n}dt\right|^p\right]\leqslant 2^{p-1}\left[\int|f(x^{(n-1)},\ x_n)|^p\,dE^{n-1}(d)\right.$$

$$\left.+\left(\int_0^b\varphi_n^{-q}(t)\,dt\right)^{\frac{p}{q}}\int\left|\varphi_n\frac{\partial f}{\partial x_n}\right|^p dE^n(b;\ d)\right].\quad (^5)$$

Integrating this inequality with respect to x_n from 0 to b, we get

$$b\,|f,\ L_p(E^{n-1}(0;\ d))|^p\leqslant 2^{p-1}\left[|f,\ L_p(E^n(b;\ d))|^p\right.$$

$$\left.+\,b\left(\int_0^b\varphi_n^{-q}(t)\,dt\right)^{\frac{p}{q}}|f,\ L^1_{p,\varphi_n}(E^n(b;\ d))|\right].$$

Dividing both members of the inequality by b and taking the pth root of both members, we get

---

(5) We have employed the inequality $(a+b)^p\leqslant 2^{p-1}(a^p+b^p)$, $a\geqslant 0$, $b\geqslant 0$, $p\geqslant 1$.

$$|f,\ L_p(E^{n-1}(0;\ d))| \leqslant 2^{\frac{1}{q}}\left[b^{-\frac{1}{p}}|f,\ L_p(E^n(b;\ d))| + \left(\int_0^b \varphi_n^{-q}(t)\,dt\right)^{\frac{1}{q}} |f,\ L^1_{p,\varphi_n}(E^n(b;\ d))|\right]. \tag{2.27}$$

Inequality (2.25) is proved. Inequality (2.26) is proved in an analogous manner. Lemma 8 is proved.

REMARK 7. If

$$|f,\ L_p(E^n(+\infty;\ d))| < +\infty, \tag{2.28}$$

then from inequality (2.27) with $b \to +\infty$ it follows that

$$|f,\ L_p(E^{n-1}(0;\ d)| \leqslant 2^{\frac{1}{q}}\left(\int_0^{+\infty} \varphi_n^{-q}(t)\right)^{\frac{1}{q}} |f,\ L^{1n}_{p,\varphi_n}(E^n(+\infty;\ d))|.$$

One must take into consideration, however, that the simplest example of a function, a nonzero constant, shows that the condition $f \in L^1_{p,\varphi_n}(E^n(+\infty; d))$ does not imply condition (2.28). Let us note also that $\int_0^{+\infty} \varphi_n^{-q}(t)dt < +\infty$ in the case of the weight function $\varphi_n(t) = t^\alpha$ is not satisfied for any α.

REMARK 8. Everything said in this section for the case of weak degeneracy at the origin carries over in a natural way also to the case of weak degeneracy at infinity.

In particular, if $f(x) \in L^{1n}_{p,\varphi_n}(\overset{+}{E}{}^n(a; d))$, $a \geqslant 0$, $0 < d \leqslant +\infty$ and if the condition (2.3) of weak degeneracy at infinity is satisfied for the weight function $\varphi_n(x)$, then for the function f there is a trace $f(x^{(n-1)}, +\infty)$, $x^{(n-1)} \in E^{n-1}(d)$, in the sense of convergence almost everywhere, which in every case, if the function φ_n depends only on x_n and satisfies condition (2.13), belongs locally to L_p and is also a trace in the sense of convergence in the mean. The analogs of the formulas and inequalities (2.21)–(2.26) are also valid in this case.

2.4. The spaces $L^{r_n}_{p,\varphi_n}(E^n_b)$ and $L^{r_n}_{p,\varphi_n}(\overset{+}{E}{}^n_a)$. In this section we shall assume that all the weight functions $\varphi_i(t)$, $i = 1, \cdots, n$, depend on a single variable.

Let r_n be a positive integer, let $0 < \eta_{r_n} < \eta_{r_n-1} < \cdots < \eta_1 < \eta_0 = b \leqslant +\infty$, let the function $\varphi_n(t)$ be defined on the interval $(0, b)$ and satisfy condition (2.9), and let

$$\left.\begin{aligned} &\varphi_{n,0}(t) = \varphi_n(t), \\ &\varphi_{n,k+1}(t) = \varphi_{n,k}^{-\frac{q}{p}}(t)\left[\int_t^{\eta_k} \varphi_{n,k}^{-q}(u)\,du\right]^{-1}, \\ &k = 0,\ 1,\ \ldots,\ r_n - 1,\ 0 < t < \eta_k; \end{aligned}\right\} \tag{2.29}$$

then each of the functions $\varphi_{n,k+1}$ also satisfies condition (2.9).

Let $f \in L^{r_n}_{p,\varphi_n}(E^n(b; d))$, $\partial^i f/\partial x_n^i \in L_p(E^n(a, b; d))$, $i = 0, 1, \cdots, r_n - 1$, $0 < a < b \leqslant +\infty$, $0 < d \leqslant +\infty$, $p \geqslant 1$, $1/p + 1/q = 1$. If $|\varphi_n^{-1}, L_q(0, c)| < +\infty$ for all $c \in (0, b)$, i.e. the condition of weak degeneracy at the origin is satisfied, then according to Lemma 3 of §2.3 the derivative $\partial^{r_n-1} f/\partial x_n^{r_n-1}$ has a trace on $E^{n-1}(0, d)$.

But if $|\varphi_n^{-1}, L_q(0, c)| = +\infty$ for all $c \in (0, b)$, i.e. the condition of strong degeneracy is satisfied, then according to Lemma 1 of §2.2 we have

$$\frac{\partial^{r_n-1} f}{\partial x_n^{r_n-1}} \in L_{p, \varphi_{n,1}}(E^n(\eta_1; d))$$

and consequently $f \in L^{(r_n-1)n}_{p,\varphi_{n,1}}(E^n(\eta_1; d))$.

It can happen that for all $k = 0, 1, \cdots, r_n - 1$ all the functions $\varphi_{n,k}$ satisfy the conditions of strong degeneracy; then for none of the derivatives $\partial^s f/\partial x_n^s$, $s = 0, 1, \cdots, r_n - 1$, can the existence of a trace on $E^{n-1}(0; d)$ be guaranteed.

If an index k_0 can be found such that for all $k = 0, 1, \cdots, k_0 - 1$ there holds

$$|\varphi_{n,k}^{-1}, L_q(0, \eta_k)| = +\infty, \qquad |\varphi_{n,k_0}^{-1}, L_q(0, \eta_{k_0})| < +\infty, \qquad 0 \leqslant k_0 < r_n, \tag{2.30}$$

then

$$\frac{\partial^{r_n-k_0} f}{\partial x_n^{r_n-k_0}} \in L_{p, \varphi_{n,k_0}}(E^n(\eta_{k_0}; d)),$$

and consequently for the derivative $\partial^{r_n-k_0-1} f/\partial x_n^{r_n-k_0-1}$ a trace on $E^{n-1}(0; d)$ exists; but if $d < +\infty$, then (see Lemma 7, §2.3)

$$\frac{\partial^{r_n-k_0-1} f}{\partial x_n^{r_n-k_0-1}} \in L_p(E^n(\eta_{k_0}; d)),$$

and hence traces will exist also for all earlier derivatives

$$\frac{\partial^s f}{\partial x_n^s}, \qquad 0 \leqslant s \leqslant s_0 = r_n - k_0 - 1. \tag{2.31}$$

Thus s_0 is the largest nonnegative integer s such that the weight φ, with which the derivative $\partial^s f/\partial x_n^s$ is pth power summable, possesses the property that $\varphi^{-1} \in L_q$.

In the case where $\varphi_n(t) = t^\alpha$ the number s_0 is defined as the largest nonnegative integer such that

$$s_0 < r_n - \alpha - \frac{1}{p}, \tag{2.32}$$

and, consequently, a condition for its existence is the inequality

$$r_n - \alpha - \frac{1}{p} > 0. \tag{2.33}$$

Independently of the strong or weak degeneracy of the weight functions $\varphi_{n,k}(t)$ the following theorem is true.

THEOREM 1. *Let the functions* $\varphi_{n,k}(t)$, $k = 0, 1, \cdots, r_n$, *be defined by the formulas* (2.29). *Then for any function* f *defined, along with its derivatives* $\partial^i f/\partial x_n^i$, $i = 1, 2, \cdots, r_n$, *on* $E^n(b; d)$, *the inequality*

$$f,\ L_{p,\varphi_n,r_n}(E^n(\eta_n;\ d))\,|$$

$$\leqslant c\left[\,|f,\ L^{r_n}_{p,\varphi_n}(E^n(b;\ d))\,| + \sum_{j=1}^{r_n-1}\left|\frac{\partial^j f}{\partial x_n^j},\ L_p(E^{n-1}(\eta_{r_n-j};\ d))\right|\right] \tag{2.34}$$

is valid, where the constant c *does not depend on* f.

This follows from Lemma 1 of §2.2 by induction. In the case of zero traces the inequality (2.34) can be strengthened in a well-defined sense with the aid of Lemma 6 of §2.3.

In order to simplify the notation we shall assume in what follows that $s_0 = r_n$ (see (2.31)), i.e. that

$$\int_0^b \varphi_n^{-q}(t)\,dt < +\infty.$$

In this case the integral

$$\int_0^b \frac{\partial^{r_n} f(x^{(n-1)},\ x_n)}{\partial x_n}\,dx_n$$

is finite for almost all $x^{(n-1)} \in E^{n-1}(d)$ (see Remark 6 in §2.3).

Everything said in this section carries over in a natural way to the case of the spaces $L^{r_n}_{p,\varphi_n}(\overset{+}{E}{}^n(a; d))$.

2.5. Properties of the traces on a hyperplane.

LEMMA 9 (FUNDAMENTAL). *For some* i $(i = 1, 2, \cdots, n-1)$ *let*

$$f(x) \in L^{r_i r_n}_{p,\varphi_i,\varphi_n}(E^n_b),\quad r_i \geqslant 1,\quad r_n \geqslant 1,\quad 0 < b \leqslant +\infty,$$

and let

$$\int_0^b x_n^{-sq}\varphi_i^{-q}(x_n)\,dx_n < +\infty,\quad \int_0^b \varphi_n^{-q}(x_n)\,dx_n < +\infty,\quad 0 \leqslant s < r_n. \tag{2.35}$$

Then if

$$\left.\begin{aligned} \Phi_s(t) &= \frac{1}{t^{s+1}}\left(\int_0^t \varphi_i^{-q}(u)\,du\right)^{\frac{1}{q}}, \\ \Psi_s(t) &= \frac{1}{t}\int_0^t\left(\int_0^y v^{(r_n-s-1)q}\varphi_n^{-q}(v)\,dv\right)^{\frac{1}{q}}dy, \end{aligned}\right\} \tag{2.36}$$

for $k \geqslant r$ *we have*

$$\left|\Delta_{h_i}^k \frac{\partial^s f}{\partial x_n^s},\ L_p(E^{n-1})\right| \leqslant c\left[|h_i|^{r_i}\Phi_s(t)\,|f,\ L_{p,\varphi_i}^{r_i}(E_b^n)| + \Psi_s(t)\,|f,\ L_{p,\varphi_n}^{r_n}(E_b^n)|\right]. \tag{2.37}$$

PROOF. Since in the inequality (2.37) a fundamental role is played by the two variables x_i and x_n (only integration is carried out with respect to the rest of the variables), we shall write $f(x_i, x_n)$ instead of $f(x) = f(x^{(n-1)}, x_n)$, when this cannot lead to ambiguities.

Using (1.18) and the Hölder inequality, we obtain

$$\begin{aligned} \left|\Delta_{h_i}^k \frac{\partial^s f(x^{(n-1)}, 0)}{\partial x_n^s}\right| &\leqslant \frac{a_s}{x_n^s}|h_i|^{r_i-1}\sum_{\nu=2}^{r_n+1}\int_0^{k|h_i|}\left|\frac{\partial^{r_i} f(x_i + v\,\mathrm{sign}\,h_i,\ \nu x_n)}{\partial x_i^{r_i}}\right|dv \\ &+ \frac{b_s}{x_n^s}\sum_{j=0}^k\left(\int_0^{(r_n+1)x_n} u^{(r_n-1)q}\varphi_n^{-q}(u)\,du\right)^{\frac{1}{q}}\left(\int_0^{(r_n+1)x_n}\varphi_n^p(u)\left|\frac{\partial^{r_n} f(x_i + jh_i,\ u)}{\partial x_n^{r_n}}\right|^p du\right)^{\frac{1}{p}} \\ &+ c_s\sum_{j=0}^k\left(\int_0^{x_n} u^{(r_n-s-1)q}\varphi_n^{-q}(u)\,du\right)^{\frac{1}{q}}\left(\int_0^{x_n}\varphi_n^p(u)\left|\frac{\partial^{r_n} f(x_i + jh_i,\ u)}{\partial x_n^{r_n}}\right|^p du\right)^{\frac{1}{p}}. \end{aligned}$$

Noting that

$$x_n^{-sq} \leqslant \left(\frac{u}{r_n+1}\right)^{-sq} \quad \text{for} \quad 0 < u < (r_n+1)x_n,$$

we combine the last two sums:

$$\begin{aligned} \left|\Delta_{h_i}^k \frac{\partial^s f(x^{(n-1)}, 0)}{\partial x_n^s}\right| &\leqslant \frac{a_s}{x_n^s}|h_i|^{r_i-1}\sum_{\nu=2}^{r_n+1}\int_0^{k|h_i|}\left|\frac{\partial^{r_i} f(x_i + v\,\mathrm{sign}\,h_i,\ \nu x_n)}{\partial x_i^{r_i}}\right|dv \\ &+ b_s'\sum_{j=0}^k\left(\int_0^{(r_n+1)x_n} u^{(r_n-s-1)q}\varphi_n^{-q}(u)\,du\right)^{\frac{1}{q}}\left(\int_0^{(r_n+1)x_n}\varphi_n^p(u)\frac{\partial^{r_n} f(x_i + jh_i,\ u)}{\partial x_n^{r_n}}\Big|^p du\right)^{\frac{1}{p}}; \end{aligned}$$

where b_s' is some constant.

Multiplying by x_n^s, we integrate the last inequality with respect to x_n^s along some segment $[0, a]$ and fix any b such that $0 < (r_n + 1)a \leqslant b$:

$$\frac{a^{s+1}}{s+1}\left|\Delta^k_{h_i}\frac{\partial^s f(x^{(n-1)},0)}{\partial x_n^s}\right| \leqslant a_s|h_i|^{r_i-1}\sum_{\nu=2}^{r_n+1}\int_0^a dx_n\int_0^{k|h_i|}\left|\frac{\partial^{r_i}f(x_i+v\operatorname{sign}h_i,\nu x_n)}{\partial x_i^{r_i}}\right|dv$$

$$+\,b'_s\sum_{j=0}^{k}\int_0^a x_n^s\left[\left(\int_0^{(r_n+1)x_n}u^{(r_n-s-1)q}\varphi_n^{-q}(u)\,du\right)^{\frac{1}{q}}\right.$$

$$\left.\times\left(\int_0^{(r_n+1)x_n}\varphi_n^p(u)\left|\frac{\partial^{r_n}f(x_i+jh_i,u)}{\partial x_n^{r_n}}\right|^p du\right)^{\frac{1}{p}}\right].$$

In the terms of the first sum we make the change of variables $y = \nu x_n$, in the terms of the second sum the change of variable $y = (r_n + 1)x_n$, and we increase the upper limit of the second integral to b and introduce the notation $t = (r_n + 1)a$. To the inequality obtained we apply the Hölder inequality:

$$\frac{a^{s+1}}{s+1}\left|\Delta^k_{h_i}\frac{\partial^s f(x^{(n-1)},0)}{\partial x_n^s}\right| \leqslant a'_s|h_i|^{r_i-1}\int_0^t dy\int_0^{k|h_i|}\left|\frac{\partial^{r_i}f(x_i+v\operatorname{sign}h_i,y)}{\partial x_i^{r_i}}\right|dv$$

$$+\,b''_s\int_0^t y^s\left(\int_0^y u^{(r_n-s-1)q}\varphi_n^{-q}(u)\,du\right)^{\frac{1}{q}}dy\sum_{j=0}^{k}\left(\int_0^b\varphi_n^p(u)\left|\frac{\partial^{r_n}f(x_i+jh_i,u)}{\partial x_n^{r_n}}\right|^p du\right)^{\frac{1}{p}}$$

$$\leqslant a''_s|h_i|^{r_i-1+\frac{1}{q}}\left(\int_0^t\varphi_i^{-q}(y)\,dy\right)^{\frac{1}{q}}\left(\int_0^b dy\int_0^{k|h_i|}\varphi_i^p(y)\left|\frac{\partial^{r_i}f(x_i+v,y)}{\partial x_i^{r_i}}\right|^p dv\right)^{\frac{1}{p}}$$

$$+\,b''_s\int_0^t y^s\left(\int_0^y u^{(r_n-s-1)q}\varphi_n^{-q}(u)\,du\right)^{\frac{1}{q}}dy\sum_{j=0}^{k}\left(\int_0^b\varphi_n^p(u)\left|\frac{\partial^{r_n}f(x_i+jh_i,u)}{\partial x_n^{r_n}}\right|^p du\right)^{\frac{1}{p}},$$

where a'_s, a''_s and b''_s are some constants.

We divide both sides of the inequality by

$$\frac{a^{s+1}}{s+1}=\frac{t^{s+1}}{(s+1)(r+1)^{s+1}},$$

increase y^s in the second term to t^s, take the $L_p(E^{n-1})$ norms of both members, and effect the changes of variables $x'_i = x_i + v \operatorname{sign} h_i$, $x_i^{(j)} = x_i + jh_i$, $(j = 1, 2, \cdots, k)$:

$$\left[\int\left|\Delta^k_{h_i}\frac{\partial^s f(x^{(n-1)},0)}{\partial x_n^s}\right|^p dE^{n-1}\right]^{\frac{1}{p}} \leqslant a'''_s|h_i|^{r_i-\frac{1}{p}}t^{-s-1}\left(\int_0^t\varphi_i^{-q}(y)\,dy\right)^{\frac{1}{q}}$$

$$\times\left(\int_0^{k|h_i|}dv\int_0^b dy\int\varphi_i^p(y)\left|\frac{\partial^{r_i}f(x_i+v\operatorname{sign}h_i,y)}{\partial x_i^{r_i}}\right|^p dE^{n-1}\right)^{\frac{1}{p}}+$$

$$+b_s''t^{-1}\int_0^t\left(\int_0^y u^{(r_n-s-1)q}\varphi_n^{-q}(u)\,du\right)^{\frac{1}{q}}dy$$

$$\times\sum_{j=0}^{k}\left(\int_0^b du\int\varphi_n^p(u)\left|\frac{\partial^{r_n}f(x_i+jh_i,\ u)}{\partial x_n^{r_n}}\right|dE^{n-1}\right)^{\frac{1}{p}}$$

$$\leqslant c\left[|h_i|^{r_i}t^{-s-1}\left(\int_0^t\varphi_i^{-q}(y)\,dy\right)^{\frac{1}{q}}\left(\int\varphi_i^p(x_n)\left|\frac{\partial^{r_i}f}{\partial x_i^{r_i}}\right|^p dE_b^n\right)^{\frac{1}{p}}\right.$$

$$\left.+t^{-1}\int_0^t dy\left(\int_0^y u^{(r_n-s-1)q}\varphi_n^{-q}(u)\,du\right)^{\frac{1}{q}}\left(\int\varphi_n^p(x_n)\left|\frac{\partial^{r_n}f}{\partial x_n^{r_n}}\right|^p dE_b^n\right)^{\frac{1}{p}}\right].$$

The lemma is proved.

The special case of formula (2.37) for $s = 0$ was published without complete proof in a paper of the author [15], and a complete formulation (also without proof) is in the note [16].

REMARK 9. If $\varphi_i(x_n) = x_n^{\alpha_i}$ and $\varphi_n(x_n) = x_n^{\alpha_n}$, then the conditions (2.35) for $b < +\infty$ are equivalent to the conditions $\alpha_i < 1/q$ and $\alpha_n < 1/q$. In this case $(r_n - s - 1 - \alpha_n)q \geqslant -\alpha_n q > -1$ $(s = 0, 1, \cdots, r_n - 1)$, and hence as $t \to +0$ we have

$$\Phi_s(t)=\frac{1}{t^{s+1}}\left(\int_0^t u^{-\alpha_i q}du\right)^{\frac{1}{q}}\asymp t^{-s-\frac{1}{p}-\alpha_i},$$

$$\Psi_s(t)=\frac{1}{t}\int_0^t\left(\int_0^y v^{(r_n-s-1-\alpha_n)q}\,dv\right)^{\frac{1}{q}}dy\asymp t^{r_n-s-\frac{1}{p}-\alpha_n},$$

and the inequality (2.37) assumes the following form:

$$\left|\Delta_{h_i}^k\frac{\partial^s f}{\partial x_n^s},\ L_p(E^{n-1})\right|\leqslant c\left[|h_i|^{r_i}t^{-s-\frac{1}{p}-\alpha_i}\left|f,\ L_{p,x_n^{\alpha_i}}^{r_i}\right|+t^{r_n-s-\frac{1}{p}-\alpha_n}\left|f,\ L_{p,x_n^{\alpha_n}}^{r_n}\right|\right].$$

THEOREM 2. *For some* $i = 1, \cdots, n-1$ *let*

$$f(x)\in L_{p,\varphi_i,\varphi_n}^{r_ir_n}(E_b^n),\qquad r_i\geqslant 1,\qquad r_n\geqslant 1,$$

$$\int_0^b\varphi_i^{-q}(x_n)\,dx_n<+\infty,\qquad\int_0^b\varphi_n^{-q}(x_n)\,dx_n<+\infty. \tag{2.38}$$

If the trace of the function f *on the hyperplane* E^{n-1} *belongs locally to the space* $L_p(E^{n-1})$:

$$\int_0^t \varphi_i^{-q}(x_n)\,dx_n = o(t^q) \quad as \quad t \to +0, \tag{2.39}$$

then the function $f(x^{(n-1)}, 0)$ *is a polynomial of degree not greater than* r_i *in the variable* x_i.

COROLLARY. *Let* $f(x) \in L^r_{p,\varphi}(E^n_b)$, $r = (r_1, \cdots, r_n)$, $r_i \geqslant 1$, $i = 1, \cdots, n$, *let* $\varphi = (\varphi_1, \cdots, \varphi_n)$ *and let*

$$\int_0^b \varphi_i^{-q}(x_n)\,dx_n < +\infty, \; i = 1, 2, \ldots, n.$$

If the trace of the function f *on the hyperplane* E^{n-1} *belongs locally to the space* $L_p(E^{n-1})$,

$$\int_0^t \varphi_i^{-q}(x_n)\,dx_n = o(t^q) \quad as \quad t \to 0, \; i = 1, 2, \ldots, n-1,$$

then the function $f(x^{(n-1)}, 0)$ *is a polynomial in the variables* $x_1, \cdots, x_{n-1}$ *of degrees, respectively, not greater than* $r_1 - 1, \cdots, r_{n-1} - 1$.

PROOF. From (2.38) it follows that

$$\lim_{y \to +0} \int_0^y v^{(r_n-1)\,q} \varphi_n^{-q}(v)\,dv = 0,$$

and hence (see (2.36)) $\lim_{t\to+0}\Psi_0(t) = 0$. If condition (2.39) is satisfied, then $\lim_{t\to+0}\Phi_0(t) = 0$. Hence, letting $t \to +0$ in (2.37), we get

$$|\Delta^k_{h_i} f, \; L_p(E^{n-1})| = 0, \qquad k \geqslant r_i.$$

Consequently $\Delta^{r_i}_{h_i} f = 0$ almost everywhere on E^{n-1}, and hence (see §1.3) the function $f(x^{(n-1)}, 0)$ is a polynomial in x_i of degree not greater than $r_i - 1$. The theorem is proved.

REMARK 10. If $\varphi_i(x_n) = x_n^{\alpha_i}$ and $\varphi_n(x_n) = x_n^{\alpha_n}$, then for $b < +\infty$ the condition (2.38) is equivalent to the condition $\alpha_i < 1/q$, $\alpha_n < 1/q$, and condition (2.39) is equivalent to a condition stronger than the first of those mentioned, namely the condition $\alpha_i < -1/p$. Hence for the power weights we have the following assertion:

If $f(x) \in L^{r_i r_n}_{p, x_n^{\alpha_i}, x_n^{\alpha_n}}(E^n_b)$, $b < +\infty$, $r_i \geqslant 1$, $r_n \geqslant n$, $\alpha_i < -1/p$, $\alpha_n < -1/p$, *then the function* $f(x^{(n-1)}, 0)$ *is a polynomial of degree no higher that* $r_i - 1$ *in* x_i.

Let us now proceed to the study of the smoothness of the trace when it is not a polynomial. As a preliminary we note that if one introduces the notation

$$G_j(t) = \frac{1}{t}\left[\int_0^t \varphi_j^{-q}(t)\,dt\right]^{\frac{1}{q}}, \qquad j = i, n, \tag{2.40}$$

(thus $G_i(t) = \Phi_0(t)$: see (2.36)), then

$$\Psi_s(t) \leqslant t^{r_n - s} G_n(t), \qquad 0 \leqslant s \leqslant r_n - 1, \tag{2.41}$$

where the function $\Psi_s(t)$ is defined by formula (2.36).

In fact, the function $\int_0^y v^{(r_n-s-1)q} \varphi_n^{-q}(v)dv$ increases monotonically, and consequently

$$\Psi_s(t) = -\frac{1}{t}\int_0^t \left[\int_0^y v^{(r_n-s-1)q}\varphi_n^{-q}(v)\,dv\right]^{\frac{1}{q}} dy$$

$$\leqslant \frac{1}{t}\left[\int_0^t v^{(r_n-s-1)q}\varphi_n^{-q}(v)\,dv\right]^{\frac{1}{q}}\int_0^t dy \leqslant t^{r_n-s-1}\left[\int_0^t \varphi_n^{-q}(v)\,dv\right]^{\frac{1}{q}};$$

from which (2.41) follows. Substituting (2.41) into (2.37), we obtain

$$\left|\Delta_{h_i}^k \frac{\partial^s f}{\partial x_n^s},\ L_p(E^{n-1})\right| \leqslant c\left[|h_i|^{r_i}\Phi_s(t) + t^{r_n-s}G_n(t)\right]\left|f,\ L_{p,\varphi_i,\varphi_n}^{r_i r_n}(E_b^n)\right|, \qquad k \geqslant r_i. \tag{2.42}$$

From this for $s = 0$ we have

$$|\Delta_{h_i}^k f,\ L_p(E^{n-1})| \leqslant c\left|f,\ L_{p,\varphi_i,\varphi_n}^{r_i r_n}(E_b^n)\right|\left[|h_i|^{r_i}G_i(t) + t^{r_n}G_n(t)\right], \qquad k \geqslant r_i. \tag{2.43}$$

Here t is a parameter. If it is chosen as a function of h_i (if this is possible), $t = t(h_i)$, in such a way that in a neighborhood of zero the equality

$$|h_i|^{r_i} G_i(t) = t^{r_n} G_n(t) \tag{2.44}$$

is satisfied, or even so that the left and right members of the equality have the same order as $t \to 0$, i.e.

$$|h_i|^{r_i} G_i(t) \asymp t^{r_n} G_n(t),$$

then from (2.43) we obtain the estimate

$$|\Delta_{h_i}^k f,\ L_p(E^{n-1})| \leqslant c\left|f,\ L_{p,\varphi_i,\varphi_n}^{r_i r_n}\right| G_i(t(h))\,|h_i|^{r_i}, \qquad k \geqslant r_i, \tag{2.45}$$

giving the smoothness characteristic of the trace of the function f.

Equation (2.44) admits a simple solution, for example, when $\varphi_i(u) = \varphi_n(u) = \varphi(u)$. Then $G_i(t) = G_n(t) = G(t)$, and hence

$$t = |h_i|^{\frac{r_i}{r_n}} \tag{2.46}$$

is a solution of equation (2.44).

Substituting (2.46) into (2.45) we obtain an explicit evaluation of the difference. We formulate this result in the form of a theorem.

THEOREM 3. *Let* $f \in L^{r_i r_n}_{p,\varphi,\varphi}(E^n_b)$, $0 < b \leqslant +\infty$, $r_i \geqslant 1$, $r_n \geqslant 1$, $k \geqslant r_i$. *Then*

$$|\Delta^k_{h_i} f, L_p(E^{n-1})| \leqslant c\,|f, L^{r_i r_n}_{p,\varphi,\varphi}(E^n_b)|\, G\left(|h_i|^{\frac{r_i}{r_n}}\right)|h_i|^{r_i}, \tag{2.47}$$

where the constant c *does not depend on* f.

COROLLARY. *If under the conditions of Theorem 3 the weight function* $\varphi(u)$ *does not decrease, then*

$$|\Delta^k_{h_i} f, L_p(E^{n-1})| \leqslant c\,|f, L^{r_i r_n}_{p,\varphi,\varphi}(E^n_b)|\, \omega^{-1}\left(|h_i|^{\frac{r_i}{r_n}}\right)|h_i|^{\frac{r_i}{r_n}\left(r_n - 1 - \frac{1}{p}\right)}, \tag{2.48}$$

where the constant c *does not depend on the function* f, *and the function* $\omega(t)$ *is determined by the formula* (1.33).

Let us prove the corollary. On the basis of the nondecreasing character of the function $\varphi(u)$ we have

$$\varphi\left(|h_i|^{\frac{r_i}{r_n}}\right)\varphi^{-1}(u) \geqslant 1, \quad 0 \leqslant u \leqslant |h_i|^{\frac{r_i}{r_n}},$$

and hence

$$|h_i|^{r_i} G\left(|h_i|^{\frac{r_i}{r_n}}\right) = |h_i|^{r_i - \frac{r_i}{r_n}} \left(\int_0^{|h_i|^{\frac{r_i}{r_n}}} \varphi^{-q}(u)\,du\right)^{\frac{1}{q}}$$

$$= |h_i|^{\frac{r_i}{r_n}(r_n-1)} \left[\int_0^{|h_i|^{\frac{r_i}{r_n}}} \varphi^q\left(|h_i|^{\frac{r_i}{r_n}}\right)\varphi^{-q}(u)\,du\right]^{-\frac{1}{p}} \varphi^{\frac{q}{p}}\left(|h_i|^{\frac{r_i}{r_n}}\right) \int_0^{|h_i|^{\frac{r_i}{r_n}}} \varphi^{-q}(u)\,du$$

$$\leqslant |h_i|^{\frac{r_i}{r_n}\left(r_n - 1 - \frac{1}{p}\right)} \omega^{-1}\left(|h_i|^{\frac{r_i}{r_n}}\right).$$

Substituting this estimate into (2.47) we get (2.48).

Equation (2.44) can be solved in other cases also; for example, when the weights are power functions: $\varphi_i(u) = u^{\alpha_i}$. In this case the desired solution is easily found in the form $t = h^\lambda$. This result will be obtained a little later by means of the solution of another simpler equation (see Remark 11).

Let us impose on the weight functions constraints of a type other than those considered above. Let there exist a constant $\epsilon > 0$ such that

$$\int_0^t \varphi_j^{-q}(u)\,du = O(t^{\varepsilon}\varphi_j^{-q}(t)) \quad \text{as} \quad t \to +0, \quad j = 1, n. \tag{2.49}$$

This condition is satisfied, for example, when $\varphi_j(u) = u^{\alpha_j}$ and $\alpha_j < 1/q$; in this case $\epsilon = 1$. It is satisfied also for $\varphi_j(u) = \ln^{-1}u$, but then it is necessary to take $\epsilon < 1$.

Substituting (2.49) into (2.43), we get

$$|\Delta_{h_i}^k f, L_p(E^{n-1})| \leqslant c\,|f, L^{r_i r_n}_{p, \varphi_i, \varphi_n}(E_b^n)|\left[|h_i|^{r_i} t^{\frac{\varepsilon}{q}-1}\varphi_i^{-1}(t) + t^{r_n + \frac{\varepsilon}{q} - 1}\varphi_n^{-1}(t)\right],$$
$$k \geqslant r_i. \tag{2.50}$$

The condition under which both terms of the right member coincide gives the equation

$$|h_i|^{r_i}\varphi_n(t) = t^{r_n}\varphi_i(t). \tag{2.51}$$

It is solvable for $t = t(h_i)$ if the function $t^{r_n}\varphi_i(t)/\varphi_n(t)$ is monotone. Substituting the indicated solution $t = t(h_i)$ into (2.50), we get in this case an estimate of the smoothness of the trace directly in terms of the weight functions φ_i and φ_n.

REMARK 11. Let us again examine the case of power weights: $\varphi_i(u) = u^{\alpha_i}$ and $\varphi_n(u) = u^{\alpha_n}$. Setting $t = h_i^{\lambda}$ in equation (2.51), we find

$$\lambda = \frac{r_i}{r_n + \alpha_i - \alpha_n}.$$

Substituting $t = h_i^{\lambda}$ with this λ and $\epsilon = 1$ into (2.50), we shall have

$$|\Delta_{h_i}^k f, L_p(E^n)| \leqslant c\left|f, L^{r_i r_n}_{p, x_n^{\alpha_i}, x_n^{\alpha_n}}(E_b^n)\right|\,|h_i|^{\mu_i},$$

where $k \geqslant r_i$ and

$$\mu_i = \frac{r_i\left(r_n - \alpha_n - \frac{1}{p}\right)}{r_n + \alpha_i - \alpha_n} \tag{2.52}$$

under the condition that $\alpha_n < 1/q$ and $\alpha_i \geqslant -1/p$ (we recall that the condition $\alpha_i < -1/p$ signifies that the trace is a polynomial). The exponents (2.52) were found earlier by S. V. Uspenskiĭ [8] in treating the case of power weights.

2.6. The characteristic of traces at infinity. Results analogous to those of the preceding section can be proved also for traces at infinity, i.e. for functions $f(x^{(n-1)}, +\infty)$. They can be obtained from the results proved above by a change of variable. For simplicity we shall restrict ourselves to the consideration of only power weights, since even in this case the specifics of the examination of traces at infinity are evident.

As a preliminary we note the following. Let the function $f(x)$ be differentiable r times for $x > 1$ and let $g(y) = f(1/y)$, $0 < y < 1$. Then

$$g^{(s)}(y) = \sum_{i=1}^{s} \frac{a_{is}}{y^{s+i}} f^{(i)}\left(\frac{1}{y}\right), \qquad s = 1, 2, \ldots, r, \tag{2.53}$$

where a_{is} is some constant.

Now let the function $f(x)$ be defined on $\overset{+}{E}{}_1^n$ and have generalized derivatives of orders up to r_i inclusively, $i = 1, \cdots, n$. Let $r = (r_1, \cdots, r_n)$ and $\beta = (\beta_1, \cdots, \beta_n)$. Let us set

$$\left| f,\ \overset{*}{L}{}^{r_n}_{p,\, x_n^{\beta_n}}(\overset{+}{E}{}_1^n) \right| = \sum_{i=1}^{r_n} \left| x_n^{\beta_n - r_n + i} \frac{\partial^i f}{\partial x_n^i},\ L_p(\overset{+}{E}{}_1^n) \right|, \tag{2.54}$$

$$x_n^\beta = (x_1^{\beta_1}, \ldots, x_n^{\beta_n})$$

and

$$\left| f,\ \overset{*}{L}{}^{r}_{p,\, x_n^{\beta}}(\overset{+}{E}{}_1^n) \right| = \sum_{j=1}^{n-1} \left| f,\ L^{r_j}_{p,\, x_n^{\beta_j}}(\overset{+}{E}{}_1^n) \right| + \left| f,\ \overset{*}{L}{}^{r_n}_{p,\, x_n^{\beta_n}}(\overset{+}{E}{}_1^n) \right|. \tag{2.55}$$

If for traces at infinity we wish to obtain results from the properties of traces on an ordinary hypersurface, we shall have to introduce into the norm of the function not only the norm of a higher derivative in some weight space, but also the norms of all lower derivatives, as was done in formula (2.54).

We denote the linear space of functions with norm (2.55) by $\overset{*}{L}{}^r_{p,x_n^\beta}(\overset{+}{E}{}_1^n)$.

THEOREM 4. *Let* $f \in \overset{*}{L}{}^r_{p,x_n^\beta}(\overset{+}{E}{}_1^n)$, $r_i \geqslant 1$, $i = 1, 2, \cdots, n$, *and let*

$$\beta_n > 2r_n - 1 - \frac{1}{p}. \tag{2.56}$$

If

$$\beta_i > -\frac{1}{p}, \qquad i = 1, 2, \ldots, n-1, \tag{2.57}$$

then the function $f(x^{(n-1)}, +\infty)$ *is a polynomial of degree not greater than* $r_i - 1$ *in the variable* x_i, $i = 1, \cdots, n-1$.

But if

$$\beta_i \leqslant -\frac{1}{p}, \tag{2.58}$$

then

$$\left| \Delta^k_{h_i} f(x^{(n-1)}, +\infty),\ L_p(E^{n-1}) \right| \leqslant c \left| f,\ \overset{*}{L}{}^r_{p,\, x_n^\beta}(\overset{+}{E}{}_1^n) \right| |h_i|^{\mu_i},$$

where

$$\mu_i = \frac{r_i\left(\beta_n + \frac{1}{p} - r_n\right)}{\beta_n - \beta_i - r_n}, \qquad i = 1, 2, \ldots, n-1.$$

PROOF. Making the change of variable $x_n = 1/t$ and setting

$$\begin{aligned} &g(x^{(n-1)}, t) = f\left(x^{(n-1)}, \frac{1}{t}\right), \\ &\alpha_j = -\beta_j' - \frac{2}{p}, \quad j = 1, 2, \ldots, n-1; \ \alpha_n = 2r_n - \beta_n - \frac{2}{p}, \end{aligned} \tag{2.59}$$

we get

$$\left|g, L^{r_j}_{p,\, t^{\alpha_j}}(E^n_1)\right| = \left|f, L^{r_j}_{p,\, x_n^{\beta_j}}(\overset{+}{E}{}^n_1)\right|, \qquad j = 1, 2, \ldots, n-1,$$

$$\left|t^{\alpha_n}\frac{\partial^{r_n} g}{\partial t^{r_n}}, L_p(E^n_1)\right| \leqslant c' \sum_{i=1}^{r_n}\left|x_n^{\beta_n - r_n + i}\frac{\partial^i f}{\partial x_n^i}, L_p(\overset{+}{E}{}^n_1)\right|$$

(we have employed formula (2.53), where c' is a constant not depending on the function f). From the inequalities obtained it follows that (see (2.55))

$$|g, L^r_{p,\, t^\alpha}(E^n_1)| \leqslant c''\left|f, \overset{+}{L}{}^r_{p,\, x_n^\beta}(\overset{+}{E}{}^n_1)\right|, \tag{2.60}$$

where c'' is a constant likewise not depending on f.

On the strength of formulas (2.59) condition (2.56) is equivalent to the condition $\alpha_n < 1 - 1/p$, (2.57) is equivalent to the condition $\alpha_i < -1/p$, and correspondingly (2.58) is equivalent to the condition $\alpha_i \geqslant -1/p$. Finally,

$$\mu_i = \frac{r_i\left(\beta_n + \frac{1}{p} - r_n\right)}{\beta_n - \beta_i - r_n} = \frac{r_i\left(r_n - \alpha_n - \frac{1}{p}\right)}{r_n + \alpha_i - \alpha_n}.$$

Since $f(x^{(n-1)}, +\infty) = g(x^{(n-1)}, 0)$, on the strength of inequality (2.60), Theorem 4 follows from Remarks 10 and 11 in §2.5. The theorem is proved.

We observe that if $\beta_n > 2r_n - 1 - 1/p$ (see (2.56)), then all the more $\beta_n - r_n + i > 1 - 1/p$, $i = 1, 2, \cdots, r_n$. The latter inequalities comprise a condition of weak degeneracy of the weight for the respective derivatives $\partial^i f/\partial x_n^i$, $i = 1, \cdots, r_n$. This guarantees the existence of a trace at infinity for the derivatives $\partial^j f/\partial x_n^j$, $j = 0, 1, \cdots, r_n - 1$. Since in this case $\alpha_n < 1 - 1/p$, the finite traces of $\partial^j g(x^{(n-1)}, 0)/\partial x_n^j$ also exist; but then (2.53) implies that

$$\frac{\partial^j f(x^{(n-1)}, +\infty)}{\partial x_n^j} = 0, \qquad j = 1, 2, \ldots, r_n - 1.$$

In conclusion we shall show that Theorem 4 implies, for example, the well-known theorem (S. V. Uspenskiĭ [4]) that a function for which the Dirichlet integral

$$D_p(f)=\int\sum_{i=1}^{n}\left|\frac{\partial f}{\partial x_i}\right|^p dE^{n-1} \tag{2.61}$$

is finite for $p<n$ converges, after modification of it on a set of n-dimensional measure zero, to one and the same constant as the point tends to infinity along a radius. Some weights do not occur in the integral (2.61); nevertheless, the roots of the indicated phenomenon lie in the characteristics of weight spaces. To prove this let us convert to spherical coordinates (s, ρ) in the integral (2.3), where $\rho = |x|$, $s = (\varphi_1, \cdots, \varphi_{n-1}) \in S^{n-1} = \{x: \rho = 1\}$, $0 \leqslant \varphi_1 \leqslant 2\pi$, $-\pi/2 \leqslant \varphi_i \leqslant \pi/2$, $i = 2, 3, \cdots, n-1$. Noting that

$$\left|\frac{\partial f}{\partial \rho}\right|\leqslant\sum_{i=1}^{n}\left|\frac{\partial f}{\partial x_i}\right|,\qquad \left|\frac{\partial f}{\partial \varphi_k}\right|\leqslant\rho\sum_{i=1}^{n}\left|\frac{\partial f}{\partial x_i}\right|,\qquad k=1,\,2,\,\ldots,\,n-1,$$

we obtain

$$\int dS^{n-1}\int_0^{+\infty}\left[\sum_{i=1}^{n}\left|\frac{1}{\rho}\frac{\partial f}{\partial \varphi_k}\right|^p+\left|\frac{\partial f}{\partial \rho}\right|^p\right]\rho^{n-1}d\rho\leqslant D_p(f),$$

where

$$dS^{n-1}=\cos\varphi_2\cos^2\varphi_3\ldots\cos^{n-2}\varphi_{n-1}d\varphi_1 d\varphi_2\ldots d\varphi_{n-1}.$$

For any $\epsilon>0$ there exists a $c(\epsilon)>0$ such that

$$\int_0^{2\pi}d\varphi_1\int_{-\frac{\pi}{2}+\varepsilon}^{\frac{\pi}{2}-\varepsilon}d\varphi_2\ldots\int_{-\frac{\pi}{2}+\varepsilon}^{\frac{\pi}{2}-\varepsilon}d\varphi_{n-1}\int_1^{+\infty}\left[\sum_{i=1}^{n}\left|\rho^{\frac{n-1}{p}-1}\frac{\partial f}{\partial \varphi_k}\right|^p+\left|\rho^{\frac{n-1}{p}}\frac{\partial f}{\partial \rho}\right|^p\right]d\rho \leqslant c(\varepsilon)D_p(f).$$

If $p<n$, then $(n-1)/p>1-1/p$, i.e. condition (2.56) is satisfied (here $r_n = 1$); but $(n-1)p-1>-1/p$, i.e. condition (2.57) is also satisfied, and consequently on the basis of Theorem 4 the function $f(x(s, +\infty))$ is constant.

BIBLIOGRAPHY

1. S. L. Sobolev, *Applications of functional analysis in mathematical physics*, Izdat. Leningrad. Gos. Univ., Leningrad, 1950; English transl., Transl. Math. Monographs, vol. 7, Amer. Math. Soc., Providence, R. I., 1963. MR 14, 565; 29 #2624.

2. S. M. Nikol'skiĭ, *Approximation of functions of several variables and imbedding theorems*, "Nauka", Moscow, 1969. (Russian)

3. S. L. Sobolev, *Density of function with compact support in the space* $L_p^{(m)}(E_n)$, Sibirsk. Mat. Ž. **4** (1963), 673–682. (Russian) MR **30** #5156.

4. S. V. Uspenskiĭ, *Imbedding theorems for weighted classes*, Trudy Mat. Inst. Steklov. **60** (1961), 282–303; English transl., Amer. Math. Soc. Transl. (2) **87** (1970), 121–145. MR **25** #440; **41** #3202.

5. V. R. Portnov, *Two imbedding theorems for the* $L_{p,b}^{(1)}(\Omega \times R_+)$ *space and their applications*, Dokl. Akad. Nauk SSSR **155** (1964), 761–764 = Soviet Math. Dokl. **5** (1964), 514–517. MR **28** #4243.

6. V. N. Sedov, *On functions reducing at infinity to a polynomial*, Imbedding Theorems and Their Applications, Proc. All-Union Sympos. Imbedding Theorems (Baku, 1966), "Nauka", Moscow, 1970, pp. 204–212. (Russian)

7. G. N. Jakovlev, *On traces of functions whose derivatives are summable with some weight*, Imbedding Theorems and Their Applications, Proc. All-Union Sympos. Imbedding Theorems (Baku, 1966), "Nauka", Moscow, 1970, pp. 225–233. (Russian)

8. S. V. Uspenskiĭ, *Imbedding and extension theorems for a certain class of functions*. II, Sibirsk. Mat. Ž. **7** (1966), 409–418 = Siberian Math. J. **7** (1966), 333–342. MR **33** #6382.

9. S. M. Nikol'skiĭ, *Some properties of differentiable functions defined on an n-dimensional open set*, Izv. Akad. Nauk SSSR Ser. Mat. **23** (1959), 213–242; English transl., Amer. Math. Soc. Transl. (2) **88** (1970), 1–40. MR **22** #3775; **41** #5135.

10. F. A. Sysoeva, *Generalization of a certain Hardy inequality*, Izv. Vysš. Učebn. Zaved. Matematika **1965**, no. 6 (49), 140–143. (Russian) MR **33** #240.

11. L. D. Kudrjavcev, *Imbedding theorems for a class of functions defined on the entire space or on a half space*. I, Mat. Sb. **69 (111)** (1966), 616–639; English transl., Amer. Math. Soc. Transl. (2) **74** (1968), 199–225; II, Mat. Sb. **70 (112)** (1966), 3–35; English transl., Amer. Math. Soc. Transl. (2) **74** (1968), 227–260. MR **34** #6522a, b.

12. ———, *Direct and inverse imbedding theorems. Applications to the solution of elliptic equations by variational methods*, Trudy Mat. Inst. Steklov. **55** (1959); English transl., Transl. Math. Monographs, Amer. Math. Soc., Providence, R. I. (to appear). MR **33** #7838.

13. V. I. Sedov, *A variational method for solving nonregular elliptic equations*, Dokl. Akad. Nauk SSSR **183** (1968), 1012–1015 = Soviet Math. Dokl. **9** (1968), 1511–1515. MR **40** #543.

14. A. Kufner, *Imbedding theorems for general Sobolev weight spaces*, Ann. Scuola Norm. Sup. Pisa (3) **23** (1969), 373–386. MR **40** #6253.

15. L. D. Kudrjavcev, *On traces of functions of several variables and direction fields which are regular at infinity for the Laplace operator*, Trudy Mat. Inst. Steklov. **112** (1971), 256–270 = Proc. Steklov Inst. Math. **112** (1971), 266–280.

16. ———, *On parametric inequalities for functions of several variables*, Dokl. Akad. Nauk SSSR **202** (1972), 1261–1264 = Soviet Math. Dokl. **13** (1972), 276–280.

Translated by R. F. RINEHART

Trudy Mat. Inst. Steklov.
117 (1972)

Proc. Steklov Inst. Math.
117 (1972)

OPERATORS CONNECTED WITH FRACTIONAL DIFFERENTIATION, AND CLASSES OF DIFFERENTIABLE FUNCTIONS

P. I. LIZORKIN

UDC 517.518.22

Abstract. This paper considers operators that are immediately connected with fractional differentiation. With their use a unified approach is given to the theory of function spaces of Sobolev's type and the generalized Hölder spaces of Nikol'skiĭ and Besov. Major attention is given to reformulating the theory of the Liouville classes $L_p^{(r)}$ in terms of the new operators. The author introduces a new family $L_{p,\theta}^{(r)}$ of function spaces which coincide with $L_p^{(r)}$ for $\theta = 2$. The proposed approach makes it possible to eliminate operations with singular integrals, and this is essential for the transition to bounded regions.

Bibliography: 10 items.

The spaces of differentiable functions most commonly encountered in analysis can ultimately be written as collections of functions which, together with their fractional derivatives of given order, belong to some "zeroth" Banach space. Examples are the $W_p^{(l)}$ spaces of Sobolev, the $H_p^{(r)}$ spaces of Nikol'skiĭ, the $B_{p,\theta}^{(r)}$ spaces of Besov, and the Liouville classes $L_p^{(r)}$. It is clear that the same construction is possible in other cases also. However, the use of fractional derivatives in such constructions is somewhat inconvenient, since it necessarily involves singular (conditionally convergent) integrals in an essential way. This is awkward in the case of bounded domains.

In the present paper we construct operators that are closely connected with fractional differentiation. By using them we can avoid having to consider conditionally convergent integrals. We also reveal a common structure for different families of function spaces ($B_{p,\theta}^{(r)}$ and $L_p^{(r)}$). In particular, there arise naturally some new frunction spaces ($L_{p,\theta}^{(r)}$) whose study seems likely to attract the attention of specialists. These questions are discussed in detail in §1.

Our principal interest is the reformulation of the theory of the Liouville classes $L_p^{(r)}$ in the language of our new operators. The basic facts of the theory of the $L_p^{(r)}$ spaces, and references, are given in §2. That section also contains some lemmas on which the subsequent proofs depend. In §3 we introduce the $L_{p,\theta}^{(r)}$ spaces and prove that $L_p^{(r)}$ and $L_{p,2}^{(r)}$ coincide up to equivalence of norm. We investigate separately the cases of coordinate-wise norm and "chain" norms. The last section (§4) discusses more

AMS (MOS) subject classifications (1970). Primary 46E15, 46E35, 26A33.

general operators of the type of fractional differentiation.

We now state a representative result.

Let numbers r *and* p *be given,* $0<r<\infty$, $1<p<\infty$, *and define the following operator on p-summable functions*:

$$D_t^r f(x) = \int_t^{2t} \frac{\Delta_h^l f(x)\, dh}{h^{1+r}}, \qquad l>r,$$

where $\Delta_h^l f(x)$ *is the difference of order* l *of* f *with step* h. *A necessary and sufficient condition for* f *to have a derivative* $D^r f \in L_p(-\infty, \infty)$ *is that*

$$X_p T_2^* [D_t^r f(x)] \equiv \left\| \left\{ \int_0^\infty | D_t^r f(x) |^2 \frac{dt}{t} \right\}^{1/2} \right\|_p$$

if finite. In this case

$$c \| D^r f \|_p \leqslant X_p T_2^* [D_t^r f(x)] \leqslant C \| D^r f \|_p,$$

where the constants c *and* C *are independent of* f.

§1. Preliminary considerations

We recall some facts about the theory of the $W_2^{(r)}(R^n) \equiv L_2^{(r)}(R^n)$ spaces of Sobolev, Slobodeckiĭ and Aronszajn. For simplicity take $n=1$ and $0<r<1$. By definition, $f \in W_2^{(r)}(R^1)$ if $f(x) \in L_2(R^1)$ and $f^{(r)}(x) \in L_2(R^1)$. Here $f^{(r)}(x)$ denotes the derivative of f of order r; it is defined by the formulas

$$f^{(r)}(x) = \frac{d}{dx} f_{(1-r)} = \frac{d}{dx} \frac{1}{\Gamma(1-r)} \int_{-\infty}^{x} \frac{f(t)\, dt}{(x-t)^2}, \tag{1}$$

$$f^{(r)}(x) = \frac{r}{\Gamma(1-r)} \int_{-\infty}^{x} \frac{f(x)-f(t)}{(x-t)^{1+r}}\, dt, \tag{2}$$

$$\overline{f^{(r)}(x)} = (i\lambda)^r \bar{f}(\lambda). \tag{3}$$

For smooth functions with compact support these formulas have the classical sense; in more general cases they have to be interpreted differently (cf., with reference to (1) and (2), the author's papers [1, 2]). In the L_2 framework the representation (3) is very natural ($\bar{f}$ is the Fourier transform of f). We shall, however, also appeal to the representation (2), since the papers just mentioned depend on it. As the initial norm in $W_2^{(r)}(R^1)$ we shall use

$$\| f \|_{W_2^{(r)}} = \left\{ \int_{R^1} (1+|\lambda|^2)^r |\bar{f}(\lambda)|^2 d\lambda \right\}^{1/2},$$

which by Parseval's theorem is finite if and only if $f \in L_2(R^1)$ and $f^{(r)} \in L_2(R^1)$. For functions in $W_2^{(r)}(R^1)$ the integral (2) is to be understood as

$$\lim_{\varepsilon\to 0}^{L_2} \frac{r}{\Gamma(1-r)} \int_{\varepsilon}^{\infty} \frac{f(x)-f(x-h)}{h^{1+r}}\,dh \overset{\text{def}}{=} D^r f, \tag{4}$$

i.e. as a conditionally convergent ($\equiv$ singular) integral. Hence without introducing the Fourier transform we can say that f belongs to $W_2^{(r)}$ if and only if the norm

$$\|f\|_2 + \|D^r f\|_2$$

is finite. However, it is convenient to work with a norm that is expressed in unconditionally convergent terms and by an absolutely convergent integral. The success of the theory of $W_2^{(r)}$ spaces depends largely on the introduction of such a norm. We now take up the reasoning leading to this norm in a form that can be generalized to the spaces $L_p^{(r)}(R^1)$.

The fractional differentiation operator $D^r f$ defined by (4) is the limit as $t\to 0$ of the operators

$$D_t^r f = a_r \int_t^{\infty} \frac{f(x)-f(x-h)}{h^{1+r}}\,dh = a_r \varphi_r(f;\ x,\ t,\ \infty).$$

It is rather natural to think of clarifying the characteristic properties of φ_r as a function of t that guarantee that f belongs to $W_2^{(r)}$ under the assumption that φ_r is square summable in x. In other words, we propose to consider φ as a square summable function of x with values in a space H determined by these characteristic properties of φ as a function of t. We shall express this by writing $\varphi\in L_2(R^1;H)$. In addition, we have to note that in describing the differential properties of f we need to consider the behavior of f only near x. Hence it is appropriate to replace $\varphi_r(f; x, t, \infty)$ by

$$\varphi_r(f;\ x,\ t,\ \tau) = \int_t^{\tau} \frac{f(x)-f(x-h)}{h^{1+r}}\,dh. \tag{5}$$

If we put $\tau = At$, where $A>1$ is a constant, then as $t\to 0$ the function $\varphi_r(f; x, t, At)$ can be diminished by shrinking the interval of integration to zero. This property has to be considered in the metric of the space H.

Put

$$\varphi_r(f;\ x,\ t,\ 2t) = \varphi_r(f;\ x,\ t) = \varphi_r(x,\ t) = \int_t^{2t} \frac{f(x)-f(x-h)}{h^{1+r}}\,dh \tag{6}$$

and take H to be the Hilbert space $L_2^*(0,\infty)$ of functions with the finite norm

$$\|\psi\|_{L^*(0,\infty)} = \left\{\int_0^{\infty} |\psi(t)|^2 \frac{dt}{t}\right\}^{1/2}.$$

For Hilbert-valued functions we have Parseval's theorem,

$$\int_{-\infty}^{\infty} |\varphi_r(x, \cdot)|_H^2\, dx = \int_{-\infty}^{\infty} |\tilde{\varphi}_r(\lambda, \cdot)|_H^2\, d\lambda, \tag{7}$$

where $\tilde{\varphi}_r(\lambda, t)$ is the Fourier transform of $\varphi_r(x, t)$ with respect to x, and this allows an exact calculation of the $L_2(R^1, H)$ norm of $\varphi_r(x, t)$ in which we are interested.

We have

$$\tilde{\varphi}_r(\lambda, t) = \tilde{f}(\lambda) \int_t^{2t} \frac{1 - e^{-i\lambda h}}{h^{1+r}}\, dh,$$

$$\int_{-\infty}^{\infty} |\tilde{\varphi}_r(\lambda, \cdot)|_H^2\, d\lambda = \int_{-\infty}^{\infty} |\tilde{f}(\lambda)|^2 \left\{ \int_0^{\infty} \left| \int_t^{2t} \frac{1 - e^{-i\lambda h}}{h^{1+r}}\, dh \right|^2 \frac{dt}{t} \right\} d\lambda \tag{8}$$

$$= \int_{-\infty}^{\infty} |\lambda|^{2r} |\tilde{f}(\lambda)|^2 \left(\int_0^{\infty} \left| \int_v^{2v} \frac{1 - e^{-iu}}{u^{1+r}}\, du \right|^2 \frac{dv}{v} \right) d\lambda = c \int_{-\infty}^{\infty} |\lambda|^{2r} |\tilde{f}(\lambda)|^2\, d\lambda.$$

It is easy to show that the expression we have obtained coincides up to a factor with

$$2 \int_{-\infty}^{\infty} dx \int_0^{\infty} \frac{|f(x) - f(x-h)|^2}{h^{1+2r}}\, dh = \int_{-\infty}^{\infty} \int_{-\infty}^{\infty} \frac{|f(x) - f(y)|^2}{|x-y|^{1+2r}}\, dx\,dy,$$

which is commonly used as a differential seminorm on $W_2^{(r)}(R^1)$. In fact, with the usual form of Plancherel's theorem (for scalar-valued functions) we have

$$\int_{-\infty}^{\infty} dx \int_0^{\infty} \frac{|f(x) - f(x-h)|^2}{h^{1+2r}}\, dh = \int_0^{\infty} \frac{dh}{h^{1+2r}} \int_{-\infty}^{\infty} |f(x) - f(x-h)|^2\, dh$$

$$= \int_0^{\infty} \frac{dh}{h^{1+2r}} \int_{-\infty}^{\infty} |\tilde{f}(\lambda)(1 - e^{i\lambda h})|^2\, d\lambda = c' \int_{-\infty}^{\infty} |\lambda|^{2r} |\tilde{f}(\lambda)|^2\, d\lambda. \tag{9}$$

Hence it follows from (6)–(9) that

$$\int_{-\infty}^{\infty} dx \left\{ \int_0^{\infty} \frac{dt}{t} \left| \int_t^{2t} \frac{f(x) - f(x-h)}{h^{1+r}}\, dh \right|^2 \right\} = \frac{c}{2c'} \int_{-\infty}^{\infty} \int_{-\infty}^{\infty} \frac{|f(x) - f(y)|^2}{|x-y|^{1+2r}}\, dx\,dy. \tag{10}$$

Although the expression on the left has a somewhat more complicated form, it is preferable. In the first place, it directly connects the "fractional" class $W_2^{(r)}(R^1)$ with the "fractional" derivative (4); in the second place – significantly – it can be generalized to the L_p-metric for $p \neq 2$, $1 < p < \infty$. In fact, the expression

$$\left[\int_{-\infty}^{\infty} |\varphi_r(x,\ \cdot)|_H^p\, dx\right]^{1/p} = \left\| \left(\int_0^{\infty} \frac{dt}{t} \left| \int_t^{2t} \frac{f(x)-f(x-h)}{h^{1+r}}\, dh \right|^2 \right)^{1/2} \right\|_p \tag{11}$$

is equivalent to the differential seminorm on $L_p^{(r)}(R^1)$, i.e.

$$\|f^{(r)}\|_p \sim \|\,|\varphi_r(\cdot,\ \cdot)|_H\|_p.\ (^1)$$

This, in particular, will appear again later (cf. Theorem 5, §3).

In addition, the same approach based on the function $\varphi_r(f; x, t)$ lets us study the more general classes $L_{p,\theta}^{(r)}(R^1)$, $1 \leqslant \theta < \infty$. These classes consist of the functions $f \in L_p$ for which $\varphi_r(f; x, t) = \varphi_r(x, t)$ has values in the space $L_\theta^*(0, \infty) = L_\theta^*$ and

$$|\varphi_r(x,\ \cdot)|_{L_\theta^*} = \left\{ \int_0^{\infty} \frac{dt}{t} \left| \int_t^{2t} \frac{f(x)-f(x-h)}{h^{1+r}}\, dh \right|^\theta \right\}^{1/\theta} \in L_p(R^1).$$

The norm in $L_{p,\theta}^{(r)}(R^1)$ is taken to be

$$\|f\|_{L_{p,\theta}^{(r)}} = \|f\|_p + \|\,|\varphi_r(x,\ \cdot)|_{L_\theta^*}\|_p.$$

It is clear that $L_{p,2}^{(r)}(R^1) = L_p^{(r)}(R^1)$. The theory of $L_{p,\theta}^{(r)}$ spaces for $\theta \neq 2$ will be given in a forthcoming paper of the author.

In this connection it is natural to raise the question of interchanging the roles of the arguments x and t of the function φ_r in forming the L_θ^*- and L_p-norms. In the cases we have considered, we first take the norm with respect to the translation, and then with respect to the space variable. Let us now introduce a norm of the form

$$\|f\|_p + |\,\|\varphi_r(\cdot,\ t)\|_p|_{L_\theta^*}.$$

It can be shown that the spaces obtained in this way coincide with the well-known Besov spaces $B_{p,\theta}^{(r)}(R^1)$ [3].

Before proceeding to the main part of the paper, we note several facts.

1. The expression on the left of (10) can be written in an equivalent form which is frequently more convenient for calculations. By Minkowski's inequality we have

$$\left\{ \int_0^{\infty} \frac{dt}{t} \left| \int_t^{2t} g(h)\, dh \right|^\theta \right\}^{1/\theta} = \left\{ \int_0^{\infty} \left| \int_1^2 g(ut)\, t\, du \right|^\theta \frac{dt}{t} \right\}^{1/\theta}$$

$$\leqslant \int_1^2 du \left(\int_0^{\infty} |g(ut)|^\theta t^{\theta-1} dt \right)^{1/\theta} = c_\theta \left(\int_0^{\infty} h^{\theta-1} |g(h)|^\theta dt \right)^{1/\theta}$$

Using this inequality, we obtain

---

(1) The equivalence relation $A(f) \sim B(f)$ means that $cA(f) \leqslant B(f) \leqslant CA(f)$, $0 < c < C < \infty$.

$$\int_{-\infty}^{\infty} dx \left\{ \int_0^{\infty} \frac{dt}{t} \left| \int_t^{2t} \frac{f(x) - f(x-h)}{h^{1+r}} dh \right|^2 \right\} \leqslant \int_{-\infty}^{\infty} dx \int_0^{\infty} \left\{ \int_t^{2t} \frac{|f(x) - f(x-h)|}{h^{1+r}} dh \right\}^2 \frac{dt}{t}$$

$$\leqslant c_2 \int_{-\infty}^{\infty} dx \int_0^{\infty} \frac{|f(x) - f(x-h)|^2}{h^{1+2r}} dh = 2c_2 \int_{-\infty}^{\infty} \int_{-\infty}^{\infty} \frac{|f(x) - f(y)|^2}{|x-y|^{1+2r}} dx dy.$$

From this and (10) we obtain the equivalence

$$\int_{-\infty}^{\infty} dx \int_0^{\infty} \frac{dt}{t} \left\{ \int_t^{2t} \frac{|f(x) - f(x-h)|}{h^{1+r}} dh \right\}^2 \sim \int_{-\infty}^{\infty} \int_{-\infty}^{\infty} \frac{|f(x) - f(y)|^2}{|x-y|^{1+2r}} dx dy. \tag{12}$$

2. A careful look at formula (8) shows that to obtain the factor $|\lambda|^{2r}$ on the right we use only the homogeneity in λ of the integral in the braces. But this homogeneity is preserved if we consider the integral

$$\int_0^{\infty} \left| \frac{1}{t^{\beta}} \int_t^{2t} \frac{1 - e^{-i\lambda h}}{h^{1+r-\beta}} dh \right|^2 \frac{dt}{t} = |\lambda|^{2r} \int_0^{\infty} \left| \frac{1}{v^{\beta}} \int_v^{2v} \frac{1 - e^{-iu}}{u^{1+r-\beta}} \right|^2 \frac{dv}{v} \tag{13}$$

Moreover, when $\beta > 0$ we may consider the integral

$$\int_0^{\infty} \left| \frac{1}{t^{\beta}} \int_0^{t} \frac{1 - e^{-i\lambda h}}{h^{1+r-\beta}} dh \right|^2 \frac{dt}{t}, \tag{14}$$

and when $\beta < 0$, the integral

$$\int_0^{\infty} \left| t^{\gamma} \int_t^{\infty} \frac{1 - e^{-i\lambda h}}{h^{1+r+\gamma}} dh \right|^2 \frac{dt}{t}, \qquad \gamma = -\beta. \tag{15}$$

These integrals lead to the consideration of the functions

$$\varphi_{r,\beta}(f;\ x,\ t,\ 2t) = \frac{1}{t^{\beta}} \int_t^{2t} \frac{f(x) - f(x-h)}{h^{1+r-\beta}} dh, \tag{13'}$$

$$\varphi_{r,\beta}(f;\ x,\ 0,\ t) = \frac{1}{t^{\beta}} \int_0^{t} \frac{f(x) - f(x-h)}{h^{1+r-\beta}} dh, \tag{14'}$$

$$\varphi_{r,\beta}(f;\ x,\ t,\ \infty) = t^{\gamma} \int_t^{\infty} \frac{f(x) - f(x-h)}{h^{1+r+\gamma}} dh \tag{15'}$$

instead of $\varphi_r(f; x, t, 2t)$. Using these functions we get the same results as by using $\varphi_r(f; x, t, 2t)$. This equivalence will be considered later. Note that for functions in $L_p^{(r)}$

the integral (14) is not singular: in view of the differential properties of f the singularity at zero is without effect.

In the preceding discussion we have considered only operators defined by first differences. In the general case it is necessary to introduce higher differences. For example, for $r = 1$ we have

$$f' = \frac{1}{2\ln 2}\int_0^\infty \frac{f(x) - 2f(x-h) + f(x-2h)}{h^2}\,dh,$$

$$\varphi_1(f;\ x,\ t,\ 2t) = \int_t^{2t} \frac{f(x) - 2f(x-h) + f(x-2h)}{h^2}\,dh.$$

In accordance with the preceding outline we consider

$$\|f'\|_p \sim \left\| \left\{ \int_0^\infty \frac{dt}{t} \left(\int_t^{2t} \frac{|f(x) - 2(x-h) + f(x-2h)|}{h^2}\,dh \right)^2 \right\}^{1/2} \right\|_p, \tag{16}$$

which is, in essence, a disguised form of Marcinkiewicz's theorem which states that

$$\|f'\|_p \sim \left\| \left\{ \int_0^\infty \frac{|f(x) - 2f(x-t) + f(x-2t)|^2}{t^3}\,dt \right\}^{1/2} \right\|. \tag{17}$$

In this connection see Remark 6 in §3.

§2. The spaces $L_p^{(\mathbf{r})}(R^n)$. Auxiliary propositions

1. Let R^n be n-dimensional Euclidean space, $x = (x_1, \cdots, x_n)$ a point in R^n, $|x| = \sqrt{x_1^2 + \cdots + x_n^2}$. Let $\mathbf{r} = (r_1, \cdots, r_n)$ be a vector with nonnegative components r_j. Writing $\mathbf{r} \geqslant 0$ ($\mathbf{r} > 0$) means that $r_j \geqslant 0$, $j = 1, \cdots, n$ ($r_j > 0$, $j = 1, \cdots, n$), and $|\mathbf{r}| = r_1 + \cdots + r_n$. If r is a real number, $[r]$ means, as usual, the integral part of r, and $[r]^- = \bar{r}$ is the integral part "by deficiency" (i.e. the largest integer less than r). Hence

$$r = \bar{r} + \alpha, \qquad 0 < \alpha \leqslant 1.$$

We consider p-summable functions $f(x_1, \cdots, x_n) = f(x)$ on R^n, i.e. functions with finite norm

$$\|f\|_p = \left\{ \int_{R^n} |f(x)|^p\,dx \right\}^{1/p}, \qquad 1 \leqslant p < \infty.$$

The norm of the element f of the Banach space B will be denoted by $\|f\|_B$ or $|f, B|$. The norm of an operator $\mathcal{K}$ from B' to B'' is denoted by $|\mathcal{K}|_{B' \to B''}$. The Fourier transform of $f(x)$ is denoted by $\tilde{f}(\lambda)$ or $(\mathfrak{F}f)(\lambda)$,

$$f(\lambda) = \frac{1}{(2\pi)^n} \int_{R^n} \tilde{f}(\lambda)\, e^{i\lambda x} d\lambda.$$

Let e_j be the basis vector for the jth axis. We consider the difference $\Delta^{l_j}_{he_j} f(x)$ of order l_j with step $-h$ of the function f in the direction of the jth axis,

$$\Delta^{l_j}_{he_j} f(x) = \sum_0^{l_j} (-1)^{k_j} C^{k_j}_{l_j} f(x - k_j h e_j).$$

As usual,

$$D^{r_j}_j f = \frac{\partial^{r_j} f}{\partial x_j^{r_j}}, \qquad D^{\mathbf{r}} f = D_1^{r_1} \ldots D_n^{r_n} f = \frac{\partial^{r_1 + \ldots + r_n} f}{\partial x_1^{r_1} \ldots \partial x_n^{r_n}}.$$

Unless the contrary is specified, these derivatives are not necessarily of integral order (cf. below). For derivatives of integral order the index r_j is usually written m_j or l_j (or k_j).

We shall often use a modified Hardy "(x)-inequality",

$$\int_0^\infty \left| \frac{1}{t} \int_{at}^{bt} g(h)\, dh \right|^\theta \frac{dt}{t^\beta} \leqslant \varkappa \int_0^\infty |g(t)|^\theta \frac{dt}{t^\beta}, \qquad 1 \leqslant \theta < \infty.$$

This follows easily from Minkowski's inequality:

$$\left\{ \int_0^\infty \left| \frac{1}{t} \int_{at}^{bt} g(h)\, dh \right|^\theta \frac{dt}{t^\beta} \right\}^{1/\theta} = \left\{ \int_0^\infty \left| \int_a^b g(\xi t)\, d\xi \right|^\theta \frac{dt}{t^\beta} \right\}^{1/\theta}$$

$$\leqslant \int_a^b \xi^{\frac{\beta - 1}{\theta}} d\xi \left(\int_0^\infty |g(\eta)|^\theta \frac{d\eta}{\eta^\beta} \right)^{1/\theta} = \varkappa_{a,b}^{1/\theta} \left(\int_0^\infty |g(t)|^\theta \frac{dt}{t^\beta} \right)^{1/\theta}$$

and holds for arbitrary β with constant

$$\varkappa_{a,b} = \left(\int_a^b \xi^{\frac{\beta-1}{\theta}} d\xi \right)^\theta, \qquad 0 < a < b < \infty.$$

Hardy's inequality in the usual form holds only for $\beta \neq \theta - 1$ and can be obtained as a special case of the (x)-inequality. In fact, if we take $a = 0$, $b = 1$ and $\beta > 1 - \theta$, then

$$\int_0^\infty \left| \frac{1}{t} \int_0^t g(h)\, dh \right|^\theta \frac{dt}{t^\beta} \leqslant \left(\frac{\theta}{\beta + \theta - 1} \right)^\theta \int_0^\infty |g(t)|^\theta \frac{dt}{t^\beta},$$

and if $a = 1$, $b = \infty$ and $\beta < 1 - \theta$, then

$$\int_0^\infty \left| \frac{1}{t} \int_t^\infty g(h)\, dh \right|^\theta \frac{dt}{t^\beta} \leqslant \left(\frac{\theta}{1 - \theta - \beta} \right)^\theta \int_0^\infty |g(t)|^\theta \frac{dt}{t^\beta}.$$

2. The original definition of the spaces $L_p^{(r_1, \ldots, r_n)}(R^n) \equiv L_p^{(\mathbf{r})}(R^n) \equiv L_p^{(\mathbf{r})}$, $1 < p < \infty$, is

$$f \in L_p^{(\mathbf{r})}(R^n) \Leftrightarrow f \in L_p(R^n), \quad \frac{\partial^{r_j} f}{\partial x_j^{r_j}} \in L_p(R^n), \qquad j = 1, \ldots, n,$$

where $\partial^{r_j} f/\partial x_j^{r_j} \equiv D_j^{r_j} f$ is the Liouville generalized (L.g.) derivative of order r_j of the function f with respect to the variable x_j [1]. For integral $r_j = l_j$ the spaces $L_p^{(\mathbf{l})}(R^n)$ are the same, as we saw in the Introduction, as the Sobolev spaces $W_p^{(l_1,\dots,l_n)}(R^n)$. The space $L_p^{(\mathbf{r})}(R^n)$ is a Banach space with the norm

$$\|f\|_{L_p^{(\mathbf{r})}} = \|f\|_p + \sum_{j=1}^{n} \|D_j^{r_j} f\|_p. \tag{I}$$

The case $r_1 = r_2 = \cdots = r_n = r$ is said to be isotropic, and in that case we use the simpler notation

$$L_p^{(\mathbf{r})} = L^{(r,\dots,r)} = L_p^{(r)}.$$

Similarly, when $r = l$,

$$W_p^{(l,\dots,l)} = W_p^{(l)}.$$

As was shown in the author's papers [2, 4], the L.g. derivatives of $f \in L_p^{(\mathbf{r})}(R^n)$ can be expressed as singular integrals of the form

$$D_j^{r_j} f(x) = a_{r_j l_j k_j} \overset{L_p}{\lim_{\varepsilon\to 0}} \int_{\varepsilon}^{\infty} \frac{\Delta_{h e_j}^{l_j} D_j^{k_j} f(x)}{h^{1+r_j-k_j}}\, dh, \tag{18}$$

where k_j and l_j are fixed integers satisfying

$$0 \leqslant k_j \leqslant r_j < r_j < l_j + k_j,$$

and $a_{r,l,k}$ are constants,

$$a_{r,l,k} = \left\{ \int_0^\infty t^{-r+k-1} (1 - e^{-t})^l \, dt \right\}^{-1}$$

Therefore the representation (18) makes it possible to eliminate the use of generalized functions in the definition of the spaces $L_p^{(\mathbf{r})}$ at the expense of using singular integrals. We have the following theorem.

THEOREM 1 [2, 4]. *A function* $f \in L_p^{(\mathbf{r})}$ *if and only if* $f \in L_p$ *and the limits* (18) *exist for* $j = 1, \cdots, n$. *In this case*

$$\|f\|_{L_p^{\mathbf{r}}} = \|f\|_p + \sum_1^n \|D_j^{r_j} f\|_p.$$

REMARK 1. If it is desirable for any reason to work with differences of at most first order, one can, for $r_j = l_j$, replace (18) by the generalized derivative $D_j^{l_j} f$ in the sense of Sobolev, and for nonintegral r_j by the limit (18) with $l_j = 1$, $k_j = [r_j]$ and $D_j^{[r_j]} f$ taken as a generalized derivative in the sense of Sobolev. However, in accordance with the traditional procedures of the theory of fractional differentiation, we shall take the canonical representation of $D_j^{r_j}f$ to be (18) with a second difference:

$$D_j^{r_j}f(x) = a_{r_j, 2, \bar{r}_j} \lim_{\varepsilon\to 0} \int_\varepsilon^\infty \frac{\Delta^2_{he_j} D_j^{\bar{r}_j} f(x)}{h^{(1+\alpha_j)}}\, dh. \tag{19}$$

The truncated differentiation operator $\bar{D}^{r_j}_{te_j}f$ is defined by

$$\bar{D}^{r_j}_{te_j}f(x) = a_{r_j, 2, \bar{r}} \int_t^{2t} \frac{\Delta^2_{he_j} D_j^{\bar{r}_j} f(x)}{h^{1+\alpha_j}}\, dh = a_{r_j, 2, \bar{r}_j} \check{D}^{r_j}_{te_j} f. \tag{20}$$

The definition itself and the criteria we have stated provide a description of the $L_p^{(\mathbf{r})}$ spaces in terms of coordinate-wise differential properties of the functions under consideration. In this description $L_p^{(\mathbf{r})}$ is the intersection of the spaces $L_p\binom{r_j}{x_j}$:

$$L_p^{(\mathbf{r})} = \bigcap_{j=1}^n L_p\binom{r_j}{x_j},$$

where

$$L_p\binom{r_j}{x_j} \Leftrightarrow f \in L_p, \qquad D_j^{r_j} f \in L_p,$$

$$\left| f,\ L_p\binom{r_j}{x_j} \right| = \frac{1}{n}\| f \|_p + \| D_j^{r_j} f \|_p.$$

We shall call $\Sigma_1^n \|D_j^{r_j}f\|_p$ the coordinate seminorm in $L_p^{(\mathbf{r})}$, and the norm (I) the coordinate norm in $L_p^{(\mathbf{r})}$.

3. There is also interest in a description of $L_p^{(\mathbf{r})}$ that makes fuller use of mixed differential properties of f. One such description is close to (18) in principle and is obtained by considering n-dimensional singular integrals containing differences; a second one characterizes $L_p^{(\mathbf{r})}$ as a space of functions represented by integrals of potential type whose densities are in L_p.

Let $a = (a_1, \cdots, a_n)$, $a_j > 0$, $a_1 + \cdots + a_n = n$, and let $\rho(h_1, \cdots, h_n) = \rho(h)$ be an a-homogeneous function of degree 1, i.e.

$$\rho(t^a h) \equiv \rho(t^{a_1} h_1, \ldots, t^{a_n} h_n) = t\rho(h).$$

For example, we can take the positive solution of the equation

$$\sum_{j=1}^n \frac{h_j^2}{\rho^{2a_j}} = 1.$$

In the isotropic case, i.e. when $a_1 = a_2 = \cdots = a_n = 1$, we have $\rho(h) = |h|$. For $\mathbf{r} > 0$ we put

$$\frac{1}{r^*} = \frac{1}{n} \sum_{j=1}^{n} \frac{1}{r_j}, \qquad a_j = \frac{r^*}{r_j}.$$

Let $\Delta_h^l f$ be the difference of order l of the function f with vector step $-h = (-h_1, \cdots, -h_n)$, $\Delta_h^l f(x) = \Sigma_{k=0}^l (-1)^k C_l^k f(x - kh)$.

THEOREM 2 [4]. *A necessary and sufficient condition for a p-summable function, $1 < p < \infty$, to belong to $L_p^{(\mathbf{r})}$ is that the limit*

$$\lim_{\varepsilon \to 0} \int_{\rho(h) > \varepsilon} \frac{\Delta_h^l f(x)\, dh}{[\rho(h)]^{n+r^*}} \overset{\text{def}}{=} (T^{\mathbf{r}} f)(x) \tag{21}$$

exists in the L_p sense, with $l > \max r_j$. In this case we have the equivalences

$$\| T^{\mathbf{r}} f \|_p \sim \sum_1^n \| D_j^{r_j} f \|_p, \tag{22}$$

$$\| f \|_p + \| T^{\mathbf{r}} f \|_p \sim \| f \|_{L_p^{(\mathbf{r})}}. \tag{23}$$

Consequently the expression

$$\| f \|_p + \| T^{\mathbf{r}} f \|_p \tag{II}$$

can be taken as a norm in $L_p^{(\mathbf{r})}$. We call it the *chain norm* in $L_p^{(\mathbf{r})}$. We call $\|T^{\mathbf{r}} f\|_p$ the *chain seminorm* in $L_p^{(\mathbf{r})}$. By a *differential seminorm* in $L_p^{(\mathbf{r})}$ we shall mean either $\|T^{\mathbf{r}} f\|_p$, or $\Sigma_1^n \|D_j^{r_j} f\|_p$, or any equivalent expression.

REMARK 2. We shall consider in detail the isotropic case $r_1 = \cdots = r_n = r$. In this case the operator $T^{\mathbf{r}} f = T^r f$ is, up to a scalar factor, the inverse of the operator known as the Riesz potential $\mathcal{R}_{-r}$. For a summable function $g(x)$ we have for $0 < r < n$

$$\mathcal{R}_{-r} g(x) = \mathcal{A}_{n,-r} \int \frac{g(y)\, dy}{|x - y|^{n-r}}, \qquad \mathcal{B}_{n,r,l} T^r \mathcal{R}_{-r} g = g.\text{(}^2\text{)}$$

For arbitrary r the operator $\mathcal{R}_r$ is to be interpreted as convolution with the kernel $\mathcal{A}_{n,r} |x|^{-r-n}$ in a suitable space of generalized functions (for example, in the space Φ' [2]). We wish to stress that for functions $f \in L_p^{(r)}$ the operator $(\mathcal{R}_r)^{-1}$ is represented, up to a factor, by the singular integral $T^{(r)} f$. For even $r = 2m$ the integral $T^{2m} f$ is a multiple of the operator $\Delta^m f$, where Δ is the Laplacian. All the statements we have enumerated are in essence reflections of the fact that in the Fourier representation

---

(2) $\mathcal{A}_{n,r}$ and $\mathcal{B}_{n,r,l}$ are normalizing constants, defined by (24) (see Appendex 2).

$$\widetilde{R_{-r}g} = |\lambda|^{-r}\hat{g}, \qquad \mathcal{B}_{n,r,l}\widetilde{T^r f} = |\lambda|^r \hat{f}, \tag{24}$$

as is easily obtained by a discussion in terms of the spaces Φ and Φ' [2].

By the methods of [4] it is easy to obtain for $L_p^{(r)}$ spaces the following results, which reveal an analogy between the Liouville derivatives $D_j^r f$ and the Riesz derivatives $T^r f$ (in the Fourier representation the first corresponds to multiplication by $(i\lambda_j)^r$ and the second to multiplication by $\mathcal{B}_{n,r,l}|\lambda|^r$).(3)

THEOREM 3. *A necessary and sufficient condition that the p-summable function* $f(x)$ *belongs to* $L_p^{(r)}$, $1 < p < \infty$, *is that for fixed* k *and* l, $0 \leqslant k \leqslant \bar{r} < r < l + k$, *the limit*

$$T^{r,l,\mathbf{m}} f = \underset{\varepsilon \to 0}{\overset{L_p}{\lim}} \int_{|h|>\varepsilon} \frac{\Delta_h^l D^{\mathbf{m}} f(x)\, dh}{|h|^{n+r-k}}, \qquad m_1 + \ldots + m_n = k,$$

exists in the sense of convergence in L_p *In this case the quantity*

$$\sum_{|\mathbf{m}|=k} \| T^{r,l,\mathbf{m}} f \|_p$$

is, for all admissible k *and* l, *equivalent to the differential seminorm of* f *in* $L_p^{(r)}$.

By this theorem if r is nonintegral we can take

$$\|f\|_{L_p} + \sum_{|\mathbf{m}|=[r]} \| T^{r,1,\mathbf{m}} f \|_p \tag{II$'$}$$

as the norm in $L_p^{(r)}$; when $r = l$ is an integer, as we have already said, $L_p^{(l)} = W_p^{(l)}$.

4. To a certain extent the most complete characterization of the spaces $L_p^{(\mathbf{r})}$ is contained in the integral representation given below. Let a_j, r^* and ρ have the same meanings as above. In R^n consider the function

$$\frac{1}{[1+\rho^2(\lambda)]^{r^*/2}}.$$

Let its inverse Fourier transform be $G_{\mathbf{r}}(x)$:

$$G_{\mathbf{r}}(x) = F^{-1}([1+\rho^2(\lambda)]^{-r^*/2}).$$

THEOREM 4 [2]. *A necessary and sufficient condition that* $f(x)$ *belongs to* $L_p^{(\mathbf{r})}$, $\mathbf{r} > 0$, $1 < p < \infty$, *is that* $f(x)$ *can be represented in the form*

$$f(x) = G_{\mathbf{r}} * g = \int_{R^n} G_{\mathbf{r}}(x-y)\, g(y)\, dy, \tag{25}$$

where $g(x)$ *is an element of* L_p. *In this case the norm of* f *in* $L_p^{(\mathbf{r})}$ *is equivalent to the norm of* g *in* L_p.

(3) See the proof of Theorem 7 and Appendix 2.

It was shown in [2] that the kernel $G_{\mathbf{r}}(x)$ is infinitely differentiable for $x \neq 0$ and summable over R^n, and various estimates for it were obtained. The isomorphic image of L_p obtained by applying the convolution operator (25) is called $L_p^{\mathbf{r}}$, $|f, L_p^{\mathbf{r}}| = \|g\|_p$. The spaces $L_p^{(\mathbf{r})}$ and $L_p^{\mathbf{r}}$ coincide up to equivalence of norm. In the isotropic case $r_1 = \cdots = r_n = r$ the spaces $L_p^{\mathbf{r}} \equiv L_p^r$ were thoroughly investigated by Aronszajn, Calderón and others under the name of Bessel potentials. In this case $G_{\mathbf{r}} \equiv G_r = \mathfrak{F}^{-1}([1 + |\lambda|^2]^{-r/2})$.

5. Our object in the present section is to provide $L_p^{(\mathbf{r})}$ with new norms that are expressed by absolutely convergent integrals. Such norms have an advantage over (I) and (II) for the passage to bounded regions. We need a number of auxiliary considerations.

Let $L_p(R^n, H) = L_p(H)$ be the set of functions $f(x)$, $x \in R^n$, with values in Hilbert space H (with norm $|\cdot|_H$), summable with their pth powers over R^n;

$$|f, L_p(H)| = \left\{ \int_{R^n} |f(x)|_H^p \, dx \right\}^{1/p}.$$

For functions $f \in L_1(H)$ the Fourier transform $\widetilde{f}(\lambda)$ is defined by the usual formula, where the integral is taken in the sense of Bochner. It is extended to $L_2(H)$ in the usual way; Plancherel's theorem remains valid, and in particular

$$\int_{R^n} |f(x)|_H^2 \, dx = \int_{R^n} |\widetilde{f}(\lambda)|_H^2 \, d\lambda. \tag{26}$$

The generalized Fourier transform is introduced in the natural way, and in particular the Fourier transform for $f \in L_p(H)$, and the concept of a p-multiplier (cf. [5]).

We shall make use of the following proposition. Let H' and H'' be Hilbert spaces; we denote by $S(H)$ the Schwartz space of infinitely differentiable functions that decrease at ∞ (together with their derivatives) faster than any power function (with values in H).

THEOREM A. 1) *Let $k(x)$ be a function, defined almost everywhere in R^n, whose values are bounded operators from H' to H''. In addition, let $k(x)$ be the inverse Fourier transform of a function $\Phi(\lambda)$ which is continuous in the operator norm for $\lambda_1 \cdots \lambda_n \neq 0$ together with the expressions*

$$\lambda^{\mathbf{m}} D^{\mathbf{m}} \Phi = \lambda_1^{m_1} \ldots \lambda_n^{m_n} \frac{\partial^{m_1 + \ldots + m_n} \Phi}{\partial \lambda_1^{m_1} \ldots \partial \lambda_n^{m_n}}, \qquad m_j = 0, 1. \tag{27}$$

In addition, let

$$|\lambda^{\mathbf{m}} D^{\mathbf{m}} \Phi|_{H' \to H''} \leqslant M. \tag{27'}$$

Then the transformation $Af = k * \varphi$ (defined in the Fourier representation for functions $\varphi \in S(H')$ by the equation $\widetilde{Af} = \widetilde{\Phi\varphi}$) can be extended to a continuous trans-

formation of $L_p(H')$ *into* $L_p(H'')$, *and then*

$$|Af,\ L_p(H'')| \leqslant C\,|f,\ L_p(H')|,$$

where C *is independent of* f.

2) *Furthermore let there exist positive constants* $a_1, \cdots, a_n$ *such that the family of functions*

$$\Phi_s(\lambda) = \Phi(s^{a_1}\lambda_1, \ldots, s^{a_n}\lambda_n) \equiv \Phi(s^a\lambda), \qquad 0 < s < \infty, \tag{28}$$

is equicontinuous with respect to the operator norm in the annulus $1 < |\lambda| < 2$. *Then, if* A *has a bounded inverse as an operator from* $L_2(H')$ *into* $L_2(H'')$,

$$|f,\ L_p(H'')| \leqslant c\,|Af,\ L_p(H')|, \qquad 1 < p < \infty,$$

where c *is independent of* f.

A proof of the first part of the theorem for $H' = H'' = H$ is outlined in the author's paper [6]. The stated generalization follows from the supplementary remark that the operator A_2 in [6] maps $L_p(l_2(H'))$ continuously into $L_p(l_2(H''))$. The second part is a corollary of the first part together with Theorem 4 of [5] (the latter is established there for $a_1 = \cdots = a_n = 1$, but the same proof applies also for arbitrary $a_j > 0$).

We shall also need to consider functions with values in a Banach space B. Let $L_p(R^n, B)$ denote the space of functions that have values in B and are p-summable over R.

THEOREM B. *Let* B' *and* B'' *be two Banach spaces, and let the operator* A *be defined on functions* $f \in L_p(R^n, B')$, $1 < p < \infty$, *by*

$$Af = \int_{R^n} k(x-y)\,f(y)\,dy,$$

where $k(x)$ *is a bounded linear operator from* B' *to* B''. *If*

$$\|Af\|_{L_2(R^n,\,B'')} \leqslant c\,\|f\|_{L_2(R^n,\,B')} \tag{29}$$

and for some constant b

$$\int_{R^n - Q_y} |k(x) - k(x-y)|_{B' \to B''}\,dx \leqslant c, \tag{30}$$

where $Q_y = \{x;\ |x_j| < b^{a_j}|y_j|,\ j = 1, \cdots, n\}$, *then*

$$\|Af\|_{L_p(R^n,\,B'')} \leqslant c\,\|f\|_{L_p(R^n,\,B')}.$$

Theorem B is proved in [5] for $a_1 = \cdots = a_n = 1$, and in [7] in the general case.

§3. The spaces $L^{(\mathbf{r})}_{p,\theta}$. Relations between the spaces $L^{(\mathbf{r})}_{p,2}$ and $L^{(\mathbf{r})}_{p}$

1. We shall consider functions $\varphi(x, t)$, $x \in R^n$, $t \in (0, \infty)$, for which the norm

$$\|\varphi(x, \cdot)\|_{L^*_\theta} = \left\{ \int_0^\infty |\varphi(x, t)|^\theta \frac{dt}{t} \right\}^{1/\theta}, \quad 1 \leqslant \theta < \infty,$$

is finite for almost all x. Then φ as a function of x has values in L^*_θ. The space $L_p(R^n, L^*_\theta) = L_p(L^*_\theta) = L_{p,\theta}$ is defined as the set of functions $\varphi(x, t)$ with finite norm

$$\|\varphi\|_{L_{p,\theta}} \equiv |\varphi, L_{p,\theta}| = \left\| \left\{ \int_0^\infty |\varphi(x, t)|^\theta \frac{dt}{t} \right\}^{1/\theta} \right\|_p, \quad 1 \leqslant p \leqslant \infty.$$

Following Hörmander, the norm of $\varphi(x, t) \in L_{p,\theta}$ will also be denoted by $X_p T^*_\theta[\varphi(x, t)]$ or $Y_p U^*_\theta[\varphi(y, u)]$, etc. This notation is sometimes very flexible since it exhibits the structure of the norm. Correspondingly,

$$T^*_\theta \varphi(x, t) \equiv \|\varphi(x, \cdot)\|_{L^*_\theta}.$$

DEFINITION 1. *A function $f(x)$ in L_p belongs to $L^{(\mathbf{r})}_{p,\theta}(R^n) \equiv L^{(\mathbf{r})}_{p,\theta}$, $\mathbf{r} > 0$, $1 \leqslant p \leqslant \infty$, $1 \leqslant \theta < \infty$, if it has generalized derivatives $\partial^{\bar{r}_j} f/\partial x_j^{\bar{r}_j}$, $j = 1, \cdots, n$, in the sense of Sobolev, and the expressions*(4)

$$\mathscr{L}^{r_j}_{p,\theta}(f) = \left\| \left\{ \int_0^\infty \frac{dt}{t} \left| \int_t^{2t} \frac{\Delta^2_{he_j} D^{\bar{r}_j}_j f(x)}{h^{1+\alpha_j}} dh \right|^\theta \right\}^{1/\theta} \right\|_p$$

are finite.

It is easily verified that the quantity(5)

$$\|f\|_{L^{(\mathbf{r})}_{p,\theta}} \overset{\text{def}}{=} \|f\|_p + \sum_1^n \mathscr{L}^{r_j}_{p,\theta}(f) \equiv \|f\|_p + \sum_{j=1}^n X_p T^*_\theta \left[D^{r_j}_{te_j} f(x) \right] \tag{31}$$

is a norm in $L^{(\mathbf{r})}_{p,\theta}$. It can be shown that the space $L^{(\mathbf{r})}_{p,\theta}$ is complete. From the (x)-inequality and then the generalized Minkowski inequality for $p \geqslant \theta$ it follows that

(4) The use of differences with positive step evidently leads to an equally inclusive definition. We recall that $r_j = \bar{r}_j + \alpha_j$, $0 < \alpha_j \leqslant 1$.

(5) The symbol $D^{r_j}_{te_j} f$ is used in this section to denote the truncated derivative $\bar{D}^{r_j}_{te_j} f$, sometimes multiplied by a constant.

$$\mathscr{L}^{r_j}_{p,\theta}(f) \leqslant c \left\{ \int_{R^n} \left(\int_0^\infty \frac{|\Delta^2_{te_j} D^{\bar{r}_j}_j f(x)|^\theta}{t^{\alpha_j \theta}} \frac{dt}{t} \right)^{p/\theta} dx \right\}^{1/p}$$

$$\leqslant c \left\{ \int_0^\infty \frac{dt}{t^{1+\theta\alpha_j}} \left(\int_{R^n} |\Delta^2_{te_j} D^{\bar{r}_j}_j f(x)|^p \, dx \right)^{\theta/p} \right\}^{1/\theta} = c \left\{ \int_0^\infty \frac{\|\Delta^2_{te_j} D^{\bar{r}_j}_j f(x)\|^\theta_p}{t^{1+\theta\alpha_j}} dx \right\}^{1/\theta}.$$

Therefore for $p \geqslant \theta$ we have the imbedding

$$B^{(\mathbf{r})}_{p,\theta}(R^n) \to L^{(\mathbf{r})}_{p,\theta}(R^n).$$

It can be shown that when $p = \theta$ the spaces $B^{(\mathbf{r})}_{p,\theta}(R^n)$ and $L^{(\mathbf{r})}_{p,\theta}(R^n)$ coincide up to norm equivalence. In the present paper we shall, however, be interested only in the spaces $L^{(\mathbf{r})}_{p,1}$ and their connections with the spaces $L^{(\mathbf{r})}_p$. We shall show that these spaces.coincide.

2. We begin with the one-dimensional case $n = 1$.

THEOREM 5. *The spaces* $L^{(r)}_p(R^1)$ *and* $L^{(r)}_{p,2}(R^1)$ *coincide up to norm equivalence.*

PROOF. Let $f(x) \in L^{(r)}_p(R^1)$. We approximate $f(x)$ by functions of a base space Φ (in the one-dimensional case Φ is the subspace of S consisting of the functions $u(x) \in S$ which are orthogonal to the polynomials [1, 8]; Φ is dense in $L^{(r)}_p$; cf. Appendix 1). Let $u_\nu(x) \in \Phi$, $u_\nu(x) \xrightarrow{L^{(r)}_p} f$. Then we have [2]

$$u^{(\bar{r})}_\nu = \frac{1}{\Gamma(\alpha)} \int_{-\infty}^x \frac{v_\nu(\xi)\, d\xi}{(x-\xi)^{1-\alpha}}, \qquad v_\nu(x) = D^r u_\nu(x). \tag{32}$$

It is important that $v_\nu \in \Phi$ and hence the indicated integral is absolutely convergent. We introduce an operator A by the equation

$$Av_\nu = \int_t^{2t} \frac{u^{(\bar{r})}_\nu(x) - 2u^{(\bar{r})}_\nu(x-h) + u^{(\bar{r})}_\nu(x-2h)}{h^{1+\alpha}} dh$$

$$= \frac{1}{\Gamma(\alpha)} \int_{-\infty}^x k(x-\xi, t)\, v_\nu(\xi)\, d\xi, \tag{33}$$

$$k(x, t) = \int_t^{2t} \frac{1}{h^{1+\alpha}} \left(x_+^{\alpha-1} - 2(x-h)_+^{\alpha-1} + (x-2h)_+^{\alpha-1}\right) dh, \qquad x_+^{\alpha-1} = \begin{cases} \dfrac{1}{x^{1-\alpha}}, & x > 0, \\ 0, & x < 0. \end{cases}$$

We are going to appeal to Theorem A. The kernel $k(x, t)$ can be considered as an operator-valued function $k(x, \cdot)$ with values in the space $\mathfrak{L}(H', H'')$ of bounded linear transformations from the Hilbert space $H' = Z$ (= space of complex numbers) to the Hilbert space $H'' = L^*_2(0, \infty)$. Then the operator norm of $k(x, \cdot)$ has the form

$$\begin{aligned}|k(x,\,\cdot)|^2_{H'\to H''}&=\int_0^\infty |k(x,\,t)|^2\frac{dt}{t}\\&=\int_0^\infty\frac{dt}{t}\left|\int_t^{2t}\frac{1}{h^{1+\alpha}}\left(x_+^{\alpha-1}-2\,(x-h)_+^{\alpha-1}+(x-2h)_+^{\alpha-1}\right)dh\right|^2.\end{aligned}\tag{34}$$

We note that $|k(x,\,\cdot\,)|_{H'\to H''}$ is a homogeneous function of degree -1, i.e.

$$|k(\tau x,\,\cdot)|_{H'\to H''}=\frac{1}{\tau}\,|k(x,\,\cdot)|_{H'\to H''},\quad \tau>0.$$

It is clear that we have $|k(x,\,\cdot\,)|_{H'\to H''}=0$ for $x<0$. In fact, in this case $x_+^{\alpha-1}-2(x-h)_+^{\alpha-1}+(x-2h)_+^{\alpha-1}\equiv 0$ by the definition of the function $x_+^{\alpha-1}$ and since $h>0$. For $x>0$ put $\tau=1/x$; then

$$|k(x,\,\cdot)|_{H'\to H''}=\frac{1}{x}\,|k(1,\,\cdot)|_{H'\to H''}=\frac{1}{x}\left(\int_0^\infty\frac{\varphi^2(t)\,dt}{t}\right)^{1/2},$$

$$\varphi(t)=\int_t^{2t}\left[1-2\,(1-h)_+^{\alpha-1}+(1-2h)_+^{\alpha-1}\right]\frac{dh}{h^{1+\alpha}}.$$

It is easy to see that there are simple estimates for $\varphi(t)$.

1) For $0<t<1/5$, if we use the representation of a second difference by a second derivative and the boundedness of the second derivative of the function $1/(1-h)^{1-\alpha}$ for $0<h<4/5$, we obtain

$$|\varphi(t)|\leqslant\int_t^{2t}\left|1-2\,(1-h)_+^{\alpha-1}+(1-2h)_+^{\alpha-1}\right|\frac{dh}{h^{1+\alpha}}\leqslant c\int_0^t\frac{h^2dh}{h^{1+\alpha}}=c_1t^{2-\alpha}.$$

2) For $1/5<t<1$,

$$|\varphi(t)|\leqslant\int_t^{2t}\left(1+2\,|1-h|^{\alpha-1}+|1-2h|^{\alpha-1}\right)\frac{dh}{h^{1+\alpha}}\leqslant\int_{\frac{1}{5}}^{2}(\ldots)\frac{dh}{h^{1+\alpha}}=\text{const}.$$

3) For $t>1$,

$$\varphi(t)=\int_t^{2t}\frac{dh}{h^{1+\alpha}}\leqslant\int_t^\infty\frac{dh}{h^{1+\alpha}}=\alpha^{-1}\frac{1}{t^\alpha}.$$

It follows from these estimates that for $x>0$

$$|k(x,\,\cdot)|_{H'\to H''}\leqslant\frac{c}{x},$$

and therefore the kernel $k(x, \cdot)$ actually has, for $x \neq 0$, values in the space of bounded operators from H' to H''.

Taking account of (32) and (33), we have for the Fourier transforms

$$\overline{Av_\nu} = \Phi(\lambda, t)\overline{v_\nu}, \qquad \Phi(\lambda, t) = \int_t^{2t} \frac{(1 - e^{-i\lambda h})^2}{(i\lambda)^\alpha h^\alpha} \frac{dh}{h},$$

where

$$(i\lambda)^\alpha = \begin{cases} e^{i\frac{\pi}{2}\alpha}\lambda^\alpha & \text{for } \lambda > 0, \\ e^{-i\frac{\pi}{2}\alpha}|\lambda|^\alpha & \text{for } \lambda < 0. \end{cases}$$

Consequently

$$\Phi(\lambda, t) = \begin{cases} e^{-i\frac{\pi}{2}\alpha}[\varphi(|\lambda|t) + i\psi(|\lambda|t)], & \lambda > 0, \\ e^{i\frac{\pi}{2}\alpha}[\varphi(|\lambda|t) - i\psi(|\lambda|t)], & \lambda < 0, \end{cases}$$

where

$$\varphi(\xi) = -4\int_\xi^{2\xi} \frac{\cos u \sin^2 \frac{u}{2}\, du}{u^{1+\alpha}}, \quad \psi(\xi) = -4\int_\xi^{2\xi} \frac{\sin u \sin^2 \frac{u}{2}\, du}{u^{1+\alpha}}.$$

We shall see that $\Phi(\lambda, t)$ satisfies the hypotheses of Theorem A. Using the (x)-inequality, we have

$$|\Phi(\lambda, \cdot)|_{H' \to H''} = \int_0^\infty |\Phi(\lambda, t)|^2 \frac{dt}{t} = \int_0^\infty (|\varphi(|\lambda|t)|^2 + |\psi(|\lambda|t)|^2) \frac{dt}{t}$$

$$= \int_0^\infty (|\varphi(\xi)|^2 + |\psi(\xi)|^2) \frac{d\xi}{\xi} \leqslant c \int_0^\infty \frac{\sin^4 \frac{\xi}{2}}{\xi^{1+2\alpha}} du = c_1,$$

$$\left|\lambda \frac{\partial \Phi}{\partial \lambda}\right|_{H' \to H''} = \int_0^\infty (|\xi\varphi'(\xi)|^2 + |\xi\psi'(\xi)|^2) \frac{d\xi}{\xi}$$

$$= c\int_0^\infty \frac{1}{\xi^{1+2\alpha}} \left\{ \left[\frac{\cos 2\xi \sin^2 \xi}{2^{1+\alpha}} - \cos \xi \sin^2 \frac{\xi}{2} \right]^2 \right.$$

$$\left. + \left[\frac{\sin 2\xi \sin^2 \xi}{2^{1+\alpha}} - \sin \xi \sin^2 \frac{\xi}{2} \right]^2 \right\} d\xi$$

$$= c\int_0^\infty \left[\frac{\sin^4 \xi}{2^{1+\alpha}} + \sin^4 \frac{\xi}{2} - \frac{1}{2^\alpha} \cos \xi \cdot \sin^2 \xi \cdot \sin^2 \frac{\xi}{2} \right] \frac{d\xi}{\xi^{1+2\alpha}} = c_2.$$

It is easy to verify the continuity of Φ and $\partial\Phi/\partial\lambda$ in the operator norm; this also follows from the existence of the derivatives. Therefore the hypotheses of the first part of Theorem A are satisfied, and hence

$$\| Av_\nu \|_{L_p(H'')} \leqslant \| v_\nu \|_{L_p(H')},$$

i.e.

$$\left\| \left\{ \int_0^\infty \frac{dt}{t} \left| \int_t^{2t} \frac{\Delta_h^2 u_\nu^{(\bar r)}(x)}{h^{1+\alpha}} dh \right|^2 \right\}^{1/2} \right\|_p \leqslant cM \| D^r u_\nu \|_p.$$

This inequality can be extended by continuity to every function in $L_p^{(r)}$, and therefore

$$L_{p,2}^{(r)} \to L_p^{(r)},$$

i.e. it follows from the fact that f belongs to $L_p^{(r)}$ that f belongs to $L_{p,2}^{(r)}$, and then

$$| f, L_{p,2}^{(r)} | \leqslant c \, | f, L_p^{(r)} |.$$

We have established somewhat more, namely an inequality between differential seminorms,

$$X_p T_2^* [D_t^r f] \leqslant c \| D^r f \|_p. \tag{35}$$

Conversely, let $f \in L_{p,2}^{(r)}$, i.e.

$$f \in L_p, \quad \left\| \left\{ \int_0^\infty \frac{dt}{t} \left| \int_t^{2t} \frac{\Delta_h^2 D^{\bar r} f(x)}{h^{1+\alpha}} dh \right|^2 \right\}^{1/2} \right\|_p < \infty.$$

We show that $f \in L_p^{(r)}$ and satisfies

$$| f, L_p^{(r)} | \leqslant c \, | f, L_{p,2}^{(r)} |.$$

Take an averaging kernel $g_\beta(x) > 0$ such that for $\beta > 0$

a) $g_\beta(x) \in S$ (the Schwartz space of test functions);

b) $g_\beta(x) = \int_{R^1} G_r(x-y)\, \omega_\beta(y)\, dy, \quad \omega_\beta \in L_1;$

c) $f * g_\beta \xrightarrow{L_p} f \in L_p$ for $\beta \to 0$.

For example, we could take $g_\beta(x) = (1/\sqrt{2\beta})\, e^{-x^2/4\beta}$.

Under these hypotheses we have

$$f_\beta = f * g_\beta = \int G_r(x-y) f * \omega_\beta dy \in L_p^{(r)}, \quad f_\beta \in C^\infty.$$

In addition,

$$\left(\int_0^\infty \frac{dt}{t}\left|\int_t^{2t} \frac{\Delta_h^2 D^{\bar r}(f*g_\beta)}{h^{1+\alpha}}\,dh\right|^2\right)^{1/2} = \left(\int_0^\infty \frac{dt}{t}\left|\left(\int_t^{2t} \frac{\Delta_h^2 D^{\bar r} f}{h^{1+\alpha}}\,dh\right) * g_\beta\right|^2\right)^{1/2}$$

$$\leqslant \int_{R^1} |g_\beta|\,dy \left\{\int_0^\infty \frac{dt}{t}\left|\int_t^{2t} \frac{\Delta_h^2 D^r f}{h^{1+\alpha}}\,dh\right|^2\right\}^{1/2}.$$

Therefore

$$f_\beta \in L_p, \quad X_p T_2^*[D_t^r f_\beta(x)] < \infty.$$

We now verify that the operator A has a bounded inverse. By Theorem A we have to establish the equicontinuity of the functions $\Phi(s\lambda,\,\cdot\,)$, $0 < s < \infty$, in the region $1 < |\lambda| < 2$, as functions with values in $L_2^*(0, \infty)$. For this it is enough to show that the derivatives $\partial\psi_s/\partial\lambda$ are uniformly bounded in this region. This is equivalent to the boundedness of $|\lambda\partial\Phi_s/\partial\lambda|_{H'\to H''}$.

For $1 < \lambda < 2$ we have

$$\left|\lambda\frac{\partial\Phi_s}{\partial\lambda}\right|_{H'\to H''} = \int_0^\infty (|\lambda st\varphi'(\lambda st)|^2 + |\lambda st\psi'|^2)\frac{dt}{t} = C_2.$$

Hence the inverse of A exists and is continuous, and hence

$$\|D^r f_\beta\|_p \leqslant C X_p T_2^*(D_t^r f_\beta(x)). \tag{36}$$

This implies the convergence of $D^r f_\beta$ in the L_p norm.

In fact, we have

$$D_t^r f - D_t^r f_\beta = \int_{R^1} [D_t^r f(x) - D_t^r f(x-y)]\,g_\beta(y)\,dy.$$

Taking the $X_p T_2^*$ norm on both sides, we obtain

$$X_p T_2^*(D_t^r f - D_t^r) \leqslant \int_{R^1} \|\,|D_t^r f(x) - D_t^r f(x-y)|_H\|_p\, g_\beta(y)\,dy$$

$$= \int_{|y|<\delta} |||D_t^r f(x) - D_t^r f(x-y)|||\, g_\beta(y)\,dy + \int_{|y|>\delta} |||\cdots|||\, g_\beta(y)\,dy$$

$$\leqslant \sup_{|y|<\delta} |||F(x,t) - F(x-y,t)||| + 2\,|||F|||\int_{|y|>\delta} g_\beta(y)\,dy,$$

$$F(x,t) = D_t^r f(x).$$

Using the continuity under translation of functions in $L_p\{R^1; H\}$, we can make the first term as small as desired by choice of δ, and then the second, by choice of β. Hence $D_t^r f_\beta \to D_t^r f$ in the $L_p(R^1, L_2^*)$ norm as $\beta \to 0$. Consequently $D_t^r f_\beta$ converges in the $X_p T_2^*$ norm. But then by (36) the sequence $D^r f_\beta$ converges as $\beta \to 0$

in the L_p norm, and since L_p is complete and the L.g. differentiation operator is closed,

$$D^r f_\beta \xrightarrow{L_p} D^r f \in L_p.$$

Therefore $f \in L_p^{(r)}$. Taking limits in (36), we obtain

$$\| D^r f \|_p \leqslant C X_p T_2^* (D_t^r f(x)). \tag{37}$$

Hence it follows trivially that

$$| f, L_p^{(r)} | \leqslant C | f, L_{p,2}^{(r)} |.$$

The theorem is established. We have actually proved not only the equivalence of the norms, but also the equivalence of the corresponding differential seminorms (cf. (35) and (37)).

REMARK 3. In the preceding proof we could have replaced $\mathfrak{L}^r_{p,2}(f)$ by the expression

$$\mathscr{L}_{p,2}^{r,l,k}(f) = \left\| \left\{ \int_0^\infty \frac{dt}{t} \left| \int_t^{2t} \frac{\Delta_h^l D^k f(x)}{h^{1+r-k}} \, dh \right|^2 \right\}^{1/2} \right\|_p,$$

$0 \leqslant k \leqslant \bar{r} < r < l + k$, $\mathfrak{L}^r_{p,2}(f) = \mathfrak{L}^{r,2,\bar{r}}_{p,2}(f)$. By considerations similar to those used above, we would have obtained the equivalence

$$\mathscr{L}_{p,2}^{r,l,h}(f) \sim \| D^r f \|_p, \quad (f \in L_p).$$

Therefore the quantities $\mathfrak{L}^{r,l,k}_{p,2}(f)$ are equivalent to each other for $0 \leqslant k \leqslant \bar{r} < r < l + k$. In particular, when r is integral we can replace $\mathfrak{L}^r_{p,2}$ by the equivalent quantity

$$\mathscr{L}_{p,2}^{r,1,\bar{r}}(f) = \left\| \left\{ \int_0^\infty \frac{dt}{t} \left| \int_t^{2t} \frac{D^{\bar{r}} f(x) - D^{\bar{r}} f(x-h)}{h^{1+\alpha}} \, dh \right|^2 \right\}^{1/2} \right\|_p.$$

REMARK 4. Theorem 5 can be somewhat strengthened to prove (for $f \in L_p$) the equivalence of

$$\| D^r f \|_p \quad \text{and} \quad \left\| \left\{ \int_0^\infty \frac{dt}{t} \left(\int_t^{2t} \frac{| \Delta_h^l D^k f(x) |}{h^{1+r-k}} \, dh \right)^2 \right\}^{1/2} \right\|_p \equiv \bar{\mathscr{L}}_{p,2}^{r,l,k}(f).$$

By Remark 1 and the definitions of $\mathfrak{L}^{r,l,k}_{p,2}(f)$ and $\bar{\mathfrak{L}}^{r,l,k}_{p,2}(f)$ we evidently have

$$\| D^r f \|_p \leqslant C_1 \mathscr{L}_{p,2}^{r,l,k}(f) \leqslant C_1 \bar{\mathscr{L}}_{p,2}^{r,l,k}(f).$$

Hence it is enough to obtain the estimate

$$\mathfrak{L}^{r,\,l,\,k}_{p,\,2}(f)\leqslant C_2\|D^r f\|_p.$$

For simplicity we confine ourselves to the case of nonintegral r. In this case (cf. the preceding remark) the problem reduces to estimating $\overline{\mathfrak{L}}^{r,1,\overline{r}}_{p,2}(f)$.

We apply Theorem B, taking B' to be the space z of complex numbers, B'' to be the space $L_2^*(0, \infty; L_1)$ of functions $\varphi(t, u)$ with norm

$$T_2^* U_1(\varphi)=\left\{\int_0^\infty \frac{dt}{t}\left(\int_{-\infty}^\infty |\varphi(t, u)|\,du\right)^2\right\}^{1/2},$$

and $k(x, \cdot)$ to be the function

$$k(x;\ \cdot)=k(x;\ t,\ u)=\begin{cases}\dfrac{1}{u^{1+\alpha}}[x_+^{\alpha-1}-(x-u)_+^{\alpha-1}], & t<u<2t,\\ 0, & u\in R^1\setminus[t,\ 2t].\end{cases}$$

Then

1) $$|k(x;\ \cdot)|_{B'\to B''}=\left\{\int_0^\infty \frac{dt}{t}\left(\int_t^{2t} \frac{1}{u^{1+\alpha}}\,|x_+^{\alpha-1}-(x-u)_+^{\alpha-1}|\,du\right)^2\right\}^{1/2}<\infty.$$

2) On a set of functions $v\in\Phi$, dense in L_p, (29) is satisfied. In fact, we have

$$Av=\int_{-\infty}^\infty k(x-y;\ t,\ u)\,v(y)\,dy,$$

$$\begin{aligned}\|Av\|_{L_p(R^1,\,B'')}&=\left\|\left\{\int_0^\infty \frac{dt}{t}\left(\int_{-\infty}^\infty |Av|\,du\right)^2\right\}^{1/2}\right\|_p\\ &=\left\|\left\{\int_0^\infty \frac{dt}{t}\left(\int_t^{2t}\left|\int_{-\infty}^\infty \frac{(x-y)_+^{\alpha-1}-(x-y-u)^{\alpha-1}}{u^{1+\alpha}}\,v(y)\,dy\right|du\right)^2\right\}^{1/2}\right\|_p\\ &=\left\|\left\{\int_0^\infty \frac{dt}{t}\left(\int_t^{2t}\left|\int_{-\infty}^x \frac{v(y)\,dy}{(x-y)^{1-\alpha}}-\int_{-\infty}^{x-u} \frac{v(y)\,dy}{(x-y-u)^{1-\alpha}}\right|\frac{du}{u^{1+\alpha}}\right)^2\right\}^{1/2}\right\|_p.\end{aligned} \tag{38}$$

Between the absolute value signs we have a difference $v_{(\alpha)}(x)-v_{(\alpha)}(x-u)$, where $v_{(\alpha)}$ is the integral of v of order α. Applying the (x)-inequality with $p=2$ and going over to Fourier transforms by Plancherel's theorem, we obtain

$$\|Av\|^2_{L_2(R^1,\,B')}\leqslant\int_{-\infty}^\infty dx\int_0^\infty \frac{|v_{(\alpha)}(x)-v_{(\alpha)}(x-t)|^2}{t^{1+2\alpha}}\,dt$$

$$=\int_0^\infty \frac{dt}{t^{1+2\alpha}}\int_{-\infty}^\infty |\overline{v}(\lambda)|^2\,\frac{|1-e^{-i\lambda t}|^2}{|\lambda|^{2\alpha}}=\int_{-\infty}^\infty |\overline{v}(\lambda)|^2\,d\lambda\int_0^\infty \frac{|1-e^{-i\lambda t}|^2}{|t\lambda|^{2\alpha}}\frac{dt}{t}=C\|v\|_2^2.$$

3. Hypothesis (30) of Theorem B is satisfied. Under our hypotheses it reduces to verifying the boundedness of

$$\int\limits_{|x|>4|y|} |k(x;\ \cdot) - k(x-y;\ \cdot)|_{B'\to B''} dx$$

$$= \int\limits_{|x|>4|y|} dx \left\{ \int_0^\infty \frac{dt}{t} \left(\int_t^{2t} \frac{1}{u^{1+\alpha}} |x_+^{\alpha-1} - (x-u)_+^{\alpha-1} - (x-y)_+^{\alpha-1} \right.\right. \tag{39}$$

$$\left.\left. + (x-y-u)_+^{\alpha-1}|\, du \right)^2 \right\}^{1/2}.$$

An elementary calculation shows that this is bounded.

Hence Theorem B applies, and consequently

$$\| Av \|_{L_p(R^1,\, B'')} \leqslant C \| v \|_p. \tag{40}$$

If we put $v = D^r f$, then $v_{(\alpha)} = D^{\bar r} f$, and by (38) and (40) we obtain

$$\left\| \left\{ \int_0^\infty \frac{dt}{t} \left(\int_t^{2t} \frac{|D^{\bar r} f(x) - D^{\bar r} f(x-u)|}{u^{1+\alpha}} du \right)^2 \right\}^{1/2} \right\|_p \leqslant C \| D^r f \|_p.$$

It remains to consider the equivalence of $\mathcal{L}_{p,2}^{r,1,\bar r}(f)$ and $\|D^r f\|_p$.

A similar discussion and calculation apply in the general case, and we obtain

$$\mathcal{L}_{p,2}^{r,l,k}(f) \sim \| D^r f \|_p \sim \bar{\mathcal{L}}_{p,2}^{r,l,k}(f) \qquad (f \in L_p). \tag{41}$$

On the basis of the preceding discussion we conclude that for nonintegral r we may take the norm on $L_p^{(r)}(R^1)$ to be

$$\| f \|_p + \bar{\mathcal{L}}_{p,2}^r(f),$$

where

$$\bar{\mathcal{L}}_{p,2}^r(f) = \mathcal{L}_{p,2}^{r,1,\bar r}(f) = \left\| \left\{ \int_0^\infty \frac{dt}{t} \left(\int_t^{2t} \frac{|\Delta_h D^{\bar r} f(x)|}{h^{1+\alpha}} dh \right)^2 \right\}^{1/2} \right\|_p.$$

We shall frequently use this norm since it is comparatively simple, but for integral r it is more convenient to use the orginal norm in L_p^r since it is expressed directly in terms of the derivative $D^r f$.

REMARK 5. All the proofs that we have discussed or mentioned go through also for differences with positive steps. In fact, the corresponding quantities formed with differences having positive or negative steps are mutually equivalent. For norms containing differences inside the absolute value signs we can use central differences.

REMARK 6. Let us say a few words about Marcinkiewicz's theorem that was mentioned at the end of §1. Relation (16) was obtained above (cf. Remark 4). Relation (17) can be obtained more simply, in fact along the lines of the proof of Theorem 5, if we introduce the operator A by the relation (cf. (33))

$$Av = \frac{u^{\bar r}(x) - 2u^{\bar r}(x-t) + u^{\bar r}(x-2t)}{t^{\alpha}} = \int_{-\infty}^{x} k(x-\xi, t)\, v(\xi)\, d\xi,$$

$$k(x, t) = \frac{1}{t^{\alpha}}\left(x_{+}^{\alpha-1} - 2(x-t)_{+}^{\alpha-1} + (x-2t)_{+}^{\alpha-1}\right).$$

However, in this approach we have to have the restriction $\alpha > ½$ in estimating $|k(x, \cdot)|_{H' \to H''}$. The rest of the proof goes through without change, and we obtain

$$\| Av \|_{L_p(H'')} \sim \| v \|_{L_p(H')},$$

i.e.

$$\left\| \left\{ \int_0^{\infty} \frac{|u^{\bar r}(x) - 2u^{\bar r}(x-t) + u^{\bar r}(x-2t)|^2}{t^{2\alpha}} \frac{dt}{t} \right\}^{1/2} \right\|_p \sim \| D^r u \|_p .$$

For $r = 1$ this coincides with Marcinkiewicz's theorem.

If we vary the orders of the differences and the orders of the derivatives that are differenced, we can show that

$$\left\| \left\{ \int_0^{\infty} \frac{|\Delta_t^l D^k f(x)|^2}{t^{2(r-k)}} \frac{dt}{t} \right\}^{1/2} \right\|_p \sim \| D^r u \|_p . \tag{42}$$

This holds for $0 \leqslant k \leqslant \bar r < r < l + k$, $r - k > ½$. The last restriction is essential whenever (42) is considered for the full range of p, $1 < p < \infty$. For $k = 0$ (42) implies a relation that holds for functions in $L_p^{(r)}$ with $r > ½$; in particular, when $½ < r < 1$ we obtain

$$\left\| \left\{ \int_0^{\infty} \frac{|f(x) - f(x-t)|^2}{t^{2r}} \frac{dt}{t} \right\}^{1/2} \right\|_p \sim \| D^r f \|_p .$$

We also note that if we apply the (x)-inequality to the right-hand side of (16) we obtain the estimate

$$\left\| \left\{ \int_0^{\infty} \frac{dt}{t} \left(\int_t^{2t} \frac{|f(x) - 2f(x-h) + f(x-2h)|}{h^2} dh \right)^2 \right\}^{1/2} \right\|_p$$

$$\leqslant C \left\| \left\{ \int_0^{\infty} \frac{|f(x) - 2f(x-t) + f(x-2t)|^2}{t^3} \right\}^{1/2} \right\|_p .$$

As follows from what we have said, these two quantities are in fact equivalent, i.e. the converse estimate also holds.

4. For arbitrary n there is a similar theorem.

THEOREM 6. *The spaces* $L_p^{(\mathbf{r})}(R^n)$ *and* $L_{p,2}^{(\mathbf{r})}(R^n)$ *coincide up to norm equivalence, and the corresponding differential seminorms are also equivalent.*

PROOF. If f belongs to one of these classes, then by Fubini's Theorem it belongs to the corresponding class $L_p^{r_j}(R^1_{x_j})$ or $L_{p,2}^{(r_j)}(R^1_{x_j})$ on almost all lines $R^1_{x_j}$ in the jth coordinate direction. But the conclusions and estimates of the preceding theorems all hold for one-dimensional classes. We obtain the required estimates by integrating over the subspace $R^{n-1}_{x_j}$ complementary to R_{x_j}. All the remarks following Theorem 1 can be extended in a similar way. We therefore obtain, in particular, the equivalence

$$\sum_{j=1}^n \|D_j^{r_j} f\| \sim \sum_{j=1}^n \left\| \left\{ \int_0^\infty \frac{dt}{t} \left(\int_t^{2t} \frac{|\Delta^2_{he_j} D_j^{\bar r_j} f(x)|}{h^{1+\alpha_j}} dh \right)^2 \right\}^{1/2} \right\|_p, \tag{43}$$

where in accordance with our conventions $r_j = \bar r_j + \alpha_j$, $\bar r_j$ is an integer, and $0 < \alpha_j \leqslant 1$. If we desire, we can simplify the right-hand side as follows: for nonintegral r_j replace the corresponding second difference by a first difference, and for integral $r_j = l_j$ replace the corresponding jth term by $\|D_j^{l_j} f\|_p$. The expression for the norm can be rewritten similarly. For example, for nonintegral r_j, $j = 1, \cdots, n$, the norm in $L_p^{(\mathbf{r})}(R^n)$ can be taken to be

$$\|f\|_p + \sum_{j=1}^n \left\| \left\{ \int_0^\infty \frac{dt}{t} \left(\int_t^{2t} \frac{|D_j^{\bar r_j} f(x) - D_j^{\bar r_j} f(x-h)|}{h^{1+\alpha_j}} dh \right)^2 \right\}^{1/2} \right\|_p.$$

5. We shall also call the norm (31) in $L_{p,\theta}^{(\mathbf{r})}(R^n)$ a coordinate norm (by analogy with $L_p^{(\mathbf{r})}$). It is appropriate to emphasize here that the index r_j in the notation $\mathfrak{L}^{r_j}_{p,\theta}(f)$ does double duty: it is both a number and the jth component of a vector. Even if $r_1 = \cdots = r_n = r$, we still write $\mathfrak{L}^{r_j}_{p,\theta}(f)$, indicating in this way that the second difference of $\Delta_j^{\bar r} f$ is taken in the jth coordinate direction. In $L_{p,\theta}^{(\mathbf{r})}$ we can also introduce the chain norm by using the expression

$$\mathscr{L}^{\mathbf{r}}_{p,\theta}(f) = \left\| \left\{ \int_0^\infty \frac{dt}{t} \left| \int_{t<\rho(h)<2t} \frac{\Delta_h^l f(x)\, dh}{[\rho(h)]^{n+r^*}} \right|^\theta \right\}^{1/\theta} \right\|_p, \tag{44}$$

where $\rho(h)$, r^* and l have the meanings assigned in §2 (cf. Theorem 2).

We note that the seminorm (44) is constructed by the same prescription as in the one-dimensional case. We illustrate this for the isotropic case $r_1 = \cdots = r_n = r$. As we have already said, the operator $T^r f$ reduces, up to a factor, to the operator of Riesz integration, and we call it the Riesz derivative of f of order r. The quantity

$$T_t^r f = \int_{t<|h|<2t} \frac{\Delta_h^l f(x)\, dh}{|h|^{n+r}}, \qquad l > r,$$

is called the truncated Riesz derivative. Then the seminorm (44) can be written in the form

$$X_p T_2^*(\bar{T}_t^r f(x)) = \left\| \left\| \int_{t<|h|<2t} \frac{\Delta_h^l f(x)\, dh}{|h|^{n+r}} \right\|_{L_\theta^*} \right\|_p . \tag{44'}$$

Restricting ourselves to the value $\theta = 2$, we put

$$\mathscr{L}_{p,2}^{r,l,\mathbf{m}}(f) = \left\| \left\{ \int_0^\infty \frac{dt}{t} \left| \int_{t<|h|<2t} \frac{\Delta_h^l D^{\mathbf{m}} f(x)\, dh}{|h|^{n+r-k}} \right|^2 \right\}^{1/2} \right\|_p .$$

We shall show that for $f \in L_p$ the quantity $\Sigma_{|\mathbf{m}|=k} \mathfrak{L}_{p,2}^{r,l,\mathbf{m}}(f)$ is, for $0 \leqslant k \leqslant \bar{r} < r < l+k$, equivalent to the differential seminorm in $L_p^{(r)}$. From this and Theorem 6 it will follow, for example, that

$$\sum_{|\mathbf{m}|=k} \mathscr{L}_{p,2}^{r,l,\mathbf{m}}(f) \sim \sum_1^n \mathscr{L}_{p,2}^{r_j}(f). \tag{45}$$

The result is unchanged if the summation on the left is only over the unmixed derivatives of order k.

It is enough to prove (45) for an arbitrary differential seminorm in $L_p^{(r)}$. In this situation it is convenient to take as such a quantity (cf. Theorem 2 and Remark 2)

$$\| T^{\mathbf{r}} f \|_p = \| T^r f \|_p = \left\| \lim \int_{|h|>\varepsilon} \frac{\Delta_h^s f(x)}{|h|^{n+r}} dh \right\|_p , \qquad s > r.$$

THEOREM 7. *For functions in* L_p *the quantities* $\|T^r f\|_p$ *and* $\Sigma_{|\mathbf{m}|=k} \mathfrak{L}_{p,2}^{r,l,\mathbf{m}}(f)$, $0 \leqslant k \leqslant \bar{r} < l+k$, *are equivalent.*

The proof proceeds along the lines of Theorem 5, and we shall be brief. If $\|T^r f\| < \infty$, then $f \in L_p^{(r)}$. Let $u_\nu(x) \in \Phi$, $u_\nu(x) \xrightarrow{L_p} f$. The operator T^r carries Φ into Φ and has the group property: $T^{r_1} T^{r_2} = T^{r_1+r_2}$; as the inverse of T^r the operator R_{-r} of Riesz integration stands out; $T^0 = R_0$ is the identity operator (cf. Appendix 2). For the present let $r \neq n + 2k$, $k = 0, 1, \cdots$. Then

$$u_\nu = C_{n,r} \int_{R^n} \frac{T^r u_\nu(y)\, dy}{|x-y|^{n-r}} = C_{n,r} \int_{R^n} \frac{v_\nu(y)\, dy}{|x-y|^{n-r}} .$$

Introduce an operator A by the equation

$$Av_\nu = \int_{t<|h|<2t} \frac{\Delta_h^l D^{\mathbf{m}} u_\nu(x)\, dh}{|h|^{n+r-k}} = \int_{R^n} k(x-y,\, t)\, v_\nu(y)\, dy,$$

$$k(x,t) = \int_{t<|h|<2t} \Delta_h^l D^{\mathbf{m}} \left(\frac{1}{|x|^{n-r}} \right) \frac{dh}{|h|^{n+r-k}} .$$

Let H' and H'' have the same meaning as in Theorem 5. We return again to Theorem A. Let us verify that for $x \neq 0$

$$|k(x,\cdot)|^2_{H'\to H''}=\int_0^\infty |k(x,t)|^2\frac{dt}{t}<\infty. \tag{46}$$

We divide the interval of integration with respect to t into the two intervals $0<t<\gamma|x|$ and $\gamma|x|<t<\infty$. We take γ so small that the points $x-jh$ that occur in the integral

$$\int_0^{\gamma|x|}\frac{dt}{t}\left|\int_{t<|h|<2t}\Delta_h^l D^{\mathbf{m}}\left(\frac{1}{|x|^{n-r}}\right)\frac{dh}{|h|^{n+r-k}}\right|^2 \tag{47}$$

avoid a neighborhood of zero.

For example, if $\gamma=1/(2l+1)$, then $|x-jh|\geqslant|x|-j|h|\geqslant|x|-l|h|\geqslant|x|-l2\gamma|x|=\gamma|x|$; therefore all the poles of the integrand in (46) lie outside the domain of integration. Using the representation of the difference $\Delta_h^l D^{\mathbf{m}}(1/|x|^{n-r})$ by a derivative of order l (in the h direction) and the boundedness of the latter, we have the estimate

$$\left|\int_{t<|h|<2t}\Delta_h^l D^{\mathbf{m}}\left(\frac{1}{x^{n-r}}\right)\frac{dh}{|h|^{n+r-k}}\right|\leqslant C\int_{|h|<2t}\frac{dh}{|h|^{n+r-k-l}}=C_1t^{l+k-r}.$$

This guarantees the convergence of (47).

The rest of the integral can be estimated directly:

$$\begin{aligned}&\int_{\gamma|x|}^\infty\frac{dt}{t}\left|\int_{t<|h|<2t}\Delta_h^l D^{\mathbf{m}}\left(\frac{1}{|x|^{n-r}}\right)\frac{dh}{|h|^{n+r-k}}\right|^2\\ &\quad\leqslant C\sum_{j=0}^{l}\int_{\gamma|x|}^\infty\frac{dt}{t}\left|\int_{t<|h|<2t}\frac{1}{|x-jh|^{n-r+k}}\frac{dh}{h^{n+r-k}}\right|^2\\ &\quad=C|x|^{-n}\sum\int_\gamma^\infty\frac{d\xi}{\xi}\left|\int_{\xi<|\eta|<2\xi}\frac{1}{(\theta-j\eta)^{n-r+k}}\frac{d\eta}{\eta^{n+r-k}}\right|^2=C_1|x|^{-n}.\end{aligned}$$

Here θ denotes a point on the unit sphere, and the convergence of the integrals is easily verified. Therefore (46) is established for $r\neq n+2k$, $k=0,1,\cdots$. In the excluded cases the estimates are obtained similarly.

We turn to the verfication of hypothesis (27′) of Theorem A:

$$\Phi(\lambda,t)=\tilde{k}(\lambda,t)=C\int_{t<|h|<2t}\frac{(1-e^{i\lambda h})^l\lambda^{\mathbf{m}}}{|\lambda|^r|h|^{n+r-k}}dh.$$

Differentiating Φ with respect to λ_j and then multiplying by λ_j preserves homogeneity, and hence (27′) is verified by a single method and without effort. We have, for example,

$$|\Phi(\lambda,\cdot)|^2_{H'\to H''}=\int_0^\infty|\Phi(\lambda,t)|^2\frac{dt}{t}=\int_0^\infty\frac{d\xi}{\xi}\left|\int_{\xi<|\eta|<2\xi}\frac{(1-e^{i\theta u})^l\,\theta^{\mathbf{m}}}{|u|^r}\frac{du}{|u|^n}\right|^2\leqslant M.$$

Therefore Theorem A applies and

$$\|Av_\nu\|_{L_p(H'')}\leqslant C\|v_\nu\|_{L_p(H')}. \tag{48}$$

The inequality can be extended by continuity to $L_p^{(r)}$, i.e. for $f\in L_p^{(r)}$

$$\sum_{|\mathbf{m}|=k}\mathcal{L}_{p,2}^{r,l,\mathbf{m}}(f)\leqslant C\|T^rf\|_p \tag{49}$$

Turning to the proof of the converse inequality, we consider the operator $\mathbf{A}v_\nu$ with components

$$A_jv_\nu=\int_{t<|h|<2t}\frac{\Delta_h^lD_j^ku_\nu(x)\,dh}{|h|^{n+r-k}}=\int_{R^n}k_j(x-y,t)\,v_\nu(y)\,dy,$$

$$k_j(x,t)=\int_{t<|h|<2t}\Delta_h^lD_j^k\left(\frac{1}{|x|^{n-r}}\right)\frac{dh}{h^{n+r-k}},\qquad j=1,\ldots,n.$$

By what has been established, the kernel $\mathbf{k}_j(x,t)=(k_1(x,t),\cdots,k_n(x,t))$ represents a bounded operator from the Hilbert space $\mathbf{H}'$, whose elements are vectors $a=(a_1,\cdots,a_n)$ with complex components a_j and scalar product $(a,b)=\Sigma_1^na_j\bar b_j$, into the Hilbert space H'' In addition,

$$\|\mathbf{A}v_\nu\|_p\leqslant C\|v_\nu\|_p,$$

which leads to the inequality

$$\sum_{\mathbf{m}=(0,\ldots,k,\ldots,0)}\mathcal{L}_{p,2}^{r,l,\mathbf{m}}(f)\leqslant C\|T^rf\|_p, \tag{50}$$

which is apparently weaker that (49) (the summation on the left extends over the unmixed derivatives). In fact (49) and (50) are equivalent (each implies the other), as the following considerations show.

For $p=2$ we invert the operator $\mathbf{A}$ (by using Plancherel's theorem it is easy to verify that $\|\widetilde{Av_\nu}\|_2\sim\|v_\nu\|_2$). In addition, in the Fourier representation $\mathbf{A}$ reduces to multiplication by $\Phi(\lambda,t)=\widetilde{\mathbf{k}(x,t)}$ and the function $\Phi(\lambda,\cdot)$ satisfies hypothesis 2) of Theorem A. Proceeding as in the proof of Theorem 5, we obtain

$$\|T^rf\|_p\leqslant C\sum_{\mathbf{m}=(0,\ldots,k,\ldots,0)}\mathcal{L}_{p,2}^{r,l,\mathbf{m}}(f). \tag{51}$$

From (49), (50) and (51) there follows the equivalence of the differential seminorms

$$\sum_{|\mathbf{m}|=k}\mathcal{L}_{p,2}^{r,l,\mathbf{m}}(f)\sim\sum_{\mathbf{m}=(0,\ldots,k,\ldots,0)}\mathcal{L}_{p,2}^{r,l,\mathbf{m}}(f)\sim\|T^2f\|_p,$$

and then the equivalence of the corresponding norms in $L_p^{(\mathbf{r})}$. This completes the proof of the theorem.

REMARK 7. Theorem 7 can be strengthened at the expense of using Theorem B along the lines of Remark 4. This leads to

$$\|f\|_{L_p^{(r)}} \sim \|f\|_p + \sum_{|\mathbf{m}|=k} \left\| \left\{ \int_0^\infty \frac{dt}{t} \left(\int_{t<|h|<2t} \frac{|\Delta_h^l D^{\mathbf{m}} f(x)|}{|h|^{n+r-k}} dh \right)^2 \right\}^{1/2} \right\|_p, \tag{52}$$

where the second term on the right is equivalent to the differential seminorm of f in $L_p^{(\mathbf{r})}$ for $0 \leqslant k \leqslant \bar{r} \leqslant r < k+l$. Here the equivalence is not affected if the summation is extended only over the unmixed derivatives of order k.

In conclusion we note the following theorem for the nonisotropic case.

THEOREM 8.[6] *A necessary and sufficient condition for the p-summable function $f(x)$ to belong to $L_p^{(\mathbf{r})}(R^n)$, $1 < p < \infty$, $\mathbf{r} > 0$, is that the function $U^{\mathbf{r}}f$ defined by*

$$(U^{\mathbf{r}}f)(x) = \left\{ \int_0^\infty \frac{dt}{t} \left(\int_{t<\rho(h)<2t} \frac{|\Delta_h^l f(x)|}{[\rho(h)]^{n+r^*}} dh \right)^2 \right\}^{1/2} \tag{53}$$

belongs to L_p. In this case

$$\|U^{\mathbf{r}}f\|_p \sim \|T^{\mathbf{r}}f\|_p. \tag{54}$$

REMARK 8. We shall call seminorms or norms expressed in terms of the moduli of corresponding differences strong seminorms (or norms). Examples are the seminorms $\mathfrak{L}_{p,2}^{r,l,k}(f)$ (cf. Remark 4), $\|U^{\mathbf{r}}f\|_p$, etc. Strong seminorms admit simplification. It is clear, for example, that (53) can be written in the equivalent forms

$$\left\{ \int_0^\infty \frac{dt}{t^{1+2n+2r^*}} \left(\int_{t<\rho(h)<2t} |\Delta_h^l f(x)|\, dh \right)^2 \right\}^{1/2}, \tag{55}$$

$$\left\{ \int_0^\infty \frac{dt}{t^{1+2r^*}} \left(\int_{1<|u|<2} |\Delta_{t^a u}^l f(x)|\, du \right)^2 \right\}^{1/2}, \tag{56}$$

where $t^a u = (t^{a_1}u_1, t^{a_2}u_2, \cdots, t^{a_n}u_n)$.

§4. Operators of the type of truncated differentiation

1. The operators $\bar{D}_t^{\mathbf{r}}$ and $\bar{T}_t^r$ have the property that in the Fourier representation they are generated by multiplication by a function whose L_2^*-norm is a homogeneous function of degree r. The class of operators satisfying such a condition is rather extensive, and its complete characterization in abstract terms presents a certain interest. Our present objective is more modest: to indicate an "operator of the type

[6] $\rho(h)$, r^* and l have the same meanings as in Theorem 2.

of truncated differentiation" for which the theorems of the preceding section are preserved. These operators may turn out to be more convenient for investigations in bounded regions. We begin again with the one-dimensional case. We introduce the operator B by the equation

$$(Bf)(x, t) = \frac{1}{\omega_{\beta, r}(q, Q)} \frac{1}{t^{\beta}} \int_{qt}^{Qt} \frac{\Delta_h^l f_{(x)}^{(k)}}{h^{1+r-k-\beta}} dh, \quad 0 \leqslant k < r < k + l, \tag{57}$$

where $\omega_{\beta,r}^{-1}(q, Q)$ is a normalizing factor which will be made explicit below, q and Q are constants, $0 < q < Q < \infty$, and β is an arbitrary real number. For $\beta = 0$, $q = 1$, $Q = 2$ we obtain, up to a factor, the operator considered in Theorem 5 of the preceding section. As we shall see, (57) can also be interpreted (with some supplementary stipulations) in the following limiting cases: 1) $q \to 0$, 2) $Q \to \infty$, 3) $q \to Q \to 1$. The last case corresponds to Marcinkiewicz's theorem (cf. Remark 6 and subsection 3 below).

THEOREM 5′. *For functions in* L_p *we have the equivalence*

$$\| D^r f \|_p \sim X_p T_2^* [Bf(x, t)], \quad 1 < p < \infty.$$

The proof proceeds precisely along the lines of the proof of Theorems 5 and 7. We confine ourselves to a formal verification of the fact (which is the key to the proof) that the operator $Av = B(D^r)^{-1} v$ is a bounded invertible operator from L_2 into $L_2(H)$, $H = L_2^*(0, \infty)$. We have

$$\| Av \|_{L_2(H)}^2 = \| \overline{Av} \|_{L_2(H)}^2 = \frac{1}{\omega_{\beta, r}^2(q, Q)} \int | \bar{f}(\lambda) |^2 d\lambda \int_0^{\infty} \left| \frac{1}{t^{\beta}} \int_{qt}^{Qt} \frac{(1 - e^{-h\nu})^l \, dh}{h^{1+r-\beta}} \right|^2 \frac{dt}{t} = \| f \|_2^2.$$

Here we have again used Plancherel's theorem and put

$$\omega_{\beta, r}^2(q, Q) = \int_0^{\infty} \left| \frac{1}{\eta^{\beta}} \int_{q\eta}^{Q\eta} \frac{(1 - e^{-iu})^l}{u^{r-\beta}} \frac{du}{u} \right|^2 \frac{d\eta}{\eta}. \tag{58}$$

The verification that the sufficient conditions of Theorem A are satisfied proceeds in a similar way to what was done in Theorems 5 and 7. The various auxiliary considerations are also similar. We shall therefore consider Theorem 5′ as established.

A stronger form of Theorem 5′ can be proved along the lines of Remark 4 of §3:

$$\| D^r f \| \sim \left\| \left\{ \int_0^{\infty} \left| \frac{1}{t^{\beta}} \int_{qt}^{Qt} \frac{| \Delta_h^l f^{(k)}(x) |}{h^{1+r-k-\beta}} dh \right| \frac{dt}{t} \right\}^{1/2} \right\|_p .$$

The indicated proof applies if 1) $q = 0$ and $Q = 1$ for $\beta > 0$; 2) $q = 1$, $Q = \infty$ for $\beta < 0$.

Just as in §3, Theorem 5′ immediately implies the corresponding result in the n-dimensional case, which asserts the equivalence of certain coordinate norms in $L_p^{(\mathbf{r})}(R^n)$. We state it in terms of second differences and in a stronger form, putting $\beta_j = 1 + \alpha_j$, $q_j = 0$ and $Q_j = 1$, $j = 1, \cdots, n$.

THEOREM 6′. *The space* $L_p^{(\mathbf{r})}(R^n)$ *admits the norm (equivalent to the original one)*

$$\|f\|_{L_p^{(\mathbf{r})}} = \|f\|_p + \sum_{j=1}^{n} \left\| \left\{ \int_0^\infty \left[\frac{1}{t^{1+\alpha_j}} \int_0^t |\Delta_h^2 D_j^{\bar{r}_j} f(x)|\, dh \right]^2 \frac{dt}{t} \right\} \right\|_p . \tag{59}$$

For $\alpha_j < 1$ the second difference can be replaced by a first difference; for $\alpha_j = 1$ the corresponding term can be replaced by $\|D_j^{\bar{r}_j} f\|_p$.

2. In the n-dimensional isotropic case, instead of (57) the following operators naturally stand out:(7)

$$B^{\mathbf{m}} f = \frac{1}{\bar{\omega}_{\beta, r}(q, Q)} \frac{1}{t^\beta} \int_{qt < |h| < Qt} \frac{\Delta_h^l D^{\mathbf{m}} f(x)}{|h|^{n+r-k-\beta}}\, dh, \qquad 0 \leqslant k < r < k + l; \tag{60}$$

with the aid of those we obtain, along the lines of Theorem 7 and Remark 7,

$$\begin{aligned} \|T^r f\|_p &\sim \sum_{|\mathbf{m}|=k} X_p T_2^* [(B^{\mathbf{m}} f)(x, t)] \\ &\sim \sum_{\mathbf{m}=k} \left\| \left\{ \int_0^\infty \left(\frac{1}{t^\beta} \int_{qt<|h|<Qt} \frac{|\Delta_h^l D^{\mathbf{m}} f(x)|}{|h|^{n+r-k-\beta}}\, dh \right)^2 \frac{dt}{t} \right\}^{1/2} \right\|_p . \end{aligned} \tag{61}$$

Here again, when $\beta > 0$ we may take $q = 0$, $Q = 1$, and when $\beta < 0$ we may take $q = 1$, $Q = \infty$. Relation (61) was obtained in [9] for special values of the parameters ($k = 0$, $l \leqslant 2$, $0 < r < 1$, $q = 0$, $Q = 1$, $\beta = n + r$).

There is a similar result for the space $L_p^{(\mathbf{r})}$ in the nonisotropic case (on the right-hand side of (60), take $k = 0$, and replace $|h|$ by $\rho(h)$, r by r^*).

3. Let us consider in more detail the operator (60) in the limiting case $q \to Q \to 1$. For $n \geqslant 2$ write

$$B^{\mathbf{m}} f = \frac{Q-q}{\bar{\omega}_{\beta, r}(q, Q)} \frac{1}{t^{\beta-1}} \frac{1}{Qt - qt} \int_{qt}^{Qt} d\rho \int_{\Omega_\rho} \frac{\Delta_h^l D^{\mathbf{m}} f(x)}{\rho^{n+r-k-\beta}}\, d\omega_\rho,$$

(7) The constants $\bar{\omega}_{\beta,r}(q, Q) = \bar{\omega}_{\beta,r}^{\mathbf{m},l}(q, Q)$ are determined by the conditions $X_2 T_2^*(B^{\mathbf{m}} f) = |\lambda|^{r-k} |(i\lambda)^{\mathbf{m}}|\, \|f\|_2$.

where $\rho = |h|$, Ω_ρ is the sphere of radius ρ, $d\omega_\rho = \rho^{n-1} d\omega$ is the surface element on Ω_ρ, $d\omega$ is the surface element of the unit sphere Ω, and ω is a point of Ω. With the indicated approach to the limit we obtain

$$B^{\mathbf{m}} f = \frac{1}{\bar{\omega}_{\beta, r}} \frac{1}{t^{n-1}} \int_\Omega \frac{\Delta^l_{t\omega} D^{\mathbf{m}} f(x)}{t^{r-k}} d\omega, \tag{62}$$

which represents spherical symmetrization of the difference quotient over the sphere of radius t. In the one-dimensional case we obtain correspondingly

$$\frac{1}{\bar{\omega}_{\beta, r}} \frac{\Delta^l_t D^k f(x) - \Delta^l_{-t} D^k f(x)}{t^{r-k}}.$$

The one-dimensional case was already discussed, in essentials, in Remark 6. It turns out that in the n-dimensional case the same restrictions on r appear.

Consider, for example, the operator (62) with $k = 0$, $0 < r < 1$, $l = 1$:

$$Bu = \frac{1}{t^{n-1}} \int_\Omega \frac{u(x) - u(x - t\omega)}{t^r} d\omega.$$

Let $u \in L^r_p(R^n)$; then

$$u = C \int \frac{v(y)\, dy}{|x - y|^{n-r}}, \quad v \in L_p,$$

and we pass to the operator A, as usual:

$$Av = C \int k(x - y) v(y)\, dy,$$

$$k(x, t) = \frac{1}{t^{n-1}} \int_\Omega \left(\frac{1}{|x|^{n-r}} - \frac{1}{|x - t\omega|^{n-r}} \right) d\omega.$$

With our usual notation we have

$$|k(x, \cdot)|^2_{H' \to H''} = \int_0^\infty \frac{dt}{t} \left| \frac{1}{t^{n-1}} \int_\Omega \left(\frac{1}{|x|^{n-r}} - \frac{1}{|x - t\omega|^{n-r}} \right) d\omega \right|^2.$$

It is easily seen that the expression on the left is finite only for $r > 1/2$. With this restriction we can prove, for example, that

$$\| T^r f \|_p \sim \left\| \left\{ \int_0^\infty \frac{dt}{t} \left(\frac{1}{t^{n-1}} \int_\Omega \frac{|\Delta^l_{t\omega} f(x)|}{t^r} d\omega \right)^2 \right\}^{1/2} \right\|_p.$$

The average (62) can be replaced by other averages. In particular, we could average the difference quotient along line segments and finally consider the difference quotients by themselves. By modifying Theorems A and B somewhat and following the same lines we could show, for example, that when $r > n/2$

$$\| T^r f \|_p \sim \left\| \left\{ \int_{R^n} \frac{dh}{|h|^n} \left| \frac{\Delta_h^l f(x)}{|h|^r} \right|^2 \right\}^{1/2} \right\|_p, \quad l > r, \quad 1 < p < \infty.$$

We shall not consider this in detail.

Appendices

1. Let us recall some properties of the space Φ [1]. An infinitely differentiable function φ belongs to the Schwartz space S if it (together with its derivatives) decreases at ∞ faster than every power of $|x|$:

$$\varphi \in \Phi \Leftrightarrow \varphi \in S, \qquad \int_{-\infty}^{\infty} x_j^k \varphi(x)\, dx_j = 0, \qquad j = 1, \ldots, n, \quad k = 0, 1, 2, \ldots.$$

According to this definition a function $\varphi(x)$ in S belongs to Φ if and only if all its moments in the coordinate directions are zero. The Fourier transform of a function $\varphi \in \Phi$ is a function $\overline{\varphi} = \Psi(\lambda) \in S$ which, together with its derivatives, vanishes at the origin of the coordinate hyperplanes. The set of functions $\Psi(\lambda)$ with this property forms the space Ψ; the Fourier transformation $\mathfrak{F}$ establishes an isomorphism between Φ and Ψ: $\mathfrak{F}\Phi = \Psi$, $\mathfrak{F}^{-1}\Psi = \Phi$. The space Ψ becomes a countably-normed space if we consider it under the system of seminorms [1, 8]:

$$_k\| \Psi \| = \sup_{\substack{|\mathbf{m}| \leqslant k \\ \lambda \in R^n}} M_k(\lambda) \left| D^{\mathbf{m}} \Psi(\lambda) \right|, \qquad k = 0, 1, 2, \ldots,$$

$$M_k(\lambda) = [M(\lambda)]^k, \qquad M(\lambda) = \max\left((1 + |\lambda|^2)^{1/2}, \frac{1}{|\lambda_1|}, \ldots, \frac{1}{|\lambda_n|} \right).$$

An example of a function $\varphi \in \Phi$ is the inverse Fourier transform of the function

$$e^{-\beta^2 \left(|\lambda|^2 + \sum_{j=1}^{n} \frac{1}{\lambda_j^2} \right)} = \prod_{j=1}^{n} e^{-\beta^2 \left(\lambda_j^2 + \frac{1}{\lambda_j^2} \right)},$$

i.e.

$$\mathcal{K}_\beta(x) = \frac{1}{(2n)^{n/2}} \int_{R^n} e^{-\beta^2 \left(|\lambda|^2 + \sum_{j=1}^{n} \frac{1}{\lambda_j^2} \right)} \cos \lambda x \, d\lambda. \tag{A.1}$$

The space Φ is dense in $L_p^{\mathfrak{r}}(R^n)$. As was done in the one-dimensional case in [2] (cf. Chapter II, §4, Supplement to Theorem 1), we can see that if $\overline{f}$ is p-summable and of compact support, the convolution $\mathcal{K}_\beta * \overline{f} = (\overline{f})_\beta$ converges to $\overline{f}$ in the L_p norm, and then $(\overline{f})_\beta \in \Phi$ (the Fourier transform of $\overline{f}$ is an entire bounded function, whence $\overline{\mathcal{K}_\beta} \cdot \overline{f} \in \Psi$; and this is equivalent to having $(\overline{f})_\beta \in \Phi$). For an arbitrary function $f \in L_p$ we write

$$f(x)=f(x)-\bar{f}(x)+\bar{f}(x)-(\bar{f})_\beta+(\bar{f})_\beta,$$

where $\bar{f}(x)$ is a function of compact support approximating f arbitrarily closely in the L_p norm. By choice of β we can ensure that $\|\bar{f}-(\bar{f})_\beta\|_p$ is arbitrarily small; consequently $\|f-(\bar{f})_\beta\|_p$ can be made arbitrarily small. Therefore Φ is dense in L_p. To prove that Φ is dense in $L_p^{\mathfrak{r}}$ it remains only to use the integral representation (25).

We note that in the isotropic case it follows immediately from the integral representation (25) that S is dense in L_p^r (since S is dense in L_p). Finally, even in the nonisotropic case S is dense in $L_p^{(\mathfrak{r})}$, although the preceding discussion is not sufficient to establish this (in the nonisotropic case $\rho^2(\lambda)$ is not an infinitely differentiable function) and it is necessary to resort to different integral representations, with kernels constructed by using Bessel derivatives [2].

2. In the space Ψ multiplication by $|\lambda|^r$ is defined. This makes it possible to follow the usual plan (cf. [10], vol. 2) of defining the convolution

$$R_r*f;\qquad f\in\Phi',\qquad R_r=\mathcal{F}^{-1}(|\lambda|^r),$$

in the space Φ' of generalized functions over Φ. The operators R_r* form an additive group of transformations of the space Φ', since by the convolution theorem we have

$$\mathcal{F}(R_r*R_s)=\mathcal{F}(R_r)\mathcal{F}(R_1)=|\lambda|^r|\lambda|^s=|\lambda|^{r+s}=\mathcal{F}(R_{r+s}).$$

For the generalized function R_r we can obtain an explicit expression by calculating the generalized Fourier transform of the function $|\lambda|^r\in\Psi'$. This expression coincides with the usual one ([10], vol. 1, Russian pp. 224–225) if we disregard polynomial terms orthogonal to Φ:

$$R_r(x)=\begin{cases}2^{r+\frac{n}{2}}\dfrac{\Gamma\left(\frac{r+n}{2}\right)}{\Gamma\left(-\frac{r}{2}\right)}|x|^{-r-n} & \text{for}\quad r\neq 2k,\ -n-2k\quad (k=0,\ 1,\ \ldots),\\ (2\pi)^{\frac{n}{2}}(-\Delta)^k\delta(x), & r=2k\quad (k=0,\ 1,\ \ldots),\\ \dfrac{(-1)^k|x|^{2k}\ln|x|}{2^{\frac{n}{2}+2k-1}\pi^{\frac{n}{2}}\Gamma\left(\frac{n}{2}+k\right)k!}, & r=-n-2k\quad (k=0,\ 1,\ \ldots).\end{cases}$$

Here R_r* acts on functions $\varphi\in\Phi$ according to the following rule. When $s<0$, we evidently have

$$R_s*\varphi=\frac{1}{(2\pi)^{n/2}}\int_{R^n}R_s(x-y)\varphi(y)\,dy,\qquad \varphi\in\Phi. \tag{A.2}$$

For $s \neq -n-2k$, $k=0, 1, \cdots$, the convolution coincides with the Riesz potential

$$R_s * \varphi = A_{n,s} \int_{R^n} \frac{\varphi(y)\,dy}{|x-y|^{n+s}} \qquad (s<0). \tag{A.2'}$$

For the excluded values $-n-2k$ we have the corresponding logarithmic potentials.

If $0<r<l$, we obtain

$$R_r * \varphi = B_{n,r,l} \int_{R^n} \frac{\Delta_h^l \varphi(x)}{h^{n+r}}\,dh = B_{n,r,l} T^r \varphi, \tag{A.3}$$

where the coefficient $B_{n,r,l}$ is easily calculated from the condition $B_{n,r,l}\widetilde{T^r\varphi} = |\lambda|^r\tilde{\varphi}$! In addition, for even $r=2k$ we need to use the relation

$$R_{2k} * \varphi = (-\Delta)^k \varphi, \qquad k=0, 1, \ldots .$$

Formulas (A.2) $(s=-r)$ and (A.3) still make sense for functions $\varphi \in L_p^{(r)}$ if they are considered as being conditionally convergent (at ∞ and zero, respectively; cf. [2, 4]). This makes it possible (in parallel with Theorems 2 and 3) to characterize $L_p^{(r)}$ as the set of functions $f \in L_p$ represented by integrals of potential type:

$$f = \int_{R^n} R_{-r}(x-y)\, g(y)\,dy, \qquad g \in L_p.$$

Here

$$\|f\|_{L_p^r} \sim \|f\|_p + \|g\|_p.$$

BIBLIOGRAPHY

1. P. I. Lizorkin, *Generalized Liouville differentiation and the function spaces* $L_p^r(E_n)$. *Imbedding theorems,* Mat. Sb. **60** (102) (1963), 325–353. (Russian) MR **27** #610.

2. ———, *Generalized Liouville differentiation and the method of multipliers in the theory of imbeddings of classes of differentiable functions,* Trudy Mat. Inst. Steklov. **105** (1969), 89–167 = Proc. Steklov Inst. Math. **105** (1969), 105–202. MR **41** #7419.

3. O. V. Besov, *Investigation of a family of function spaces in connection with theorems of imbedding and extension,* Trudy Mat. Inst. Steklov. **60** (1961), 42–81; English transl., Amer. Math. Soc. Transl. (2) **40** (1964), 85–126. MR **24** #A3501.

4. P. I. Lizorkin, *Description of the spaces* $L_p^{(r)}(R^m)$ *in terms of singular difference integrals,* Mat. Sb. **81** (123) (1970), 79–91 = Math. USSR Sb. **10** (1970), 77–90. MR **41** #2376.

5. A. Benedek, A. P. Calderón and R. Panzone, *Convolution operators on Banach space valued functions,* Proc. Nat. Acad. Sci. U.S.A. **48** (1962), 356–365. MR **24** #A3479.

6. P. I. Lizorkin, (L_p, L_q)*-multipliers of Fourier integrals,* Dokl. Akad. Nauk SSSR **152** (1963), 808–811 = Soviet Math. Dokl. **4** (1963), 1420–1423. MR **27** #4016.

7. P. Krée, *Propriétés de continuité dans* L^p *de certains noyaux,* Bol. Un. Mat. Ital. (3) **22** (1967), 330–344. MR **37** #4604.

8. K. Yoshinaga, *On Liouville's differentiation,* Bull. Kyushu Inst. Tech. Math. Natur. Sci. No. 11 (1964), 1–17. MR **30** #5071.

9. R. Strichartz, *Multipliers on fractional Sobolev spaces,* J. Math. Mech. **16** (1967), 1031–1060. MR **35** #5927.

10. I. M. Gel'fand and G. E. Šilov, *Generalized functions.* Vol. 1: *Operations on them,* Fizmatgiz, Moscow, 1958; English transl., Academic Press, New York, 1964. MR **20** #4182; **29** #3869.

———, *Generalized functions,* Vol. 2: *Spaces of fundamental functions,* Fizmatgiz, Moscow, 1958; English transl., Academic Press; Gordon and Breach, New York, 1968. MR **21** #5142a; **37** #5693.

Translated by R. P. BOAS

Trudy Mat. Inst. Steklov.
117 (1972)

Proc. Steklov Inst. Math.
117 (1972)

ASYMPTOTIC BEHAVIOR AT INFINITY OF DIFFERENTIABLE FUNCTIONS OF WEIGHT CLASSES

UDC 517.43

Ju. S. NIKOL'SKIĬ

Abstract. Functions $f(x)$, $x \in E_n$, are considered whose generalized derivatives $D_i^{l_i} f(x)$ $(i = 1, \cdots, n)$ are p-summable with a weight that degenerates at the origin and at infinity. It is shown that in a certain sense these functions or their derivatives "go over to polynomials" as $|x| \to \infty$, and that those, and only those, which "go over to zero" can be approximated by infinitely differentiable functions with compact support.

Bibliography: 14 items.

In the weight-free case questions concerning the propety of "going over into a polynomial" for differentiable functions as $|x| \to \infty$ and concerning the density of finite functions were studied in the case of isotropic spaces in the work of S. V. Uspenskiĭ [1], S. L. Sobolev [2] and V. N. Sedov [3], and for nonisotropic spaces in the work of O. V. Besov [4]. In the work of V. P. Portnov [5], T. S. Pigolkina [6], the author [7] and Sedov [3] these problems were investigated in certain weight-isotropic spaces with weights having singularities at infinity.

We shall adopt the notation conventionally used in the theory of differentiable functions of several variables (see, e.g., [8]).

Let $\alpha = (\alpha_1, \cdots, \alpha_n)$ be a vector with positive coordinates, $|\alpha| = \Sigma_1^n \alpha_i = n$. A continuous α-homogeneous function $\rho(x) = \rho(x_1, \cdots, x_n)$ of degree $m = 1$, i.e. which has the property $\rho(t^{\alpha_1} x_1, \cdots, t^{\alpha_n} x_n) = t\rho(x)$ for any $t > 0$, which is positive for $x \neq 0$, will be called an α-distance of x (from the origin).

As an example of an α-distance we give the function

$$\rho(x) = \sum_1^n |x_i|^{1/\alpha_i}, \quad \pi(x) = \max_{1 \leqslant i \leqslant n} |x_i|^{1/\alpha_i}.$$

α-distances posess the following property (see [9]). For any α-distance $\rho(x)$, Minkowski's inequality

$$\rho(x + y) \leqslant c_0 (\rho(x) + \rho(y)) \tag{1}$$

AMS (MOS) subject classifications (1970). Primary 46E35; Secondary 46E30.

is satisfied, where the constant c_0 does not depend on $x, y \in E_n$.

Any two α-distances are equivalent, i.e. for some positive numbers c' and c''

$$c'\rho_1(x) \leqslant \rho_2(x) \leqslant c''\rho_1(x). \tag{2}$$

We define the weight space $L_{p,\beta}$, $1 < p < \infty$, of functions $f(x)$ defined on E_n which have finite norm

$$\|f\|_{p,\beta} = \|[\rho(x)]^\beta f(x)\|_p$$

The space $L_{p,\beta}$ is complete.

Let $\Phi(y)$ be a piecewise continuous function concentrated in the cube $\{0 < \delta < y_i < 1,\ i = 1, \cdots, n\}$. Let $0 < \epsilon < r < \infty$. Let[1]

$$(I_{\varepsilon,r}f)(x) = \int k_{\varepsilon,r}(y) f(x+y)\,dy, \tag{3}$$

$$k_{\varepsilon,r}(y) = \int_\varepsilon^r \Phi\left(\frac{y}{h^\alpha}\right) h^{-n-s}\,dh, \tag{4}$$

where $f \in L_{p,\beta}$ $(1 < p < \infty;\ \beta < n/p',\ (1/p) + (1/p') = 1)$.

It is easily shown that the family (4) of kernels $k_{\epsilon,r}(y)$ has the following properties.

I. For any $t > 0$

$$k_{\varepsilon,r}(t^\alpha y) = t^{-n-s+1} k_{\frac{\varepsilon}{t},\frac{r}{t}}(y).$$

II. The family of kernels $k_{\epsilon,r}(y)$ is uniformly bounded on the unit ρ-sphere, $\rho(\zeta) = 1$, i.e.

$$|k_{\varepsilon,r}(\zeta)| \leqslant M,\ \rho(\zeta) = 1,\ 0 \leqslant \varepsilon < r \leqslant \infty,$$

where the constant M does not depend on ϵ or r.

LEMMA 1. *Let $1 < p \leqslant q < \infty$, $\beta < n/p'$, $0 \leqslant \mu < n/q + \beta$ and $f \in L_{p,\beta}$. Then*

$$\left\| h^{-n} \int \Phi\left(\frac{y}{h^\alpha}\right) f(x+y)\,dy \right\|_{q,\beta-\mu} \leqslant c h^{-\left(\frac{1}{p}-\frac{1}{q}\right)n-\mu} \|f\|_{p,\beta}, \tag{5}$$

where the constant c does not depend on f and $h > 0$.

PROOF. Since α-distances are equivalent in the sense of the inequality (2), it is sufficient to carry out the proof for the case $\rho(x) = \Sigma_1^n |x_i|^{1/\alpha_i}$. We set $\varphi(x) = [\rho(x)]^\beta f(x)$.

---

[1] If no region of integration Ω is indicated below the integral sign, it is to be understood that $\Omega = E_n$.

If $\rho(x) > 2c_0 a$, $\rho(y) < a$ and $a > 0$ (c_0 is the constant in inequality (1)), then by applying (1) it is easily proved that

$$c_1\rho(x) \leqslant \rho(x+y) \leqslant c_2\rho(x), \tag{6}$$

where the positive constants c_1 and c_2 do not depend on a.

By noting that $\{y: \delta h^{\alpha_i} < y_i < h^{\alpha_i},\ i = 1, \cdots, n\} \subset \{y: \rho(y) < nh\}$, using inequality (6) for $a = nh$ and the fact that the lemma is true for $\beta = \mu = 0$ (see [4]), we find

$$\begin{aligned} &\left\| h^{-n}[\rho(x)]^{-\mu} \int\limits_{\rho(y)<nh} \left|\Phi\left(\frac{y}{h^{\alpha}}\right)\right| \left[\frac{\rho(x)}{\rho(x+y)}\right]^{\beta} |\varphi(x+y)|\,dy \right\|_{L_q(\rho(x)>2c_0nh)} \\ &\leqslant ch^{-\mu}\left\| h^{-n}\int \left|\Phi\left(\frac{y}{h^{\alpha}}\right)\right| |\varphi(x+y)|\,dy\right\|_q \leqslant c_1 h^{-\left(\frac{1}{p}-\frac{1}{q}\right)n-\mu} \|f\|_{p,\beta}. \end{aligned} \tag{7}$$

By applying the Hölder inequality, noting that there exists a positive number b, which does not depend on h, such that the set $\{y: \rho(y-x) < nh, \rho(x) < 2c_0nh\} \subset \{y: \rho(y) < bnh\}$, and introducing generalized spherical coordinates (ρ, ζ) by the formulas $x_i = \rho^{\alpha_i}\zeta_i$ ($i = 1, \cdots, n$), where $\rho = \rho(x)$, $\zeta = (\zeta_1, \cdots, \zeta_n)$ is the point of intersection of the unit sphere, i.e. $\rho(\zeta) = 1$, with the α-trajectory $(t^{\alpha_1}\zeta_1, \cdots, t^{\alpha_n}\zeta_n)$, $0 < t < \infty$, which passes through the point x (see [9]), we obtain

$$\begin{aligned} \int \left|\Phi\left(\frac{y}{h^{\alpha}}\right) f(x+y)\right| dy &\leqslant \left(\int\limits_{\rho(y)<bnh} \left|\Phi\left(\frac{y-x}{h^{\alpha}}\right)\right|^{p'} \frac{dy}{[\rho(y)]^{\beta p'}}\right)^{1/p'} \|\varphi\|_p \\ &\leqslant c\left(\int\limits_0^{bnh} \frac{d\rho}{\rho^{\beta p'-n+1}}\right)^{1/p'} \|\varphi\|_p \leqslant c_1 h^{-\beta+\frac{n}{p'}} \|f\|_{p,\beta}. \end{aligned}$$

From the last inequality we find

$$\left\| h^{-n}\int \Phi\left(\frac{y}{h^{\alpha}}\right) f(x+y)\,dy \right\|_{L_{q,\beta-\mu}(\rho(x)<2c_0nh)} \leqslant ch^{-\left(\frac{1}{p}-\frac{1}{q}\right)n-\mu} \|f\|_{p,\beta}. \tag{8}$$

From inequalities (7) and (8) we obtain the desired estimate (5).

LEMMA 2. *Let* $1 < p \leqslant q < \infty$, $\beta < n/p'$, $0 \leqslant \mu < n/q + \beta$, $\kappa = 1 - (1/p - 1/q)n - \mu - s < 0$ *and* $f \in L_{p,\beta}$. *Then for* $0 < \epsilon < r \leqslant \infty$

$$\|I_{\varepsilon,r} f\|_{q,\beta-\mu} \leqslant c\varepsilon^{\kappa} \|f\|_{p,\beta}.$$

The proof follows immediately from Minkowski's inequality and Lemma 1.

LEMMA 3. *Let* $1 < p < \infty$, $\beta < n/p'$, $\kappa = 1 - n/p - \beta - s < 0$ *and* $f \in L_{p,\beta}$. *Then as* $\epsilon, r \to \infty$

$$|I_{\varepsilon, r} f| \to 0 \quad \text{uniformly on any compact set } K \subset E_n. \tag{9}$$

PROOF. Let $x \in K$ and $\rho(x) = \Sigma_1^n |x_i|^{1/\alpha_i}$. Then, by assuming that ϵ is sufficiently large, by using properties I and II of the kernel $k_{\epsilon,r}(y)$, and noting that $\{y: \delta\epsilon^{\alpha_i} < y_i < r^{\alpha_i},\ i = 1, \cdots, n\} \subset \{y: \rho(y) > b\epsilon,\ b = \Sigma_1^n (\delta)^{1/\alpha_i}\}$, after applying Hölder's inequality and introducing spherical coordinates (ρ, ζ), where $x_i = \rho^{\alpha_i}\zeta_i$ $(i = 1, \cdots, n)$, we obtain

$$\begin{aligned} |I_{\varepsilon, r} f| &\leqslant c \int\limits_{\delta\varepsilon^{\alpha_i} < y_i < r^{\alpha_i}} [\rho(y)]^{-n-s+1} |f(x+y)|\, dy \\ &\leqslant c \left\{ \int\limits_{b\varepsilon < \rho(y)} \frac{dy}{[\rho(y)]^{p'(n+\beta-1+s)}} \right\}^{1/p'} \|f\|_{p,\beta} \leqslant c_1 \varepsilon^{\varkappa} \|f\|_{p,\beta} \to 0, \quad \varepsilon, r \to \infty, \end{aligned} \tag{10}$$

which also proves relation (9).

LEMMA 4. *Let*

$$(T_\nu f)(x) = \int \frac{f(y)\, dy}{[\rho(x)]^s [\rho(y)]^t [\rho(x-y)]^{n-\nu-s-t}}.$$

If $1 < p \leqslant q < \infty$ *and* $\nu = (1/p - 1/q)n$, $s < n/q$, $t < n/p'$, *where* $s + t \geqslant 0$ *if* $p < q$, *and* $s + t > 0$ *if* $p = q$, *and if* $f \in L_p$, *then*

$$\|T_\nu f\|_q \leqslant c \|f\|_p, \tag{11}$$

where the constant c *does not depend on* f.

PROOF. With no restriction in generality we can assume that $f \geqslant 0$ and $\rho(x) = \Sigma_1^n \alpha_i |x_i|^{1/\alpha_i}$.

I. Let $s, t \geqslant 0$. By making use of the inequality relating the arithmetic and geometric mean (see [10], §2.5), we find

$$\begin{aligned} (T_\nu f)(x) &\leqslant c \int \frac{f(y)\, dy}{\prod_1^n |x_i|^{\frac{s}{n}} |y_i|^{\frac{t}{n}} |x_i - y_i|^{1 - \frac{\nu}{n} - \frac{s}{n} - \frac{t}{n}}} \\ &= \left(\prod_{i=1}^n \int \frac{dy_i}{|x_i|^{\frac{s}{n}} |y_i|^{\frac{t}{n}} |x_i - y_i|^{1 - \frac{\nu}{n} - \frac{s}{n} - \frac{t}{n}}} \right) f(y). \end{aligned} \tag{12}$$

Since the one-dimensional integral operator

$$(T_{\nu,1}u)(x)=\int_{-\infty}^{\infty}\frac{u(y)\,dy}{|x|^{s}|y|^{t}|x-y|^{1-\nu-s-t}}$$

according to a theorem of Hardy and Littlewood acts boundedly from $L_p(E_1)$ into $L_q(E_1)$ if $1<p\leqslant q<\infty$, $\nu=1/p-1/q$, $s<1/q$ and $t<1/p'$, where $s+t\geqslant 0$ if $p<q$, and $s+t>0$ if $p=q$ (see [10], Russian p. 359), we obtain the desired estimate (11) from (12).

II. Let $s, t>0$. If $\rho(y)<2\rho(x-y)$, then

$$\int\limits_{\rho(y)<2\rho(x-y)}\frac{f(y)\,dy}{[\rho(x)]^{s}[\rho(y)]^{t}[\rho(x-y)]^{n-\nu-s-t}}\leqslant c\int\limits_{\rho(y)<2\rho(x-y)}\frac{f(y)\,dy}{[\rho(x)]^{s}[\rho(x-y)]^{n-\nu-s}}$$
$$\leqslant c_1\left(\prod_{i=1}^{n}\int\frac{dy_i}{|x_i|^{\frac{s}{n}}|x_i-y_i|^{1-\frac{\nu}{n}-\frac{s}{n}}}\right)f(y). \tag{13}$$

Now let $\rho(y)>2\rho(x-y)$. Consider the function $\omega(x)=[\rho(x)]^{\alpha_0}$, $\alpha_0=\min_{1\leqslant i\leqslant n}\alpha_i$. The function $\omega(x)$ obeys the triangle inequality, i.e. $\omega(x+y)\leqslant\omega(x)+\omega(y)$ for any $x, y\in E_n$ (see [11]). Since $\omega(y)>2^{\alpha_0}\omega(x-y)$, we have $\omega(y)>2^{\alpha_0}(\omega(x)-\omega(y))$ and $\omega(y)>2^{\alpha_0}(\omega(y)-\omega(x))$. Hence we find that $1-2^{-\alpha_0}<\omega(x)/\omega(y)<1+2^{-\alpha_0}$, i.e.

$$c_1<\rho(x)/\rho(y)<c_2.$$

With the aid of the last inequality we obtain

$$\int\limits_{\rho(y)>2\rho(x-y)}\frac{f(y)\,dy}{[\rho(x)]^{s}[\rho(y)]^{t}[\rho(x-y)]^{n-\nu-s-t}}\leqslant c\int\limits_{\rho(y)>2\rho(x-y)}\frac{f(y)\,dy}{[\rho(x)]^{s+t}[\rho(x-y)]^{n-\nu-s-t}}$$
$$\leqslant c_1\left(\prod_{i=1}^{n}\int\frac{dy_i}{|x_i|^{\frac{s}{n}+\frac{t}{n}}|x_i-y_i|^{1-\frac{\nu}{n}-\frac{s}{n}-\frac{t}{n}}}\right)f(y). \tag{14}$$

From (13), (14) and Hardy and Littlewood's theorem mentioned above we find the inequality (11).

The case $s<0, t>0$ is treated in exactly the same way as case II. Thus the lemma is proved.

LEMMA 5. *Let*

$$(T_{\nu,\mu}f)(x)=\int\frac{f(y)\,dy}{[\rho(x-y)]^{n-\nu-\mu}}.$$

If $1<p\leqslant q<\infty$, $\nu=(1/p-1/q)n$, $\beta<n/p'$, *where* $0\leqslant\mu<n/q+\beta$ *if* $p<q$, *and* $0<\mu<n/q+\beta$ *if* $p=q$, *and if* $f\in L_{p,\beta}$, *then*

$$\|T_{\nu,\mu}f\|_{q,\beta-\mu}\leqslant c\|f\|_{p,\beta},$$

where the constant c *does not depend on* f.

The lemma follows immediately from Lemma 4.

LEMMA 6. *Let* $1 < p \leqslant q < \infty$, $\beta < n/p'$, $0 \leqslant \mu < (n/q) + \beta$ *if* $p < q$, *and* $0 < \mu < n/q + \beta$ *if* $p = q$; *let* $\kappa = 1 - (1/p - 1/q)n - \mu - s = 0$ *and* $f \in L_{p,\beta}$. *Then for* $0 \leqslant \epsilon < r < \infty$

$$\|I_{\varepsilon,r}f\|_{q,\beta-\mu} \leqslant c\|f\|_{p,\beta}, \tag{15}$$

$$\|I_{\varepsilon,r}f\|_{q,\beta-\mu} \to 0, \quad \varepsilon, r \to \infty, \quad \|I_{\varepsilon,r}f\|_{q,\beta-\mu} \to 0, \quad \varepsilon, r \to 0.$$

PROOF. By making use of properties I and II of the kernel $k_{\epsilon,r}(y)$, we obtain

$$|I_{\varepsilon,r}f| \leqslant M \int \frac{|f(y)|\,dy}{[\rho(y-x)]^{n-\left(\frac{1}{p}-\frac{1}{q}\right)n-\mu}}$$

whence by applying Lemma 5 we obtain the desired estimate (15).

The second assertion of the lemma follows from the first, which allows us to apply the theorem on interchanging the limit and the Lebesgue integral for functions with integrable majorant.

LEMMA 7. *Let* $\Phi(x) \in C_0^\infty(E_n)$ *be concentrated in the cube* $\{0 < \delta < y_i < 1, i = 1, \cdots, n\}$ *and*

$$\int \Phi(x)\,dx = 0. \tag{16}$$

If $1 < p < \infty$, $\kappa = 1 - s = 0$, $0 \leqslant \epsilon < r < \infty$, $f \in L_{p,\beta}$ *and* $-(n/p) < \beta < n/p'$, *then*

$$\|I_{\varepsilon,r}f\|_{p,\beta} \leqslant c\|f\|_{p,\beta},$$

where c *does not depend on* ϵ, r *and* f, *and* I_{0r}, $I_{\epsilon\infty}$ *are to be understood as limits in the sense of convergence in* $L_{p,\beta}$ *and* $L_{p,\beta}(K)$ *on each compact set* $K \subset E_n$, *respectively.*

The proof of this lemma is presented in [11].

Let $\mathbf{l} = (l_1, \cdots, l_n)$ be a vector with positive integer coordinates ($l_i > 0, i = 1, \cdots, n$). In the following it will be assumed that $\alpha_i = \lambda/l_i$, $i = 1, \cdots, n$, where $\lambda = n/\Sigma_1^n(1/l_i)$.

We define the space $L^{\mathbf{l}}_{p,\beta}$ ($1 < p < \infty$, $\beta < n/p'$) of functions $f(x)$ which have generalized derivatives of order l_i ($i = 1, \cdots, n$) in $E_n \setminus \{0\}$, in the sense of S. L. Sobolev [12], with respect to x_i if $n \geqslant 2$, and in E_1, if $n = 1$, with finite seminorm

$$\|f\|_{L^{\mathbf{l}}_{p,\beta}} = \sum_{i=1}^n \|[\rho(x)]^\beta D_i^{l_i} f(x)\|_p = \sum_{i=1}^n \|D_i^{l_i} f\|_{p,\beta}.$$

If $\beta < n/p'$, then it is easily shown that the functions $f \in L^{\mathbf{l}}_{p,\beta}$ will have generalized derivatives of order l_i ($i = 1, \cdots, n$) in E_n with respect to x_i in the sense of S. L. Sobolev. Hence it follows that for any function $f \in L^{\mathbf{l}}_{p,\beta}$ ($\beta < n/p'$) and for any vector $\mathbf{k} = (k_1, \cdots, k_n)$, k_i a nonnegative integer, we have the Il'in-Besov integral identity [8, 13]

$$D^{\mathbf{k}}f_{\varepsilon,r}(x)=D^{\mathbf{k}}f_{\varepsilon}(x)-D^{\mathbf{k}}f_{r}(x)$$
$$=(-1)^{|\mathbf{k}|}\sum_{i=1}^{n}\int_{\varepsilon}^{r}\int h^{-n-1+\lambda-(\mathbf{k},\alpha)}\Phi_i^{(\mathbf{k})}\left(\frac{y}{h^{\alpha}}\right)D_i^{l_i}f(x+y)\,dy\,dh, \tag{17}$$

where $0<\epsilon<r<\infty$ and

$$D^{\mathbf{k}}f_r(x)=(-1)^{|\mathbf{k}|}r^{-n-(\mathbf{k},\alpha)}\int\Phi_0^{(\mathbf{k})}\left(\frac{y}{r^{\alpha}}\right)f(x+y)\,dy.$$

The functions $\Phi_i(x)\in C_0^{\infty}(E_n)$, $i=0,1,\cdots,n$, are concentrated in the cube $\{0<\delta<y_i<1,\ i=1,\cdots,n\}$ and for $i=1,\cdots,n$ have property (16).

LEMMA 8. *Let $f\in L^1_{p,\beta}$, $-n/p<\beta<n/p'$ and $\kappa=\lambda-(1/p-1/q)n-\mu-(\mathbf{k},\alpha)=\lambda-(1/p-1/q)n-\mu-\lambda\Sigma_1^n k_i/l_i$. Then for $r\to\infty$, $D^{\mathbf{k}}f_r(x)$ and each of the integrals in* (17) *converges*

1) $L_{q,\beta-\mu}$ *for* $\kappa<0$, $1<p\leqslant q<\infty$, $0\leqslant\mu<n/q+\beta$, $\epsilon>0$;

2) *uniformly on any compact set* $K\subset E_n$ *for* $\kappa<0$, $1<p<\infty$, $\mu=n/q+\beta$, $\epsilon>0$;

3) $L_{q,\beta-\mu}$ *for* $\kappa=0$, $1<p\leqslant q<\infty$, $0\leqslant\mu<n/q+\beta$ *if* $p<q$, *and* $0<\mu<(n/q)+\beta$ *if* $p=q$;

4) $L_{p,\beta}(K)$ *for any compact set* $K\subset E_n$ *for* $\kappa=0$, $1<p=q<\infty$, $\mu=0$.

If $k_j\geqslant l_j$ *for some* j, *then*

5) $\|D^{\mathbf{k}}f_r\|_{q,\beta-\mu}\to 0$, $r\to\infty$, $\kappa<0$, $1<p\leqslant q<\infty$, $0\leqslant\mu<n/q+\beta$;

6) $|D^{\mathbf{k}}f_r|\to 0$, $r\to\infty$ *uniformly on any compact set* $K\subset E_n$ *for* $\kappa<0$, $1<p<\infty$, $\mu=n/q+\beta$;

7) $\|D^{\mathbf{k}}f_r\|_{p,\beta}\to 0$, $r\to\infty$, $\kappa=0$, $1<p=q<\infty$, $\mu=0$.

PROOF. Assertions 1)–4) follow from (17) and Lemmas 2, 3, 6 and 7, in which it is necessary to set $s=1-\lambda+(\mathbf{k},\alpha)$.

We set $\mathbf{k}=\mathbf{k}'+\mathbf{m}$, where $\mathbf{k}'=(0,\cdots,0,l_j,0,\cdots,0)$. Then

$$D^{\mathbf{k}}f_r(x)=(-1)^{|\mathbf{m}|}r^{-n-(\mathbf{m},\alpha)}\int\Phi_0^{(\mathbf{m})}\left(\frac{y}{r^{\alpha}}\right)D_j^{l_j}f(x+y)\,dy,$$

and as a consequence of Lemma 1, noting that $(\mathbf{k},\alpha)=(\mathbf{k}',\alpha)+(\mathbf{m},\alpha)=\lambda+(\mathbf{m},\alpha)$, we obtain

$$\|D^{\mathbf{k}}f_r(x)\|_{q,\beta-\mu}\leqslant cr^{-(m,\alpha)-\left(\frac{1}{p}-\frac{1}{q}\right)n-\mu}\|D_j^{l_j}f\|_{p,\beta}=cr^{\varkappa}\|D_j^{l_j}f\|_{p,\beta}\to 0,$$

whence assertion 5) also follows.

Let $x\in K$. Then, proceeding exactly as for (10), we find for case 6)

$$|D^{\mathbf{k}}f_r(x)|\leqslant cr^{\varkappa}\|D_j^{l_j}f\|_{p,\beta}\to 0,\qquad r\to\infty.$$

In the case 7)

$$\|D^{\mathbf{k}}f_r\|_{p,\beta}=\|D_j^{l_j}f_r\|_{p,\beta}=\|(D_j^{l_j}f)_r\|_{p,\beta}\to 0,\qquad r\to\infty,$$

since if $\varphi \in L_{p,\beta}$ $(-n/p<\beta<n/p')$, then $\|\varphi_r\|_{p,\beta} \to 0$ as $r\to\infty$ (see [11]).

We shall write $\mathbf{k}<\mathbf{l}$ if $k_i<l_i$, $i=1,\cdots,n$.

LEMMA 9. *For any function $f\in L^1_{p,\beta}$, $-n/p<\beta<n/p'$, there exists a polynomial*

$$\mathscr{P}_{\mathbf{l}-\mathbf{I}}(x;\ f)=\mathscr{P}_{\mathbf{l}-\mathbf{I},\ p,\ \beta}(x;\ f)=\sum_{\mathbf{m}\leqslant\mathbf{l}-\mathbf{I}} c_{\mathbf{m}}x^{\mathbf{m}}, \tag{18}$$

$$c_{\mathbf{m}}=0 \quad for \quad (\mathbf{m},\ \boldsymbol{\alpha})\leqslant\lambda-\frac{n}{p}-\beta, \tag{19}$$

such that for $r\to\infty$

$$D^{\mathbf{k}}f_r(x)\to D^{\mathbf{k}}\mathscr{P}_{\mathbf{l}-\mathbf{I}}(x;\ f),\quad (\mathbf{k},\ \boldsymbol{\alpha})>\lambda-\frac{n}{p}-\beta, \tag{20}$$

in the sense of conditions 1)–4) *of Lemma* 8.

PROOF. We set

$$c_{\mathbf{m}}=\lim_{r\to\infty}\frac{1}{\mathbf{m}\,!}D^{\mathbf{m}}f_r(0),\qquad (\mathbf{m},\ \boldsymbol{\alpha})>\lambda-\frac{n}{p}-\beta. \tag{21}$$

These limits exist by Lemma 8, part 2). From assertion 6) of the same lemma we have $c_{\mathbf{m}}=0$ for $\mathbf{m}\not\leqslant\mathbf{l}-\mathbf{I}$.

By applying Taylor's formula and property 6) of Lemma 8, it is easily shown that if $(k,\alpha)>\lambda-n/p-\beta$, then $D^kf_r(x)\to D^k\mathcal{P}_{1-I}(x;f)$ as $r\to\infty$ uniformly on any compact set $K\subset E_n$, i.e. (20) is satisfied in the sense of 2) of Lemma 8. Consequently the right-hand side of (20) is also the limit of the left-hand side in the sense of conditions 1), 2) and 4) of Lemma 8.

THEOREM 1. *For any function $f\in L^1_{p,\beta}$, $-n/p<\beta<n/p'$, there exists a polynomial $\mathcal{P}_{1-1}(x;f)$ of the form* (18), (19), (21), *which depends linearly on f, such that for any nonnegative integer vector $\mathbf{k}$ for which $(\mathbf{k},\alpha)>\lambda-n/p-\beta$ we have*

$$D^{\mathbf{k}}[f_\varepsilon(x)-\mathscr{P}_{\mathbf{l}-\mathbf{I}}(x;\ f)]=(-1)^{|\mathbf{k}|}\sum_{i=1}^{n}\int_\varepsilon^\infty\int h^{-n-1+\lambda-(\mathbf{k},\ \boldsymbol{\alpha})}\Phi_i^{(\mathbf{k})}\left(\frac{y}{h^\alpha}\right)D_i^{l_i}f(x+y)\,dy\,dh, \tag{22}$$

where $\epsilon\geqslant 0$ if $\lambda-n/p-\beta<(\mathbf{k},\alpha)\leqslant\lambda$; $\epsilon>0$ if $\lambda<(\mathbf{k},\alpha)$; and

$$D^{\mathbf{k}}f_0(x)=D^{\mathbf{k}}f(x),\quad \lambda-\frac{n}{p}-\beta<(\mathbf{k},\ \boldsymbol{\alpha})\leqslant\lambda.$$

In addition, the outer integral on the right-hand side of (22) *in the case $\epsilon=0$ $(\lambda-n/p-\beta<(\mathbf{k},\alpha)\leqslant\lambda)$ converges to zero in the sense of $L_{q(\mathbf{k}),\beta-\mu(\mathbf{k})}$, where $q=q(\mathbf{k})\in[p,\infty)$ and $\mu=\mu(\mathbf{k})\in[0,\ n/p\ +\beta)$ are given by the condition $(\mathbf{k},\alpha)=\lambda-(1/p\ -\ 1/q)n-\mu$. The outer integral on the right-hand side of* (22) *converges at infinity in the sense of conditions* 1)–4) *of Lemma* 8.

The proof of this theorem consists of applying Lemmas 6–9 to the representation (17).

THEOREM 2. *Let* $f \in L^{\mathbf{l}}_{p,\beta}$, $-n/p<\beta<n/p'$, $1<p\leqslant q<\infty$, $\kappa=\lambda-(1/p-1/q)n-\mu-(\mathbf{k},\alpha)$ *and* $0\leqslant\mu<n/q+\beta$.

a) *If* $\kappa=0$, *then for the polynomial* $\mathcal{P}_{\mathbf{l}-\mathbf{I}}(x;f)$ *in* (18), (19), (21) *we have*

$$\| D^{\mathbf{k}} [f(x) - \mathcal{P}_{\mathbf{l}-\mathbf{I}}(x;\ f)] \|_{q,\ \beta-\mu} \leqslant c \| f \|_{L^{\mathbf{l}}_{p,\ \beta}}, \tag{23}$$

where the constant c *does not depend on* f.

b) *If* $\kappa<0$, *then for any* $\epsilon>0$

$$\| D^{\mathbf{k}} [f_\varepsilon(x) - \mathcal{P}_{\mathbf{l}-\mathbf{I}}(x;\ f)] \|_{q,\ \beta-\mu} \leqslant c\varepsilon^{\varkappa} \| f \|_{L^{\mathbf{l}}_{p,\ \beta}}, \tag{24}$$

where the constant c *does not depend on* f *or* ϵ.

The inequality (23), as well as Theorem 1, was proven for $p=0$ and $\mu=0$ by O. V. Besov [4]. We note that the inequality (23) for $\beta=0,\mu=0,p=q$ was given by Besov in an earlier paper [13], and independently by S. V. Uspenskiĭ [14].

PROOF. In view of (22) it is sufficient to estimate the integrals on the right-hand sides. For $p<q$ and $0\leqslant\mu<n/q+\beta$, or $p=q$ and $0<\mu<n/q+\beta$, (23) follows from Lemma 6. For $p=q$, $\mu=0$ this inequality follows from Lemma 7. The inequality (24) follows from Lemma 2.

The theorem just proved shows that the derivatives defined for the functions $f(x)$ or $f_\epsilon(x)$ close to "infinity" correspond to the derivatives of a certain polynomial $\mathcal{P}_{\mathbf{l}-\mathbf{I}}(x;f)$ in the sense that the left-hand sides of (23) and (24) are finite. In this case we shall simply say that the function $f\in L^{\mathbf{l}}_{p,\beta}$ goes over at infinity into the polynomial $\mathcal{P}_{\mathbf{l}-\mathbf{I}}(x;f)$. If for the function $\varphi\in L^{\mathbf{l}}_{p,\beta}$ the polynomial $\mathcal{P}_{\mathbf{l}-\mathbf{I}}(x;f)\equiv 0$, we shall say that the function φ goes over into zero at infinity.

If $\lambda-n/p-\beta\geqslant 0$, the polynomial $\mathcal{P}_{\mathbf{l}-\mathbf{I}}(x;f)$ does not contain "low order terms" (does not contain powers for which $(\mathbf{m},\alpha)\leqslant\lambda-n/p-\beta$), as is clear from (19). However, the addition of "low order terms" does not influence the inequalities (23) and (24), since they disappear on differentiating the left-hand sides of these inequalities.

Note that for a function $f\in L^{\mathbf{l}}_{p,\beta}$ there exists only a single polynomial $\mathcal{P}_{\mathbf{l}-\mathbf{I}}(x;f)$ without "low order terms", into which it goes over at infinity. Indeed, for any other polynomial $\mathcal{P}^*_{\mathbf{l}-\mathbf{I}}(x;f)=\Sigma_{\mathbf{m}\leqslant\mathbf{l}-\mathbf{I}}c^*_{\mathbf{m}}x^{\mathbf{m}}$, $c^*_{\mathbf{m}}=0$ for $(\mathbf{m},\alpha)\leqslant\lambda-n/p-\beta$, the embedding

$$D^{\mathbf{k}} [\mathcal{P}_{\mathbf{l}-\mathbf{I}}(x;\ f) - \mathcal{P}^*_{\mathbf{l}-\mathbf{I}}(x;\ f)] \in L_{q,\ \beta-\mu},\quad \varkappa \leqslant 0,$$

should be realized, and this is possible only if $c^*_{\mathbf{m}}=c_{\mathbf{m}}$ for all $\mathbf{m}\leqslant\mathbf{l}-\mathbf{I}$.

We shall let $W^{\mathbf{l}}_{p,\beta}$ denote the space of functions with norm

$$\| f \|_{W^{\mathbf{l}}_{p,\ \beta}} = \| f \|_{L_p(1<|x|<2)} + \| f \|_{L^{\mathbf{l}}_{p,\ \beta}}. \tag{25}$$

The space $W^{\mathbf{l}}_{p,\beta}$ is complete.

We shall study the problem of approximating functions of $W^1_{p,\beta}$ by infinitely differentiable finite functions.

Since a finite function $\varphi(x)\in C_0^\infty(E_n)$ goes over into zero at infinity, it is easily shown that any function $f(x)$ which can be approximated to any degree in the norm of $W^1_{p,\beta}$ by functions in $C_0^\infty(E_n)$ also goes over into zero at infinity. We shall prove below that also, conversely, any function $f(x)\in W^1_{p,\beta}$ which goes over into zero at infinity can be approximated to any precision in the norm of $W^1_{p,\beta}$ by functions of $C_0^\infty(E_n)$. Thus it will be established that the family of functions of $W^1_{p,\beta}$ which go over into zero at infinity coincides with the closure in $W^1_{p,\beta}$ of the set of functions belonging to $C_0^\infty(E_n)$.

Below we make use of the following lemma due to O. V. Besov [4], which we reformulate in more convenient terms.

LEMMA 10. *For* $\delta = \lambda - n/p - \beta \geqslant 0, \beta > - n/p$ *for a sufficiently smooth function* $f(x)$ *there exists a projection operator*

$$Q(x;\ f)=\sum_{(\mathbf{m},\,\alpha)\leqslant\delta} c_{\mathbf{m}}x^{\mathbf{m}}$$

such that

$$\| f-Q(x;\ f)\|_{L_p(1<|x|<2)}\leqslant c\sum_{\delta<(\mathbf{k},\,\alpha)<2\lambda}\| D^{\mathbf{k}}f\|_{L_p\left(\frac{1}{2}<|x|<3\right)}.$$

LEMMA 11. *For* $(\mathbf{m},\alpha)\leqslant\lambda - n/p - \beta$ *any monomial* x^m *can be approximated to arbitrary precision in* $W^1_{p,\beta}$ *by functions of* $C_0^\infty(E_n)$.

PROOF. Let $\rho(x)=\sqrt{x_1^{2/\alpha_1}+\cdots+x_n^{2/\alpha_n}}$ and let $\psi(t)$ be an infinitely differentiable function of the variable $t\in E_1$, where $\psi(t)=1$ for $t<½$, and $\psi(t)=0$ for $t>1$. We shall show that as $\eta\to+\infty$ the norm in $W^1_{p,\beta}$ of the difference

$$g_\eta(x)=x^{\mathbf{m}}-x^{\mathbf{m}}\psi\left(\frac{\ln\rho}{\ln\eta}\right)$$

tends to zero. This will prove the lemma.

Since for $\eta>0$ sufficiently large we have $\|g_\eta(x)\|_{L_p(1<|x|<2)}=0$, it is only necessary to estimate the term $\|D_i^{l_i}g_\eta(x)\|_{p,\beta}$ of the norm (25). We note that $D_i^{l_i}x^{\mathbf{m}}=0$.

It is easily shown that

$$\left|\frac{d^s\psi\left(\frac{\ln\rho}{\ln\eta}\right)}{dx_i^s}\right|\leqslant\frac{c}{\rho^{\alpha_i s}\ln\eta},\qquad s=1,\ 2,\ \ldots,\qquad\sqrt{\eta}<\rho<\eta.$$

By using this inequality and the fact that $\ln\eta>\ln\rho$ for $\sqrt{\eta}<\rho<\eta$, we find

$$\left|D_i^{l_i}\left[x^{\mathbf{m}}\psi\left(\frac{\ln \rho}{\ln \eta}\right)\right]\right| = \left|\frac{x^{\mathbf{m}}}{x_i^{m_i}}\sum_{s=0}^{m_i} c_s x_i^{m_i-s}\frac{d^{l_i-s}}{dx_i^{l_i-s}}\psi\left(\frac{\ln \rho}{\ln \eta}\right)\right| \leqslant c\frac{|x^{\mathbf{m}}|}{\ln \rho}\sum_{s=0}^{m_i}\frac{1}{|x_i|^s \rho^{(l_i-s)\alpha_i}}, \quad \sqrt{\eta}<\rho<\eta. \tag{26}$$

By noting that the functions $D_i^{l_i}g_\eta(x)$, $i = 1, \cdots, n$, are equal to zero outside the region $\sqrt{\eta} < \rho < \eta$, applying (26) and going to the generalized spherical coordinates (ρ, ζ) defined by $x_j = \rho^{\alpha_j}\zeta_j$ $(j = 1, \cdots, n)$, we find that as $\eta \to +\infty$

$$\|D_i^{l_i}g_\eta(x)\|_{p,\beta} = \|D_i^{l_i}g_\eta(x)\|_{L_{p,\beta}(\sqrt{\eta}<\rho<\eta)} \leqslant c\sum_{s=0}^{m_i}\left\|\frac{x^{\mathbf{m}}}{x_i^s\rho^{(l_i-s)\alpha_i}\ln\rho}\right\|_{L_{p,\beta}(\rho>\sqrt{\eta})} \leqslant c_1\left\{\int_{\sqrt{\eta}}^{\infty}\frac{d\rho}{\rho^{[\lambda-(\mathbf{m},\alpha)-\beta]p-n+1}(\ln\rho)^p}\right\}^{1/p} \to 0,$$

since $\lambda - (\mathbf{m}, \alpha) - \beta - n/p \geqslant 0$.

THEOREM 3. *Any function* $f \in W^1_{p,\beta}$ *which goes over into zero at infinity can be approximated arbitrarily closely in the norm of* $W^1_{p,\beta}$ *by functions of* $C_0^\infty(E_n)$.

PROOF. Let $f \in W^1_{p,\beta}$ go over into zero at infinity, i.e. $\mathcal{P}_{1-1}(x; f) \equiv 0$.

Since $\|u - u_\epsilon\|_{p,\beta} \to 0$ and $\epsilon \to 0$ for $u \in L_{p,\beta}$ $(1<p<\infty, -n/p<\beta<n/p$ (see [11]), it is easily shown that

$$\|f - f_\epsilon\|_{W^1_{p,\beta}} \to 0, \quad \epsilon \to 0. \tag{27}$$

By making use of (17), noting that if $\mathcal{P}_{1-1}(x; f) \equiv 0$ for $f \in W^1_{p,\beta}$, then also $\mathcal{P}_{1-1}(x; f) \equiv 0$, and applying (23) of Theorem 2 and part 7 of Lemma 8, we find for given $\epsilon > 0$ and $r \to \infty$ that

$$\sum_{\mathbf{k}\in A}\|D^{\mathbf{k}}(f_\epsilon - f_{\epsilon,r})\|_{q(\mathbf{k}),\beta-\mu(\mathbf{k})} = \sum_{\mathbf{k}\in A}\|D^{\mathbf{k}}f_r\|_{q(\mathbf{k}),\beta-\mu(\mathbf{k})} \leqslant c\sum_{i=1}^{n}\|D_i^{l_i}f_r\|_{p,\beta} \to 0,$$

where $A = \{\mathbf{k}: \lambda - n/p - \beta < (\mathbf{k}, \alpha) \leqslant \lambda\}$, $q = q(\mathbf{k}) \in [p, \infty)$ and $\mu = \mu(\mathbf{k}) \in [0, n/p + \beta)$ are determined by the condition

$$(\mathbf{k}, \alpha) = \lambda - \left(\frac{1}{p} - \frac{1}{q}\right)n - \mu.$$

By using inequality (24) of Theorem 2 for $p = q$ and $\mu = 0$, we find that for $(\mathbf{k}, \alpha) > \lambda$

$$\|D^{\mathbf{k}}(f_\epsilon - f_{\epsilon,r})\|_{p,\beta} = \|D^{\mathbf{k}}f_r\|_{p,\beta} \leqslant cr^{\lambda-(\mathbf{k},\alpha)}\|f\|_{L^1_{p,\beta}} \to 0, \quad r \to \infty.$$

Thus the function $f_\epsilon(x)$ can be approximated arbitrarily closely in the seminorm

$$\sum_{\mathbf{k}\in A}\|D^{\mathbf{k}}u\|_{q(\mathbf{k}),\beta-\mu(\mathbf{k})} + \sum_{\lambda<(\mathbf{k},\alpha)<2\lambda}\|D^{\mathbf{k}}u\|_{p,\beta} \tag{28}$$

by the functions

$$f_{\varepsilon, r}(x)=\sum_{i=1}^{n}\int_{\varepsilon}^{r}\int h^{-n-1+\lambda}\Phi_i\left(\frac{y}{h^{\alpha}}\right)D_i^{l_i}f(x+y)\,dydh.$$

Consider the functions $f_{\epsilon,r}^{[R]}(x)$ which differ from $f_{\epsilon,r}(x)$ only by the replacement of $D_i^{l_i}f(x)$ by $\chi(x;R)D_i^{l_i}f(x)$ under the integral sign, where $\chi(x;R)$ is the characteristic function of the sphere $|x|<R$. For given $\epsilon>0, r<\infty$ and $R\to\infty$, we have from Lemmas 6 and 7 (for $\mathbf{k}\in A$) and Lemma 2 (for $(\mathbf{k},\alpha)>\lambda$), in which it is necessary to set $p=q$ and $\mu=0$,

$$\|D^{\mathbf{k}}(f_{\varepsilon, r}-f^{[R]}_{\varepsilon, r})\|_{q(\mathbf{k}),\,\beta-\mu(\mathbf{k})}\leqslant c\|f\|_{L^1_{p,\,\beta}(|x|>R)}\to 0,$$

$$\|D^{\mathbf{k}}(f_{\varepsilon, r}-f^{[R]}_{\varepsilon, r})\|_{p,\,\beta}\leqslant c\varepsilon^{\lambda-(\mathbf{k},\,\alpha)}\|f\|_{L^1_{p,\,\beta}(|x|>R)}\to 0,\qquad(\mathbf{k},\ \alpha)>\lambda.$$

Hence we see that in the seminorm (28)

$$f^{[R]}_{\varepsilon, r}(x)\to f_{\varepsilon, r}(x),\qquad \varepsilon>0,\quad r<\infty,\quad R\to\infty.$$

It is clear from the forgoing calculation that the seminorm (28) of the difference $f_{\epsilon}(x)-f_{\epsilon,r}^{[R]}(x)$ for any $\epsilon>0$ becomes as small as desired with the proper choice of $r<\infty$ and $R<\infty$. By making use of Lemma 10 and the fact that $c'\leqslant\rho(x)<c''$ for $\frac{1}{2}<|x|<3$, we shall have

$$\begin{aligned}
&\|f_{\varepsilon}-f^{[R]}_{\varepsilon, r}-Q(x;\ f_{\varepsilon}-f^{[R]}_{\varepsilon, r})\|_{L_p(1<|x|<2)}\\
&\leqslant c\sum_{\lambda-\frac{n}{p}-\beta<(\mathbf{k},\alpha)<2\lambda}\|D^{\mathbf{k}}(f_{\varepsilon}-f^{[R]}_{\varepsilon, r})\|_{L_p\left(\frac{1}{2}<|x|<3\right)}\\
&\leqslant c_1\Big\{\sum_{\mathbf{k}\in A}\|D^{\mathbf{k}}(f_{\varepsilon}-f^{[R]}_{\varepsilon, r})\|_{\substack{q(\mathbf{k}),\,\beta-\mu(\mathbf{k})\\(1/2<|x|<3)}}+\sum_{\lambda<(\mathbf{k},\alpha)<2\lambda}\|D^{\mathbf{k}}(f_{\varepsilon}-f^{[R]}_{\varepsilon, r})\|_{L_{p,\beta}\left(\frac{1}{2}<|x|<3\right)}\Big\}\\
&\leqslant c_1\Big\{\sum_{\mathbf{k}\in A}\|D^{\mathbf{k}}(f_{\varepsilon}-f^{[R]}_{\varepsilon, r})\|_{q(\mathbf{k}),\,\beta-\mu(\mathbf{k})}+\sum_{\lambda<(\mathbf{k},\,\alpha)<2\lambda}\|D^{\mathbf{k}}(f_{\varepsilon}-f^{[R]}_{\varepsilon, r})\|_{p,\,\beta}\Big\}.
\end{aligned}\tag{29}$$

In view of (27) and (29), for a suitable choice of $\epsilon>0, r<\infty$ and $R<\infty$ the norm

$$\|f-f^{[R]}_{\varepsilon, r}-Q(x;\ f_{\varepsilon}-f^{[R]}_{\varepsilon, r})\|_{W^1_{p,\beta}}$$

can be made arbitrarily small. By the construction of the $f_{\epsilon,r}^{[R]}\in C_0^{\infty}(E_n)$ and on the basis of Lemma 11 the polynomial $Q(x;f_{\epsilon}-f_{\epsilon,r}^{[R]})$ can be approximated with arbitrary precision by functions of $C_0^{\infty}(E_n)$ in the norm of $W^1_{p,\beta}$. Hence we conclude that any function $f\in W^1_{p,\beta}$ which goes over into zero at infinity can be approximated to arbitrary precision in the norm of $W^1_{p,\beta}$ by functions of $C_0^{\infty}(E_n)$.

REMARK. These considerations have been carried out with reference to functions $f(x)$ defined over the entire space E_n. The results obtained are true as well for more the general regions considered in [13] which possess the property of "accessibility from within an infinite horn", thanks to the possibility, proved in [11], of extending a function which is determined in such a region onto the entire space E_n while preserving its differential properties and the corresponding inequalities satisfied by its norm.

BIBLIOGRAPHY

1. S. V. Uspenskiĭ, *Imbedding theorems for weighted classes,* Trudy Mat. Inst. Steklov. **60** (1961), 282–303; English transl., Amer. Math. Soc. Transl. (2) **87** (1970), 121–145. MR **25** #440; **41** #3202.

2. S. L. Sobolev, *Density of functions with compact support in the space* $L_p^m(E_n)$, Dokl. Akad. Nauk SSSR **149** (1963), 40–43 = Soviet Math. Dokl. **4** (1963), 313–316. MR **26** #6758.

3. V. N. Sedov, *On functions reducing at infinity to a polynomial,* Imbedding Theorems and Their Applications, Proc. All-Union Sympos. Imbedding Theorems (Baku, 1966), "Nauka", Moscow, 1970, pp. 204–212. (Russian)

4. O. V. Besov, *Behavior of differentiable functions at infinity and the density of functions with compact support,* Trudy Mat. Inst. Steklov. **105** (1969), 3–14 = Proc. Steklov Inst. Math. **105** (1969), 1–15. MR **42** #6603.

5. V. R. Portnov, *A theorem on the density of functions with compact support in weight classes,* Dokl. Akad. Nauk SSSR **160** (1965), 545–548 = Soviet Math. Dokl. **6** (1965), 592–595. MR **31** #597.

6. T. S. Pigolkina, *The density of functions with compact support in the weight classes,* Mat. Zametki **2** (1967), 53–60 = Math. Notes **2** (1967), 518–522. MR **36** #3121.

7. Ju. S. Nikol'skiĭ, *Boundary values of functions from weight classes,* Dokl. Akad. Nauk SSSR **164** (1965), 503–506 = Soviet Math. Dokl. **6** (1965), 1260–1263. MR **33** #1763.

8. V. P. Il'in, *Integral representations of differentiable functions and their applications to questions of extension of functions of the* $W_p^{(1)}(G)$ *classes,* Sibirsk. Mat. Ž. **8** (1967), 573–586 = Siberian Math. J. **8** (1967), 421–432. MR **36** #5683.

9. O. V. Besov and P. I. Lizorkin, *Singular integral operators and sequences of convolutions in* L_p*-spaces,* Mat. Sb. **73** (**115**) (1967), 65–88 = Math. USSR Sb. **2** (1967), 57–76. MR **36** #1936.

10. G. H. Hardy, J. E. Littlewood and G. Pólya, *Inequalities,* Cambridge Univ. Press, New York, 1934; Russian transl., IL, Moscow, 1948. MR **18**, 722.

11. Ju. S. Nikol'skiĭ, *Extension of weight spaces of differentiable functions,* Sibirsk. Mat. Ž. **12** (1971), 158–170 = Siberian Math. J. **12** (1971), 113–122. MR **44** #4513.

12. S. L. Sobolev, *Applications of functional analysis in mathematical physics,* Izdat. Leningrad. Gos. Univ., Leningrad, 1950; English transl., Transl. Math. Monographs, vol. 7, Amer. Math. Soc., Providence, R. I., 1963. MR **14**, 565; **29** #2624.

13. O. V. Besov, *The continuation of functions of* L_p^1 *and* W_p^1, Trudy Mat. Inst. Steklov. **89** (1967), 5–17 = Proc. Steklov Inst. Math. **89** (1967), 1–15. MR **35** #5920.

14. S. V. Uspenskiĭ, *Differential properties of solutions to quasi-elliptic equations in unbounded domains,* Dokl. Akad. Nauk SSSR **181** (1968), 562–564 = Soviet Math. Dokl. **9** (1968), 900–903. MR **38** #1390.

Translated by M. L. GLASSER

Trudy Mat. Inst. Steklov.
117 (1972)

Proc. Steklov Inst. Math.
117 (1972)

INVESTIGATION OF CERTAIN CLASSES OF FUNCTIONS BY "ANGULAR" APPROXIMATION

UDC517.518.22

M. K. POTAPOV

Abstract. In this article we consider certain classes of functions whose mixed smoothness moduli have various properties. These classes of functions satisfy direct and inverse "angular" approximation theorems. Using these constructive characteristics for our function classes, we obtain several equivalent definitions, as well as imbedding theorems, interpolation theorems, and other theorems.

Bibliography: 11 items.

There has been a recent increase in the study of classes of functions whose mixed smoothness moduli have various properties (see, for example, [1–3]). We shall show that the investigations of such function classes can be conveniently carried out using "angular" approximation.

§1. Definitions, notation, auxiliary lemmas

Suppose we are given a natural number n and a number p in the interval $1 \leqslant p \leqslant \infty$. We shall say that $f \in \mathfrak{L}_p([0, 2\pi]^n)$ if $f = f(x_1, \cdots, x_n)$ is a measurable function of n variables which is 2π-periodic in each variable and which belongs to $\mathfrak{L}_p$.

Henceforth we shall not distinguish between equivalent functions, i.e. functions which differ on a set of n-dimensional measure zero.

LEMMA 1 (SEE [4], THEOREM 3). *Every function $f \in \mathfrak{L}_p([0, 2\pi]^n)$ is representable (in fact, uniquely) in the form*

$$\begin{aligned} f = F_{1\,2\ldots n} &+ \sum_{i_1=1}^{n} F_{1\ldots i_1-1\; i_1+1\ldots n} + \sum_{i_2=2}^{n}\sum_{i_1=1}^{i_2-1} F_{1\ldots i_1-1\; i_1+1\ldots i_2-1\; i_2+1\ldots n} \\ &+ \sum_{i_3=3}^{n}\sum_{i_2=2}^{i_3-1}\sum_{i_1=1}^{i_2-1} F_{1\ldots i_1-1\; i_1+1\ldots i_2-1\; i_2+1\ldots i_3-1\; i_3+1\ldots n} \\ &+ \cdots + \sum_{j=2}^{n}\sum_{i=1}^{j-1} F_{ij} + \sum_{i=1}^{n} F_i + F_0 \equiv \sum_{\substack{1\leqslant i_j\leqslant n \\ 1\leqslant j\leqslant m\leqslant n}} F_{i_1\ldots i_j\ldots i_m} + F_0, \end{aligned} \tag{1}$$

AMS (MOS) subject classifications (1970). Primary 42A08, 46E35; Secondary 42A92.

where F_0 *is a constant, and each function* $F_{i_1 \cdots i_m}$ *only depends on the* m *variables* x_{i_j} *with indices* $i_1, \cdots, i_m$ *and has mean value over all these variables equal to zero, i.e.*

$$\int_0^{2\pi} F_{i_1 \dots i_m} dx_{i_j} = 0, \quad 1 \leqslant j \leqslant m,$$

and where, in addition, $F_{i_1 \cdots i_m} \in \mathfrak{L}_p([0, 2\pi]^m)$. *Here the Fourier series of the function* f *is the sum of the Fourier series of the functions* $F_{i_1 \cdots i_m}$.

The Fourier series of the function $F \in \mathfrak{L}_p([0, 2\pi]^n)$ is usually written in the form

$$\sum_{\nu_1=-\infty}^{\infty} \cdots \sum_{\nu_n=-\infty}^{\infty} c_{\nu_1 \dots \nu_n} e^{i \sum_{j=1}^{n} \nu_j x_j} \equiv \sum_{\nu_1=-\infty}^{\infty} \cdots \sum_{\nu_n=-\infty}^{\infty} B_{\nu_1 \dots \nu_n},$$

where

$$c_{\nu_1 \dots \nu_n} = \frac{1}{(2\pi)^n} \int_0^{2\pi} \cdots \int_0^{2\pi} f(t_1, \dots, t_n) e^{-i \sum_{j=1}^{n} t_j \nu_j} dt_1 \dots dt_n.$$

However, we shall find it more convenient to write this series in the form

$$\sum_{\nu_1=-\infty}^{\infty} \cdots \sum_{\nu_n=-\infty}^{\infty} A_{\nu_1 \dots \nu_n}(x_1, \dots, x_n), \tag{2}$$

where

$$A_{\nu_1 \dots \nu_n} = \sum_{j=1}^{n} (B_{\nu_1, \dots, \nu_{j-1} \nu_j \nu_{j+1} \dots \nu_n} + B_{\nu_1 \dots \nu_{j-1} \nu_j^* \nu_{j+1} \dots \nu_n}), \quad \nu_j^* = -\nu_j.$$

With this notation, the Fourier series of $F_{i_1 \cdots i_m}$ in representation (1) has the form

$$\sum_{\nu_{i_1}=1}^{\infty} \cdots \sum_{\nu_{i_m}=1}^{\infty} A_{\nu_{i_1} \dots \nu_{i_m}} \quad (x_{i_1}, \dots, x_{i_m}), \tag{3}$$

where $A_{\nu_{i_1} \cdots \nu_{i_m}} = A_{l_1 \cdots l_n}$, with $l_{i_j} = \nu_{i_j}$ for $1 \leqslant j \leqslant m$, and $l_{i_j} = 0$ for $m + 1 \leqslant j \leqslant n$.

We shall let $C, C_1, C_2, \cdots$ denote constants, which are generally different in different formulas and which depend on certain fixed parameters; we shall usually indicate what the given constant does and does not depend on.

We let $\beta_i(t)$ designate a function which is nonnegative and continuous on $[0, 1]$ and which satisfies the following conditions:

1. $\beta_i(t_1) \leqslant C_1 \beta_i(t_2)$, if $0 \leqslant t_1 \leqslant t_2 \leqslant 1$,
2. $\beta_i(2t_1) \leqslant C_2 \beta_i(t_1)$, if $0 \leqslant t_1 \leqslant ½$,

where C_1 and C_2 do not depend on t_1 or t_2. Any modulus of continuity is an example of such a function.

We say that the function $\beta_i(t)$ satisfies the τ_i-condition if there exists a real number τ_i such that for all t in the interval $(0, 1)$

$$\int_t^1 \frac{\beta_i(u)}{u^{1+\tau_i}}\,du \leqslant C_3 \frac{\beta_i(t)}{t^{\tau_i}},$$

where C_3 does not depend on t.

We say that the function $\beta_i(t)$ satisfies the ρ_i-condition if there exists a real number ρ_i such that for all t in the interval $(0, 1)$

$$\int_0^t \frac{\beta_i(u)}{u^{1+\rho_i}}\,du \leqslant C_4 \frac{\beta_i(t)}{t^{\rho_i}},$$

where C_4 does not depend on t.

LEMMA 2. *If $\beta_i(t)$ satisfies the ρ_i-condition, then for any $\kappa \leqslant \rho_i$*

$$I_N = \sum_{\nu=N}^{\infty} 2^{\kappa\nu} \beta_i\left(\frac{1}{2^\nu}\right) \leqslant C_5 2^{\kappa N} \beta_i\left(\frac{1}{2^N}\right),$$

$$I = \sum_{\nu=0}^{\infty} (\nu+1)^{\kappa-1} \beta_i\left(\frac{1}{\nu+1}\right) < \infty,$$

where C_5 does not depend on N.

PROOF. Using property 2 of the function $\beta_i(t)$, we have

$$I_N \leqslant C_6 \sum_{\nu=N}^{\infty} 2^{\kappa\nu} \beta_i\left(\frac{1}{2^{\nu+1}}\right) \leqslant C_7 \sum_{\nu=N}^{\infty} \beta_i\left(\frac{1}{2^{\nu+1}}\right) \int_\nu^{\nu+1} 2^{\kappa u}\,du.$$

Using property 1 of the function $\beta_i(t)$, we can write

$$I_N \leqslant C_8 \sum_{\nu=N}^{\infty} \int_\nu^{\nu+1} \beta_i\left(\frac{1}{2^u}\right) 2^{\kappa u}\,du = C_8 \int_N^{\infty} \beta_i\left(\frac{1}{2^u}\right) 2^{\kappa u}\,du$$

$$\leqslant C_9 \int_0^{1/2^N} \frac{\beta_i(t)}{t^{\kappa+1}}\,dt = C_9 \int_0^{1/2^N} \frac{\beta_i(t)}{t^{1+\rho_i}} t^{\rho_i-\kappa}\,dt.$$

Now taking into account that $\rho_i - \kappa \geqslant 0$ and that $\beta_i(t)$ satisfies the β_i-condition, we have successively

$$I_N \leqslant C_9 2^{N(\kappa-\rho_i)} \int_0^{1/2^N} \frac{\beta_i(t)}{t^{1+\rho_i}}\,dt \leqslant C_{10} 2^{N\kappa} \beta_i\left(\frac{1}{2^N}\right),$$

and the first inequality of Lemma 2 is proved.

We now prove the second inequality. Using property 1 of the function $\beta_i(t)$, we can write

$$I \leqslant C_{11} \sum_{\mu=0}^{\infty} \sum_{\nu=2^{\mu}-1}^{2^{\mu+1}} (\nu+1)^{\varkappa-1} \beta_i\left(\frac{1}{\nu+1}\right)$$

$$\leqslant C_{12} \sum_{\mu=0}^{\infty} \beta_i\left(\frac{1}{2^{\mu}}\right) \sum_{\nu=2^{\mu}-1}^{2^{\mu+1}} (\nu+1)^{\varkappa-1} \leqslant C_{13} \sum_{\mu=0}^{\infty} \beta_i\left(\frac{1}{2^{\mu}}\right) 2^{\varkappa\mu}$$

Using the first inequality of Lemma 2, which we have already proved, we obtain $I \leqslant C_{14}\beta_i(1)$. Since the function $\beta_i(t)$ is continuous on $[0, 1]$, we have $\beta_i(1) = C_{15}$. Thus we have found that $I \leqslant C_{16} < \infty$, and Lemma 2 is completely proved.

We shall let $\alpha_i(t)$ designate a function which is measurable on $[0, 2\pi]$, which is summable on $[\delta, 2\pi]$ for any positive $\delta < 2\pi$, and for which there exists a positive constant C_1 not depending on t such that $\alpha_i(t) \geqslant C_1$ for all $t \in [0, 2\pi]$.

We set

$$\mu_i(\nu) = \int_{1/2^{\nu}}^{1/2^{\nu-1}} \alpha_i(t)\, dt, \qquad \nu = 1, 2, \ldots,$$

$$\mu_i(0) = \int_{1}^{2\pi} \alpha_i(t)\, dt.$$

We say that the function $\alpha_i(t)$ satisfies the σ_i-condition if there exists a real number σ_i such that for all positive $\delta \leqslant \pi$

$$\int_{0}^{\delta} \alpha_i(t)\, t^{\sigma_i}\, dt \leqslant C_2 \delta^{\sigma_i} \int_{\delta}^{2\delta} \alpha_i(t)\, dt,$$

where C_2 does not depend on δ.

We say that the function $\alpha_i(t)$ satisfies the λ_i-condition if there exists a real number λ_i such that for all positive $\delta \leqslant \pi$

$$\int_{\delta}^{2\pi} \alpha_i(t)\, t^{\lambda_i} dt \leqslant C_3 \delta^{\lambda_i} \int_{\delta}^{2\delta} \alpha_i(t)\, dt,$$

where C_3 does not depend on δ.

We say that the function $\alpha_i(t)$ satisfies the γ-condition if for all positive $\lambda \in \pi$

$$\int_{\delta}^{2\pi} \frac{\alpha_i(t)}{t}\, dt \leqslant C_4 \alpha_i(\delta),$$

where C_4 does not depend on δ.

LEMMA 3. *If the function $\alpha_i(t)$ satisfies the λ_i-condition, then*

$$2^{n\lambda_i} \leqslant C_5\mu_i(n), \quad \sum_{\nu=1}^{n} \mu_i(\nu)\, 2^{-\nu\lambda_i} \leqslant C_6\mu_i(n)\, 2^{-n\lambda_i}, \qquad n=1,\ 2,\ \ldots,$$

where C_5 *and* C_6 *do not depend on* n.

This lemma was proved in [5].

We now consider a function of two variables $f(x, y) \in \mathfrak{L}_p([0, 2\pi]^2)$ which satisfies the conditions

$$\int_0^{2\pi} f(x,\ y)\, dx = \int_0^{2\pi} f(x,\ y)\, dy = 0.$$

Then its series has the form (3), i.e. it can be written in the form

$$\sum_{\nu_1=1}^{\infty} \sum_{\nu_2=1}^{\infty} A_{\nu_1\nu_2}(x,\ y).$$

We introduce the notation

$$\Delta_{l_1 l_2} = \sum_{\nu_1=[2^{l_1-1}]+1}^{2^{l_1}} \ \sum_{\nu_2=[2^{l_2-1}]+1}^{2^{l_2}} A_{\nu_1\nu_2}(x,\ y),$$

where $[2^{l_i-1}] = 0$ for $l_i = 0$ and $[2^{l_i-1}] = 2^{l_i-1}$ for $l_i = 1, 2, 3, \cdots$.

With this notation we have the following lemmas.

LEMMA 4 (SEE [6], P. 63). a) *Suppose that* $f(x, y) \in \mathfrak{L}_p([0, 2\pi]^2)$ *for some* p *in the interval* $1 < p < \infty$, *and that*

$$\int_0^{2\pi} f(x,\ y)\, dx = \int_0^{2\pi} f(x,\ y)\, dy = 0.$$

Then

$$\|f\|_p \leqslant C_1 \left(\int_0^{2\pi}\int_0^{2\pi} \left[\sum_{l_1=0}^{\infty} \sum_{l_2=0}^{\infty} \Delta_{l_1 l_2}^2 \right]^{p/2} dx dy \right)^{1/p} \leqslant C_2 \|f\|_p,$$

where C_1 *and* C_2 *do not depend on* f.

b) *Suppose that for some* p *in the interval* $1 < p < \infty$ *and for some trigonometric series*

$$\sum_{\nu_1=1}^{\infty} \sum_{\nu_2=1}^{\infty} A_{\nu_1\nu_2}$$

the inequality

$$I_p = \left(\int_0^{2\pi}\int_0^{2\pi} \left[\sum_{l_1=0}^{\infty} \sum_{l_2=0}^{\infty} \Delta_{l_1 l_2}^2 \right]^{p/2} dx dy \right)^{1/p} < \infty$$

holds. Then this series is the Fourier series of some function $\varphi(x, y) \in \mathfrak{L}_p([0, 2\pi]^2)$, *where* $\|\varphi\|_p \leqslant C_3 I_p$, *with* C_3 *not depending on* φ *or* f.

LEMMA 5 (MARCINKIEWICZ; SEE [6], P. 67). *Suppose that* $f(x, y) \in \mathfrak{L}_p([0, 2\pi]^2)$ *for some* p *in the interval* $1 < p < \infty$, *and that*

$$\int_0^{2\pi} f(x, y)\,dx = \int_0^{2\pi} f(x, y)\,dy = 0.$$

Suppose we are given a sequence of numbers $\{\lambda_{\nu_1\nu_2}\}$ *satisfying the conditions* $|\lambda_{\nu_1\nu_2}| \leqslant M$ *and*

$$\sum_{l_1=[2^{\nu_1-1}]}^{2^{\nu_1}-1} \sum_{l_2=[2^{\nu_2-1}]}^{2^{\nu_2}-1} |\lambda_{l_1+1\, l_2+1} - \lambda_{l_1+1\, l_2} - \lambda_{l_1 l_2+1} + \lambda_{l_1 l_2}|$$
$$+ \sum_{l_1=[2^{\nu_1-1}]}^{2^{\nu_1}-1} |\lambda_{l_1+1\, 2^{\nu_2}} - \lambda_{l_1 2^{\nu_2}}| + \sum_{l_2=[2^{\nu_2-1}]}^{2^{\nu_2}-1} |\lambda_{2^{\nu_1} l_2+1} - \lambda_{2^{\nu_1} l_2}| \leqslant M$$

for all $\nu_1, \nu_2 = 0, 1, 2, \cdots$. *Then the trigonometric series*

$$\sum_{\nu_1=1}^{\infty} \sum_{\nu_2=1}^{\infty} \lambda_{\nu_1\nu_2} A_{\nu_1\nu_2}(x, y)$$

is the Fourier series of some function $F(x, y) \in \mathfrak{L}_p([0, 2\pi]^2)$, *where* $\|F\|_p \leqslant C_4 M \|f\|_p$, *with* C_4 *not depending on* f, M *or* F.

§2. "Angular" approximation

The results of this section were the content of my report at the conference on Constructive Function Theory (24 August–3 September 1969, Budapest, Hungary) and were published in [7]. They are included here because they are a fundamental device in proving almost all the theorems in this paper.

Suppose that $T_{l_{i_j}} \in \mathfrak{L}_p([0, 2\pi]^n)$ is a trigonometric polynomial of order l_{i_j} in the variable x_{i_j}, $1 \leqslant i_j \leqslant n$, $1 \leqslant j \leqslant m \leqslant n$.

We call

$$Y_{l_{i_1} \ldots l_{i_m}}(f)_p = \inf_{\substack{T_{l_{i_j}} \\ 1 \leqslant j \leqslant m}} \left\| f - \sum_{j=1}^{m} T_{l_{i_j}} \right\|_p, \quad l = 0, 1, 2, \ldots,$$

the best m-dimensional "angular" approximation of the function f in the variables $x_{i_1}, \cdots, x_{i_m}$.

We let $\Delta_{h_i}^{k_i} f$ denote the k_ith difference with increment h_i with respect to the variable x_i of the function $F(x_1, \cdots, x_n)$, i.e.

$$\Delta_{h_i}^{k_i} F = \sum_{\nu_i=0}^{k_i} (-1)^{k_i-\nu_i} C_{k_i}^{\nu_i} F(x_1, \ldots, x_{i-1}, x_i + \nu_i h_i, x_{i+1}, \ldots, x_n).$$

We shall define the mixed difference as follows:

$$\Delta_{h_{i_1} h_{i_2} \ldots h_{i_j}}^{k_{i_1} k_{i_2} \ldots k_{i_j}} F = \Delta_{h_{i_1}}^{k_{i_1}} \varphi,$$

where

$$\varphi = \Delta_{h_{i_2} \ldots h_{i_j}}^{k_{i_2} \ldots k_{i_j}} F.$$

We let $\omega_{k_{i_1} \cdots k_{i_m}}(f, \delta_{i_1}, \cdots, \delta_{i_m})_p$ designate the mixed m-dimensional $(m \leqslant n)$ smoothness modulus of order k_{i_j} in the variable x_{i_j} of the function $f \in \mathfrak{L}_p([0, 2\pi]^n)$, i.e.

$$\omega_{k_{i_1} \ldots k_{i_m}}(f, \delta_{i_1}, \ldots, \delta_{i_m})_p = \sup_{\substack{|h_{i_j}| \leqslant \delta_{i_j} \\ 1 \leqslant j \leqslant m}} \left\| \Delta_{h_{i_1} \ldots h_{i_m}}^{k_{i_1} \ldots k_{i_m}} f \right\|_p.$$

LEMMA 6. *If* $f \in \mathfrak{L}_p([0, 2\pi]^n)$, *and* k_{i_j} $(1 \leqslant i_j \leqslant n, 1 \leqslant j \leqslant m \leqslant n)$ *are given natural numbers, then*

$$Y_{l_{i_1} \ldots l_{i_m}}(f)_p \leqslant C_1 \omega_{k_{i_1} \ldots k_{i_m}}\left(f, \frac{1}{l_{i_1}+1}, \ldots, \frac{1}{l_{i_m}+1}\right)_p$$
$$\leqslant C_2 \prod_{j=1}^{m} (l_{i_j}+1)^{-k_{i_j}} \sum_{\nu_{i_1}=0}^{l_{i_1}} \cdots \sum_{\nu_{i_m}=0}^{l_{i_m}} \prod_{j=1}^{m} (\nu_{i_j}+1)^{k_{i_j}-1} Y_{\nu_{i_1} \ldots \nu_{i_m}}(f)_p,$$

where $l_{i_j} = 0, 1, 2, \cdots$, *and* C_1 *and* C_2 *do not depend on* f *or* l_{i_j}.

We easily obtain the following two lemmas as corollaries of this lemma.

LEMMA 7. *Suppose given a collection of indices* $i_1, \cdots, i_m$ *such that* $1 \leqslant i_j \leqslant n, 1 \leqslant j \leqslant m \leqslant n$, *a collection of real numbers* r_{i_j} *and* τ_{i_j}, *a collection of natural numbers* k_{i_j} *such that* $k_{i_j} \geqslant r_{i_j} + \tau_{i_j}$, *and a collection of functions* $\beta_{i_j}(t)$ *each of which satisfies the* τ_{i_j}-*condition. Then in order for a function* $f \in \mathfrak{L}_p([0, 2\pi]^n)$ *to have an m-dimensional mixed smoothness modulus of order* k_{i_j} *in the variable* x_{i_j} *satisfying the condition*

$$\omega_{k_{i_1} \ldots k_{i_m}}(f, \delta_{i_1}, \ldots, \delta_{i_m})_p \leqslant C_3 \prod_{j=1}^{m} \delta_{i_j}^{r_{i_j}} \beta_{i_j}(\delta_{i_j}),$$

it is necessary and sufficient that the m-dimensional "angular" approximations of this function in the variables x_{i_j} *satisfy the following inequality for any* $l_{i_j} = 0, 1, 2, \cdots$:

$$Y_{l_{i_1}\dots l_{i_m}}(f)_p \leqslant C_4 \prod_{j=1}^{m} (l_{i_j}+1)^{-r_{i_j}} \beta_{i_j}\left(\frac{1}{l_{i_j}+1}\right),$$

where C_3 *and* C_4 *do not depend on* f, δ_{i_j} *or* l_{i_j}.

LEMMA 8. *Suppose that* $f \in \mathfrak{L}_p([0, 2\pi]^n)$ *and* θ *is some number in the interval* $1 \leqslant \theta < \infty$; *suppose given a collection of indices* $i_1, \cdots, i_m$ *such that* $1 \leqslant i_j \leqslant n$, $1 \leqslant j \leqslant m \leqslant n$, *a collection of real numbers* σ_{i_j}, *a collection of natural numbers* k_{i_j} *such that* $k_{i_j} \geqslant \sigma_{i_j}/\theta$, *and a collection of functions* $\alpha_{i_j}(t)$ *each of which satisfies the* σ_{i_j}*-condition. Then, in order for the function* f *to have m-dimensional mixed smoothness modulus of order* k_{i_j} *in the variable* x_{i_j} *satisfying the condition*

$$\int_0^{2\pi} \cdots \int_0^{2\pi} \omega^{\theta}_{k_{i_1}\dots k_{i_m}}(f, t_{i_1}, \dots, t_{i_m})_p \prod_{j=1}^{m} \alpha_{i_j}(t_{i_j})\, dt_{i_1} \dots dt_{i_m} < \infty,$$

it is necessary and sufficient that the m-dimensional "angular" approximations of this function in the same variables x_{i_j} $(1 \leqslant j \leqslant m)$ *satisfy the inequality*

$$\sum_{\nu_{i_1}=0}^{\infty} \cdots \sum_{\nu_{i_m}=0}^{\infty} \prod_{j=1}^{m} \mu_{i_j}(\nu_{i_j})\, Y^{\theta}_{[2^{\nu_{i_1}}-1]\dots[2^{\nu_{i_m}}-1]}(f)_p < \infty.$$

We shall later need the following lemmas, which are also proved in [7].

LEMMA 9. *Suppose that* $f \in \mathfrak{L}_p([0, 2\pi]^n)$, $\int_0^{2\pi} f(x_1, \cdots, x_n)dx_i = 0$ *for* $i = 1, \cdots, n$; *suppose given a collection of real numbers* r_i *and* τ_i, *a collection of natural numbers* k_i *such that* $k_i \geqslant r_i + \tau_i$, *and a collection of functions* $\beta_i(t)$ *each of which satisfies the* τ_i*-condition. Then, if the n-dimensional mixed smoothness modulus of the function* f *of order* k_i *in the variable* x_i *satisfies the inequality*

$$\omega_{k_1\dots k_n}(f, \delta_1, \dots, \delta_n)_p \leqslant C_5 \prod_{i=1}^{n} \delta_i^{r_i} \beta_i(\delta_i),$$

it follows that, for any natural numbers s *and* i_j *such that* $1 \leqslant i_j \leqslant n$, $1 \leqslant j \leqslant s \leqslant n$, *we have the inequality*

$$\omega_{k_{i_1}\dots k_{i_s}}(f, \delta_{i_1}, \dots, \delta_{i_s})_p \leqslant C_6 \prod_{j=1}^{s} \delta_{i_j}^{r_{i_j}} \beta_{i_j}(\delta_{i_j}),$$

where C_5 *and* C_6 *do not depend on* f *or* δ_i.

LEMMA 10. *Suppose that* $f \in \mathfrak{L}_p([0, 2\pi]^n)$, $\int_0^{2\pi} f(x_1, \cdots, x_n)dx_i = 0$ *for* $i = 1, \cdots, n$, *and* θ *is some number in the interval* $1 \leqslant \theta < \infty$; *suppose given a collection of real numbers* σ_i, *a collection of natural numbers* k_i *such that* $k_i \geqslant \sigma_i/\theta$, *and a collection of functions* $\alpha_i(t)$ *each of which satisfies the* σ_i*-condition. Then, if the n-dimensional mixed smoothness modulus of the function* f *of order* k_i

in the variable x_i *satisfies the inequality*

$$\int_0^{2\pi} \cdots \int_0^{2\pi} \omega^{\theta}_{k_1 \dots k_n}(f, t_1, \dots, t_n)_p \prod_{i=1}^{n} \alpha_i(t_i)\, dt_1 \dots dt_n < \infty,$$

it follows that, for any natural numbers s *and* i_j *such that* $1 \leqslant i_j \leqslant n$, $1 \leqslant j \leqslant s \leqslant n$, *we have the inequality*

$$\int_0^{2\pi} \cdots \int_0^{2\pi} \omega^{\theta}_{k_{i_1} \dots k_{i_s}}(f, t_{i_1}, \dots, t_{i_s})_p \prod_{j=1}^{s} \alpha_{i_j}(t_{i_j})\, dt_{i_1} \dots dt_{i_s} < \infty.$$

LEMMA 11. *Suppose that* $f \in \mathfrak{L}_p([0, 2\pi]^n)$, $\int_0^{2\pi} f(x_1, \cdots, x_n)dx_i = 0$ *for* $i = 1, \cdots, n$, $1 \leqslant p \leqslant q \leqslant \infty$ *and the* l_i *are fixed nonnegative integers. Then, if the n-dimensional "angular" approximations of the function* f *satisfy the inequality*

$$\sum_{\nu_1=0}^{\infty} \cdots \sum_{\nu_n=0}^{\infty} \prod_{i=1}^{n} (\nu_i + 1)^{l_i+1/p-1/q-1} Y_{\nu_1 \dots \nu_n}(f)_p < \infty,$$

it follows that $f(x_1, \cdots, x_n)$ *has a generalized mixed derivative*

$$f^{l_1 \dots l_n} = \frac{\partial^{l_1 + \dots + l_n}}{\partial x_1^{l_1} \dots \partial x_n^{l_1}} f$$

of order l_i *in the variable* x_i, $1 \leqslant i \leqslant n$, *which belongs to* $\mathfrak{L}_q([0, 2\pi]^n)$. *In addition, the n-dimensional "angular" approximations of the function* $f^{l_1 \cdots l_n}$ *satisfy the inequality*

$$Y_{2^{N_1-1} \dots 2^{N_n-1}}\left(f^{l_1 \dots l_n}\right)_q \leqslant C_7 \sum_{\nu_1=N_1}^{\infty} \cdots \sum_{\nu_n=N_n}^{\infty} \prod_{i=1}^{n} 2^{\nu_i(l_i+1/p-1/q)} Y_{[2^{\nu_1-1}] \dots [2^{\nu_n-1}]}(f)_p,$$

where C_7 *does not depend on* f *or* N_i.

Finally, following Nikol'skiĭ (see [6], §6.4), we define the trace of a function $f \in \mathfrak{L}_p([0, 2\pi]^n)$ on the subspace R_{n-s}.

We first introduce the following notation. Suppose we are given a fixed collection of indices $i_1, \cdots, i_s$ such that $1 \leqslant i_j \leqslant n$, $1 \leqslant j \leqslant s < n$. We define the subspace $R_{n-s} \subset R_n$ as follows: $R_{n-s} = \{0 \leqslant x_{i_j} \leqslant 2\pi,\ s + 1 \leqslant j \leqslant n\}$. We write $f \in \mathfrak{L}_p(R_{n-s}, x^0_{i_j}, 1 \leqslant j \leqslant s)$ if for fixed $x^0_{i_j}$, $1 \leqslant j \leqslant s$, the function f as a function of the $n - s$ variables x_{i_j}, $s + 1 \leqslant j \leqslant n$, belongs to $\mathfrak{L}_p([0, 2\pi]^{n-s})$. If $x^0_{i_j} = 0$ for $1 \leqslant j \leqslant s$, then we write $f = f(0, \cdots, 0, x_{i_{s+1}}, \cdots, x_{i_n})$.

We say that the function $\varphi = \varphi(x_{i_{s+1}}, \cdots, x_{i_n})$ is the trace of the function $f \in \mathfrak{L}_p([0, 2\pi]^n)$ on R_{n-s} if f can be modified on a set of n-dimensional measure zero so that the following conditions then hold:

1. $f \in \mathfrak{L}_p(R_{n-s}, x^0_{i_j}, 1 \leqslant j \leqslant s)$ for any fixed x_{i_j}, $1 \leqslant j \leqslant s$, such that $\sum_{j=1}^s x_{i_j}^2 < \delta$, where δ is sufficiently small;

2. $f(0, \cdots, 0, x_{i_{s+1}}, \cdots, x_{i_n}) = \varphi(x_{i_{s+1}}, \cdots, x_{i_n})$;

3. $\|f - \varphi\|_{\mathfrak{L}_p(R_{n-s}, x^0_{i_j}, 1 \leqslant j \leqslant s)} \to 0$ for $\sum_{j=1}^s x_{i_j}^2 \to 0$.

Nikol'skiĭ showed (see [6], §6.4) that the trace of the function f on the subspace R_{n-s} defined as above is unique to within R_{n-s}-equivalence.

LEMMA 12. *Suppose that* $f \in \mathfrak{L}_p([0, 2\pi]^n)$, $\int_0^{2\pi} f(x_1, \cdots, x_n)\, dx_i = 0$ *for* $i = 1, \cdots, n$; *suppose that* $i_1, \cdots, i_s$ *is a fixed collection of indices such that* $1 \leqslant i_j \leqslant n$, $1 \leqslant j \leqslant s$. *Then, if the n-dimensional "angular" approximations of the function* f *satisfy the inequality*

$$I = \sum_{\nu_1=0}^{\infty} \cdots \sum_{\nu_n=0}^{\infty} \prod_{j=1}^{s} (\nu_{i_j} + 1)^{1/p-1} \prod_{j=s+1}^{n} (\nu_{i_j} + 1)^{-1} Y_{\nu_1 \ldots \nu_n}(f)_p < \infty,$$

it follows that $f(x_1, \cdots, x_n)$ *has a trace* $\varphi = \varphi(x_{i_{s+1}}, \cdots, x_{i_n})$ *on* R_{n-s} *belonging to* $\mathfrak{L}_p([0, 2\pi]^{n-s})$. *Here the* $(n-s)$*-dimensional "angular" approximations of the function* φ *satisfy the inequality*

$$Y_{2^{N_{s+1}}-1 \ldots 2^{N_n}-1}(\varphi)_{\mathscr{L}_p([0, 2\pi]^{n-s})}$$

$$\leqslant C_8 \sum_{\nu_{i_1}=0}^{\infty} \cdots \sum_{\nu_{i_s}=0}^{\infty} \sum_{\nu_{i_{s+1}}=N_{i_{s+1}}}^{\infty} \cdots \sum_{\nu_{i_n}=N_{i_n}}^{\infty} \prod_{j=1}^{s} 2^{\nu_{i_j}/p}\, Y_{[2^{\nu_1}-1] \ldots [2^{\nu_n}-1]}(f)_{\mathscr{L}_p([0, 2\pi]^n)},$$

where C_8 *does not depend on* φ, f *or* N_i.

We conclude §2 with a lemma for a function of two variables, which was also proved in [7].

We let $D_l(u)$ denote the Dirichlet kernel, i.e.

$$D_l(u) = \frac{\sin\left(l + \frac{1}{2}\right)u}{2 \sin \frac{u}{2}},$$

we let $S_{l_1 \infty}(f)$ denote the l_1th order partial sum with respect to x of the Fourier series of $f(x, y)$, i.e.

$$S_{l_1 \infty}(f) = \frac{1}{\pi} \int_{-\pi}^{\pi} f(x + t, y)\, D_{l_1}(t)\, dt,$$

and we let $S_{\infty l_2}(f)$ denote the l_2th order partial sum with respect to y of the Fourier series of $f(x, y)$, i.e.

$$S_{\infty l_2}(f) = \frac{1}{\pi} \int_{-\pi}^{\pi} f(x, y+u) D_{l_2}(u)\, du.$$

Finally, we let $S_{l_1 l_2}(f)$ denote the partial sum of order l_1 in x and of order l_2 in y of the Fourier series of $f(x, y)$, i.e.

$$S_{l_1 l_2}(f) = \frac{1}{\pi^2} \int_{-\pi}^{\pi} \int_{-\pi}^{\pi} f(x+t, y+u) D_{l_1}(t) D_{l_2}(u)\, dt du.$$

LEMMA 13. *Suppose that* $f(x, y) \in \mathfrak{L}_p([0, 2\pi]^2)$ *for some* p *in the interval* $1 \leqslant p < \infty$, *and*

$$u^*_{l_1 l_2}(f) = S_{l_1 \infty}(f) + S_{\infty l_2}(f) - S_{l_1 l_2}(f).$$

Then

$$\| f - u^*_{l_1 l_2}(f) \|_p \leqslant C_9 Y_{l_1 l_2}(f)_p,$$

where C_9 *depends only on* p *and not on* f, l_1 *or* l_2.

§3. The function class $S^r_p H(\beta, \tau, n)$

Suppose that we are given a natural number n, a number p in the interval $1 \leqslant p \leqslant \infty$, the vectors $r = \{r_1, \cdots, r_n\}$, $\tau = \{\tau_1, \cdots, \tau_n\}$ and $\beta = \{\beta_1, \cdots, \beta_n\}$, where r_i and τ_i are real numbers, and the function $\beta_i = \beta_i(t)$ is defined in §1.

We say that $f \in S^r_p H(\beta, \tau, n)$, if

1) $f \in \mathfrak{L}_p([0, 2\pi]^n)$;

2) each function $\beta_i(t)$ satisfies the τ_i-condition;

3) for all natural numbers m and i_j such that $1 \leqslant i_j \leqslant n$, $1 \leqslant j \leqslant m \leqslant n$,

we have the inequalities

$$\omega_{k_{i_1} \ldots k_{i_m}}(f, \delta_{i_1}, \ldots, \delta_{i_m})_p \leqslant C \prod_{j=1}^{m} \delta_{i_j}^{r_{i_j}} \beta_{i_j}(\delta_{i_j}), \tag{4}$$

where the k_{i_j} are fixed natural numbers satisfying the condition $k_{i_j} \geqslant r_{i_j} + \tau_{i_j}$, and C is a constant not depending on f or δ_i.

We make several remarks concerning this definition.

1. It is obvious that this function class depends *formally* on the selection of the k_{i_j}. However, applying Lemma 7, we easily show that the choice of the k_{i_j}

$(k_{i_j} \geqslant r_{i_j} + \tau_{i_j})$ effects only the constant in inequality (4). Since we shall not be interested in the constants, we shall assume that these function classes coincide for any k_{ν_i} and k_{μ_i} satisfying the conditions $k_{\nu_i} \geqslant r_i + \tau_i$ and $k_{\mu_i} \geqslant r_i + \tau_i$.

2. The condition $k_i \geqslant r_i + \tau_i$ was introduced for two reasons. In the first place, this condition removes from consideration the case $\beta_i \equiv 1, k_{i_j} = r_{i_j}$, which corresponds to the class of functions having derivatives $f^{r_{i_1} \cdots r_{i_m}}$, about which we only know that they are bounded in the $\mathfrak{L}_p$-metric. In the second place, this condition is introduced for the other function classes so that the classes $S_p^r H(\beta, \tau, n)$ are "meaningful." For example, if we take $k_1 < r_1 + \tau_1$ for $n = 1, \beta(t) \equiv 1, p = \infty$ $(\mathfrak{L}_\infty \equiv C)$, $r_1 \geqslant 1$, i.e. if we take $k_1 = 1$, then the function class $S_p^r H(\beta, \tau, n)$ is the class of functions satisfying the condition

$$|f(x+h) - f(x)| \leqslant C |h|^{r_1},$$

and for $r_1 = 1$ this is the class of functions having a bounded derivative, while for $r_1 > 1$ this function class consists only of constants.

We now give a constructive characterization of the function class $S_p^r H(\beta, \tau, n)$. As follows from Lemma 7, we have

THEOREM 1. *In order that $f \in S_p^r H(\beta, \tau, n)$, it is necessary and sufficient that for all natural numbers m and l_j such that $1 \leqslant i_j \leqslant n, 1 \leqslant j \leqslant m \leqslant n$, the following inequalities hold*:

$$Y_{l_{i_1} \ldots l_{i_m}}(f)_p \leqslant C \prod_{j=1}^{m} (l_{i_j} + 1)^{-r_{i_j}} \beta_{i_j}\left(\frac{1}{l_{i_j}+1}\right),$$

where $l_{i_j} = 0, 1, 2, \cdots$, and C does not depend on f or l_{i_j}.

We introduce one further function class. We say that $f \in \Sigma S_p^r H^0(\beta, \tau, n)$, if

1) $f \in \mathfrak{L}_p([0, 2\pi]^n)$;

2) each function $\beta_i(t)$ satisfies the τ_i-condition;

3) each function $F_{i_1 \cdots i_m}$ in representation (1) (see §1, Lemma 1) satisfies the condition

$$\omega_{k_{i_1} \ldots k_{i_m}}(F_{i_1 \ldots i_m}, \delta_{i_1}, \ldots, \delta_{i_m})_p \leqslant C \prod_{j=1}^{m} \delta_{i_j}^{r_{i_j}} \beta_{i_j}(\delta_{i_j}), \tag{5}$$

where the k_{i_j} are fixed natural numbers such that $k_{i_j} \geqslant r_{i_j} + \tau_{i_j}$, and C is a constant which does not depend on $F_{i_1 \cdots i_m}$, f or δ_i.

Here we also easily show that the selection of the k_{i_j} only affects the constant inequality (5). We shall hence assume that these function classes coincide for any k_{ν_i} and k_{μ_i} such that $k_{\nu_i} \geqslant r_i + \tau_i$ and $k_{\mu_i} \geqslant r_i + \tau_i$.

Later in this section we shall consider functions depending on different numbers of variables. However, despite this, we shall always take the norm of a function over all n variables:

$$\|F\|_p=\left(\int_0^{2\pi}\cdots\int_0^{2\pi}|F|^p\,dx_1\ldots dx_n\right)^{1/p}$$

It is clear here that in the cases when F depends on a smaller number of variables we have

$$\|F\|_p=(2\pi)^{\frac{n-m}{p}}\left(\int_0^{2\pi}\cdots\int_0^{2\pi}|F|^p\,dx_{i_1}\ldots dx_{i_m}\right)^{1/p}.$$

THEOREM 2. *The function classes $S_p^rH(\beta,\tau,n)$ and $\Sigma S_p^rH^0(\beta,\tau,n)$ coincide.*

PROOF. If $f\in\Sigma S_p^rH^0(\beta,\tau,n)$, we show that $f\in S_p^rH(\beta,\tau,n)$. For any fixed choice of indices $i_1,\cdots,i_s$ $(1\leqslant s\leqslant n)$ we choose natural numbers k_{i_j} so that $k_{i_j}\geqslant r_{i_j}+\tau_{i_j}$. Taking into account representation (1), we find that

$$\omega_{k_{i_1}\ldots k_{i_s}}(f,\delta_{i_1},\ldots,\delta_{i_s})_p\equiv\omega_{k_{i_1}\ldots k_{i_s}}(\Phi_{i_1\ldots i_s};\delta_{i_1},\ldots,\delta_{i_s})_p,$$

where $\Phi_{i_1\cdots i_s}$ is the sum of all functions in representation (1) which contain all the x_{i_j} for the indices chosen. It is clear that almost all of these functions also depend on other variables. Hence the smoothness modulus of the function f does not exceed the sum of the smoothness moduli of the functions $F_{i_1\cdots i_s i_{s+1}\cdots i_m}$, where $1\leqslant s\leqslant m\leqslant n$. Since each function $F_{i_1\cdots i_m}$ satisfies inequality (5) and $\int_0^{2\pi}F_{i_1\cdots i_m}dx_{i_j}=0$ for $1\leqslant j\leqslant m$, it follows on applying Lemma 9 that

$$\omega_{k_{i_1}\ldots k_{i_s}}(F_{i_1\ldots i_s i_{s+1}\ldots i_m},\delta_{i_1},\ldots,\delta_{i_s})\leqslant C_1\prod_{j=1}^{s}\delta_{i_j}^{r_{i_j}}\beta_{i_j}(\delta_{i_j})$$

where C_1 does not depend on F or δ_i. But this means that

$$\omega_{k_{i_1}\ldots k_{i_s}}(f,\delta_{i_1},\ldots,\delta_{i_s})_p\leqslant C_2\prod_{j=1}^{s}\delta_{i_j}^{r_{i_j}}\beta_{i_j}(\delta_{i_j}),$$

where C_2 does not depend on f or δ_{i_j}. Since the last inequality was proved for any collection of indices $i_1,\cdots,i_s$, $1\leqslant s\leqslant n$, it follows that $f\in S_p^rH(\beta,\tau,n)$. In other words, we have proved the inclusion

$$\sum S_p^rH^0(\beta,\tau,n)\subset S_p^rH(\beta,\tau,n).$$

It remains to prove the opposite inclusion.

Now if $f\in S_p^rH(\beta,\tau,n)$, we show that $f\in\Sigma S_p^rH^0(\beta,\tau,n)$. For simplicity, set $n=3$. Then by Lemma 1 this function has the representation

$$f = f(x_1, x_2, x_3) = F_{123}(x_1, x_2, x_3) + F_{12}(x_1, x_2) + F_{13}(x_1, x_3) + F_{23}(x_2, x_3) + F_1(x_1) + F_2(x_2) + F_3(x_3) + F_0.$$

It is first of all clear that

$$\omega_{k_1k_2k_3}(F_{123}, \delta_1, \delta_2, \delta_3)_p \equiv \omega_{k_1k_2k_3}(f, \delta_1, \delta_2, \delta_3)_p.$$

Hence, taking into account that $f \in S_p^r H(\beta, \tau, n)$, we obtain

$$\omega_{k_1k_2k_3}(F_{123}, \delta_1, \delta_2, \delta_3)_p \leqslant C_3 \prod_{j=1}^{3} \delta_j^{r_j} \beta_j(\delta_j), \tag{6}$$

where C_3 does not depend on F_{123}, f or δ_j.

It is also easy to see that

$$\omega_{k_1k_2}(F_{12}, \delta_1, \delta_2)_p \equiv \omega_{k_1k_2}(f - F_{123}, \delta_1, \delta_2)_p \leqslant \omega_{k_1k_2}(f, \delta_1, \delta_2)_p + \omega_{k_1k_2}(F_{123}, \delta_1, \delta_2)_p.$$

Since $f \in S_p^r H(\beta, \tau, n)$, we have

$$\omega_{k_1k_2}(f, \delta_1, \delta_2)_p \leqslant C_4 \delta_1^{r_1} \delta_2^{r_2} \beta_1(\delta_1) \beta_2(\delta_2),$$

where C_4 does not depend on f or δ_j.

Since F_{123} satisfies inequality (6) and $\int_0^{2\pi} F_{123}\, dx_i = 0$ for $i = 1, 2, 3$, it follows by Lemma 9 that

$$\omega_{k_1k_2}(F_{123}, \delta_1, \delta_2)_p \leqslant C_5 \delta_1^{r_1} \delta_2^{r_2} \beta_1(\delta_1) \beta_2(\delta_2),$$

where C_5 does not depend on F_{123}, f or δ_i.

Adding the last two inequalities, we obtain

$$\omega_{k_1k_2}(F_{12}, \delta_1, \delta_2)_p \leqslant C_6 \delta_1^{r_1} \delta_2^{r_2} \beta_1(\delta_1) \beta_2(\delta_2), \tag{7}$$

where C_6 does not depend on F_{123}, F_{12}, f or δ_i.

It is similarly shown that

$$\left.\begin{aligned} \omega_{k_2k_3}(F_{23}, \delta_2, \delta_3)_p &\leqslant C_7 \delta_2^{r_2} \delta_3^{r_3} \beta_2(\delta_2) \beta_3(\delta_3), \\ \omega_{k_1k_3}(F_{13}, \delta_1, \delta_3)_p &\leqslant C_8 \delta_1^{r_1} \delta_3^{r_3} \beta_1(\delta_1) \beta_3(\delta_3), \end{aligned}\right\} \tag{8}$$

where C_7 and C_8 do not depend on F_{123}, F_{23}, F_{13}, f or δ_i.

Finally, we consider

$$\omega_{k_1}(F_1, \delta_1)_p \equiv \omega_{k_1}(f - F_{123} - F_{12} - F_{13}, \delta_1)_p \leqslant \omega_{k_1}(f, \delta_1)_p + \omega_{k_1}(F_{123}, \delta_1)_p + \omega_{k_1}(F_{12}, \delta_1)_p + \omega_{k_1}(F_{13}, \delta_1)_p.$$

Since $f \in S_p^r H(\beta, \tau, 3)$, we have

$$\omega_{k_1}(f_1, \delta_1)_p \leqslant C_9 \delta_1^{r_1} \beta_1^{r_1}(\delta_1),$$

where C_9 does not depend on f or δ_1.

Since F_{12} satisfies inequality (7) and $\int_0^{2\pi} F_{12}\,dx_l = 0$ for $l = 1, 2$, it follows by Lemma 9 that

$$\omega_{k_1}(F_{12}, \delta_1)_p \leqslant C_{10}\delta_1^{r_1}\beta_1(\delta_1),$$

where C_{10} does not depend on F_{12}, f or δ_1.

Analogous estimates also hold for $\omega_{k_1}(F_{13}, \delta_1)_p$ and $\omega_{k_1}(F_{123}, \delta_1)_p$. It follows from all these estimates that

$$\omega_{k_1}(F_1, \delta_1)_p \leqslant C_{11}\delta_1^{r_1}\beta_1(\delta_1), \tag{9}$$

where C_{11} does not depend on $F_1, F_{12}, F_{13}, F_{123}, f$ and δ_1. It is similarly shown that

$$\left.\begin{aligned} \omega_{k_2}(F_2, \delta_2)_p &\leqslant C_{12}\delta_2^{r_2}\beta_2(\delta_2), \\ \omega_{k_3}(F_3, \delta_3)_p &\leqslant C_{13}\delta_3^{r_3}\beta_3(\delta_3), \end{aligned}\right\} \tag{10}$$

where C_{12} and C_{13} do not depend on $F_2, F_3, F_{12}, F_{13}, F_{23}, F_{123}, f$ and δ_i. But inequalities (6)–(10) mean that $f \in \Sigma S_p^r H^0(\beta, \tau, n)$. In other words, we have proved the inclusion

$$S_p^r H(\beta, \tau, n) \subset \Sigma S_p^r H^0(\beta, \tau, n),$$

and Theorem 2 is thereby proved.

Now suppose we are given a natural number n, a number p in the interval $1 \leqslant p \leqslant \infty$, and vectors

$$r = \{r_1, \ldots, r_n\}, \quad \tau = \{\tau_1, \ldots, \tau_n\}, \quad l = \{l_1, \ldots, l_n\} \text{ and } \beta = \{\beta_1, \ldots, \beta_n\},$$

where r_i and τ_i are real numbers, l_i are nonnegative integers, and $\beta_i = \beta_i(t)$ are the functions defined in §1.

We say that $f \in S_p^r W^l H(\beta, \tau, n)$, if

1) $f \in L_p([0, 2\pi]^n)$;

2) each function $\beta_i(t)$ satisfies the τ_i-condition;

3) for all natural numbers m and i_j such that $1 \leqslant i_j \leqslant n$, $1 \leqslant j \leqslant m \leqslant n$, the function f has generalized partial derivatives

$$f^{l_m} = \frac{\partial^{l_{i_1} + \cdots + l_{i_m}}}{\partial x_{i_1}^{l_{i_1}} \cdots \partial x_{i_m}^{l_{i_m}}} f$$

of order l_{i_j} in the variable x_{i_j};

4) for each such derivative we have the inequality

$$\omega_{k_{i_1} \ldots k_{i_m}}(f^{l_m}, \delta_{i_1}, \ldots, \delta_{i_m})_p \leqslant C \prod_{j=1}^m \delta_{i_j}^{r_{i_j}} \beta_{i_j}(\delta_{i_j}),$$

where the k_{i_j} are certain fixed natural numbers satisfying the condition $k_{i_j} \geqslant r_{i_j} + \tau_{i_j}$, and C is a constant not depending on f or δ_i.

THEOREM 3. *If each of the functions* $\beta_i(t)$ *satisfies the* ρ_i*-condition and* $\rho_i + r_i \geqslant 0$, *then for any vector* $l = \{l_1, \cdots, l_n\}$ *such that* l_i *are nonnegative integers satisfying the conditions* $l_i \leqslant \rho_i + r_i$ *the function classes* $S_p^r H(\beta, \tau, n)$ *and* $S_p^{r-l} W^l H(\beta, \tau, n)$ *coincide.*

PROOF. If all the $l_i = 0$, $1 \leqslant i \leqslant n$, then Theorem 3 has no content (although it is true). Hence we shall assume that $l_i > 0$ for at least one i, $1 \leqslant i \leqslant n$.

Now suppose that for some vector l the function $f \in S_p^{r-l} W^l H(\beta, \tau, n)$. Then, for any collection of i_j such that $1 \leqslant i_j \leqslant n$, $1 \leqslant j \leqslant m \leqslant n$, and any collection of k_{i_j} such that $k_{i_j} \geqslant r_{i_j} + \tau_{i_j}$, we may write that

$$\omega_{k_{i_1}\ldots k_{i_m}}(f, \delta_{i_1}, \ldots, \delta_{i_m})_p \leqslant C_1 \prod_{j=1}^{m} \delta_{i_j}^{l_{i_j}} \omega_{k_{i_1}-l_{i_1}\ldots k_{i_m}-l_{i_m}}(f^{l_m}, \delta_{i_1}, \ldots, \delta_{i_m})_p$$

$$\leqslant C_2 \prod_{j=1}^{m} \delta_{i_j}^{l_{i_j}} \prod_{j=1}^{m} \delta_{i_j}^{r_{i_j}-l_{i_j}} \beta_{i_j}(\delta_{i_j}) = C_2 \prod_{j=1}^{m} \delta_{i_j}^{r_{i_j}} \beta_{i_j}(\delta_{i_j}),$$

where C_1 and C_2 do not depend on f or δ_i. But this means that $f \in S_p^r H(\beta, \tau, n)$. In other words, we have proved the inclusion

$$S_p^{r-l} W^l H(\beta, \tau, n) \subset S_p^r H(\beta, \tau, n).$$

We show that the opposite inclusion holds. Let $f \in S_p^r H(\beta, \tau, n)$. Then, by Theorem 2, $f \in \Sigma S_p^r H^0(\beta, \tau, n)$. In other words, each function $F = F_{i_1 \ldots i_m}$ of the representation (1) is such as to satisfy inequality (5).

Now applying Lemma 6, we find that for any $\nu_{i_j} = 0, 1, 2, \cdots$ we have the inequality

$$Y_{\nu_{i_1}\ldots\nu_{i_s}}(F)_p \leqslant C_3 \prod_{j=1}^{s} (\nu_{i_j} + 1)^{-r_{i_j}} \beta_{i_j}\left(\frac{1}{\nu_{i_j}+1}\right),$$

where C_3 does not depend on F or ν_i.

We now take a vector $l = \{l_1, \cdots, l_n\}$ such that $l_i \leqslant r_i + \rho_i$. We fix the number m $(1 \leqslant m \leqslant s)$ and choose the numbers l_{i_j} so that l_{i_j} are the corresponding components of the vector l for $1 \leqslant j \leqslant m$ and $l_{i_j} = 0$ for $m < j \leqslant s$. We consider the sum

$$I_1 = \sum_{\nu_{i_1}=0}^{\infty} \cdots \sum_{\nu_{i_s}=0}^{\infty} \prod_{j=1}^{s} (\nu_{i_j} + 1)^{l_{i_j}-1} Y_{\nu_{i_1}\ldots\nu_{i_s}}(F)_p.$$

Substituting the estimate for $Y_{\nu_{i_1}\cdots\nu_{i_s}}(F)_p$, we have

$$I_1 \leqslant C_4 \prod_{j=1}^{s} \sum_{\nu_{i_j}=0}^{\infty} (\nu_{i_j} + 1)^{l_{i_j}-r_{i_j}-1} \beta_{i_j}\left(\frac{1}{\nu_{i_j}+1}\right).$$

Taking into account that $l_{i_j} \leqslant r_{i_j} + \rho_{i_j}$, and applying Lemma 2, we find that $I_1 < \infty$. But then, by Lemma 11, the function F has a generalized derivative F^{l_m} of order

l_{i_j} in x_{i_j}, $1 \leqslant j \leqslant m$, and, furthermore, it satisfies the inequality

$$Y_{2^{N_{i_1}-1}\ldots 2^{N_{i_s}-1}}(F^{l_m})_p \leqslant C_5 \prod_{j=1}^{s} \sum_{\nu_{i_j}=N_{i_j}}^{\infty} 2^{\nu_{i_j}(l_{i_j}-r_{i_j})}\beta_{i_j}\left(\frac{1}{2^{\nu_{i_j}}}\right),$$

where C_5 does not depend on F or N_i.

Again applying Lemma 2, we obtain

$$Y_{2^{N_{i_1}-1}\ldots 2^{N_{i_s}-1}}(F^{l_m})_p \leqslant C_6 \prod_{j=1}^{s} 2^{N_{i_j}(l_{i_j}-r_{i_j})}\beta_{i_j}\left(\frac{1}{2^{N_{i_j}}}\right),$$

where C_6 does not depend on F or N_i.

We now take natural numbers k_{i_j} such that $k_{i_j} \geqslant r_{i_j} + \tau_{i_j} - l_{i_j}$. Applying Lemma 7, and then Lemma 9, we arrive at the conclusion that

$$\omega_{k_{i_1}\ldots k_{i_s}}(F^{l_m}, \delta_{i_1}, \ldots, \delta_{i_s})_p \leqslant C_7 \prod_{j=1}^{s} \delta_{i_j}^{r_{i_j}-l_{i_j}}\beta_{i_j}(\delta_{i_j}), \tag{11}$$

where C_7 does not depend on F^{l_m} or δ_i.

We now choose a collection of indices $i_1, \cdots, i_m$. It then follows from the above that each function $F_{i_1\cdots i_m i_{m+1}\cdots i_s}$, $s \geqslant m$, of representation (1) depending on all the variables x_{i_j} with the given indices (and perhaps also depending on other variables) has a generalized derivative of order l_{i_j} in x_{i_j}, $1 \leqslant j \leqslant m$, which satisfies inequality (11). At the same time it is obvious that each function $F_{i_1\cdots i_t i_{m+1}\cdots i_s}$, $1 \leqslant t \leqslant m$, of representation (1) which does not depend on all the x_{i_j} with the given indices has a generalized derivative $F^{l_m}_{i_1\cdots i_t i_{m+1}\cdots i_s}$ of order l_{i_j} in x_{i_j}, $1 \leqslant j \leqslant m$, satisfying the condition

$$\omega_{k_{i_1}\ldots k_{i_s}}(F^{l_m}_{i_1\ldots i_t i_{m+1}\ldots i_s})_p \equiv 0.$$

Hence our function f has a generalized derivative f^{l_m} of order l_{i_j} in x_{i_j}, $1 \leqslant j \leqslant m$, and the smoothness modulus of the function f^{l_m} does not exceed the sum of the smoothness moduli of all the functions $F^{l}_{i_1\cdots i_m i_{m+1}\cdots i_s}$, which depend on all the variables x_{i_j} with the given indices.

We have thereby found that for any collection of indices $i_1, \cdots, i_m$ and any vector $l = \{l_1, \cdots, l_n\}$, where $l_i \leqslant r_i + \rho_i$, the function f has a generalized derivative f^{l_m}, and this derivative satisfies the inequality

$$\omega_{k_{i_1}\ldots k_{i_m}}(f^{l_m}, \delta_{i_1}, \ldots, \delta_{i_m})_p \leqslant C_8 \prod_{j=1}^{m} \delta^{r_{i_j}-l_{i_j}}\beta_{i_j}(\delta_{i_j}),$$

where $k_{i_j} \geqslant r_{i_j} + \tau_{i_j} - l_{i_j}$ and C_8 does not depend on f or δ_i. But this means that $f \in S_p^{r-l}\, W^l H(\beta, \tau, n)$. In other words, we have proved the inclusion

$$S_p^r H(\beta, \tau, n) \subset S_p^{r-l} W^l H(\beta, \tau, n),$$

and have thereby completed the proof of Theorem 3.

THEOREM 4 (IMBEDDING THEOREM). *Suppose that each function* $\beta_i(t)$ *satisfies the* ρ_i*-condition; suppose that* $1 \leqslant p \leqslant q \leqslant \infty$ *and* $1/p - 1/q \leqslant \rho_i + r_i$ *for all* i, $1 \leqslant i \leqslant n$. *Then, if* $f \in S_p^r H(\beta, \tau, n)$, *it follows that* $f \in S_q^{r^*}(\beta, \tau, n)$, *where* $r^* = \{r_1^*, \cdots, r_n^*\}$ *and* $r_i^* = r_i - 1/p + 1/q$. *In other words, we have the following imbedding*:

$$S_p^r H(\beta, \tau, n) \subset S_q^{r^*} H(\beta, \tau, n).$$

PROOF. Suppose that some function $f \in S_p^r H(\beta, \tau, n)$. Then, by Theorem 2, $f \in \Sigma S_p^r H^0(\beta, \tau, n)$, i.e. each function $F = F_{i_1 \cdots i_m}$ of representation (1) satisfies inequality (5).

Now, applying Lemma 6, we find that each function F satisfies the following condition as well:

$$Y_{\nu_{i_1} \ldots \nu_{i_m}}(F)_p \leqslant C_1 \prod_{j=1}^{m} (\nu_{i_1} + 1)^{-r_{i_j}} \beta_{i_j}\left(\frac{1}{\nu_{i_j} + 1}\right),$$

where C_1 does not depend on F or ν_i.

We consider the sum

$$I_2 = \sum_{\nu_{i_1}=0}^{\infty} \cdots \sum_{\nu_{i_m}=0}^{\infty} \prod_{j=1}^{m} (\nu_{i_j} + 1)^{1/p - 1/q - 1} Y_{\nu_{i_1} \ldots \nu_{i_m}}(F)_p.$$

It is clear that

$$I_2 \leqslant C_2 \prod_{j=1}^{m} \sum_{\nu_{i_j}=0}^{\infty} (\nu_{i_j} + 1)^{1/p - 1/q - 1 - r_{i_j}} \beta_{i_j}\left(\frac{1}{\nu_{i_j} + 1}\right).$$

Since, by assumption, $1/p - 1/q - r_i \leqslant \rho_i$, it follows by Lemma 2 that $I_2 < \infty$. But according to Lemma 11 this condition means that $F \in L_q([0, 2\pi]^m)$ and, in addition, that

$$Y_{2^{N_{i_1}-1} \ldots 2^{N_{i_m}-1}}(F)_q \leqslant C_3 \prod_{j=1}^{m} \sum_{\nu_{i_j}=N_{i_j}}^{\infty} 2^{\nu_{i_j}(1/p - 1/q - r_{i_j})} \beta_{i_j}\left(\frac{1}{2^{\nu_{i_j}}}\right).$$

where C_3 does not depend on F or N_i.

Again applying Lemma 2, we have

$$Y_{2^{N_{i_1}-1} \ldots 2^{N_{i_m}-1}}(F)_q \leqslant C_4 \prod_{j=1}^{m} 2^{N_{i_j}(1/p - 1/q - r_{i_j})} \beta_{i_j}\left(\frac{1}{2^{N_{i_j}}}\right),$$

where C_4 does not depend on F or N_i.

Finally, applying Lemma 7, we obtain

$$\omega_{k_{i_1}\cdots k_{i_m}}(F)_q \leqslant C_5 \prod_{j=1}^{m} \delta^{1/p-1/q-r_{ij}}\beta_{i_j}(\delta_{i_j}), \tag{12}$$

where C_5 does not depend on F or δ_i, and the k_{i_j} are natural numbers such that $k_{i_j} \geqslant r_{i_j} + \tau_{i_j}$. But inequality (12) and the condition $f \in L_q([0, 2\pi]^n)$ mean that $f \in \Sigma S_q^{r^*} H^0(\beta, \tau, n)$. Now applying Theorem 2, we find that $f \in S_q^{r^*} H(\beta, \tau, n)$, and Theorem 4 is thereby proved.

THEOREM 5 (ON THE TRACE OF A FUNCTION). *Suppose that each function $\beta_i(t)$ satisfies the ρ_i-condition; suppose given a collection of indices $j_1, \cdots, j_s$ such that $1 \leqslant j_l \leqslant n$, $1 \leqslant l \leqslant s < n$; finally, suppose that $1/p \leqslant \rho_{j_l} + r_{j_l}$ for $1 \leqslant l \leqslant s$ and $\rho_{i_l} + r_{i_l} \geqslant 0$ for $s + 1 \leqslant l \leqslant n$. Then, if the function $f \in S_p^r H(\beta, \tau, n)$, it follows that it has a trace*

$$\Phi(x_{i_{s+1}}, \ldots, x_{i_n}) \in S_p^{r^*} H(\beta^*, \tau^*, n - s),$$

where

$$r^* = \{r_{i_{s+1}}, \ldots, r_{i_n}\}, \qquad \beta^* = \{\beta_{i_{s+1}}, \ldots, \beta_{i_n}\}, \qquad \tau^* = \{\tau_{i_{s+1}}, \ldots, \tau_{i_n}\}.$$

In other words, the following imbedding holds:

$$S_p^r H(\beta, \tau, n) \subset S_p^{r^*} H(\beta^*, \tau^*, n - s).$$

PROOF. If the function $f \in S_p^r H(\beta, \tau, n)$, then by Theorem 2, $f \in \Sigma S_p^r H^0(\beta, \tau, n)$. Each function $F_{i_1 \cdots i_m}$ of representation (1) lies in one of the following three classes of functions:

1) the class of functions $F_{i_1 \cdots i_m}$ which do not depend on the variables x_{j_l} with the given indices;

2) the class of functions $F_{j_1 \cdots j_t}$ which only depend on all or part of the variables x_{j_l} with the given indices $(t \leqslant s)$ and do not depend on the other variables;

3) the class of functions $F_{j_1 \cdots j_t, i_{s+1} \cdots i_m}$, which depend on all or part of the variables x_{j_l} with the given indices $(t \leqslant s)$ and which also must depend on variables x_{i_l} with other indices $i_{s+1}, \cdots, i_m$.

We are interested in the question of the trace of each funciton $F_{i_1 \cdots i_m}$ on R_{n-s}.

1. It is completely clear that the trace of any function $F_{i_1 \cdots i_m}$ of the first class of R_{n-s} coincides with the function itself.

2. Applying Lemma 11 to any function $F_{j_1\cdots j_t}$ of the second class, we find that, after modification on a set of t-dimensional $(t \leqslant s < n)$ measure zero, this function becomes continuous (at least in a neighborhood of the point $(0, \cdots, 0)$). But this means that $F_{j_1\cdots j_t}(0, \cdots, 0) = C_{j_1\cdots j_t}$, where $C_{j_1\cdots j_t}$ is a constant, and here

$$|F_{j_1\ldots j_t}(x_{j_1}, \ldots, x_{j_t}) - C_{j_1\ldots j_t}| \to 0 \quad \text{as} \quad \sum_{l=1}^{t} x_{j_l}^2 \to 0.$$

It is natural to assume that the constant $C_{j_1\cdots j_t}$ is actually the trace of the function $F_{j_1\cdots j_t}$ on R_{n-s}.

3. We now consider an arbitrary function $F_{j_1\cdots j_t i_{s+1}\cdots i_m}$ of the third class. Since $f \in \Sigma S_p^r H^0(\beta, \tau, n)$, we have the following inequality for $F = F_{j_1\cdots j_t i_{s+1}\cdots i_m}$:

$$\omega_{k_{j_1}\ldots k_{j_t} k_{i_{s+1}}\ldots k_{i_m}}(F, \delta_{j_1}, \ldots, \delta_{j_t}, \delta_{i_{s+1}}, \ldots, \delta_{i_m})_p$$

$$\leqslant C_1 \prod_{l=1}^{t} \delta_{j_l}^{r_{j_l}} \beta_{j_l}(\delta_{j_l}) \prod_{l=s+1}^{m} \delta_{i_l}^{r_{i_l}} \beta_{i_l}(\delta_{i_l}),$$

where C_1 does not depend on F or δ_i and $k_{i_j} \geqslant r_{i_j} + \tau_{i_j}$.

Applying Lemma 6, we obtain

$$Y_{\nu_{j_1}\ldots\nu_{j_t}\nu_{i_{s+1}}\ldots\nu_{i_m}}(F)_p \leqslant C_2 \prod_{l=1}^{t} (\nu_{j_l}+1)^{-r_{j_l}} \beta_{j_l}\left(\frac{1}{\nu_{j_l}+1}\right)$$

$$\times \prod_{l=s+1}^{m} (\nu_{i_l}+1)^{-r_{i_l}} \beta_{i_l}\left(\frac{1}{\nu_{i_l}+1}\right),$$

where C_2 does not depend on F or ν_i.

We now consider the sum

$$I_3 = \sum_{\nu_{j_1}=0}^{\infty} \cdots \sum_{\nu_{j_t}=0}^{\infty} \sum_{\nu_{i_{s+1}}=0}^{\infty} \cdots \sum_{\nu_{i_m}=0}^{\infty} \prod_{l=1}^{t} (\nu_{jt}+1)^{1/p-1}$$

$$\times \prod_{l=s+1}^{m} (\nu_{i_l}+1)^{-1} Y_{\nu_{j_1}\ldots\nu_{j_t}\nu_{i_{s+1}}\ldots\nu_{i_m}}(F)_p.$$

It is clear that

$$I_3 \leqslant C_2 \prod_{l=1}^{s} \sum_{\nu_{j_l}=0}^{\infty} (\nu_{j_l}+1)^{1/p-1-r_{j_l}} \beta_{j_l}\left(\frac{1}{\nu_{j_l}+1}\right)$$

$$\times \prod_{l=s+1}^{m} \sum_{\nu_{i_l}=0}^{\infty} (\nu_{i_l}+1)^{-1-r_{i_l}} \beta_{j_l}\left(\frac{1}{\nu_{i_l}+1}\right).$$

Since, by assumption, $1/p - r_{j_l} = \rho_{j_l}$ $(1 \leqslant l \leqslant t)$ and $\rho_{i_l} \geqslant -r_{i_l}$ $(s+1 \leqslant l \leqslant m)$, it follows by Lemma 2 that $I_3 < \infty$. But according to Lemma 12 this condition means that, after modification on a set of $(m-s+t)$-dimensional measure zero

$(m - s + t < n)$, the function $F_{j_1 \dots j_t i_{s+1} \dots i_m}$ has a trace $\varphi = \varphi_{i_{s+1} \dots i_m}$ on R_{m-t-s}, and, in addition, the function φ satisfies the inequality

$$Y_{2^{N_{i_{s+1}}-1} \dots 2^{N_{i_m}-1}} (\varphi)_p \leqslant C_3 \prod_{l=1}^{t} \sum_{\nu_{j_l}=0}^{\infty} 2^{\nu_{j_l}(1/p - r_{j_l})} \beta_{j_l}\left(\frac{1}{2^{\nu_{j_l}}}\right) \times \prod_{l=s+1}^{m} \sum_{\nu_{i_l}=N_{i_l}}^{\infty} 2^{-\nu_{i_l} r_{i_l}} \beta_{i_l}\left(\frac{1}{2^{\nu_{i_l}}}\right),$$

where C_3 does not depend on φ and ν_i.

Again applying Lemma 2, we obtain

$$Y_{2^{N_{i_{s+1}}-1} \dots 2^{N_{i_m}-1}} (\varphi)_p \leqslant C_4 \prod_{l=s+1}^{m} 2^{-N_{i_l} r_{i_l}} \beta_{i_l}\left(\frac{1}{2^{N_{i_l}}}\right),$$

where C_4 does not depend on φ or N_i.

Finally, using Lemma 7, we have

$$\omega_{k_{i_{s+1}} \dots k_{i_m}} (\varphi, \delta_{i_{s+1}}, \dots, \delta_{i_m})_p \leqslant C_5 \prod_{l=s+1}^{m} \delta_{i_l}^{r_{i_l}} \beta_{i_l}(\delta_{i_l}), \tag{13}$$

where C_5 does not depend on φ or δ_i and $k_i \geqslant r_i + \tau_i$. Thus we have shown that each function $F_{j_1 \dots j_t i_{s+1} \dots i_m}$ of the third class has on R_{n-s} a trace $\varphi = \varphi_{i_{s+1} \dots i_m}$, satisfying condition (13).

We now consider the function Φ, which is equal to the sum of the traces of all the functions from the three classes, and we show that Φ is the trace of the function f on R_{n-s}. The first condition of the definition of the trace of a function is fulfilled in an obvious way. The second condition follows because f has representation (1) and because all the modifications of the functions of the second and third classes result in modifying f on a set of n-dimensional measure zero. Finally, the third condition follows from the fact that $\|f - \Phi\|_{\mathfrak{L}_p(R_{n-s})}$ does not exceed the sum of the norms of the differences of the functions of representation (1) and their traces.

Thus Φ is the trace of f on R_{n-s}. Since Φ is equal to the sum of all traces of the functions in representation (1), it follows that the function itself is a representation of the same type, where each function $\varphi_{i_1 \dots i_l}$ of this representation satisfies the inequality

$$\omega_{k_{i_1} \dots k_{i_l}} (\varphi_{i_1, \dots, i_l}, \delta_{i_1}, \dots, \delta_{i_l})_p \leqslant C_6 \prod_{j=1}^{l} \delta_{i_j}^{r_{i_j}} \beta_{i_j}(\delta_{i_j}),$$

where C_6 does not depend on φ or δ_i. But this means that $\Phi \in \Sigma S_p^{r^*} H^0(\beta^*, \tau^*, n - s)$. Now applying Theorem 2, we obtain $\Phi \in S_p^{r^*} H(\beta^*, \tau^*, n - s)$, which completes the proof of Theorem 5.

THEOREM 6 (EXTENSION THEOREM). *If* $f \in S_p^r H(\beta, \tau, n)$, *then for any* $m > n$ *this function can be extended onto* R_m *so that*

1) *the extended function* $F \in S_p^{r^*} H(\beta^*, \tau^*, m)$, *where* $r^* = \{r_1, \cdots, r_n, r_{n+1}, \cdots, r_m\}$, $\beta^* = \{\beta_1, \cdots, \beta_n, \beta_{n+1}, \cdots, \beta_m\}$, $\tau^* = \{\tau_1, \cdots, \tau_n, \tau_{n+1}, \cdots, \tau_m\}$ *and the* r_i *are any real numbers for* $n + 1 \leqslant i \leqslant m$, *and the* $\beta_i(t)$ *are any functions defined in* §1 *and satisfying the* τ_i-*conditions for* $n + 1 \leqslant i \leqslant m$; *and*

2) $F(x_1, \cdots, x_n, 0, \cdots, 0) \equiv f$.

PROOF. We take a function $\psi(t)$ which is λ times continuously differentiable, where $\lambda > r_i + \tau_i$ for all i: $n + 1 \leqslant i \leqslant m$, and is such that $\psi(0) = 1$ and $\int_0^{2\pi} |\psi^k(t)|^p\, dt \leqslant C$ for all $k = 0, 1, \cdots, \lambda$. It follows from these properties of the function ψ that for $k_i \geqslant \lambda$

$$\omega_{k_i}(\psi(x_i), \delta_i)_p \leqslant C_1 \delta_i^\lambda \| \psi^\lambda(x_i) \|_p \leqslant C_2 \delta_i^\lambda \leqslant C_2 \delta_i^{r_i + \tau_i},$$

where C_1 and C_2 do not depend on ψ and δ_i ($\delta_i \leqslant 1$).

We now show that $\delta_i^{\tau_1} \leqslant C_3 \beta_i(\delta)$, where C_3 does not depend on δ. Letting $t \leqslant 1/2$, we may then write that

$$\int_t^1 \frac{\beta_i(u)}{u^{1+\tau_i}}\, du \geqslant \int_{1/2}^1 \frac{\beta_i(u)}{u^{1+\tau_i}}\, du \geqslant C_4 \int_{1/2}^1 \beta_i(u)\, du,$$

where $C_4 = 1$ for $\tau_i \geqslant -1$ and $C_4 = 2^{1+\tau_i}$ for $\tau_i < -1$. Since $\beta_i(u)$ is continuous on $[0, 1]$, we have $\int_{1/2}^1 \beta_i(u) du = C_5$, where C_5 naturally does not depend on t. We may hence write the following inequality which holds for $t \leqslant 1/2$:

$$C_6 \leqslant \int_t^1 \frac{\beta_i(u)}{u^{1+\tau_i}}\, du,$$

where C_6 does not depend on t. Taking into account that $\beta_i(u)$ satisfies the τ_i-condition, we hence obtain the inequality $t^{\tau_i} \leqslant C_7 \beta_i(t)$ for $t \leqslant 1/2$, where C_7 does not depend on t.

Now if $1/2 \leqslant t \leqslant 1$, then, setting $t_1 = t/2$, we have $t_1 \leqslant 1/2$. By what has just been proved, $t_1^{\tau_1} \leqslant C_7 \beta_i(t_1)$ or $(t/2)^{\tau_i} \leqslant C_7 \beta_i(t/2)$. Since $t/2 \leqslant t$, it follows by property 2 of the function $\beta_i(t)$ that $t^{\tau_i} \leqslant C_8 \beta_i(t)$, where C_8 does not depend on t.

Thus we have shown that $\delta^{\tau_i} \leqslant C_3 \beta_i(\delta_i)$ for any $\delta_i < 1$, where C_3 does not depend on the δ_i. We may hence write that

$$\omega_{k_i}(\psi(x_i), \delta_i)_p \leqslant C_9 \delta^{r_i} \beta_i(\delta_i),$$

where C_9 does not depend on ψ or δ_i.

We now take the function

$$F(x_1, \ldots, x_m) = f(x_1, \ldots, x_n) \prod_{j=n+1}^m \psi(x_i).$$

It is completely obvious that $F \in S_p^{r^*}H(\beta^*, \tau^*, m)$ and $F(x_1, \cdots, x_n, 0, \cdots, 0) \equiv f(x_1, \cdots, x_n)$, and Theorem 6 is thereby proved.

We note that the idea of proving the extension theorems in this way belongs to Bugrov [8].

THEOREM 7 (INTERPOLATION). *If* $f \in S_p^r H(\beta, \tau, n)$ *and* $f \in S_p^{\bar{r}} H(\bar{\beta}, \bar{\tau}, n)$, *then* $f \in S_p^{r^*} H(\beta^*, \tau^*, n)$, *where*

$$r^* = \{r_1^*, \ldots, r_n^*\}, \qquad \beta^* = \{\beta_1^*, \ldots, \beta_n^*\}, \qquad \tau^* = \{\tau_1^*, \ldots, \tau_n^*\},$$

$$r_i^* = r_i\lambda_i + \bar{r}_i\mu_i, \qquad \beta_i^* = \beta_i^*(t) = \{\beta_i(t)\}^{\lambda_i}\{\bar{\beta}_i(t)\}^{\mu_i}$$

for any λ_i *and* μ_i *such that* $\lambda_i \geqslant 0, \mu_i \geqslant 0, \lambda_i + \mu_i \leqslant 1$, *with each function* $\beta_i^*(t)$ *satisfying the* τ_i^**-condition, where* $\tau_i^* = \tau_i\lambda_i + \bar{\tau}_i\mu_i$.

PROOF. We choose natural numbers $k_1, \cdots, k_n$ so that $k_i \geqslant \max\{r_i + \tau_i, \bar{r}_i + \bar{\tau}_i\}$. Then for the m and i_j chosen, but for the same k_{i_j}, we have the inequalities

$$\omega_{k_{i_1}\ldots k_{i_m}}(f, \delta_{i_1}, \ldots, \delta_{i_m})_p \leqslant C_1 \prod_{j=1}^m \delta_{i_j}^{r_{i_j}} \beta_{i_j}(\delta_{i_j}),$$

$$\omega_{k_{i_1}\ldots k_{i_m}}(f, \delta_{i_1}, \ldots, \delta_{i_m})_p \leqslant C_2 \prod_{j=1}^m \delta^{\bar{r}_{i_j}} \bar{\beta}_{i_j}(\delta_{i_j}),$$

where C_1 and C_2 do not depend on f or δ_i.

It is completely clear that, by examining the different cases, we can easily show that these inequalities imply the inequalities

$$\omega_{k_{i_1}\ldots k_{i_m}}(f, \delta_{i_1}, \ldots, \delta_{i_m})_p \leqslant C_3 \prod_{j=1}^m \delta_{ij}^{r_{i_j}\lambda_{i_j} + \bar{r}_{i_j}\mu_{i_j}} [\beta_{i_j}(\delta_{i_j})]^{\lambda_{i_j}} [\bar{\beta}_{i_j}(\delta_{i_j})]^{\mu_{i_j}},$$

where $C_3 = \max(C_1, C_2)$ and hence does not depend on f or δ_i. It is equally easy to show that the function $\beta_i^*(t)$ satisfies the τ_i^*-condition, and Theorem 7 is proved.

§4. The function class $S_pB(\alpha, \sigma, \theta, n)$

Suppose we are given a natural number n, a number p in the interval $1 \leqslant p \leqslant \infty$, a number θ in the interval $1 \leqslant \theta < \infty$, and vectors $\sigma = \{\sigma_1, \cdots, \sigma_n\}$ and $\alpha = \{\alpha_1, \cdots, \alpha_n\}$, where the σ_i are real numbers, and $\alpha_i = \alpha_i(t)$, the functions defined in §1.

We say that $f \in S_pB(\alpha, \sigma, \theta, n)$, if

1) $f \in L_p([0, 2\pi]^n)$;

2) each function $\alpha_i(t)$ satisfies the σ_i-condition;

3) for all natural numbers m and i_j such that $1 \leqslant i_j \leqslant n, 1 \leqslant j \leqslant m \leqslant n$, we have the inequalities

$$\int_0^{2\pi} \cdots \int_0^{2\pi} \omega^\theta_{k_{i_1} \ldots k_{i_m}} (f, t_{i_1}, \ldots, t_{i_m})_p \prod_{j=1}^m \alpha_{i_j}(t_{i_j}) \, dt_{i_1} \ldots dt_{i_m} < \infty,$$

where the k_{i_j} are certain fixed natural numbers satisfying the condition $k_{i_j} \geqslant \sigma_{i_j}/\theta$.

REMARK. Like the function class $S^r_p H(\beta, \tau, n)$, *formally* the function class $S_pB(\alpha, \sigma, \theta, n)$ also depend on the choice of natural numbers k_{i_j}. However, applying Lemma 8, we easily show that, actually, the function class $S_pB(\alpha, \sigma, \theta, n)$ does not depend on the choice of the k_{i_j}, if we only require that the k_{i_j} be sufficiently large: $k_{i_j} \geqslant \sigma_{i_j}/\theta$. The condition $k_{i_j} \geqslant \sigma_{i_j}/\theta$ is also introduced so that the function class $S_pB(\alpha, \sigma, \theta, n)$ is "meaningful" (the condition $k_{i_j} \geqslant \sigma_{i_j}/\theta$ is discussed in more detail for functions of one variable in [5]).

We now give a constructive characterization of the function class $S_pB(\alpha, \sigma, \theta, n)$. It follows from Lemma 8 that we have

THEOREM 8. *In order that $f \in S_pB(\alpha, \sigma, \theta, n)$, it is necessary and sufficient that for all natural numbers i_j and m such that $1 \leqslant i_j \leqslant n$,*

$$\sum_{\nu_{i_1}=0}^{\infty} \cdots \sum_{\nu_{i_m}=0}^{\infty} \prod_{j=1}^m \mu_{i_j}(\nu_{i_j}) Y^\theta_{[2^{\nu_{i_1}-1}] \ldots [2^{\nu_{i_m}-1}]} (f)_p < \infty.$$

We further say that $f \in \Sigma S_pB^0(\alpha, \sigma, \theta, n)$ if

1) $f \in \mathfrak{L}_p([0, 2\pi]^n)$;
2) each function $\alpha_i(t)$ satisfies the σ_i-condition;
3) each function $F_{i_1 \ldots i_m}$ in representation (1) satisfies the inequality

$$\int_0^{2\pi} \cdots \int_0^{2\pi} \omega_{k_{i_1} \ldots k_{i_m}} (F_{i_1 \ldots i_m}, t_{i_1}, \ldots, t_{i_m})_p \prod_{j=1}^m \alpha_{i_j}(t_{i_j}) \, dt_{i_1} \ldots dt_{i_m} < \infty, \tag{14}$$

where the k_{i_j} are fixed natural numbers satisfying the condition $k_{i_j} \geqslant \sigma_{i_j}/\theta$.

Here it is also easy to show that the choice of k_{i_j} does not effect the definition of the function class $\Sigma S_pB^0(\alpha, \sigma, \theta, n)$, if we only require that the k_{i_j} be sufficiently large: $k_{i_j} \geqslant \sigma_{i_j}/\theta$.

THEOREM 9. *The function classes $S_pB(\alpha, \sigma, \theta, n)$ and $\Sigma S_pB^0(\alpha, \sigma, \theta, n)$ coincide.*

The proof of this theorem repeats almost word for word the proof of Theorem 2, except that the references to Lemma 9 must be replaced by references to Lemma 10; hence we shall not give the proof here.

Now suppose that we are given a natural number n, a number p in the interval $1 \leqslant p \leqslant \infty$, a number θ in the interval $1 \leqslant \theta < \infty$, vectors $\sigma = \{\sigma_1, \cdots, \sigma_n\}$,

$l = \{l_1, \cdots, l_n\}$ and $\alpha = \{\alpha_1, \cdots, \alpha_n\}$, where the σ_i are real numbers, and the l_i are nonnegative integers, and the functions $\alpha_i = \alpha_i(t)$ are defined in §1.

We say that $f \in S_p W^l B(\alpha, \sigma, \theta, n)$ if

1) $f \in \mathfrak{L}_p([0, 2\pi]^n)$;

2) each function $\alpha_i(t)$ satisfies the σ_i-condition;

3) for all natural numbers m and i_j such that $1 \leqslant i_j \leqslant n, 1 \leqslant j \leqslant m \leqslant n$, the function f has generalized partial derivatives

$$f^{l_m} = \frac{\partial^{l_{i_1}}}{\partial x_{i_1}^{l_{i_1}}} \cdots \frac{\partial^{l_{i_m}}}{\partial x_{i_m}^{l_{i_m}}} f$$

of order l_{i_j} in the variable x_{i_j};

4) for each such derivative we have the inequality

$$\int_0^{2\pi} \cdots \int_0^{2\pi} \omega^\theta_{k_{i_1} \cdots k_{i_m}} (f^{l_m}, t_{i_1}, \ldots, t_{i_m})_p \prod_{j=1}^m \alpha_{i_j}(t_{i_j})\, dt_{i_1} \ldots dt_{i_m} < \infty,$$

where the k_{i_j} are fixed natural numbers satisfying $k_{i_j} \geqslant \sigma_{i_j}/\theta$.

THEOREM 10. *If each of the functions $\alpha_i(t)$ satisfies the λ_i-condition and $\lambda_i > 0$, then, for any vector $l = \{l_1, \cdots, l_n\}$ such that the l_i are nonnegative integers satisfying the conditions $l_i < \lambda_i/\theta$, the function classes $S_p B(\alpha, \sigma, \theta, n)$ and $S_p W^l B(\alpha^*, \sigma^*, \theta, n)$ coincide, where $\alpha^* = \{\alpha_1^*, \cdots, \alpha_n^*\}$, $\sigma^* = \{\sigma_1^*, \cdots, \sigma_n^*\}$, $\alpha_i^* = \alpha_i(t) t^{l_i \theta}$ and $\sigma_i^* = \sigma_i - l_i \theta$.*

PROOF. The inclusion $S_p W^l B(\alpha^*, \sigma^*, \theta, n) \subset S_p B(\alpha, \sigma, \theta, n)$ is proved in the same way as the analogous inclusion in Theorem 3.

We show the opposite inclusion. If $f \in S_p B(\alpha, \sigma, \theta, n)$, then, according to Theorem 9, $f \in \Sigma S_p B^0(\alpha, \sigma, \theta, n)$, i.e. each function $F = F_{i_1 \cdots i_s}$ of representation (1) is such that inequality (14) holds.

Now applying Lemma 8, we arrive at the inequality

$$I_1 = \sum_{\nu_{i_1}=0}^{\infty} \cdots \sum_{\nu_{i_s}=0}^{\infty} \prod_{j=1}^{s} \mu_{i_j}(\nu_{i_j})\, Y^\theta_{[2^{\nu_{i_1}}-1] \ldots [2^{\nu_{i_s}}-1]}(F_p) < \infty.$$

Since $\alpha_i(t)$ satisfies the λ_i-condition, it follows by Lemma 3 that $2^{n\lambda_i} \leqslant C_1 \mu(n)$, where C_1 does not depend on n. Hence for $l_i < \lambda_i/\theta$ the following series converges:

$$\sum_{\nu_{i_1}=0}^{\infty} \cdots \sum_{\nu_{i_s}=0}^{\infty} \prod_{j=1}^{s} 2^{l_{i_j} \nu_{i_j}} Y_{[2^{\nu_{i_1}}-1] \ldots [2^{\nu_{i_s}}-1]}(F)_p.$$

Now applying Lemma 11, we find that the function F has a generalized derivative

F^{l_m} of order l_{i_j} in x_{i_j} for $1 \leqslant j \leqslant m$ and of order $l_{i_j} = 0$ for $m < j \leqslant s$, which satisfies the inequality

$$Y_{2^{N_{i_1}}-1 \ldots 2^{N_{i_s}}-1}(F^{l_m})_p \leqslant C_2 \sum_{\nu_{i_1}=N_{i_1}}^{\infty} \cdots \sum_{\nu_{i_s}=N_{i_s}}^{\infty} \prod_{j=1}^{s} 2^{l_{i_j}\nu_{i_j}} Y_{[2^{\nu_{i_1}}-1] \ldots [2^{\nu_{i_s}}-1]}(F)_p,$$

where C_2 does not depend on F^{l_m} or N_i.

We now consider

$$I_2 = \sum_{\nu_{i_1}=0}^{\infty} \cdots \sum_{\nu_{i_s}=0}^{\infty} \prod_{j=1}^{s} \mu^*_{i_j}(\nu_{i_j}) Y^{\theta}_{[2^{\nu_{i_1}}-1] \ldots [2^{\nu_{i_s}}-1]}(F^l)_p,$$

where $\mu^*_{i_j}(0) = \mu_{i_j}(0)$,

$$\mu^*_{i_j}(\nu_{i_j}) = \int_{1/2^{\nu_{i_j}}}^{1/2^{\nu_{i_j}-1}} \alpha^*_{i_j}(t)\,dt \leqslant C_3 2^{-\nu_{i_j} l_{i_j}\theta} \mu_{i_j}(\nu_{i_j}), \quad \nu_{i_j} = 1, 2, \ldots,$$

and C_3 does not depend on ν_{i_j}.

Taking into account the estimates for $\mu^*_{i_j}(\nu_{i_j})$ and $Y_{2^{N_{i_1}}-1 \ldots 2^{N_{i_s}}-1}(F^{l_m})_p$, we obtain

$$I_2 \leqslant C_4 \sum_{\nu_{i_1}=0}^{\infty} \cdots \sum_{\nu_{i_s}=0}^{\infty} \prod_{j=1}^{s} 2^{-\nu_{i_j} l_{i_j}\theta} \mu_{i_j}(\nu_{i_j}) A^{\theta},$$

where

$$A = \sum_{r_{i_1}=\nu_{i_1}}^{\infty} \cdots \sum_{r_{i_s}=\nu_{i_s}}^{\infty} \prod_{j=1}^{s} 2^{\lambda_{i_j} r_{i_j}/\theta} Y_{[2^{r_{i_1}}-1] \ldots [2^{r_{i_s}}-1]}(F)_p \prod_{j=1}^{s} 2^{r_{i_j}\left(l_{i_j} - \frac{\lambda_{i_j}}{\theta}\right)}.$$

Now applying Hölder's inequality to A and taking into account that $l_{i_j} - \lambda_{i_j}/\theta < 0$, we have

$$A^{\theta} \leqslant C_5 \prod_{j=1}^{s} 2^{\nu_{i_j}\left(l_{i_j} - \frac{\lambda_{i_j}}{\theta}\right)\theta} \sum_{r_{i_1}=\nu_{i_1}}^{\infty} \cdots \sum_{r_{i_s}=\nu_{i_s}}^{\infty} \prod_{j=1}^{s} 2^{\lambda_{i_j} r_{i_j}} Y^{\theta}_{[2^{r_{i_1}}-1] \ldots [2^{r_{i_s}}-1]}(F)_p,$$

where C_5 does not depend on ν_{i_j}.

We may now write that

$$I_2 \leqslant C_6 \sum_{\nu_{i_1}=0}^{\infty} \cdots \sum_{\nu_{i_s}=0}^{\infty} \cdots \prod_{j=1}^{s} 2^{-\nu_{i_j}\lambda_{i_j}} \mu_{i_j}(\nu_{i_j}) \times$$

$$\times \sum_{r_{i_1}=\nu_{i_1}}^{\infty} \cdots \sum_{r_{i_s}=\nu_{i_s}}^{\infty} \prod_{j=1}^{s} 2^{\lambda_{i_j} r_{i_j}} Y^{\theta}_{[2^{r_{i_1}}-1]\ldots[2^{r_{i_s}}-1]}(F)_p$$

$$= C_6 \sum_{r_{i_1}=0}^{\infty} \cdots \sum_{r_{i_s}=0}^{\infty} \prod_{j=1}^{s} 2^{\lambda_{i_j} r_{i_j}} Y^{\theta}_{[2^{r_{i_1}}-1]\ldots[2^{r_{i_s}}-1]}(F)_p \prod_{j=1}^{s} \sum_{\nu_{i_j}=0}^{r_{i_j}} 2^{-\nu_{i_j}\lambda_{i_j}} \mu_{i_j}(\nu_{i_j}).$$

Applying Lemma 3, we have

$$I_2 \leqslant C_7 \sum_{r_{i_1}=0}^{\infty} \cdots \sum_{r_{i_s}=0}^{\infty} \prod_{j=1}^{s} \mu_{i_j}(r_{i_j}) Y^{\theta}_{[2^{r_{i_1}}-1]\ldots[2^{r_{i_s}}-1]}(F)_p,$$

i.e. $I_2 \leqslant C_7 I_1 < \infty$.

Taking into account that $\alpha_i^*(t)$ satisfies the σ_i^*-condition and that $I_2 < \infty$, and applying Lemma 8, we obtain

$$\int_0^{2\pi} \cdots \int_0^{2\pi} \omega^{\theta}_{k_{i_1}\ldots k_{i_s}}(F^{l_m}, t_{i_1}, \ldots, t_{i_s})_p \prod_{j=1}^{s} \alpha^*_{i_j}(t_{i_j})\, dt_{i_1} \ldots dt_{i_s} < \infty, \tag{15}$$

where the k_{i_j} are natural numbers and $k_{i_j} \geqslant \sigma^*_{i_j}/\theta$.

Thus we have obtained that each function $F_{i_1 \ldots i_s}$ in representation (1) has a generalized derivative $F^{l_m}_{i_1 \ldots i_s}$ of order l_{i_j} in x_{i_j} $(1 \leqslant j \leqslant m)$ which satisfies inequality (15). The rest of the proof of Theorem 10 is analogous to the last part of the proof of Theorem 3, and will not be given here.

We say that $f \in S_p\Delta(\alpha, \sigma, \theta, n)$ if

1) $f \in \mathfrak{L}_p[(0, 2\pi]^n)$:

2) each function $\alpha_i(t)$ satisfies the σ_i-condition;

3) for all natural numbers m and i_j such that $1 \leqslant i_j \leqslant n$, $1 \leqslant j \leqslant m \leqslant n$, we have the inequalities

$$\int_0^{2\pi} \cdots \int_0^{2\pi} \left\| \Delta^{k_{i_1} \ldots k_{i_m}}_{t_{i_1} \ldots t_{i_m}} f \right\|_p^{\theta} \prod_{j=1}^{m} \alpha_{i_j}(t_{i_j})\, dt_{i_1} \ldots dt_{i_m} < \infty, \tag{16}$$

where the k_{i_j} are fixed natural numbers satisfying $k_{i_j} \geqslant \sigma_{i_j}/\theta$.

THEOREM 11. *If each of the functions $\alpha_i(t)$ satisfies the γ-condition, then the function classes $S_pB(\alpha, \sigma, \theta, n)$ and $S_p\Delta(\alpha, \sigma, \theta, n)$ coincide.*

PROOF. If some function $f \in S_pB(\alpha, \sigma, \theta, n)$, then it is obvious that $f \in S_p\Delta(\alpha, \sigma, \theta, n)$, i.e. we have the obvious inclusion

$$S_pB(\alpha, \sigma, \theta, n) \subset S_p\Delta(\alpha, \sigma, \theta, n).$$

We now show the opposite inclusion.

Suppose that $f \in S_p\Delta(\alpha, \sigma, \theta, n)$, i.e. f satisfies (16).

Now fix i_j $(1 \leqslant j \leqslant m)$ and k_{i_j} $(k_{i_j} \geqslant \sigma_{i_j}/\theta)$. We choose a natural number k_0 satisfying the condition $k_0 \geqslant \max(k_{i_j}/2 + 1, (k_{i_j}\theta + 1)/2)$ and for every $n = 0, 1, \cdots$ we determine the natural number m from the inequality

$$\frac{n}{2k_0} < m \leqslant \frac{n}{2k_0} + 1.$$

We consider the function

$$K_n(t) = b_m \left(\frac{\sin \frac{mt}{2}}{\sin t/2} \right)^{2k_0},$$

where b_m is chosen so that $\int_{-\pi}^{\pi} K_n(t)dt = 1$.

In [7] it is shown that for any $l_{i_j} = 0, 1, 2, \cdots,$

$$Y_{l_{i_1}\ldots l_{i_m}}(f)_p \leqslant \int_{-\pi}^{\pi} \cdots \int_{-\pi}^{\pi} \prod_{j=1}^{m} K_{l_{i_j}}(t_{i_j}) \left\| \Delta_{t_{i_1}\ldots t_{i_m}}^{k_{i_1}\ldots k_{i_m}} f \right\|_p dt_{i_1} \ldots dt_{i_m}.$$

Applying Hölder's inequality to the right side of this relation, we find that for $1 \leqslant \theta < \infty$

$$Y_{l_{i_1}\ldots l_{i_m}}^{\theta}(f)_p \leqslant \int_{-\pi}^{\pi} \cdots \int_{-\pi}^{\pi} \prod_{j=1}^{m} K_{l_{i_j}}(t_{i_j}) \left\| \Delta_{t_{i_1}\ldots t_{i_m}}^{k_{i_1}\ldots k_{i_m}} f \right\|_p^{\theta} dt_{i_1} \ldots dt_{i_m}.$$

It is then easy to show that

$$Y_{l_{i_1}\ldots l_{i_m}}^{\theta}(f)_p \leqslant C_1 \int_0^{\pi/2} \cdots \int_0^{\pi/2} \prod_{j=1}^{m} K_{l_{i_j}}(2t_{i_j}) \left\| \Delta_{2t_{i_1}\ldots 2t_{i_m}}^{k_{i_1}\ldots k_{i_m}} f \right\|_p^{\theta} dt_{i_1} \ldots dt_{i_m}.$$

We consider

$$I_3 = \sum_{\nu_{i_1}=0}^{\infty} \cdots \sum_{\nu_{i_m}=0}^{\infty} \prod_{j=1}^{m} \mu_{i_j}(\nu_{i_j}) Y_{[2^{\nu_{i_1}-1}]\ldots[2^{\nu_{i_m}-1}]}^{\theta}(f)_p.$$

It is clear that

$$I_3 \leqslant C_2 \int_0^{\pi/2} \cdots \int_0^{\pi/2} \left\| \Delta_{2t_{i_1}\ldots 2t_{i_m}}^{k_{i_1}\ldots k_{i_m}} f \right\|_p^{\theta} \prod_{j=1}^{m} \sum_{\nu_{i_j}=0}^{\infty} \mu_{i_j}(\nu_{i_j}) K_{[2^{\nu_{i_j}-1}]}(2t_{i_j}) dt_{i_1} \ldots dt_{i_m}.$$

We now estimate

$$I_4 = \sum_{\nu=0}^{\infty} \mu_i(\nu) K_{[2^{\nu-1}]}(2t).$$

Since $K_{[2^{\nu-1}]}(2t) = 2\pi$ for $\nu = 0$ and $\nu = 1$, we have

$$I_4 \leqslant C_3 + \sum_{\nu=2}^{\infty} \mu_i(\nu) K_{2^{\nu-1}}(2t) = C_3 + \sum_{\nu=1}^{\infty} \mu_i(\nu+1) K_{2^\nu}(2t).$$

Since the function $\alpha_i(t)$ satisfies the σ_i-condition, it is easy to show that $\mu_i(\nu+1) \leqslant C_4 \mu_i(\nu)$. Hence

$$I_4 \leqslant C_5 \left[1 + \sum_{\nu=1}^{\infty} \mu_i(\nu) K_{2^\nu}(2t)\right].$$

In [5] it is shown that, if the function $\alpha_i(t)$ satisfies the γ- and σ_i-conditions, then

$$\sum_{\nu=1}^{\infty} \mu_i(\nu) K_{2^\nu}(2t) \leqslant C_6 \alpha_i(2t)$$

for every fixed t in the interval $0 < t \leqslant \pi/2$. Further, taking into account that $\alpha_i(t)$ satisfies the condition $1 \leqslant C_7 \alpha_i(t)$ for any t, we finally obtain

$$I_4 \leqslant C_8 \alpha_i(2t).$$

We have thus shown that

$$I_3 \leqslant C_9 \int_0^{\pi/2} \cdots \int_0^{\pi/2} \left\| \Delta_{2t_{i_1} \ldots 2t_{i_m}}^{k_{i_1} \ldots k_{i_m}} f \right\|_p^\theta \prod_{j=1}^{m} \alpha_{i_j}(2t_{i_j})\, dt_{i_1} \ldots dt_{i_m} < \infty.$$

Hence for any natural numbers m, i_j and k_{i_j} such that $1 \leqslant i_j \leqslant n$, $1 \leqslant j \leqslant m \leqslant n$, $k_{i_j} \geqslant \sigma_{i_j} \geqslant \sigma_{i_j}/\theta$, we prove the inequalities

$$\sum_{\nu_{i_1}=0}^{\infty} \cdots \sum_{\nu_{i_m}=0}^{\infty} \prod_{j=1}^{m} \mu_{i_j}(\nu_{i_j})\, Y^\theta_{[2^{\nu_{i_1}}-1] \ldots [2^{\nu_{i_m}}-1]}(f)_p < \infty.$$

But by Theorem 8 this means that $f \in S_p B(\alpha, \sigma, \theta, n)$. In other words, we have $S_p \Delta(\alpha, \sigma, \theta, n) \subset S_p B(\alpha, \sigma, \theta, n)$, and Theorem 11 is proved.

THEOREM 12 (IMBEDDING THEOREM). *Suppose that each function $\alpha_i(t)$ satisfies the λ_i-condition; suppose that $1 \leqslant p \leqslant q \leqslant \infty$ and $1/p - 1/q < \lambda_i/\theta$ for all i $(1 \leqslant i \leqslant n)$. Then, if $f \in S_p B(\alpha, \sigma, \theta, n)$, it follows that $f \in S_q B(\alpha^*, \sigma^*, \theta, n)$. In other words, the following imbedding holds:*

$$S_p B(\alpha, \sigma, \theta, n) \subset S_q B(\alpha^*, \sigma^*, \theta, n),$$

where

$$\alpha^* = \{\alpha_1^*, \ldots, \alpha_n^*\}, \qquad \sigma^* = \{\sigma_1^*, \ldots, \sigma_n^*\},$$

$$\alpha_1^*(t) = \alpha_i(t)\, t^{(1/p-1/q)\theta}, \qquad \sigma_i^* = \sigma_i - \left(\frac{1}{p} - \frac{1}{q}\right)\theta.$$

PROOF If some function $f \in S_pB(\alpha, \sigma, \theta, n)$, then, by Theorem 8, $f \in \Sigma S_pB^0(\alpha, \sigma, \theta, n)$, i.e. each function $F = F_{i_1 \cdots i_m}$ of representation (1) satisfies inequality (14). Applying Lemma 8, we find that

$$I_5 = \sum_{\nu_{i_1}=0}^{\infty} \cdots \sum_{\nu_{i_m}=0}^{\infty} \prod_{j=1}^{m} \mu_{i_j}(\nu_{i_j}) Y^{\theta}_{[2^{\nu_{i_1}}-1] \ldots [2^{\nu_{i_m}}-1]}(F)_p < \infty.$$

Since $\alpha_i(t)$ satisfies the λ_i-condition, it follows by Lemma 3 that $2^{n\lambda_i} \leqslant C_1\mu_1(n)$, where C_1 does not depend on n. Hence the following series converges for $1/p - 1/q < \lambda_{i_j}/\theta$:

$$\sum_{\nu_{i_1}=0}^{\infty} \cdots \sum_{\nu_{i_m}=0}^{\infty} \prod_{j=1}^{m} 2^{\nu_{i_j}\left(\frac{1}{p}-\frac{1}{q}\right)} Y_{[2^{\nu_{i_1}}-1] \ldots [2^{\nu_{i_m}}-1]}(F)_p.$$

Now applying Lemma 11, we find that $F \in \mathfrak{L}_q([0, 2\pi]^n)$ and

$$Y_{2^{N_{i_1}}-1 \ldots 2^{N_{i_m}}-1}(F)_q \leqslant C_2 \sum_{\nu_{i_1}=N_{i_1}}^{\infty} \cdots \sum_{\nu_{i_m}=N_{i_m}}^{\infty} \prod_{j=1}^{m} 2^{\nu_{i_j}\left(\frac{1}{p}-\frac{1}{q}\right)} Y_{[2^{\nu_{i_1}}-1] \ldots [2^{\nu_{i_m}}-1]}(F)_p,$$

where C_2 does not depend on N_i.

We now consider

$$I_6 = \sum_{l_{i_1}=0}^{\infty} \cdots \sum_{l_{i_m}=0}^{\infty} \prod_{j=1}^{m} \mu^*_{i_j}(l_{i_j}) Y^{\theta}_{[2^{l_{i_j}}-1] \ldots [2^{l_{i_m}}-1]}(F)_q,$$

where

$$\mu^*_i(l_i) = \int_{{}^1\!/_2{}^{l_i}}^{{}^1\!/_2{}^{l_i-1}} \alpha^*_i(t)\,dt \leqslant C_3 2^{-l_i\theta\left(\frac{1}{p}-\frac{1}{q}\right)} \mu_i(l_i), \qquad l_i \geqslant 1,$$

$$\mu^*_i(0) = \mu_i(0),$$

where C_3 does not depend on l_i.

Taking into account the estimates for $\mu^*_i(l_i)$ and $Y_{2^{N_{i_1}-1} \ldots 2^{N_{i_m}-1}}(F)_q$, we obtain

$$I_6 \leqslant C_4 \sum_{l_{i_1}=0}^{\infty} \cdots \sum_{l_{i_m}=0}^{\infty} \prod_{j=1}^{m} 2^{-l_{i_j}\left(\frac{1}{p}-\frac{1}{q}\right)\theta} \mu_{i_j}(l_{i_j}) B^{\theta},$$

where

$$B = \sum_{\nu_{i_1}=l_{i_1}}^{\infty} \cdots \sum_{\nu_{i_m}=l_{i_m}}^{\infty} \prod_{j=1}^{m} 2^{\nu_{i_j}\lambda_{i_j}/\theta} Y_{[2^{\nu_{i_1}}-1] \ldots [2^{\nu_{i_m}}-1]}(F)_p \prod_{j=1}^{m} 2^{\nu_{i_j}\left(\frac{1}{p}-\frac{1}{q}-\frac{\lambda_{i_j}}{\theta}\right)}.$$

Applying Hölder's inequality to B and taking into account that $1/p - 1/q - \lambda_{i_j}/\theta < 0$, we have

$$B^\theta \leqslant C_5 \prod_{j=1}^m 2^{l_{i_j}\left(\frac{1}{p}-\frac{1}{q}-\frac{\lambda_{i_j}}{\theta}\right)\theta} \sum_{\nu_{i_1}=l_{i_1}}^\infty \cdots \sum_{\nu_{i_m}=l_{i_m}}^\infty \prod_{j=1}^m 2^{\nu_{i_j}\lambda_{i_j}} Y^\theta_{[2^{\nu_{i_1}}-1]\ldots[2^{\nu_{i_m}}-1]}(F)_p,$$

where C_5 does not depend on l_i.

Substituting this estimate in the estimate for I_6, we obtain

$$I_6 \leqslant C_6 \sum_{l_{i_1}=0}^\infty \cdots \sum_{l_{i_m}=0}^\infty \prod_{j=1}^m 2^{-l_{i_j}\lambda_{i_j}} \mu_{i_j}(l_{i_j})$$

$$\times \sum_{\nu_{i_1}=l_{i_1}}^\infty \cdots \sum_{\nu_{i_m}=l_{i_m}}^\infty \prod_{j=1}^m 2^{\nu_{i_j}\lambda_{i_j}} Y^\theta_{[2^{\nu_{i_1}}-1]\ldots[2^{\nu_{i_m}}-1]}(F)_p$$

$$= C_6 \sum_{\nu_{i_1}=0}^\infty \cdots \sum_{\nu_{i_m}=0}^\infty \prod_{j=1}^m 2^{\nu_{i_j}\lambda_{i_j}} Y^\theta_{[2^{\nu_{i_1}}-1]\ldots[2^{\nu_{i_m}}-1]}(F)_p \prod_{j=1}^m \sum_{l_{i_j}=0}^{\nu_{i_j}} 2^{-l_{i_j}\lambda_{i_j}} \mu_{i_j}(l_{i_j}).$$

Now applying Lemma 3, we arrive at the conclusion that $I_6 \leqslant C_7 I_5 < \infty$. Taking into account that $\alpha_i^*(t)$ satisfies the σ_i^*-condition and applying Lemma 8, we find that each function $F_{i_1\cdots i_m}$ in representation (1) satisfies the inequality

$$\int_0^{2\pi} \cdots \int_0^{2\pi} \omega^\theta_{k_{i_1}\ldots k_{i_m}}(F_{i_1\ldots i_m}, t_{i_1}, \ldots, t_{i_m})_q \prod_{j=1}^m \alpha^*_{i_j}(t_{i_j})\, dt_{i_1} \ldots dt_{i_m} < \infty,$$

where $k_{i_j} \geqslant \sigma^*_{i_j}/\theta$. But this means that $f \in \Sigma S_q B^0(\alpha^*, \sigma^*, \theta, n)$. Applying Theorem 8, we find that $f \in S_q B(\alpha^*, \sigma^*, \theta, n)$, and Theorem 12 is proved.

THEOREM 13 (ON THE TRACE OF A FUNCTION). *Suppose that each function $\alpha_i(t)$ satisfies the λ_i-condition, $\lambda_i > 0$; suppose given a collection of indices $j_1, \cdots, j_s$ such that $1 \leqslant j_l \leqslant n$, $1 \leqslant l \leqslant s < n$; finally, suppose that $1/p < \lambda_{i_l}/\theta$ for $1 \leqslant l \leqslant s$. Then, if $f \in S_p B(\alpha, \sigma, \theta, n)$, it follows that the function has a trace $\varphi(x_{i_{s+1}}, \cdots, x_{i_n}) \in S_p B(\alpha^*, \sigma^*, \theta, n-s)$. In other words, the following imbedding holds*:

$$S_p B(\alpha, \sigma, \theta, n) \subset S_p B(\alpha^*, \sigma^*, \theta, n-s),$$

where $\alpha^* = \{\alpha_{i_{s+1}}, \cdots, \alpha_{i_n}\}$, $\sigma^* = \{\sigma_{i_{s+1}}, \cdots, \sigma_{i_n}\}$.

The proof of this theorem almost completely repeats the proof of Theorem 5. The only difference is in the proof of the existence of the trace of a function $F_{j_1\cdots j_r i_{s+1},\cdots,i_m}$ of the third class. Hence here we only give that part of the proof of Theorem 13, and omit the rest.

Since $f \in \Sigma S_p B^0(\alpha, \sigma, \theta, n)$, it follows that $F = F_{j_1\cdots j_r i_{s+1}\cdots i_m}$ satisfies the inequality

$$\int_0^{2\pi}\cdots\int_0^{2\pi}\omega^\theta_{k_{j_1}\ldots k_{j_r}k_{i_{s+1}}\ldots k_{i_m}}(F,\ t_{j_1},\ \ldots\ t_{j_r},\ t_{i_{s+1}}\ldots\ t_{i_m})_p$$
$$\times\prod_{l=1}^r\alpha_{j_l}(t_{j_l})\prod_{l=s+1}^m\alpha_{i_l}(t_{i_l})\,dt_{j_1}\ldots dt_{j_r}dt_{i_{s+1}}\ldots dt_{i_m}<\infty.$$

Applying Lemma 8, we obtain

$$I_7=\sum_{\nu_{j_1}=0}^\infty\cdots\sum_{\nu_{j_r}=0}^\infty\sum_{\nu_{i_{s+1}}=0}^\infty\cdots\sum_{\nu_{i_m}=0}^\infty\prod_{l=1}^r\mu_{j_l}(\nu_{i_l})$$
$$\times\prod_{l=s+1}^m\mu_{i_l}(\nu_{i_l})Y^\theta_{[2^{\nu_{j_1}}-1]\ldots[2^{\nu_{j_r}}-1][2^{\nu_{i_{s+1}}}-1][2^{\nu_{i_m}}-1]}(F)_p<\infty.$$

Since $\alpha_i(t)$ satisfies the λ_i-condition, it follows by Lemma 3 that $2^{n\lambda_i}\leqslant C_1\mu_i(n)$, where C_1 does not depend on n. We hence have the inequality

$$I_8=\sum_{\nu_{j_1}=0}^\infty\cdots\sum_{\nu_{j_r}=0}^\infty\sum_{\nu_{i_{s+1}}=0}^\infty\cdots\sum_{\nu_{i_m}=0}^\infty\prod_{l=1}^r2^{\nu_{j_l}\lambda_{i_l}}$$
$$\times\prod_{l=s+1}^m2^{\nu_{i_l}\lambda_{i_l}}Y^\theta_{[2^{\nu_{j_1}}-1]\ldots[2^{\nu_{j_r}}-1][2^{\nu_{i_{s+1}}}-1]\ldots[2^{\nu_{i_m}}-1]}(F)_p<\infty.$$

We consider

$$I_9=\sum_{\nu_{j_1}=0}^\infty\cdots\sum_{\nu_{j_r}=0}^\infty\sum_{\nu_{i_{s+1}}=0}^\infty\cdots\sum_{\nu_{i_m}=0}^\infty\prod_{l=1}^r2^{\nu_{j_l}/p}Y_{[2^{\nu_{j_1}}-1]\ldots[2^{\nu_{j_r}}-1][2^{\nu_{i_{s+1}}}-1]\ldots[2^{\nu_{i_m}}-1]}(F)_p.$$

Since $1/p<\lambda_{j_l}/\theta$, on applying Hölder's inequality we find that $I_9\leqslant C_1I_8<\infty$. Further applying Lemma 12, we arrive at the conclusion that, after modification on a set of $(m-s+r)$-dimensional measure zero $(m-s+r<n)$, the function F has a trace $\varphi=\varphi_{i_{s+1}\cdots i_m}$ on R_{m-s-r}, and, in addition, the function φ satisfies the inequality

$$Y_{2^{N_{i_{s+1}}-1}\ldots 2^{N_{i_m}-1}}(\varphi)_p\leqslant C_1\sum_{\nu_{j_1}=0}^\infty\cdots\sum_{\nu_{j_r}=0}^\infty\sum_{\nu_{i_{s+1}}=N_{i_{s+1}}}^\infty\cdots\sum_{\nu_{i_m}=N_{i_m}}^\infty K\prod_{l=1}^r2^{\nu_{j_l}\lambda_{j_l}/\theta}$$
$$\times\prod_{l=s+1}^m2^{\nu_{i_l}\lambda_{j_l}/\theta}Y_{[2^{\nu_{j_1}}-1]\ldots[2^{\nu_{j_r}}-1][2^{\nu_{i_{s+1}}}-1]\ldots[2^{\nu_{i_m}}-1]}(F)_p,$$

where

$$K=\prod_{l=1}^r2^{\nu_{j_l}\left(\frac{1}{p}-\frac{\lambda_{j_l}}{\theta}\right)}\prod_{l=s+1}^m2^{-\nu_{i_l}\lambda_{i_l}/\theta}$$

and C_1 does not depend on N_i or F.

Applying Hölder's inequality to the right side of this inequality and taking into account that $1/p - \lambda_{i_l}/\theta < 0$ for $1 \leqslant l \leqslant r$, we obtain

$$Y^{\theta}_{2^{N_{i_{s+1}}-1}\dots 2^{N_{i_m}-1}}(\varphi)_p$$

$$\leqslant C_2 \prod_{l=s+1}^{m} 2^{-N_{i_l}\lambda_{i_l}} \sum_{\nu_{j_1}=0}^{\infty} \dots \sum_{\nu_{j_r}=0}^{\infty} \sum_{\nu_{i_{s+1}}=N_{i_{s+1}}}^{\infty} \dots \sum_{\nu_{i_m}=N_{i_m}}^{\infty} \prod_{l=1}^{r} 2^{\nu_{j_l}\lambda_{j_l}} \prod_{l=s+1}^{m} 2^{\nu_{i_l}\lambda_{i_l}}$$

$$\times Y^{\theta}_{[2^{\nu_{j_1}}-1]\dots[2^{\nu_{j_r}}-1][2^{\nu_{i_{s+1}}}-1]\dots[2^{\nu_{i_m}}-1]}(F)_p.$$

We consider

$$I_{10} = \sum_{l_{i_{s+1}}=0}^{\infty} \dots \sum_{l_{i_m}=0}^{\infty} \prod_{j=s+1}^{m} \mu^*_{i_j}(l_{i_j})\, Y^{\theta}_{[2^{l_{i_{s+1}}}-1]\dots[2^{l_{i_m}}-1]}(\dot{\varphi})_p,$$

where $\mu^*_{i_j} = \mu_{i_j}$ for $s+1 \leqslant j \leqslant m$. Taking into account the estimate for $Y_{2^{N_{i_{s+1}}}-1\dots 2^{N_{i_m}}-1}(\varphi)_p$, after changing the order of summation we obtain

$$I_{10} \leqslant C_3 \sum_{\nu_{j_1}=0}^{\infty} \dots \sum_{\nu_{j_r}=0}^{\infty} \sum_{\nu_{i_{s+1}}=0}^{\infty} \dots \sum_{\nu_{i_m}=0}^{\infty} \prod_{l=1}^{r} 2^{\nu_{j_l}\lambda_{j_l}} \prod_{l=s+1}^{m} 2^{\nu_{i_l}\lambda_{i_l}}$$

$$\times Y^{\theta}_{[2^{\nu_{j_1}}-1]\dots[2^{\nu_{j_r}}-1][2^{\nu_{i_{s+1}}}-1]\dots[2^{\nu_{i_m}}-1]}(F)_p \cdot M,$$

where

$$M = \prod_{j=s+1}^{m} \sum_{l_{i_j}=0}^{\nu_{i_j}} \mu_{i_j}(l_{i_j})\, 2^{-l_{i_j}\lambda_{i_j}}.$$

Now applying Lemma 3, we have

$$M \leqslant C_4 \prod_{j=s+1}^{m} \mu_{i_j}(\nu_{i_j})\, 2^{-\nu_{i_j}\lambda_{i_j}},$$

where C_4 does not depend on ν_{i_j}. Further taking into account that $2^{\nu_i\lambda_i} \leqslant C_5\mu_i(\nu_i)$, where C_5 does not depend on ν_i, we obtain $I_{10} \leqslant C_6 I_8 < \infty$.

Thus, applying Lemma 8, we arrive at the conclusion that each function $F_{j_1\dots j_r i_{s+1}\dots i_m}$ of the third class has a trace $\varphi_{i_{s+1}\dots i_m}$ which satisfies the condition

$$\int_0^{2\pi} \dots \int_0^{2\pi} \omega^{\theta}_{k_{i_{s+1}}\dots k_{i_m}}(\varphi_{i_{s+1}\dots i_m}, t_{i_{s+1}}, \dots, t_{i_m})_p \prod_{j=s+1}^{m} \alpha_{i_j}(t_{i_j})\, dt_{i_{s+1}} \dots dt_{i_m} < \infty,$$

where $k_{i_j} \geqslant \sigma^*_{i_j}/\theta$ for $s+1 \leqslant j \leqslant m$.

The end of the proof of Theorem 13 proceeds in the same way as the last part of the proof of Theorem 5, and hence we will not give it here.

THEOREM 14 (EXTENSION THEOREM). *If $f \in S_pB(\alpha, \sigma, \theta, n)$, then for any $m < n$ the function can be extended onto R_m so that*

1) *the extended function* $F \in S_p B(\alpha^*, \sigma^*, \theta, m)$, *where* $\alpha_1^* = \{\alpha_1, \cdots, \alpha_n, \alpha_{n+1}, \cdots, \alpha_m\}$, $\sigma^* = \{\sigma_1, \cdots, \sigma_n, \sigma_{n+1}, \cdots, \sigma_m\}$, σ_i *are arbitrary real numbers for* $n+1 \leqslant i \leqslant m$, *and the* $\alpha_i(t)$ *are arbitrary functions defined in* §1 *and satisfying the* σ_i*-conditions for* $n+1 \leqslant i \leqslant m$; *and*

2) $F(x_1, \cdots, x_n, 0, \cdots, 0) \equiv f$.

PROOF. We consider the function $\psi(t)$ that was introduced in the proof of Theorem 6. Suppose that this function is l times continuously differentiable, where $l \geqslant \sigma_i/\theta$ for all $i, n+1 \leqslant i \leqslant m$; $\psi(0) = 1$ and, for all $k = 0, 1, \cdots, l$,

$$\int_0^{2\pi} |\psi^k(t)|^p \, dt \leqslant C.$$

As was shown in the proof of Theorem 6, for $k_i \geqslant l$ we have

$$\omega_{k_i}(\psi(x_i), \delta_i)_p \leqslant C_1 \delta_i^l,$$

where C_1 does not depend on ψ or δ_i. Since $l \geqslant \sigma_i/\theta$, we have $\omega_{k_i}(\psi(x_i), \delta_i)_p \leqslant C_1 \delta_i^{\sigma_i/\theta}$.

We now consider

$$I = \int_0^{2\pi} \omega_{k_i}^\theta(\psi(x_i), t_i)_p \, \alpha_i(t_i) \, dt_i.$$

Applying the estimate we have just obtained, we have

$$I \leqslant C_1 \int_0^{2\pi} t_i^{\sigma_i} \alpha_i(t_i) \, dt_i = C_1 \left(\int_0^{\pi} + \int_{\pi}^{2\pi} \right).$$

Since $\alpha_i(t)$ satisfies the σ_i-condition, it follows that

$$I \leqslant C_2 \left[(\pi)^{\sigma_i} \int_{\pi}^{2\pi} \alpha_i(t_i) \, dt_i + \int_{\pi}^{2\pi} \alpha_i(t_i) \, dt_i \right] \leqslant C_3 \int_{\pi}^{2\pi} \alpha_i(t_i) \, dt_i.$$

Since $\alpha_i(t)$ is summable on $[\pi, 2\pi]$, we have

$$\int_{\pi}^{2\pi} \alpha_i(t_i) \, dt_i = C_4.$$

We have thus obtained $I \leqslant C_5 < \infty$.

It is now completely clear that the function

$$F(x_1, \ldots, x_m) = f(x_1, \ldots, x_n) \prod_{i=n+1}^{m} \psi(x_i)$$

satisfies the conditions of Theorem 18.

§5. The function class $S_pW(\alpha, \theta, n)$

Suppose that we are given a natural number n, a number p in the interval $1 < p < \infty$, a number θ in the interval $1 \leqslant \theta < \infty$, and a vector $\alpha = \{\alpha_1, \cdots, \alpha_n\}$, where $\alpha_i = \alpha_i(t)$ are the functions defined in §1.

We say that $f \in S_pW(\alpha, \theta, n)$ if

1) $f \in \mathfrak{L}_p([0, 2\pi]^n)$;

2) the Fourier series of f has the form (2) (see §1), and the series

$$\sum_{\nu_1=0}^{\infty} \cdots \sum_{\nu_n=0}^{\infty} A_{\nu_1 \ldots \nu_n} \prod_{i=1}^{n} \beta_i^{1/\theta}(\nu_i)$$

is the Fourier series of some function $\varphi \in \mathfrak{L}_p[(0, 2\pi]^n)$. Here $\beta_i(0) = 1$, and for $\nu_i \geqslant 1$

$$\beta_i(\nu_i) = \int_{1/\nu_i}^{2\pi} \alpha_i(t)\, dt.$$

THEOREM 15. *If $\theta = p$ for $1 < p \leqslant 2$ or $\theta = 2$ for $2 \leqslant p < \infty$, then the condition $f \in S_pB(\alpha, \sigma, \theta, n)$ implies that $f \in S_pW(\alpha, \theta, n)$. In other words, the following imbedding holds*:

$$S_pB(\alpha, \sigma, \theta, n) \subset S_pW(\alpha, \theta, n).$$

THEOREM 16. *If $\theta = 2$ for $1 < p \leqslant 2$ or $\theta = p$ for $2 \leqslant p < \infty$, then the condition $f \in S_pW(\alpha, \theta, n)$ and the condition that each function $\alpha_i(t)$ satisfies the σ_i-condition imply that $f \in S_pB(\alpha, \sigma, \theta, n)$. In other words, the following imbedding holds*:

$$S_pW(\alpha, \theta, n) \subset S_pB(\alpha, \sigma, \theta, n).$$

PROOF OF THEOREM 15. For simplicity, let $n = 2$. We then write $f = f(x, y)$. By Theorem 9 the condition $f \in S_pB(\alpha, \sigma, \theta, 2)$ implies that $f \in \Sigma S_pB^0(\alpha, \sigma, \theta, 2)$. In other words, the function f is representable in the form (1) (see Lemma 1), i.e.

$$f = F_{12} + F_1 + F_2 + F_0,$$

and the functions F_1, F_2 and F_{12} satisfy the inequalities

$$\int_0^{2\pi} \alpha_1(t_1)\,\omega_{k_1}^{\theta}(F_1, t_1)_p\,dt_1 < \infty,$$

$$\int_0^{2\pi} \alpha_2(t_2)\,\omega_{k_2}^{\theta}(F_2, t_2)_p\,dt_2 < \infty, \tag{17}$$

$$I_1 = \int_0^{2\pi}\int_0^{2\pi} \alpha_1(t_1)\,\alpha_2(t_2)\,\omega_{k_1k_2}^{\theta}(F_{12}, t_1, t_2)_p\,dt_1dt_2 < \infty,$$

where $k_i \geqslant \sigma_i/\theta$. In addition, by Lemma 1 the Fourier series of f is equal to the sum of the Fourier series of the functions F_1, F_2 and F_{12}. We note that the Fourier series of F_1, F_2 and F_{12} have the form (3) (see §1), i.e. they can be written in the form

$$F_1 \sim \sum_{\nu_1=1}^{\infty} A_{\nu_1}(x), \quad F_2 \sim \sum_{\nu_2=1}^{\infty} A_{\nu_2}(y), \quad F_{12} \sim \sum_{\nu_1=1}^{\infty}\sum_{\nu_2=1}^{\infty} A_{\nu_1\nu_2}(x, y).$$

To prove the theorem, we must show that the series

$$\sum_{\nu_1=1}^{\infty} A_{\nu_1}(x)\,\beta_1^{1/\theta}(\nu_1), \tag{18}$$

$$\sum_{\nu_2=1}^{\infty} A_{\nu_2}(y)\,\beta_2^{1/\theta}(\nu_2), \tag{19}$$

$$\sum_{\nu_1=1}^{\infty}\sum_{\nu_2=1}^{\infty} A_{\nu_1\nu_2}(x, y)\,\beta_1^{1/\theta}(\nu_1)\,\beta_2^{1/\theta}(\nu_2) \tag{20}$$

are the Fourier series of certain functions

$$\varphi_1(x) \in \mathscr{L}_p([0, 2\pi]), \ \varphi_2(y) \in \mathscr{L}_p([0, 2\pi]) \text{ and } \varphi_{12}(x, y) \in \mathscr{L}_p([0, 2\pi]^2).$$

The fact that the series (18) and (19) are the Fourier series of certain functions $\varphi_1(x) \in \mathfrak{L}_p([0, 2\pi])$ and $\varphi_2(x) \in \mathfrak{L}_p([0, 2\pi])$ is proved in [9]. It remains to show that the series (20) is the Fourier series of some function $\varphi_{12} \in \mathfrak{L}_p([0, 2\pi]^2)$.

Since F_{12} satisfies condition (17), it follows by Lemma 8 that

$$\sum_{\nu_1=0}^{\infty}\sum_{\nu_2=0}^{\infty} \mu_1(\nu_1)\,\mu_2(\nu_2)\,Y^{\theta}_{[2^{\nu_1}-1][2^{\nu_2}-1]}(F_{12})_p < \infty.$$

By Lemma 13,

$$\|F_{12} - u^*_{[2^{\nu_1}-1][2^{\nu_2}-1]}(F_{12})\|_p \leqslant C_1 Y_{[2^{\nu_1}-1][2^{\nu_2}-1]}(F_{12})_p,$$

where C_1 does not depend on F_{12}, ν_1 or ν_2, and

$$u^*_{[2^{\nu_1}-1][2^{\nu_2}-1]}(F_{12}) = S_{[2^{\nu_1}-1]\infty}(F_{12}) + S_{\infty[2^{\nu_2}-1]}(F_{12}) - S_{[2^{\nu_1}-1][2^{\nu_2}-1]}(F_{12}).$$

It is clear that the function $F_{12} - u^*_{[2^{\nu_1-1}][2^{\nu_2-1}]}(F_{12})$ has the Fourier series

$$\sum_{l_1=[2^{\nu_1-1}]+1}^{\infty}\ \sum_{l_2=[2^{\nu_2-1}]+1}^{\infty} A_{l_1l_2}(x,\ y).$$

Applying Lemma 4 to this function, we obtain

$$\left(\int_0^{2\pi}\int_0^{2\pi}\left[\sum_{l_1=\nu_1}^{\infty}\sum_{l_2=\nu_2}^{\infty}\Delta^2_{l_1l_2}\right]^{p/2} dxdy\right)^{1/p} \leqslant C_2\left\| F_{12}-u^*_{[2^{\nu_1-1}][2^{\nu_2-1}]}(F_{12})\right\|_p,$$

where C_2 does not depend on F_{12}, ν_1 or ν_2.

Hence it follows from condition (17) that

$$I_2=\sum_{\nu_1=0}^{\infty}\sum_{\nu_2=0}^{\infty}\mu_1(\nu_1)\mu_2(\nu_2)\left(\int_0^{2\pi}\int_0^{2\pi}\left[\sum_{l_1=\nu_1}^{\infty}\sum_{l_2=\nu_2}^{\infty}\Delta^2_{l_1l_2}\right]^{p/2} dxdy\right)^{\theta/p}\leqslant C_3I_1<\infty,$$

where C_3 does not depend on F_{12}.

We now prove the inequality

$$I_3=\int_0^{2\pi}\int_0^{2\pi}\left\{\sum_{l_1=0}^{\infty}\sum_{l_2=0}^{\infty}[\Delta_{l_1l_2}\beta_1^{1/\theta}(2^{l_1})\beta_2^{1/\theta}(2^{l_2})]^2\right\}^{p/2} dxdy\leqslant C_4I_1^{p/\theta}<\infty, \tag{21}$$

where C_4 does not depend on F_{12}.

We prove this separately for $1<p\leqslant 2$ and $2\leqslant p<\infty$.

a) If $1<p\leqslant 2$, then $\theta=p$. We first consider

$$I_4=\left\{\sum_{l_1=0}^{\infty}\sum_{l_2=0}^{\infty}[\Delta_{l_1l_2}\beta_1^{1/p}(2^{l_1})\beta_2^{1/p}(2^{l_2})]^2\right\}^{p/2}.$$

It is clear that

$$I_4\leqslant\left\{\sum_{l_1=0}^{\infty}\sum_{l_2=0}^{\infty}\left[\sum_{\nu_1=0}^{l_1}\sum_{\nu_2=0}^{l_2}|\Delta_{l_1l_2}|^p\mu_1(\nu_1)\mu_2(\nu_2)\right]^{2/p}\right\}^{p/2}.$$

Applying the generalized Minkowski inequality to this sum, we obtain

$$I_4\leqslant\sum_{\nu_1=0}^{\infty}\sum_{\nu_2=0}^{\infty}\mu_1(\nu_1)\mu_2(\nu_2)\left\{\sum_{l_1=\nu_1}^{\infty}\sum_{l_2=\nu_2}^{\infty}\Delta^2_{l_1l_2}\right\}^{p/2}.$$

But this implies the inequality $I_3\leqslant I_1C_4$.

b) If $2\leqslant p<\infty$, then $\theta=2$. It is clear that

$$I_3^{2/p}\leqslant\left(\int_0^{2\pi}\int_0^{2\pi}\left\{\sum_{l_1=0}^{\infty}\sum_{l_2=0}^{\infty}\Delta^2_{l_1l_2}\sum_{\nu_1=0}^{l_1}\sum_{\nu_2=0}^{l_2}\mu_1(\nu_1)\mu_2(\nu_2)\right\}^{p/2} dxdy\right)^{2/p}$$
$$=\left(\int_0^{2\pi}\int_0^{2\pi}\left\{\sum_{\nu_1=0}^{\infty}\sum_{\nu_2=0}^{\infty}\mu_1(\nu_1)\mu_2(\nu_2)\sum_{l_1=\nu_1}^{\infty}\sum_{l_1=\nu_2}^{\infty}\Delta^2_{l_1l_2}\right\}^{p/2} dxdy\right)^{2/p}.$$

Applying the generalized Minkowski inequality, we obtain

$$I_3^{2/p} \leqslant \sum_{\nu_1=0}^{\infty} \sum_{\nu_2=2}^{\infty} \mu_1(\nu_1)\mu_2(\nu_2)\left(\int_0^{2\pi}\int_0^{2\pi}\left\{\sum_{l_1=\nu_1}^{\infty}\sum_{l_2=\nu_2}^{\infty} \Delta_{l_1 l_2}^2\right\}^{p/2} dxdy\right)^{2/p} \leqslant C_5 I_1 < \infty.$$

Thus inequality (21) is proved for all p $(1 < p < \infty)$. We now consider the trigonometric series

$$\sum_{l_1=1}^{\infty}\sum_{l_2=1}^{\infty} A_{l_1 l_2}(x, y)\gamma_{l_1 l_2},$$

where $\gamma_{l_1 l_2}^{\theta} = \beta_1(2^{\nu_1})\beta_2(2^{\nu_2})$ for $[2^{\nu_i - 1}] + 1 \leqslant l_i \leqslant 2^{\nu_i}, \nu_i = 0, 1, 2, \cdots$. The condition $I_3 < 0$ together with Lemma 4 imply that this series is the Fourier series of some function $g(x, y) \in \mathfrak{L}_p([0, 2\pi]^2)$, where $\|g\|_p \leqslant CI_3^{1/\theta}$.

We now consider the sequence of numbers

$$\lambda_{l_1 l_2}^{\theta} = \frac{\beta_1(l_1)\beta_2(l_2)}{\gamma_{l_1 l_2}}.$$

We easily verify that this sequence satisfies the conditions of Lemma 5, and thus the trigonometric series

$$\sum_{l_1=1}^{\infty}\sum_{l_2=1}^{\infty} A_{l_1 l_2}\gamma_{l_1 l_2}\lambda_{l_1 l_2} \equiv \sum_{l_1=1}^{\infty}\sum_{l_2=1}^{\infty} A_{l_1 l_2}\beta_1^{1/\theta}(l_1)\beta_2^{1/\theta}(l_2)$$

is the Fourier series of some function $\varphi_{12}(x, y) \in \mathfrak{L}_p([0, 2\pi]^2)$, where $\|\varphi_{12}\|_p \leqslant CI_1^{1/\theta}$, and Theorem 15 is proved.

PROOF OF THEOREM 16. Again let $n = 2$ and $f = f(x, y)$. By Theorem 9 the conditions $f \in S_p B(\alpha, \sigma, \theta, 2)$ and $f \in \Sigma S_p B^0(\alpha, \sigma, \theta, 2)$ are equivalent. Hence to prove Theorem 16 we must show that $f \in \Sigma S_p B^0(\alpha, \sigma, \theta, 2)$, i.e. we must prove the inequalities

$$I_4 = \int_0^{2\pi} \alpha_1(t_1)\omega_{k_1}^{\theta}(F_1, t_1)_p\, dt_1 < \infty,$$

$$I_5 = \int_0^{2\pi} \alpha_2(t_2)\omega_{k_2}^{\theta}(F_2, t_2)_p\, dt_2 < \infty,$$

$$I_6 = \int_0^{2\pi}\int_0^{2\pi} \alpha_1(t_1)\alpha_2(t_2)\omega_{k_1 k_2}^{\theta}(F_{12}, t_1, t_2)_p\, dt_1 dt_2 < \infty.$$

We recall that the condition $f \in S_p W(\alpha, \theta, 2)$ means that $f \in \mathfrak{L}_p([0, 2\pi]^2)$ has a Fourier series

$$\sum_{\nu_1=0}^{\infty}\sum_{\nu_2=0}^{\infty} A_{\nu_1 \nu_2}(x, y),$$

and the series

$$\sum_{\nu_1=0}^{\infty}\sum_{\nu_2=0}^{\infty} A_{\nu_1\nu_2}(x,\ y)\,\beta_1^{1/\theta}(\nu_1)\,\beta_2^{1/\theta}(\nu_2)$$

is the Fourier series of some function $\varphi \in \mathfrak{L}_p([0,\ 2\pi]^2)$.

Applying Lemma 1 to the function f, we find that f is representable in the form

$$f = F_{12} + F_1 + F_2 + F_0,$$

where the functions $F_1 \in \mathfrak{L}_p$, $F_2 \in \mathfrak{L}_2$ and $F_{12} \in \mathfrak{L}_p([0,\ 2\pi]^2)$ have the Fourier series

$$F_1 \sim \sum_{\nu_1=1}^{\infty} A_{\nu_1}(x),\quad F_2 \sim \sum_{\nu_2=1}^{\infty} A_{\nu_2}(y),\quad F_{12} \sim \sum_{\nu_1=1}^{\infty}\sum_{\nu_2=1}^{\infty} A_{\nu_1\nu_2}(x,\ y).$$

Applying Lemma 1 to the function $\varphi(x,\ y)$, we find that φ is representable in the form

$$\varphi = \varphi_{12} + \varphi_1 + \varphi_2 + \varphi_0,$$

where the functions $\varphi_1 \in \mathfrak{L}_p$, $\varphi_2 \in \mathfrak{L}_p$ and $\varphi_{12} \in \mathfrak{L}_p([0,\ 2\pi]^2)$ have the Fourier series

$$\varphi_1 \sim \sum_{\nu_1=1}^{\infty} A_{\nu_1}(x)\,\beta_1^{1/\theta}(\nu_1),\qquad \varphi_2 \sim \sum_{\nu_2=1}^{\infty} A_{\nu_2}(y)\,\beta_2^{1/\theta}(\nu_2),$$

$$\varphi_{12} \sim \sum_{\nu_1=1}^{\infty}\sum_{\nu_2=1}^{\infty} A_{\nu_1\nu_2}(x,\ y)\,\beta_1^{1/\theta}(\nu_1)\,\beta_2^{1/\theta}(\nu_2).$$

In addition, it follows from the definition of the class $S_pW(\alpha,\ \theta,\ 2)$ that $\varphi_0 = F_0$.

It is now completely clear that to prove Theorem 16 we must prove the following assertions:

1) If $F_1 \in \mathfrak{L}_p$ and $\varphi_1 \in \mathfrak{L}_p$, then $I_4 < \infty$;

2) if $F_2 \in \mathfrak{L}_p$ and $\varphi_2 \in \mathfrak{L}_p$, then $I_5 < \infty$;

3) if $F_{12} \in \mathfrak{L}_p$ and $\varphi_{12} \in \mathfrak{L}_p$, then $I_6 < \infty$.

The first two assertions are proved in [9]. It remains to prove the third.

We consider the series

$$\sum_{\nu_1=1}^{\infty}\sum_{\nu_2=1}^{\infty} A_{\nu_1\nu_2}(x,\ y)\,\beta_1^{1/\theta}(\nu_1)\,\beta_2^{1/\theta}(\nu_2)\,\lambda_{\nu_1\nu_2}, \tag{22}$$

where

$$\lambda_{\nu_1\nu_2}^{\theta} = \frac{\beta_1(2^{l_1})\,\beta_2(2^{l_2})}{\beta_1(\nu_1)\,\beta_2(\nu_2)}$$

for $[2^{l_1-1}] + 1 \leqslant \nu_1 \leqslant 2^{l_1}$, $[2^{l_2-1}] + 1 \leqslant \nu_2 \leqslant 2^{l_2}$, $l_i = 0,\ 1,\ 2,\ \cdots$. Taking into account that $\alpha_i(t)$ satisfies the σ_i-condition, we easily show that the sequence $\lambda_{\nu_1\nu_2}$ satisfies the conditions of Lemma 5, and hence the series (22) is the Fourier series of some function $g_{12} \in \mathfrak{L}_p([0, 2\pi]^2)$, with $\|g_{12}\|_p \leqslant C_1 \|\varphi_{12}\|_p$, where C_1 does not depend on φ_{12} or g_{12}.

Now applying Lemma 4 to the function g_{12}, we obtain

$$I_7 = \left\{\int_0^{2\pi}\int_0^{2\pi}\left[\sum_{l_1=0}^{\infty}\sum_{l_2=0}^{\infty}\Delta_{l_1 l_2}^2 \beta_1^{2/\theta}(2^{l_1})\,\beta_2^{2/\theta}(2^{l_2})\right]^{p/2} dxdy\right\}^{1/p} \leqslant C_2\|\varphi_{12}\|_p,$$

where C_2 does not depend on φ_{12}.

We now show that

$$I_8 = \sum_{\nu_1=0}^{\infty}\sum_{\nu_2=0}^{\infty}\mu_1(\nu_1)\,\mu_2(\nu_2)\left\{\int_0^{2\pi}\int_0^{2\pi}\left[\sum_{l_1=\nu_1}^{\infty}\sum_{l_2=\nu_2}^{\infty}\Delta_{l_1 l_2}^2\right]^{p/2} dxdy\right\}^{\theta/p} \leqslant I_7^\theta < \infty.$$

First let $2 \leqslant p < \infty$. Then $\theta = p$, and hence

$$I_8 = \int_0^{2\pi}\int_0^{2\pi}\left\{\sum_{\nu_1=0}^{\infty}\sum_{\nu_2=0}^{\infty}\left[\sum_{l_1=\nu_1}^{\infty}\sum_{l_2=\nu_2}^{\infty}\Delta_{l_1 l_2}^2 \mu_1^{2/p}(\nu_1)\,\mu_2^{2/p}(\nu_2)\right]^{p/2}\right\} dxdy.$$

Applying the generalized Minkowski inequality to the sum in braces, we have

$$I_8 \leqslant \int_0^{2\pi}\int_0^{2\pi}\left[\sum_{l_1=0}^{\infty}\sum_{l_2=0}^{\infty}\Delta_{l_1 l_2}^2\left\{\sum_{\nu_1=0}^{l_1}\sum_{\nu_2=0}^{l_2}\mu_1(\nu_1)\,\mu_2(\nu_2)\right\}^{2/p}\right]^{p/2} dxdy =: I_7^p,$$

since $\Sigma_{\nu_1=0}^{l_1}\mu_i(\nu_i) = \beta_i(2^{l_i})$, $i = 1, 2$.

Now let $1 < p \leqslant 2$. Then $\theta = 2$, and hence

$$I_8^{p/2} = \left(\sum_{\nu_1=0}^{\infty}\sum_{\nu_2=0}^{\infty}\left\{\int_0^{2\pi}\int_0^{2\pi}\left[\sum_{l_1=\nu_1}^{\infty}\sum_{l_2=\nu_2}^{\infty}\Delta_{l_1 l_2}^2 \mu_1(\nu_1)\,\mu_2(\nu_2)\right]^{p/2} dxdy\right\}^{2/p}\right)^{p/2}$$

Applying the generalized Minkowski inequality, we obtain

$$I_8^{p/2} \leqslant \int_0^{2\pi}\int_0^{2\pi}\left(\sum_{\nu_1=0}^{\infty}\sum_{\nu_2=0}^{\infty}\left[\sum_{l_1=\nu_1}^{\infty}\sum_{l_2=\nu_2}^{\infty}\Delta_{l_1 l_2}^2\,\mu_1(\nu_1)\,\mu_2(\nu_2)\right]\right)^{p/2} dxdy$$

$$= \int_0^{2\pi}\int_0^{2\pi}\left(\sum_{l_1=0}^{\infty}\sum_{l_2=0}^{\infty}\Delta_{l_1 l_2}^2\left[\sum_{\nu_1=0}^{l_1}\sum_{\nu_2=0}^{l_2}\mu_1(\nu_1)\,\mu_2(\nu_2)\right]\right)^{p/2} dxdy = I_7^p.$$

Thus for all p in the interval $1 < p \leqslant \infty$ we have shown that $I_8 \leqslant I_7^\theta < \infty$. We note that we already showed when proving Theorem 15 that

$$Y_{[2^{\nu_1}-1]\,[2^{\nu_2}-1]}(f)_p \leqslant C_3\left(\int_0^{2\pi}\int_0^{2\pi}\left[\sum_{l_1=\nu_1}^{\infty}\sum_{l_2=\nu_2}^{\infty}\Delta_{l_1 l_2}^2\right]^{p/2} dxdy\right)^{1/p},$$

where C_3 does not depend on φ_{12}. Hence this inequality and the inequality $I_8 \leqslant I_7^\theta < \infty$ imply that

$$\sum_{\nu_1=0}^{\infty} \sum_{\nu_2=0}^{\infty} \mu_1(\nu_1)\mu_2(\nu_2) Y^\theta_{[2^{\nu_1}-1]\,[2^{\nu_2}-1]}(f)_p \leqslant C_4 I_7^\theta \leqslant C_5 \|\varphi_{12}\|_p^\theta < \infty,$$

where C_4 and C_5 do not depend on φ_{12}.

Now taking into account that the function $\alpha_i(t)$ satisfies the σ_i-condition, and applying Lemma 8, we obtain $I_6 \leqslant C_6 \|\varphi_{12}\|_p^\theta < \infty$, which completes the proof of Theorem 16.

§6. Corollaries

1. Setting $\beta_i \equiv 1$ and $r_i > 0$ for all $i = 1, \cdots, n$ in the definition of the function class $S_p^r H(\beta, \tau, n)$, we arrive at the function class $S_p^r H$ which was introduced by Nikol'skiĭ in [1]. In this case:

a) Theorems 4–7 of §3 coincide with corresponding Theorems in [1]; and

b) Theorems 1–3 of §3, along with results in [1], give new definitions of the function classes $S_p^r H$, which are equivalent to the definition in [1].

2. Setting $\alpha_i(t) = t^{-r_i\theta - 1}$ and $r_i > 0$ for all $i = 1, \cdots, n$ in the definition of the function class $S_p B(\alpha, \sigma, \theta, n)$, we arrive at the function class $S^{\mathbf{r}}_{p*\theta} B$ $(0 \leqslant x_{i_j} \leqslant 2\pi,\ j = 1, 2, \cdots, n)$, which was considered by Amanov in [3]. In this case:

a) Theorems 12–14 of §4 coincide with corresponding theorems in [3]; and

b) Theorems 8–11 of §4, along with results in [3], give new definitions of the function classes $S^{\mathbf{r}}_{p*\theta} B$, which are equivalent to the definition in [3].

3. Setting $\alpha_i(t) = t^{-r_i\theta - 1}$ and $r_i > 0$ for all $i = 1, \cdots, n$, we obtain the fundamental result in [10] as a corollary to Theorem 15.

In conclusion, we note that in certain special cases "angular" approximation was already used in practice in [10] and [11]. In [11] (see p. 47) an "angular" approximation was found for functions having generalized mixed derivative bounded in the $\mathfrak{L}_p$-metric. In [10] the first inequality of Lemma 6 was proved, and an "angular" approximation was used to prove the fundamental result of the paper.

BIBLIOGRAPHY

1. S. M. Nikol'skiĭ, *Functions with dominant mixed derivative satisfying a multiple Hölder condition*, Sibirsk. Mat. Ž. 4 (1963), 1342–1364; English transl., Amer. Math. Soc. Transl. (2) 102 (1973), 27–51. MR 28 #3322.

2. ———, *On a certain class of differentiable functions in several variables*, Studies Contemporary Problems Constructive Theory of Functions (Proc. Second All-Union Conf., Baku, 1962), Izdat. Akad. Nauk Azerbaĭdzan. SSR, Baku, 1965, pp. 502–510. (Russian) MR 35 #716.

3. T. I. Amanov, *Representation and imbedding theorems for the function spaces* $S^r_{p,\theta}B(R_n)$ *and* $S^r_{p^*\theta}B(0 \leqslant x_j \leqslant 2\pi; j = 1, \cdots, n)$, Trudy Mat. Inst. Steklov. **77** (1965), 3–34 = Proc. Steklov Inst. Math. **77** (1965), 3–36. MR **33** #1718.

4. P. I. Lizorkin and S. M. Nikol'skiĭ, *A classification of differentiable functions in some fundamental spaces with dominant mixed derivative*, Trudy Mat. Inst. Steklov. **77** (1965), 143–167 = Proc. Steklov Inst. Math. **77** (1965), 160–187. MR **33** #450.

5. M. K. Potapov, *The imbedding and equality of certain classes of functions*, Izv. Akad. Nauk SSSR Ser. Mat. **33** (1969), 840–860 = Math. USSR Izv. **3** (1969), 795–814. MR **40** #6254.

6. S. M. Nikol'skiĭ, *Approximation of functions of several variables and imbedding theorems*, "Nauka", Moscow, 1969. (Russian)

7. M. K. Potapov, *On "angular" approximation*, Proc. Conf. Constructive Function Theorey (Budapest, 1969), Akad. Kiadó, Budapest, 1972, pp. 371–399. RŽ Mat **1972** #8Б120.

8. Ja. S. Bugrov, *Continuation of functions*, Dokl. Akad. Nauk SSSR 150 (1963), 1191–1194 = Soviet Math. Dokl. **4** (1963), 826–830. MR **27** #6112.

9. M. K. Potapov, *The interconnection of some classes of functions*, Mat. Zametki 2 (1967), 361–372 = Math. Notes 2 (1967), 706–714. MR **36** #3046.

10. ———, *Certain conditions for mixed derivatives to belong to* L_p, Mathematica (Cluj) 10(33) (1968), 355–367. (Russian) MR **39** #4605.

11. Ja. S. Bugrov, *Approximation of functions of several variables by trigonometric polynomials*, Trudy Naučn. Ob"ed. Prepodavat. Fiz.-Mat. Fak. Ped. Inst. Dal'n. Vostok., tom 1, Khabarovsk, 1962, pp. 28–49. (Russian) RŽ Mat **1964** #4Б89.

Translated by N. KOBLITZ

Trudy Mat. Inst. Steklov.
117 (1972)

Proc. Steklov. Inst. Math.
117 (1972)

ON THE REPRESENTATION OF FUNCTIONS DEFINED BY A CLASS OF HYPOELLIPTIC OPERATORS

UDC 517.518.22

S. V. USPENSKIĬ

Abstract. In this paper an integral representation is constructed of functions differentiable on the entire Euclidean space. The kernel of this representation is related with the fundamental solution of a class of hypoelliptic operators, which gives the possibility of investigating the behavior at infinity of the solutions of hypoelliptic equations and of obtaining various a priori estimates.

Bibliography: 18 items.

In this paper a representation is constructed of a class of functions defined on the whole space E_n. This representation is closely related to the solutions of quasi-elliptic equations with constant coefficients, which gives the possibility of investigating, in a sufficiently complete manner, the differential properties and the behavior of the solutions at infinity. The construction of such a "good" representation is, essentially, equivalent to the construction of a fundamental solution with best properties, both locally and at infinity. The procedures for the construction of fundamental solutions of equations with constant coefficients are sufficiently developed (see [1, 2]). The most general results are due to Hörmander [2] and Trèves [1], who have constructed fundamental solutions for general differential polynomials with constant coefficients by using the method of multidimensional contour integration. However, their utilization for the investigation of the solutions of the corresponding equations poses serious difficulties. Moreover, the most refined a priori estimates, for example, for elliptic equations (see [3]) have been carried out by using Radon's representation [4, 5], associated with the polyharmonic operator. From other papers, where fundamental solutions have been obtained for special classes of operators, we mention [6–12].

For the construction of the fundamental solution of the class of differential operators under consideration, we have made use of the well-known method connected with the integration of the fundamental solution of a certain parabolic equation by generalizing it to the case of quasielliptic operators (see, for example, [13]). This representation turned out to be more convenient if written in the form of the limit of a sequence of analytic functions, subject to some inequalities of the Bernstein type. In this connection we follow the ideas developed in S. M. Nikol′skiĭ's papers [14, 15].

We introduce some definitions. We denote by E_n the Euclidean space of the points $x(x_1, \cdots, x_n)$,

---

AMS (MOS) subject classifications (1970). Primary 47F05, 35C15; Secondary 35E05.

$$|x|=\left(\sum_{i=1}^{n} x_i^2\right)^{1/2}, \qquad x^\beta = x_1^{\beta_1}\ldots x_n^{\beta_n},$$

$$|l|=\sum_{i=1}^{n} l_i, \qquad l\alpha=\sum_{i=1}^{n} l_i\alpha_i,$$

$$D^\beta = D^{\beta_1}\ldots D^{\beta_n} = \frac{\partial^{\beta_1}}{\partial x_1^{\beta_1}}\cdots\frac{\partial^{\beta_n}}{\partial x_n^{\beta_n}},$$

$$\Delta(f, he_i) = f(x_1, \ldots, x_i, \ldots, x_n) - f(x_1, \ldots, x_i+h, \ldots, x_n),$$

$$\mathfrak{M}_v[f(x)] = \prod_{i=1}^{n}\frac{1}{v_i}\int_0^{v_1}\int_0^{v_n} f(x+u)\,du, \tag{1}$$

$$\|f, L_p(g)\|^p = \int_g |f|^p\,dg, \qquad g\subset E_n, \quad 1<p<\infty, \tag{2}$$

$$\|f, C\| = \sup_{x\in E_n}|f(x)|, \tag{3}$$

$$\|f, C^\mu\| = \sum_{i=1}^{n}\sup_{\substack{x\in E_n\\ |h|\leqslant 1}}\left|\frac{\Delta(f, he_i)}{h^{\mu_i}}\right| + \|f, C\|, \tag{4}$$

$$G_0(x) = \exp - \sum_{i=1}^{n}(2m\alpha_i)^{-1}x_i^{2m}, \tag{5}$$

where m is a positive integer and the α_i are positive real numbers.

Below we will examine the differential operators $P(D)$, which satisfy the following conditions:

a) The characteristic polynomial $P(is)$ with constant real coefficients has no real zeros for $|s|\neq 0$.

b) The polynomial $P(is)$ is quasihomogeneous, i.e. for every $\lambda>0$ we have the equality

$$P(is\lambda^\alpha) = \lambda P(is), \tag{6}$$

where $\alpha=(\alpha_1\cdots\alpha_n)$, and the α_i are positive real numbers.

The vector α will be called the *homogeneity exponent* of the operator $P(D)$. We note that from the conditions a) and b) it follows that the polynomial $P(is)$ is hypoelliptic and that we have the estimate

$$|P(is)| \geqslant C\sum_{i=1}^{n}|s_i|^{1/\alpha_i} \tag{7}$$

(see [2], Russian p. 142).

As usual, the Fourier transform of the function f will be denoted by $\hat{f}$, and the convolution of f with the function φ, by $f*\varphi$. We introduce the averaging operator

$$A_h f = \frac{h^{-|\alpha|}}{(2\pi)^n} \int\limits_{E_n} \hat{G}_0\left(\frac{t-x}{h^\alpha}\right) f(t)\, dt, \tag{8}$$

which will be called the *parabolic mean*, where $\alpha = (\alpha_1 \cdots \alpha_n)$, $\alpha_i > 0$, $h > 0$ and $i = 1, \cdots, n$. It is easy to see that for $h > 0$ the operator A_h is defined for every function f which is summable on every compactum $g \subset E_n$ and which satisfies the condition

$$\overline{\lim_{|v| \to \infty}} \, |\mathfrak{M}_v [f(x)]| < \infty. \tag{9}$$

We consider some properties of the operator A_h, which we formulate as lemmas.

LEMMA 1. *Let* $f \in L_p^{\mathrm{loc}}$, $1 < p < \infty$ *(continuous), be such that*

$$\lim_{|v| \to \infty} \mathfrak{M}_v [f(x)] = 0, \qquad x \in E_n.$$

Then for almost all $x \in E_n$ *(for all* $x \in E_n$*)*

$$\lim_{h \to 0} A_h f = f, \tag{10}$$

$$\lim_{h \to \infty} A_h f = 0. \tag{11}$$

PROOF. By virtue of the Fourier integral theorem (see, for example, [16]) we have

$$\int\limits_{E_n} \hat{G}_0(t)\, dt = (2\pi)^n$$

and consequently

$$\frac{h^{-|\alpha|}}{(2\pi)^n} \int\limits_{E_n} \hat{G}_0\left(\frac{t-x}{h^\alpha}\right) dt = 1. \tag{12}$$

For arbitrary N we have, taking (5) into account,

$$|D^m \hat{G}_0(t)| \leqslant C_N \frac{1}{(1+|t|)^N}, \qquad 0 \leqslant m < \infty. \tag{13}$$

Then, integrating by parts, we obtain

$$|A_h f - f| = \left| \frac{1}{(2\pi^n)} \int\limits_{E_n} \prod_{i=1}^n t_i D^n \hat{G}_0(t)\, \mathfrak{M}_t [\Delta(f(x),\ h^\alpha ue)]\, dt \right|$$

(taking (13) into account)

$$\leqslant \sup_{|t| \geqslant N} \frac{C}{(|t|+1)^k} |\mathfrak{M}_{h^{\alpha}t}[f] - f(x)| + \sup_{|t| \leqslant N} C\, |\mathfrak{M}_{h^{\alpha}t}[f] - f(x)|. \tag{14}$$

Since $f \in L_p^{\mathrm{loc}}$, $1 < p < \infty$, it follows from (14) that for sufficiently large k we have for almost all $x \in E_n$ (for all $x \in E_n$) (see [17])

$$\overline{\lim_{h\to\infty}}\, |A_h f - f| \leqslant \varepsilon$$

or

$$\lim_{h\to\infty} |A_h f - f| = 0.$$

We also have, integrating by parts,

$$|A_h f| \leqslant \left| \frac{h^{-2|\alpha|}}{(2\pi)^n} \int\limits_{E_n} \prod_{i=1}^{n} t_i D^n \hat{G}_0\left(\frac{t}{h^\alpha}\right) \mathfrak{M}_t[f(x)]\, dt \right|$$

(taking (13) into account)

$$\leqslant CN^n h^{-|\alpha|} \sup_{|t|\leqslant N} |\mathfrak{M}_t[f(x)]| + C \sup_{|t|\geqslant N} |\mathfrak{M}_t[f(x)]|. \tag{15}$$

Choosing N sufficiently large and taking the limit for $h \to \infty$, we obtain for almost all $x \in E_n$ (for all $x \in E_n$)

$$\lim_{h\to\infty} A_h f = 0.$$

LEMMA 2. *Let* $\|f, C^\lambda\| < \infty$ *and*

$$\lim_{|v|\to\infty} \mathfrak{M}_v[f(x)] = 0$$

uniformly with respect to $x \in E_n$. *Then*

$$\lim_{h\to 0} \| A_h f - f, \quad C\| = 0, \tag{16}$$

$$\lim_{h\to\infty} \| A_h f, \qquad C\| = 0. \tag{17}$$

PROOF. Since

$$|\mathfrak{M}_v[f(x)] - f(x)| \leqslant C \max |v_i|^\lambda \|f, C^\lambda\|, \tag{18}$$

$$|\mathfrak{M}_v[f(x)]| \leqslant \|f, C\|, \tag{19}$$

the assertions of Lemma 2 follow from the estimates (14) and (15).

LEMMA 3. *Let* $\|f, L_p(E_n)\| < \infty$, $1 < p < \infty$. *Then*

$$\lim_{h\to\infty} \| A_h f - f, \ L_p(E_n)\| = 0, \tag{20}$$

$$\lim_{h\to\infty} \| A_h f, \qquad L_p(E_n)\| = 0. \tag{21}$$

PROOF. Carrying out the estimates as in (14), we have

$$\|A_h f - f,\ L_p\| \leqslant \sup_{|t|\geqslant N} \frac{C}{(|t|+1)^m} \|\mathfrak{M}_{h^{\alpha}t}[f] - f,\ L_p\|$$

$$+ \sup_{|t|\leqslant N} C(|t|+1)^{-m} \|\mathfrak{M}_{h^{\alpha}t}[f] - f,\ L_p\|$$

(putting $\alpha = \min \alpha_i$)

$$\leqslant \frac{C}{(N+1)^m} \|f,\ L_p\| + \sup_{|v|\leqslant Nh^{\alpha}} \|\mathfrak{M}_v[f] - f,\ L_p\|.$$

From this, taking the limit for $h \to 0$, due to the arbitrariness of N we obtain (20).

Let us prove (21). Let φ_ϵ be finite outside the rectangle

$$\square_N = (|x_i| < N,\ i = 1 \dots n),$$
$$\|f - \varphi_\varepsilon,\ L_p| < \varepsilon.$$

For N sufficinetly large and depending on ϵ, this is always possible. Then

$$\|A_h f,\ L_p\| \leqslant C_0 \|\varphi_\varepsilon - f,\ L_p\| + \|A_h \varphi_\varepsilon,\ L_p\|$$

$$\leqslant C_0 \varepsilon + Ch^{-|\alpha|} \left\| \int_{E_n} \hat{G}_0\left(\frac{t-x}{h^{\alpha}}\right) \varphi_\varepsilon dt,\ L_p \right\|$$

(Minkowski's inequality)

$$\leqslant C_0 \varepsilon + Ch^{-|\alpha|/p'} \|\hat{G}_0,\ L_p\| \, \|\varphi_\varepsilon,\ L_1\|$$
$$\leqslant C_0 \varepsilon + Ch^{-|\alpha|/p'} (\|f,\ L_p\| + \varepsilon).$$

Since $1/p' = 1 - 1/p > 0$ and C_0 does not depend on h, f or ϵ, we obtain, by taking the limit for $h \to \infty$,

$$\overline{\lim_{h\to\infty}} \|A_h f,\ L_p\| \leqslant \varepsilon$$

or

$$\lim_{h\to\infty} \|A_h f,\ L_p\| = 0.$$

We put

$$G(x) = G_0(x) \sum_{i=1}^{n} x_i^{2m} [P(ix)]^{-1}, \tag{22}$$

where the characteristic polynomial $P(ix)$ satisfies the conditions a) and b) and has homogeneity exponent $\boldsymbol{\alpha} = (\alpha_1 \cdots \alpha_n)$.

LEMMA 4. *Let $f \in L_p^{\mathrm{loc}}$, $1 < p < \infty$ (continuous), be such that*

$$\lim_{|v|\to\infty} \mathfrak{M}_v[f(x)] = 0$$

for almost all $x \in E_n$ (for all $x \in E_n$). Then for almost all $x \in E_n$ (for all $x \in E_n$)

we have the following representation:

$$A_h[f]-A_{h^{-1}}[f]=\frac{1}{(2\pi)^n}\int_h^{h^{-1}} v^{-|\alpha|}\int_{E_n} P\left(\frac{\partial}{\partial x}\right)\hat{G}\left(\frac{t-x}{v^\alpha}\right) f dt dv. \tag{23}$$

PROOF. We consider the function

$$I(v, x)=v^{-|\alpha|}P(-D)\,\hat{G}\left(\frac{x}{v^\alpha}\right).$$

We have, taking into account the α-homogeneity of $P(is)$,

$$I(v, x)=\sum_{\nu=1}^n v^{-|\alpha|-1}\int_{E_n} t_\nu^{2m}\exp-\left[\sum_{s=1}^n \frac{ix_s t_s}{v^{\alpha_s}}+(2a_s m)^{-1}t_s^{2m}\right]dE_n$$

$$=\sum_{\nu=1}^n \alpha_\nu v^{-\alpha_\nu-1}\int_{-\infty}^{\infty}\left[t_\nu \exp-\frac{ix_\nu t_\nu}{v^{\alpha_\nu}}\right]\alpha_\nu^{-1}t_\nu^{2m-1}\exp-(2\alpha_\nu m)^{-1}t_\nu^{2m}dt$$

$$\times\prod_{\substack{s=1\\ s\neq\nu}}^n\int_{-\infty}^{\infty} v^{-\alpha_s}\exp-\frac{ix_s t_s}{v^{\alpha_s}}\exp-(2\alpha_s m)^{-1}t_s^{2m}dt_s$$

(integrating by parts)

$$=\sum_{\nu=1}^n \alpha_\nu v^{-\alpha_\nu-1}\int_{-\infty}^{\infty}\exp-\frac{ix_\nu t_\nu}{v^{\alpha_\nu}}-(2\alpha_\nu m)^{-1}t_\nu^{2m}$$

$$\times\left(1-\frac{it_\nu x_\nu}{v^{\alpha_\nu}}\right)dt_\nu\prod_{\substack{s=1\\ s\neq\nu}}^n\int_{-\infty}^{\infty} v^{-\alpha_s}\exp-\frac{ix_s t_s}{v^{\alpha_s}}-(2\alpha_s m)^{-1}t_s^{2m}dt_s$$

$$=\sum_{\nu=1}^n\left[\int_{-\infty}^{\infty}\alpha_\nu v^{-\alpha_\nu-1}\exp-\frac{ix_\nu t_\nu}{v^{\alpha_\nu}}-(2\alpha_\nu m)^{-1}t_\nu^{2m}dt_\nu\right.$$

$$\left.-\int_{-\infty}^{\infty} v^{-\alpha_\nu}\alpha_\nu v^{-\alpha_\nu-1}it_\nu x_\nu\exp-\frac{ix_\nu t_\nu}{v^{\alpha_\nu}}-(2\alpha_\nu m)^{-1}t_\nu^{2m}dt_\nu\right]$$

$$\times\prod_{\substack{s=1\\ s\neq\nu}}^n\int_{-\infty}^{\infty} v^{-\alpha_s}\exp-\frac{ix_s t_s}{v^{\alpha_s}}-(2\alpha_s m)^{-1}t_\nu^{2m}dt_s=-\frac{d}{dv}v^{-|\alpha|}\hat{G}_0\left(\frac{x}{v^\alpha}\right). \tag{24}$$

Taking into account the uniform convergence for $h>0$ of the integral

$$\int_h^{h^{-1}}\int_{E_n} v^{-|\alpha|}\hat{G}_0\left(\frac{t-x}{v^\alpha}\right)f(t)\,dt dv,$$

we have (see, for example, [17])

$$\frac{1}{(2\pi)^n}\int_h^{h^{-1}} v^{-|\alpha|}\int_{E_n} P\left(\frac{\partial}{\partial x}\right)\hat{G}\left(\frac{t-x}{v^\alpha}\right) f dt dv$$
$$= -\frac{1}{(2\pi)^n}\int_h^{h^{-1}} \frac{d}{dv} v^{-|\alpha|}\int_{E_n}\hat{G}_0\left(\frac{t-x}{v^\alpha}\right) f dt dv = A_h[f]-A_{h^{-1}}[f].$$

COROLLARY 1. *If the conditions of Lemma 4 hold, then for almost all* $x \in E_n$ (*for all* $x \in E_n$)

$$f=\lim_{h\to 0}\frac{1}{(2\pi)^n}\int_h^{h^{-1}} v^{-|\alpha|}\int_{E_n} P\left(\frac{\partial}{\partial x}\right)\hat{G}\left(\frac{t-x}{v^\alpha}\right) f dt dv. \tag{25}$$

Corollary 1 follows from Lemmas 1 and 4.

COROLLARY 2. *Assume that* $\|f, C^\lambda\| < \infty$ *and* $\lim_{|\nu|\mapsto\infty} M_\nu |f(x)| = 0$ *uniformly for* $x \in E_n$. *Then relation* (25) *holds, the convergence for* $h \to 0$ *being uniform.*

This assertion follows from Lemmas 2 and 4.

COROLLARY 3. *Let* $\|f, L_p\| < \infty$, $1 < p < \infty$. *Then relation* (25) *holds, the convergence for* $h \to 0$ *being in* L_p.

Corollary 3 follows from Lemmas 3 and 4.

We introduce some definitions. Assume that the operator $P(D)$ satisfies conditions a) and b), and let α be its homogeneity exponent. The function f will be said to be regular with respect to the operator $P(D)$ if on every compactum $g \subset E_n$ it satisfies the following conditions:

a) $\|D^\rho f, L_p(g)\| < \infty$ for some $p \in (1, \infty)$ and for all ρ satisfying the condition $0 \leqslant \rho\alpha \leqslant 1$.

b) There exists a natural number N, depending on f, such that for almost all $x' = (x_1, \cdots, x_{i-1}, x_{i+1}, \cdots, x_n)$ one has

$$\lim_{|x_i|\to\infty}(1+|x_i|^N|D^\rho f|=0,\ 0\leqslant\rho\alpha<1. \tag{26}$$

We define the kernel G by (22), putting

$$2m > N^n \max_{1\leqslant i\leqslant n}\alpha_i^{-1}.$$

Let us prove that

$$|D^j\hat{G}(x)| \leqslant C\prod_{\nu=1}^n (1+|x_\nu|)^{-N}. \tag{27}$$

Denoting $j = \Sigma_1^n j_s$, $|x_i| > 1$, $i = 1, \cdots, n$, we have

$$|D^j\hat{G}(x)| = \left| \int_{E_n} e^{-itx} \prod_{s=1}^{n} t_s^{j_s} \left(\sum_{s=1}^{n} t_s^{2m} \right) [P(it)]^{-1} \exp \sum_{s=1}^{n} (2m\alpha_s)^{-1} t_s^{2m} dE_n \right|$$

(integrating by parts Nn times)

$$= \prod_{\nu=1}^{n} \frac{1}{|x_\nu|^N} \left| \int_{E_n} e^{-itx} \frac{\partial^{Nn}}{\partial t_\nu^{Nn}} \left[\left(\exp - \sum_{s=1}^{n} (2m\alpha_s)^{-1} t_s^{2m} \right) \frac{\prod_{s=1}^{n} t_s^{j_s} \sum_{s=1}^{n} t_s^{2m}}{P(it)} \right] dE_n \right|.$$

Then the estimate (27) reduces to the proof of the boundedness in the neighborhood of zero of the function

$$\frac{\sum_{s=1}^{n} t_s^{2m}}{|P(it)|^{Nn}},$$

which follows from the estimate

$$|P(it)| \geqslant C \sum_{s=1}^{n} |t_s|^{1/\alpha_s}$$

and the condition $2m > N^n \max_{1 \leqslant s \leqslant n} \alpha_s^{-1}$ (see [2]). For $|x_i| \leqslant 1$ the estimate (27) is obvious.

Thus for every N one can always construct a kernel $\hat{G}$ for which the estimate (27) holds.

THEOREM 1. *Assume that the operator* $P(D)$ *satisfies the conditions* a) *and* b), *and let* $\alpha = (\alpha_1 \cdots \alpha_n)$ *be its homogeneity exponent. If the function* f *is regular with respect to the operator* $P(D)$ *and*

$$\lim_{|v| \to \infty} \mathfrak{M}_v[f] = 0$$

for almost all $x \in E_n$ *(for all* $x \in E_n$*), then for almost all* $x \in E_n$ *(for all* $x \in E_n$*) one has*

$$f = \lim_{h \to 0} \frac{1}{(2\pi)^n} \int_h^{h^{-1}} \int_{E_n} v^{-|\alpha|} \hat{G}\left(\frac{t-x}{v^\alpha} \right) P(D) f dt dv.$$

If $\|f, C^\lambda\| < \infty$, *then the convergence is uniform; for* $\|f, L_p\|$, $1 < p < \infty$, *the convergence is in* $L_p(E_n)$.

The assertions of the theorem follow from relation (23) of Lemma 4 and the properties of the kernel $\hat{G}$.

BIBLIOGRAPHY

1. F. Trèves, *Solutions élémentaires d'équations aux dérivées partielles dépendent d'une paramètre*, C. R. Acad. Sci. Paris **242** (1956), 1250–1252. MR 19, 552.

2. L. Hörmander, *Linear partial differential operators*, Die Grundlehren der math. Wissenschaften, Band 116, Academic Press, New York; Springer-Verlag, Berlin, 1963; Russian transl., "Mir", Moscow, 1965. MR **28** #4221; **37** #6595.

3. S. A. Agmon, A. Douglis and L. Nirenberg, *Estimates near the boundary for solutions of elliptic partial differential equations satisfying general boundary conditions.* I, Comm. Pure Appl. Math. **12** (1959), 623–727; Russian transl., IL, Moscow, 1962. MR **23** #A2610.

4. J. Radon, *Über die Bestimmung von Funktionen durch ihre Integralwerte längs gewisser Mannigfaltigkeiten*, Ber. Verh. Sächs. Ges. Wiss. Leipzig Math.-Phys. Kl. **69** (1917), 262–277.

5. F. John, *Plane waves and spherical means applied to partial differential equations*, Interscience Tracts in Pure and Appl. Math., vol. 2, Interscience, New York, 1955; Russian transl., IL, Moscow, 1958. MR **17**, 746.

6. V. P. Palamodov, *Linear differential operators with constant coefficients*, "Nauka", Moscow, 1967; English transl., Die Grundlehren der math. Wissenschaften, Band 168, Springer-Verlag, Berlin and New York, 1970. MR **39** #4517; **41** #8793.

7. L. Brusotti, *Sur la solution élémentaire d'une équation linéaire aux dérivées partielles d'ordre quatre et à trois variables indépendantes*, Acad. Roy. Belg. Bull. Cl. Sci. (5) **33** (1947), 473.

8. I. Fredholm, *Sur l'intégrale fondamentale d'une équation différentielle elliptique à coefficients constants*, Rend. Circ. Mat. Palermo **25** (1908), 346–351.

9. G. Herglotz, *Über die Integration linearer partieller Differentialgleichungen mit konstanten Koeffizienten*, Ber. Verh. Sächs. Akad. Wiss. Leipzig Math.-Phys. Kl. **78** (1926), 93–126, 287–318.

10. I. G. Petrovskiĭ, *On the diffusion of waves and the lacunas for hyperbolic equations*, Mat. Sb. **17** (59) (1945), 289–370. (Russian) MR **8**, 79.

11. P. Mader, *Über die Darstellung von Punktfunktionen in n-dimensionalen euklidisch Raum durch Ebenenintegrale*, Math. Z. **26** (1927), 646.

12. G. Girand, *Généralisations des problèmes sur les opérations du type elliptique*, Bull. Sci. Math. **56** (1932), 248–272, 281–312, 316–352.

13. S. D. Èĭdel'man, *Parabolic systems*, "Nauka", Moscow, 1964; English transl., Noordhoff, Groningen; North-Holland, Amsterdam, 1969. MR **29** #4998; **40** #6023.

14. S. M. Nikol'skiĭ, *Inequalities for entire functions of exponential type and their application to the theory of differentiable functions of several variables*, Trudy Mat. Inst. Steklov. 38 (1951), 244–278; English transl., Amer. Math. Soc. Transl. (2) **80** (1969), 1–38. MR **14**, 32.

15. ———, *Properties of certain classes of functions of several variables on differentiable manifolds*, Mat. Sb. **33** (75) (1953), 261–326; English transl., Amer. Math. Soc. Transl. (2) 80 (1969), 39–118. MR **16**, 453.

16. S. Bochner, *Vorlesungen über Fouriersche Integrale*, Akad. Verlag., Leipzig, 1932; English transl., Ann. of Math. Studies, no. 42, Princeton Univ. Press, Princeton, N. J., 1959; Russian transl., Fizmatgiz, Moscow, 1962. MR **21** #5851.

17. A. Zygmund, *Trigonometric series*. Vol. II, 2nd rev. ed., Cambridge Univ. Press, New York, 1959; Russian transl., "Mir", Moscow, 1965. MR **21** #6498; **31** #2554.

18. S. L. Sobolev, *Equations of mathematical physics*, 3rd ed., GITTL, Moscow, 1954; English transl., Pergamon Press, London; Addison-Wesley, Reading, Mass., 1964. MR **16**, 1027; **31** #2478.

Translated by E. DEUTSCH

Trudy Mat. Inst. Steklov.
117 (1972)

Proc. Steklov Inst. Math.
117 (1972)

INTEGRAL ESTIMATES OF GENERALIZED DERIVATIVES OF SOLUTIONS OF SECOND ORDER ELLIPTIC EQUATIONS IN THE L_p-METRICS, AND SOME RELATED IMBEDDING THEOREMS

A. S. FOHT

UDC 517.518.22

Abstract. In the paper we consider a solution u of a second order linear elliptic equation with variable coefficients. This equation arises as the Euler equation which corresponds to the functional

$$D(u, u) = \int_G \sum_{i,j} a_{ij}(x) \frac{\partial u}{\partial x_i} \frac{\partial u}{\partial x_j} dG < +\infty,$$

where G is a bounded domain with a sufficiently smooth boundary Γ, $G \subset R_n$. We show that for $p \geqslant 2$ the following estimate of any generalized derivative

$$D_\nu^s u = \frac{\partial^s u}{\partial x_1^{\nu_1} \ldots \partial x_n^{\nu_n}}, \quad \sum_{j=1}^n \nu_j = s$$

holds:

$$\int_G |D_\nu^s u|^p t^{sp} dC \leqslant C_1 \int_G |u| p\, dG; \qquad t = \rho(x, \Gamma), \qquad x \in G.$$

For a p with $1 < p < 2$ the estimate

$$\int_G | D_\nu^s u |^p t^{sp} dG \leqslant C_2(G) \left[\int_G |u|^{p_1} dG \right]^{\frac{p}{p_1}}$$

holds, where p_1 is no less than two and arbitrary otherwise, $C_1 > 0$ is a constant which does not depend on meas G, and $C_2(G)$ is a constant which depends on meas G.

Bibliography: 9 items.

Introduction

In this paper we consider an arbitrary generalized solution of a second order equation of elliptic type, which is given on a bounded domain $G \subset R_n$ with a sufficiently smooth boundary Γ and sufficiently smooth variable coefficients $a_{ij}(x), x = (x_1, \cdots, x_n)$.

This equation is given in the form of a functional which generates it as the corresponding Euler equation. The functional has the form

$$D(u, u) = \int_G \sum_{i,j} a_{ij}(x) \frac{\partial u}{\partial x_i} \frac{\partial u}{\partial x_j} dG < +\infty. \tag{0.1}$$

AMS (MOS) subject classifications (1970). Primary 35J15, 35B45, 35D99; Secondary 35J20, 46E35.

The ellipticity condition states that

$$\sum_{i,j} a_{ij}(x)\,\xi_i\xi_j \geqslant k^2 \sum_i \xi_i^2 \qquad (k^2>0). \tag{0.2}$$

The existence of all arbitrary high-order generalized partial derivatives for such an equation (with $a_{ij}(x)$ sufficiently smooth) follows, e.g., from the results of L. Hörmander [1], or from [2]–[4]. Therefore here we shall assume that this fact has been already established.

In this paper we shall prove that for the generalized partial derivatives

$$D_\nu^s u = \frac{\partial^s u}{\partial x_1^{\nu_1}\partial x_2^{\nu_2}\dots\partial x_n^{\nu_n}}, \qquad \sum_{j=1}^n \nu_j = s,$$

the following integral estimates hold:

$$\int_G |D_\nu^s u|^p t^{sp} dG \leqslant C_{s,p} \int_G |u|^p dG, \tag{0.3}$$

where $2<p<\infty$ and $C_{s,p}>0$ is a constant which does not depend on u, or the domain G, or on x, $t=t(x)=\rho(x,\Gamma)$.

Earlier [3], [4] we obtained a similar estimate for solutions of elliptic equations of any order $2l$, but only for $p=2$.

For p_1 within the interval $1<p_1<2$ the following estimate holds:

$$\int_G |D_\nu^s u|^{p_1} t^{sp_1} dG$$

(by virtue of Hölder inequality)

$$\leqslant (\text{meas } G)^{\frac{p-p_1}{p}} \left(\int_G |D_\nu^s u|^p t^{sp} dG\right)^{p_1/p}$$

(by virtue of estimate (0.3), where $2<p<\infty$, or a similar estimate of [3], [4], where $p=2$)

$$\leqslant C_{s,p}(G)\left(\int_G |u|^p dG\right)^{p_1/p}. \tag{0.4}$$

Here $C_{s,p}(G)=C_{s,p}(\text{meas } G)^{(p-p_1)/p}$, $C_{s,p}>0$ is a constant which depends neither on x nor on the domain G, and p is an arbitrary number in the interval $2\leqslant p<\infty$. Thus, assuming that $\int_G |u|^p\,dG<+\infty$ $(2\leqslant p<+\infty)$, we have obtained estimate (0.4) for any p_1 with $1<p_1<2$.

In the L_p-metric (for any p with $1<p<\infty$) we obtained an estimate similar to (0.3) only for polyharmonic functions (the paper is in press).

The basic tool for obtaining estimate (0.3) consisted of the weight imbedding theorems obtained in [3], [4] and of their generalizations developed in this paper. We note also that estimate (0.3) does not follow from the previous papers. In conclusion, we indicate the following estimate in the L_p-metric for harmonic functions, which is close in character to our estimates and was obtained earlier by S. M. Nikol'skiĭ [5] (for $1 \leqslant p \leqslant \infty$):

$$\|u\|_{L_p^{(s)}(\Gamma_\delta)} < \frac{C_2 \|u\|_{L_p(G)}}{\delta^{s+\frac{1}{p}}}, \tag{0.5}$$

where

$$\|u\|_{L_p^{(s)}(\Gamma_\delta)} = \left(\int_{\Gamma_\delta} |u^{(s)}|^p \, d\Gamma_\delta\right)^{1/p}, \quad \|u\|_{L_p(G)} = \left(\int_G |u|^p \, dG\right)^{1/p} < +\infty,$$

$u^{(s)}$ is an arbitrary s-order partial derivative with respect to u, Γ_δ is the surface which lies at the distance δ along the normals from Γ ($0 < \delta < \delta_*$, where δ_* is a sufficiently small number which depends on Γ), and $C_2 > 0$ is a constant.

We also note the papers [6] and [7], where estimates of the type of (0.5) were obtained for polyharmonic functions.

§1. Auxiliary theorems

First we shall prove a number of auxiliary lemmas and theorems.

LEMMA 1. *Let it be known that all $b_j \geqslant 0$ and $r > 1$, $j = 1, 2, \cdots, n$. Then the following inequality holds:*

$$\left(\sum_{j=1}^n b_j\right)^r \leqslant n^{r-1} \left(\sum_{j=1}^n b_j^r\right). \tag{1.1}$$

PROOF. Without loss of generality we shall assume that $b_n \neq 0$. We introduce $x_j = b_j/b_n$ ($x_j \geqslant 0$, $j = 1, 2, \cdots, n-1$). Then inequality (1.1) takes on the form

$$\left(1 + \sum_{j=1}^{n-1} x_j\right)^r \leqslant n^{r-1} \left(\sum_{j=1}^{n-1} x_j^r + 1\right). \tag{1.2}$$

We consider the function

$$\gamma(\bar{x}) = \frac{\left(1 + \sum_{j=1}^{n-1} x_j\right)^r}{1 + \sum_{j=1}^{n-1} x_j^r}, \quad \bar{x} = (x_1, x_2, \ldots, x_{n-1}); \tag{1.3}$$

$\gamma(\bar{x}) > 0$ for all admissible values of $x_j \geqslant 0$, and its maximum is attained at the point $\bar{x}_0 = (1, 1, 1, \cdots, 1)$, $\gamma_0 = \gamma(\bar{x}_0) = n^{r-1}$, which was to be proved.

We consider the expression $[\varphi(x)]^p$, where $0 < p < \infty$, and $\varphi(x)$ is a real valued function which can have positive, zero, and negative values. For an arbitrary p from the interval indicated, the expression $[\varphi(x)]^p$ is, generally speaking, many-valued. We agree to assume that $[\varphi(x)]^p \geqslant 0$. If $\varphi(x) < 0$, then we shall choose an arbitrary single complex value $(-1)_*^p$ of the number $(-1)^p$, and we agree to assume that

$$[\varphi(x)]^p = (-1)^p_* \, |\varphi(x)|^p. \tag{1.4}$$

DEFINITION. *If* $\varphi(x) < 0$, *then a fixed number* $e^{i\gamma}$ *whose product with* $(-1)^p_*$ *is one*:

$$(-1)^p_* \, e^{i\gamma} = 1 \tag{1.5}$$

will be denoted by $\operatorname{sign}^*[\varphi(x)]^p$. *For* $\varphi(x) \geqslant 0$, *we set* $\operatorname{sign}^*[\varphi(x)]^p = +1$.

LEMMA 2. *Let a function* $\varphi(x) = \varphi(x_k, \widetilde{x}), x = (x_1, \cdots, x_n) = (x_k, \widetilde{x}), \widetilde{x} = (x_1, \cdots, x_{k-1}, x_{k+1}, \cdots, x_n)$, *be continuous in the variable* x_k, *and let it have the first generalized derivative* $\partial\varphi/\partial x_k$ *with respect to this variable on* (x_k', x_k'') (*for any* $\widetilde{x}$). *Then the function*

$$w = (\varphi)^{p-1} \operatorname{sign}^* (\varphi)^{p-2} \qquad (p > 2) \tag{1.6}$$

has the generalized derivative $\partial w/\partial x_k$ *on the interval* (x_k', x_k''):

$$\frac{\partial w}{\partial x_k} = (p-1)(\varphi)^{p-2} \frac{\partial \varphi}{\partial x_k} \operatorname{sign}^* (\varphi)^{p-2}. \tag{1.7}$$

PROOF. We introduce into consideration the set E_+ of all points $x_k \in (x_k', x_k'')$ for which $\varphi(x_k, x) \geqslant 0$, and the set E_- of all points $x_k \in (x_k', x_k'')$ for which $\varphi(x_k, x) < 0$. As a result, for any addmissible function $\psi(x_k, \widetilde{x}) \in \mathfrak{M}^{(1)}(x_k', x_k'')$ [8] we obtain

$$\int_{x_k'}^{x_k''} [\operatorname{sign}^* (\varphi)^{p-2}] (\varphi)^{p-1} \frac{\partial \psi}{\partial x_x} dx_k = \int_{E_+} (\varphi)^{p-1} \frac{\partial \psi}{\partial x_k} dx_k + (-1)^{p-2}_* \int_{E_-} (\varphi)^{p-1} \frac{\partial \psi}{\partial x_k} dx_k$$

(integrating by parts)

$$\begin{aligned} &= \varphi^{p-1}\psi \,|_{x_k \in \Gamma_{E_+}} + (-1)^{p-2}_* \varphi^{p-1}\psi \,|_{x_k \in \Gamma_{E_-}} \\ &\quad - (p-1) \int_{E_+} (\varphi)^{p-2} \frac{\partial \varphi}{\partial x_k} \psi dx_k - (-1)^{p-2}_* \int_{E_-} (\varphi)^{p-2} \frac{\partial \varphi}{\partial x_k} \psi dx_k \\ &= -(p-1) \int_{x_k'}^{x_k''} (\varphi)^{p-2} \operatorname{sign}^* (\varphi)^{p-2}] \frac{\partial \varphi}{\partial x_k} \psi dx_k. \end{aligned} \tag{1.8}$$

Hence it follows that

$$\frac{\partial w}{\partial x_k} = (p-1)(\varphi)^{p-2} \frac{\partial \varphi}{\partial x_k} \operatorname{sign}^* (\varphi)^{p-2};$$

where Γ_{E_+} and Γ_{E_-} denote the boundaries of E_+ and E_-, respectively. By virtue of the continuity of the function $\varphi(x_k, \widetilde{x})$ and the boundedness of $\psi(x_k, \widetilde{x})$ it follows that $\varphi^{p-1}\psi|_{x_k \in \Gamma_{E_+}} = 0$ and $\varphi^{p-1}\psi|_{x_k \in \Gamma_{E_-}} = 0$, since these are points at which $\varphi = 0$ by continuity. The lemma is proved.

REMARK. The class of (admissible) functions with the aid of which the first generalized derivative is constructed is defined (according to S. G. Mihlin [8]) as the class of functions which are continuously differentiable on the interval $[x_k', x_k'']$ and such that $\psi(x_k', \tilde{x}) = \psi(x_k'', \tilde{x}) = 0$ ($\tilde{x}$ is considered to be a parameter).

LEMMA 3. *Let it be known that a function* $\varphi(x) = \varphi(x_k, \tilde{x})$, $\tilde{x} = (x_1, \cdots, x_{k-1}, x_{k+1}, \cdots, x_n)$, $x = (x_k, \tilde{x})$, *has the generalized derivative* $\partial\varphi/\partial x_k$ *on a bounded domain* $G \subset R_n$ *with a sufficiently smooth boundary* Γ (*nothing is said with respect to the continuity of* $\varphi(x)$). *Then also the function*

$$w = (\varphi)^{p-1} \operatorname{sign}^* (\varphi)^{p-2} \qquad (p > 2) \tag{1.9}$$

has the generalized derivative

$$\frac{\partial w}{\partial x_k} = (p-1)(\varphi)^{p-2} \frac{\partial \varphi}{\partial x_k} \operatorname{sign}^* (\varphi)^{p-2} \tag{1.10}$$

on G.

PROOF. As was shown in [9] by S. M. Nikol'skiĭ, if there exists a generalized (in the sense of S. L. Sobolev) derivative $\partial\varphi/\partial x_k$ on G, then the initial function $\varphi(x) = \varphi(x_1, \cdots, x_n)$ must be alsolutely continuous in x_k. This means that, no matter which interval $[x_k^0, x_k] \subset G$ we choose, the following representation must hold almost everywhere with respect to $\tilde{x} = (x_1, \cdots, x_{k-1}, x_{k+1}, \cdots, x_n)$ in G:

$$\varphi(x_k, \tilde{x}) = \varphi(x_k^0, \tilde{x}) + \int_{x_k^0}^{x_k} \frac{\partial \varphi(t, \tilde{x})}{\partial t} dt. \tag{1.11}$$

Of course, the continuity of the function $\varphi(x_k, \tilde{x})$ in the variable x_k follows from this formula.

Thus we obtain that (1.8) holds almost everywhere with respect to $\tilde{x}$ on the domain G. Further,

$$\int_G \operatorname{sign}^* (\varphi)^{p-2} (\varphi)^{p-1} \frac{\partial \psi}{\partial x_k} dG = \int_{\tilde{G}} \left[\int_{x_k'}^{x_k''} (\operatorname{sign}^* (\varphi)^{p-2}) (\varphi)^{p-1} \frac{\partial \psi}{\partial x_k} dx_k \right] d\tilde{G}$$

(by virtue of (1.8))

$$\begin{aligned} &= -(p-1) \int_{\tilde{G}} \left[\int_{x_k'}^{k_k''} (\varphi)^{p-2} (\operatorname{sign}^* (\varphi)^{p-2}) \frac{\partial \varphi}{\partial x_k} \psi dx_k \right] d\tilde{G} \\ &= \int_G \left[-(p-1)(\varphi)^{p-2} \frac{\partial \varphi}{\partial x_k} \psi \operatorname{sign}^* (\varphi)^{p-2} \right] dG. \end{aligned} \tag{1.12}$$

Hence follows the existence of the generalized derivative $\partial w/\partial x_k$, determined by (1.10), on G. Here $d\widetilde{G} = dx_1 \cdots dx_{k-1} dx_{k+1} \cdots dx_n$, $\widetilde{G}$ is the domain of the variable $\widetilde{x}$, $G = \widetilde{G} \times (x'_k, x''_k)$, $x'_k = x'_k(\widetilde{x})$, $x''_k = x''_k(\widetilde{x})$ are the points of contact of some straight line parallel to the axis OX_k with the boundary Γ of G.

REMARK. For simplicity of calculations we assumed that (x'_k, x''_k) is a unique interval. In the general case the proof is similar.

LEMMA 4. *If the function $\varphi(x)$ has the generalized derivative $\partial\varphi/\partial x_k$ on the domain G, then also the function $f = |\varphi|$ has the generalized derivative*

$$\frac{\partial f}{\partial x_k} = \frac{\partial |\varphi|}{\partial x_k} = \frac{\partial \varphi}{\partial x_k} \operatorname{sign} \varphi \tag{1.13}$$

on G.

The proof of this lemma is entirely similar to the proof of Lemma 3.

§2. Imbedding theorems

We consider the function

$$v(x) = |D^s_\nu u|^{p-1} [\operatorname{sign}(D^s_\nu u)] (D^{s-1}_{\nu_i} u) \mu(x) \quad (\text{где } p > 2), \tag{2.1}$$

$$D^s_\nu u = \frac{\partial^s u}{\partial x_1^{\nu_1} \partial x_2^{\nu_2} \dots \partial x_n^{\nu_n}}, \tag{2.2}$$

where ν is a vector with fixed coordinates $\nu = (\nu_1, \cdots, \nu_n)$, $\Sigma_1^n \nu_j = s$,

$$D^{s-1}_{\nu_i} u = \frac{\partial^{s-1} u}{\partial x_1^{\nu_1} \dots \partial x_{i-1}^{\nu_{i-1}} \partial x_i^{\nu_i} \partial x_{i+1}^{\nu_{i+1}} \dots \partial x_n^{\nu_n}} \qquad (\nu_i \geqslant 1), \tag{2.3}$$

and $\mu(x) \geqslant 0$ is a sufficiently smooth function which is defined on a bounded domain $G \subset R_n$, vanishes on the boundary Γ of G, and satisfies the following inequalities on G:

$$C_1 t^{sp} > \mu(x) > C_2 t^{sp}, \tag{2.4}$$

$$\left| \frac{\partial \mu}{\partial x_i} \right| \leqslant C_3 t^{sp-1}. \tag{2.5}$$

Here C_1, C_2 and C_3 are positive constants, the symbol sign has its usual sense, and $t = t(x)$ is the distance from a variable point $x \in G$ to the boundary Γ of G. The construction of a function of this kind was carried out in [2]. By virtue of Lemmas 3 and 4, $v(x)$ has the generalized partial derivative

$$\begin{aligned} \frac{\partial v}{\partial x_i} &= (p-1) |D^s_\nu u|^{p-2} \frac{\partial |D^s_\nu u|}{\partial x_i} [\operatorname{sign}(D^s_\nu u)] (D^{s-1}_{\nu_i} u) \mu(x) \\ &+ |D^s_\nu u|^{p-1} [\operatorname{sign}(D^s_\nu u)] (D^{s-1}_{\nu_i} u) \frac{\partial \mu}{\partial x_i} + |D^s_\nu u|^{p-1} [\operatorname{sing}(D^s_\nu u)] (D^s_\nu u) \mu(x), \end{aligned} \tag{2.6}$$

$$\int_G \frac{\partial v}{\partial x_i} dG = 0, \tag{2.7}$$

because $v|_\Gamma = 0$ by virtue of the properties of the auxiliary function $\mu(x)$.

We substitute the right-hand side of expression (2.6) into formula (2.7) and carry out the estimate of the positive term $|D^s_\nu u|^p \mu(x)$ by the remaining terms, making use of (2.4), (2.5) and the Schwarz inequality:

$$\begin{aligned} C \int_G |D^s_\nu u|^p t^{sp} dG &\leqslant \int_G \left|\frac{\partial D^s_\nu u}{\partial x_i}\right| |D^s_\nu u|^{p-2} |D^{s-1}_{\nu_i} u| |\mu(x)| dG \\ &+ \int_G |D^s_\nu u|^{p-1} |D^{s-1}_{\nu_i} u| \left|\frac{\partial \mu}{\partial x_i}\right| dG \\ &\leqslant C_1' \left(\int_G \left|\frac{\partial D^s_\nu u}{\partial x_i}\right|^2 |D^s_\nu u|^{p-2} t^{sp+2} dG\right)^{1/2} \left(\int_G |D^s_\nu u|^{p-2} |D^{s-1}_{\nu_i} u|^2 t^{sp-2} dG\right)^{1/2} \\ &+ C_2' \left(\int_G |D^s_\nu u|^p t^{sp} dG\right)^{1/2} \left(\int_G |D^s_\nu u|^{p-2} |D^{s-1}_{\nu_i} u|^2 t^{sp-2} dG\right)^{1/2}. \end{aligned} \tag{2.8}$$

It follows from (2.8) by virtue of [3] and [4] that

$$\begin{aligned} \int_G |D^s_\nu u|^p t^{sp} dG \leqslant C_3' \Bigg[&\int_G |D^s_\nu u|^{p-2} |D^{s-1}_{\nu_i} u|^2 t^{sp-2} dG \\ &+ \left(\int_G \left|\frac{\partial D^s_\nu u}{\partial x_i}\right|^2 |D^s_\nu u|^{p-2} t^{sp+2} dG\right)^{1/2} \left(\int_G |D^s_\nu u|^{p-2} |D^{s-1}_{\nu_i} u|^2 t^{sp-2} dG\right)^{1/2} \Bigg]. \end{aligned} \tag{2.9}$$

We now introduce the notation

$$|D^s u| = \sum_{(\nu)} \left| \frac{\partial^s u}{\partial x_1^{\nu_1} \dots \partial x_n^{\nu_n}} \right| = \sum_{(\nu)} |D^s_\nu u|. \tag{2.10}$$

If we replace all the $|D^s_\nu u|$ and $|D^{s-1}_{\nu_i} u|$ in the right-hand side by $|D^m u|$, we only strengthen inequality (2.9). As a result we obtain

$$\begin{aligned} \int_G |D^s_\nu u|^p t^{sp} dG \leqslant C_3' \Bigg[&\int_G |D^s u|^{p-2} |D^{s-1} u|^2 t^{sp-2} dG \\ &+ \left(\int_G \left|\frac{\partial D^s_\nu u}{\partial x_i}\right|^2 |D^s u|^{p-2} t^{sp+2} dG\right)^{1/2} \left(\int_G |D^s u|^{p-2} |D^{s-1} u|^2 t^{sp-2} dG\right)^{1/2} \Bigg]. \end{aligned} \tag{2.11}$$

Replacing $|\partial D^s_\nu u/\partial x_i|$ by $|\partial D^s u/\partial x_i|$ and by $|D^{s+1} u|$, we obtain two more modified imbedding theorems (2.11):

$$\begin{aligned} \int_G |D^s_\nu u|^p t^{sp} dG \leqslant C_3' \Bigg[&\int_G |D^s u|^{p-2} |D^{s-1} u|^2 t^{sp-2} dG \\ &+ \left(\int_G \left|\frac{\partial D^s u}{\partial x_i}\right|^2 |D^s u|^{p-2} t^{sp+2} dG\right)^{1/2} \left(\int_G |D^s u|^{p-2} |D^{s-1} u|^2 t^{sp+2} dG\right)^{1/2} \Bigg] \end{aligned} \tag{2.12}$$

and

$$\int_G |D_\nu^s u|^p t^{sp} dG \leqslant C_3' \Bigg[\int_G |D^s u|^{p-2} |D^{s-1}u|^2 t^{sp-2}\, dG + \left(\int_G |D^{s+1}u|^2 |D^s u|^{p-2} t^{sp+2} dG \right)^{1/2} \left(\int_G |D^s u|^{p-2} |D^{s-1}u|^2 t^{sp-2} dG \right)^{1/2} \Bigg]. \tag{2.13}$$

Setting $|D_\nu^s u| = b_\nu = b_{\nu_1,\dots,\nu_n}$ and applying Lemma 1, we obtain

$$\sum_{(\nu)} |D_\nu^s u|^p = \sum_{(\nu)} |b_{\nu_1,\nu_2,\dots,\nu_n}|^p \geqslant \frac{1}{N^{p-1}} \left(\sum_\nu b_{\nu_1,\nu_2,\dots,\nu_n} \right)^p = \frac{1}{N^{p-1}} \left(\sum_\nu |D_\nu^s u| \right)^p \tag{2.14}$$

where N is the number of the combinations of ν_j of a vector $\nu = (\nu_1, \cdots, \nu_n)$ such that $\Sigma_j \nu_j = s$.

Taking (2.14) into account, we obtain as a result from (2.12) and (2.13) that

$$\int_G |D^s u|^p t^{sp} dG \leqslant N^{p-1} \int_G \sum_\nu |D_\nu^s u|^p t^{sp} dG \leqslant N^p C_3' \Bigg[\int_G |D^s u|^{p-2} |D^{s-1}u|^2 t^{sp-2} dG + \left(\int_G \left| \frac{\partial D^s u}{\partial x_i} \right|^2 |D^s u|^{p-2} t^{sp+2} dG \right)^{1/2} \left(\int_G |D^s u|^{p-2} |D^{s-1}u|^2 t^{sp-2} dG \right)^{1/2} \Bigg], \tag{2.15}$$

and

$$\int_G |D^s u|^p t^{sp} dG \leqslant C_4' \Bigg[\int_G |D^s u|^{p-2} |D^{s-1}u|^2 t^{sp-2} dG + \left(\int_G |D^{s+1}u|^2 |D^s u|^{p-2} t^{sp+2} dG \right)^{1/2} \left(\int_G |D^s u|^{p-2} |D^{s-1}u|^2 t^{sp-2} dG \right)^{1/2} \Bigg], \tag{2.16}$$

where $C_4' = N^p C_3'$.

The inequalities (2.12)–(2.16) are exactly the imbedding theorems which we wanted to derive. We note one more case of the imbedding theorem (2.16) with $s = 1$, which we shall need in the process of further investigations:

$$\int_G |D^1 u|^p t^p dG \leqslant C_s' \Bigg[\int_G |D^1 u|^{p-2} u^2 t^{p-2} dG + \left(\int_G |D^2 u|^2 |D^1 u|^{p-2} t^{p+2} dG \right)^{1/2} \left(\int_G |D^1 u|^{p-2} u^2 t^{p-2} dG \right)^{1/2} \Bigg], \tag{2.17}$$

or, in an expanded form,

$$\sum_i \int_G \left| \frac{\partial u}{\partial x_i} \right|^p t^p dG \leqslant C_s' \sum_k \sum_i \left[\int_G \left(\frac{\partial^2 u}{\partial x_i \partial x_k} \right)^2 \left| \frac{\partial u}{\partial x_i} \right|^{p-2} t^{p+2} dG \right]^{1/2} \left[\int_G \left| \frac{\partial u}{\partial x_i} \right|^{p-2} u^2 t^{p-2} dG \right]^{1/2} + \sum_i \int_G \left| \frac{\partial u}{\partial x_i} \right|^{p-2} u^2 t^{p-2} dG. \tag{2.18}$$

§3. Obtaining the estimates for solutions of equations

The equation in variations which corresponds to functional (0.1) has the form

$$\int_G \sum_{i,j} a_{i,j}(x)\left(\frac{\partial u}{\partial x_i}\frac{\partial v}{\partial x_j}+\frac{\partial u}{\partial x_j}\frac{\partial u}{\partial x_i}\right)dG=0. \tag{3.1}$$

If it is known in advance (as is the case here; see the Introduction) that the functions $a_{ij}(x)$ and $\partial u/\partial x_i$ have the required generalized derivatives, then expression (3.1) can be differentiated any number of times with respect to any combination of variables $(x_1, \cdots, x_n)$ according to the formula

$$\int_G \sum_{i,j}\left[\frac{\partial v}{\partial x_j} D_\nu^s\left(a_{ij}(x)\frac{\partial u}{\partial x_i}\right)+\frac{\partial v}{\partial x_i} D_\nu^s\left(a_{ij}(x)\frac{\partial u}{\partial x_j}\right)\right]dG=0, \tag{3.2}$$

where, as before,

$$D_\nu^s f=\frac{\partial^s f}{\partial x_1^{\nu_1}\dots\partial x_n^{\nu_n}},\quad \sum_j \nu_j=s.$$

Formula (3.2) follows from the results of [2] and [4]; therefore we shall not derive it again here. For $D_\nu^s=\partial/\partial x_k$, (3.2) takes on the form

$$\int_G \sum_{i,j} a_{ij}(x)\left(\frac{\partial^2 u}{\partial x_i\partial x_k}\frac{\partial v}{\partial x_j}+\frac{\partial^2 u}{\partial x_j\partial x_k}\frac{dv}{dx_i}\right)dG$$
$$=-\int_G \sum_{i,j}\frac{\partial a_{ij}(x)}{\partial x_k}\left(\frac{\partial u}{\partial x_i}\frac{\partial v}{\partial x_j}+\frac{\partial u}{\partial x_j}\frac{\partial v}{\partial x_i}\right)dG. \tag{3.3}$$

We set

$$v(x)=\left(\frac{\partial u}{\partial x_k}\right)^{p-1}\left[\operatorname{sign}^*\left(\frac{\partial u}{\partial x_k}\right)^{p-2}\right]\lambda(x), \tag{3.4}$$

where the function $\lambda(x)$ vanishes on the boundary Γ of the domain G and has the following properties:

$$\bar C_2 t^{p+2}\leqslant\lambda(x)\leqslant\bar C_2 t^{p+2}, \tag{3.5}$$

$$\left|\frac{\partial\lambda(x)}{\partial x_k}\right|\leqslant\bar C_3 t^{p+1}, \tag{3.6}$$

where $\bar C_1$, $\bar C_2$ and $\bar C_3$ are positive constants, and $t=t(x)$ is the distance from the point x to the boundary Γ of G. The actual construction of such a function was carried out in [2]. The symbol sign* was defined in §1 of this paper. We evaluate the generalized derivative $\partial v/\partial x_i$ of the function $v(x)$ given by formula (3.4):

$$\frac{\partial v}{\partial x_i}=(p-1)\frac{\partial^2 u}{\partial x_i\partial x_k}\left(\frac{\partial u}{\partial x_k}\right)^{p-2}\left[\operatorname{sign}^*\left(\frac{\partial u}{\partial x_k}\right)^{p-2}\right]\lambda(x)+\frac{\partial\lambda}{\partial x_i}\left(\frac{\partial u}{\partial x_k}\right)^{p-1}\operatorname{sign}^*\left(\frac{\partial u}{\partial x_k}\right)^{p-2}. \tag{3.7}$$

This generalized derivative exists by virtue of Lemma 3.

We substitute (3.7) into (3.3) and estimate the integral which contains the expression

$$\sum_{i,j}\left(\frac{\partial^2 u}{\partial x_i \partial x_k}\right)^2 \left|\frac{\partial u}{\partial x_k}\right|^{p-2} \lambda(x),$$

making use of the definition of ellipticity (0.2), of the Schwarz and Hölder inequalities, and of the estimates (3.5) and (3.6) for the function $\lambda(x)$. As a result we obtain

$$\begin{aligned}
\bar{C}\int_G \sum_i \left(\frac{\partial^2 u}{\partial x_i \partial x_k}\right)^2 \left|\frac{\partial u}{\partial x_k}\right|^{p-2} t^{p+2}dG &\leqslant 2\int_G \sum_{i,j} a_{ij}(x)\Big[\left(\frac{\partial^2 u}{\partial x_i \partial x_k}\right)^2 \left|\frac{\partial u}{\partial x_k}\right|^{p-2} \\
&+\left(\frac{\partial^2 u}{\partial x_j \partial x_k}\right)^2 \left|\frac{\partial u}{\partial x_k}\right|^{p-2}\Big]\lambda(x)\,dG \leqslant C_5\Big|\int_G \sum_{i,j} |a_{ij}|\Big[\left|\frac{\partial u}{\partial x_k}\right|^{p-1}\left|\frac{\partial^2 u}{\partial x_i \partial x_k}\right|\left|\frac{\partial \lambda}{\partial x_j}\right| \\
&+\left|\frac{\partial^2 u}{\partial x_j \partial x_k}\right|\left|\frac{\partial u}{\partial x_k}\right|^{p-1}\left|\frac{\partial \lambda}{\partial x_i}\right|\Big]dG\Big| + C_6\int_G \sum_{i,j}\left|\frac{\partial a_{ij}}{\partial x_k}\right|\Big[\left|\frac{\partial u}{\partial x_i}\right|\left|\frac{\partial^2 u}{\partial x_j \partial x_k}\right|\left|\frac{\partial u}{\partial x_k}\right|^{p-2}\lambda(x) \\
&+\left|\frac{\partial u}{\partial x_j}\right|\left|\frac{\partial^2 u}{\partial x_i \partial x_k}\right|\left|\frac{\partial u}{\partial x_k}\right|^{p-2}\lambda(x)+\left|\frac{\partial u}{\partial x_i}\right|\left|\frac{\partial u}{\partial x_k}\right|^{p-1}\left|\frac{\partial \lambda}{\partial x_j}\right| \\
&+\left|\frac{\partial u}{\partial x_j}\right|\left|\frac{\partial u}{\partial x_k}\right|^{p-1}\left|\frac{\partial \lambda}{\partial x_i}\right|\Big]dG \\
&\leqslant D_1\left(\int_G \sum_{(i)}\left(\frac{\partial^2 u}{\partial x_i \partial x_k}\right)^2\left|\frac{\partial u}{\partial x_k}\right|^{p-2} t^{p+2}dG\right)^{1/2}\left(\int_G \sum_{(k)}\left|\frac{\partial u}{\partial x_k}\right|^p t^p dG\right)^{1/2} \\
&+D_2\int_G \sum_{(i)}\left|\frac{\partial u}{\partial x_i}\right|^p t^p dG,
\end{aligned} \tag{3.8}$$

because

$$\begin{aligned}
&\int_G \sum_{(i)}\left|\frac{\partial u}{\partial x_i}\right|\left|\frac{\partial u}{\partial x_k}\right|^{p-1} |\lambda'_{x_i}|\,dG \\
&\leqslant D_3\left(\int_G \sum_i \left|\frac{\partial u}{\partial x_i}\right|^p t^p dG\right)^{1/p}\left(\int_G \sum_k \left|\frac{\partial u}{\partial x_k}\right|^p t^p dG\right)^{\frac{p-1}{p}} = D_3\int_G \sum_i \left|\frac{\partial u}{\partial x_i}\right|^p t^p dG,
\end{aligned}$$

where C_5, C_6, D_1, D_2 and D_3 are positive constants. It follows from (3.8) that

$$\int_G \sum_i \left(\frac{\partial^2 u}{\partial x_i \partial x_k}\right)^2 \left|\frac{\partial u}{\partial x_k}\right|^{p-2} t^{p+2}dG \leqslant \tilde{C}\int_G \sum_i \left|\frac{\partial u}{\partial x_i}\right|^p t^p dG, \tag{3.9}$$

where $\tilde{C} > 0$ is a constant. We transform estimate (3.9) as follows:

$$\int_G \sum_i \sum_k \left(\frac{\partial^2 u}{\partial x_k \partial x_i}\right)^2 \left|\left|\frac{\partial u}{\partial x_i}\right|^{p-2} t^{p+2}dG \leqslant \tilde{C}_n\int_G \sum_i \left|\frac{\partial u}{\partial x_i}\right|^p t^p dG, \tag{3.10}$$

where $\widetilde{C}_n = n\widetilde{C}$. By virtue of the results of [3] and [4], it follows from (3.10) and from the imbedding theorem (2.18) that

$$\int_G \sum_k \sum_i \left(\frac{\partial^2 u}{\partial x_i \partial x_k}\right)^2 \left|\frac{\partial u}{\partial x_i}\right|^{p-2} t^{p+2} dG \leqslant \widetilde{D} \int_G \sum_i \left|\frac{\partial u}{\partial x_i}\right|^p u^2 t^{p-2} dG, \tag{3.11}$$

$$\int_G \sum_i \left|\frac{\partial u}{\partial x_i}\right|^p t^p dG \leqslant \widetilde{D}_1 \int_G \sum_i \left|\frac{\partial u}{\partial x_i}\right|^{p-2} u^2 t^{p-2} dG. \tag{3.12}$$

By virtue of the Hölder inequality,

$$\int_G \left|\frac{\partial u}{\partial x_i}\right|^{p-2} u^2 t^{p-2} dG \leqslant \left(\int_G \left|\frac{\partial u}{\partial x_i}\right|^p t^p dG\right)^{\frac{p-2}{p}} \left(\int_G |u|^p \, dG\right)^{2/p}. \tag{3.13}$$

Finally, we obtain from (3.13), (3.11) and (3.12) that

$$\int_G \sum_i \left|\frac{\partial u}{\partial x_i}\right|^p t^p dG \leqslant \widetilde{D}_2 \sum_i \left(\int_G \left|\frac{\partial u}{\partial x_i}\right|^p t^p dG\right)^{\frac{p-2}{p}} \left(\int_G |u|^p \, dG\right)^{2/p}. \tag{3.14}$$

Setting

$$b_i = \left(\int_G \left|\frac{\partial u}{\partial x_i}\right|^p t^p dG\right)^{\frac{p-2}{p}},$$

we obtain by virtue of Lemma 1

$$\left[\sum_i \left(\int_G \left|\frac{\partial u}{\partial x_i}\right|^p t^p dG\right)^{\frac{p-2}{p}}\right]^r \leqslant n^{\frac{2}{p-2}} \sum_i \int_G \left|\frac{\partial u}{\partial x_i}\right|^p t^p dG \quad \left(r = \frac{p}{p-2} > 1\right). \tag{3.15}$$

Hence, raising the last inequality to the power $1/r$, we obtain

$$\sum_i \left(\int_G \left|\frac{\partial u}{\partial x_i}\right|^p t^p dG\right)^{\frac{p-2}{p}} \leqslant n^{2/p} \left(\sum_i \int_G \left|\frac{\partial u}{\partial x_i}\right|^p t^p dG\right)^{\frac{p-2}{p}}. \tag{3.16}$$

Finally, from (3.16) and (3.14) we obtain

$$\begin{aligned} \sum_i \int_G \left|\frac{\partial u}{\partial x_i}\right|^p t^p dG &\leqslant \bar{D}_2 \sum_i \left(\int_G \left|\frac{\partial u}{d x_i}\right|^p t^p dG\right)^{\frac{p-2}{p}} \left(\int_G |u|^p \, dG\right)^{2/p} \\ &\leqslant h^{2/p} \bar{D}_2 \left(\int_G |u|^p \, dG\right)^{2/p} \left(\sum_i \int_G \left|\frac{\partial u}{\partial x_i}\right|^p t^p dG\right)^{\frac{p-2}{p}}. \end{aligned} \tag{3.17}$$

FIRST STEP. It follows from (3.17) that

$$\sum_i \int_G \left| \frac{\partial u}{\partial x_i} \right|^p t^p dG \leqslant \bar{D}_3 \int_G |u|^p dG, \tag{3.18}$$

where $\bar{D}_3 = n(\bar{D}_2)^{p/2} > 0$ is a constant. It follows from (3.18) and (3.11) that

$$\sum_i \sum_k \int_G \left(\frac{\partial^2 u}{\partial x_k \partial x_i} \right)^2 \left| \frac{\partial u}{\partial x_i} \right|^{p-2} t^{p+2} dG \leqslant \bar{D}_4 \int_G |u|^p dG, \tag{3.19}$$

where $\bar{D}_4$ is a constant. The proof of formulas (3.18) and (3.19) is the first step of the proof by induction of our estimates.

SECOND STEP. Assume that the following estimates hold:

$$\int_G \left(\frac{\partial^s u}{\partial x_i \partial x_1^{k_1} \ldots \partial x_n^{k_n}} \right)^2 \left| \frac{\partial^{s-1} u}{\partial x_1^{k_1} \ldots \partial x_n^{k_n}} \right|^{p-2} t^{ps-p+2} dG \leqslant \tilde{C}_1 \int_G |u|^p dG \tag{3.20}$$

and

$$\int_G \left| \frac{\partial^{s-1} u}{\partial x_1^{k_1} \ldots \partial x_n^{k_n}} \right|^p t^{sp-p} dG \leqslant \tilde{C}_2 \int_G |u|^p dG \tag{3.21}$$

for any $k = (k_1, \cdots, k_n)$, $\Sigma_j k_j = s - 1$, and for any $i = 1, \cdots, n$.

THIRD STEP. We are required to prove the following estimates:

$$\int_G \left(\frac{\partial^{s+1} u}{\partial x_i \partial x_1^{\nu_1} \ldots \partial x_n^{\nu_n}} \right)^2 \left| \frac{\partial^s u}{\partial x_1^{\nu_1} \ldots \partial x_n^{\nu_n}} \right|^{p-2} t^{sp+2} dG \leqslant \tilde{C}_3 \int_G |u|^p dG \tag{3.22}$$

and

$$\int_G \left| \frac{\partial^s u}{\partial x_1^{\nu_1} \ldots \partial x_n^{\nu_n}} \right|^p t^{sp} dG \leqslant \tilde{C}_4 \int_G |u|^p dG. \tag{3.23}$$

In order to prove them, we consider the following relation, which follows from (3.2):

$$\begin{aligned} \int_G \sum_{i,j} a_{ij}(x) \left(\frac{\partial^{s+1} u}{\partial x_i \partial x_1^{\nu_1} \ldots \partial x_n^{\nu_n}} \frac{\partial v}{\partial x_j} + \frac{\partial^{s+1} u}{\partial x_j \partial x_1^{\nu_1} \ldots \partial x_n^{\nu_n}} \frac{\partial v}{\partial x_i} \right) dG \\ \leqslant \tilde{C}_s \sum_i \int_G |\tilde{D}^s u| \left| \frac{\partial v}{\partial x_i} \right| dG, \end{aligned} \tag{3.24}$$

where

$$|\tilde{D}^s u| = \sum_{k=0}^{s} \sum_{k_j} \left| \frac{\partial^k u}{\partial x_1^{k_1} \partial x_2^{k_2} \ldots \partial x_n^{k_n}} \right|. \tag{3.25}$$

We set

$$v(x) = \lambda(x) (D_\nu^s u)^{p-1} \operatorname{sign}^* (D_\nu^s u)^{p-2}. \tag{3.26}$$

Here the function $\lambda(x)$ vanishes on the boundary Γ of G and satisfies the estimates

$$\bar{C} t^{sp+2} \leqslant \lambda(x) \leqslant \tilde{C} t^{sp+2}, \tag{3.27}$$

$$\left| \frac{\partial \lambda}{\partial x_i} \right| \leqslant C_6 t^{sp+1} \tag{3.28}$$

(with regard to the construction of $\lambda(x)$, see [2]). The symbol sign* was defined in §1. We find the generalized derivative $\partial v/\partial x_i$ (which exists by virtue of Lemma 3) of the function $v(x)$ defined by (3.26), and substitute it into (3.24). Making use of the definition of ellipticity (0.2), of the Hölder inequality, and of the estimates (3.27) and (3.28) for the function $\lambda(x)$, we obtain

$$\begin{aligned} \bar{C} \int_G \sum_i \left(\frac{\partial D_\nu^s u}{\partial x_i} \right)^2 | D_\nu^s u |^{p-2} t^{sp+2} dG \\ \leqslant 2(p-1) \int_G \sum_{i,j} a_{ij}(x) \frac{\partial D_\nu^s u}{\partial x_j} \frac{\partial D_\nu^s u}{\partial x_i} | D_\nu^s u |^{p-2} \lambda(x)\, dG \\ \leqslant \tilde{C}_6 \sum_i \left| \int_G | D^s u | \left[\left| \frac{\partial D_\nu^s u}{\partial x_i} \right| | D_\nu^s u |^{p-2} \lambda(x) + | \lambda'_{x_i} | \, | D_\nu^s u |^{p-1} \right] dG \right| \\ + \tilde{C}_7 \int_G \sum_{i,j} | a_{ij} | \left[\left| \frac{\partial D_\nu^s u}{\partial x_i} \right| | D_\nu^s u |^{p-1} \left| \frac{\partial \lambda}{\partial x_j} \right| + \left| \frac{\partial D_\nu^s u}{\partial x_j} \right| | D_\nu^s u |^{p-1} \left| \frac{\partial \lambda}{\partial x_i} \right| \right] dC \\ + \tilde{C}_8 \sum_i \int_G | \tilde{D}^s u | \left| \frac{\partial v}{\partial x_i} \right| dG, \end{aligned} \tag{3.29}$$

where $\bar{C}, \tilde{C}, C_6, \tilde{C}_6, \tilde{C}_7$ and $\tilde{C}_8$ are positive constants.

By virtue of the induction assumption (the second step) we have

$$\int_G | \tilde{D}^k u |^p t^{pk} dG \leqslant C'_k \int | u |^p \, dG \qquad (k = 1, 2, \ldots, s-1). \tag{3.30}$$

Therefore

$$\int_G | \tilde{D}^s u |^p t^{pk} dG \leqslant C''_s \left(\int_G | u |^p \, dG + \sum_\nu \int_G | D_\nu^s u |^p t^{sp} dG \right). \tag{3.31}$$

Taking (3.31) into account, we extend estimate (3.29):

$$\int_G \sum_i \left(\frac{\partial D_\nu^s u}{\partial x_i} \right)^2 | D^s u |^{p-2} (t^{sp+2} dG) \leqslant A_1 \left[\int_G | D^s u |^p t^{sp} dG + \int_G | u |^p dG \right], \tag{3.32}$$

where, as before,

$$| D^s u | = \sum_\nu \left| \frac{\partial^s u}{\partial x_1^{\nu_1} \ldots \partial x_n^{\nu_n}} \right|, \quad \sum_j \nu_j = s.$$

It follows from (3.32) and the imbedding theorem (2.9) that

$$\int_G | D_\nu^s u |^p t^{sp} dG \leqslant A_2 \left[\int_G | D_\nu^s u |^{p-2} | D_\nu^{s-1} u |^2 t^{sp-2} dG + \int_G | u |^p \, dG \right] \tag{3.33}$$

and

$$\int_G \left| \frac{\partial D_\nu^s u}{\partial x_i} \right|^2 | D_\nu^s u |^{p-2} t^{sp+2} dG$$
$$\leqslant A_3 \left[\int_G | D_\nu^s u |^{p-2} | D_\nu^{s-1} u |^2 t^{sp-2} dG + \int_G | u |^p dG \right]. \tag{3.34}$$

We apply the Hölder inequality to (3.33). We obtain

$$\int_G | D_\nu^s u |^p t^{sp} dG \leqslant A_2 \left[\int_G | u |^p dG + \left(\int_G | D_\nu^s u |^p t^{sp} dG \right)^{\frac{p-2}{p}} \left(\int_G | D_\nu^{s-1} u |^p t^{sp-p} dG \right)^{2/p} \right]$$

(by virtue of the induction assumption, formula (3.30))

$$\leqslant A_3 \left[\int_G | u |^p dG + \left(\int_G | u |^p dG \right)^{2/p} \left(\int_G | D_\nu^s u |^p t^{sp} dG \right)^{\frac{p-2}{p}} \right], \tag{3.35}$$

where $0 < (p-2)/p < 1$. Therefore, by virtue of the lemmas of [3] or [4], we obtain from inequality (3.5)

$$\int_G | D_\nu^s u |^p t^{sp} dG \leqslant A_5 \int_G | u |^p dG \tag{3.36}$$

and

$$\int_G | D^s u |^p t^{sp} dG \leqslant A_s \int_G | u |^p dG. \tag{3.37}$$

It follows from (3.30) and (3.34) that

$$\int_G \left| \frac{\partial D_\nu^s u}{\partial x_i} \right|^2 | D_\nu^s u |^{p-2} t^{sp+2} dG \leqslant A_6 \int_G | u |^p dG \tag{3.38}$$

for any $i = 1, \cdots, n$, which was to be proved.

BIBLIOGRAPHY

1. L. Hörmander, *Linear partial differential operators,* Die Grundlehren der math. Wissenschaften, Band 116, Academic Press, New York; Springer-Verlag, Berlin, 1963; Russian transl., "Mir", Moscow, 1965. MR **28** #4221; **37** #6595.

2. A. S. Foht, *Some L_2 inequalities for solutions and derivatives of solutions of elliptic equations near the boundary,* Trudy Mat. Inst. Steklov. **77** (1965), 168–191 = Proc. Steklov Inst. Math. **77** (1965), 188–212. MR **32** #7942.

3. ————, *A lemma in the calculus of variations and its applications to imbedding theorems,* Dokl. Akad. Nauk SSSR **176** (1967), 536–537 = Soviet Math. Dokl. **8** (1967), 1154–1156. MR **37** #4390.

4. ————, *Some imbedding theorems for solution of equations of elliptic type,* Trudy Mat. Inst. Steklov. **105** (1969), 230–242 = Proc. Steklov Inst. Math. **105** (1969), 281–293. MR **41** #629.

5. S. M. Nikol'skiĭ, *On a boundary estimate for a harmonic function in n-dimensional space*, Sibirsk. Mat. Ž. **1** (1960), 78–87. (Russian) MR **23** #A2543.

6. A. S. Foht, *A certain estimate of a polyharmonic function and its derivatives near the boundary of the domain*. I, Differencial'nye Uravnenija **4** (1968), 1265–1282 = Differential Equations 4 (1968), 657–666. MR **38** #2326.

7. ———, *A certain estimate of a polyharmonic function and its derivatives near the boundary of the domain*. II, Differencial'nye Uravnenija **4** (1968), 1509–1518 = Differential Equations 4 (1968), 784–788. MR **38** #6087.

8. S. G. Mihlin, *Mathematical physics, an advanced course*, "Nauka", Moscow, 1968, p. 27; English transl., North-Holland Ser. in Appl. Math. and Mech., vol. 11, North-Holland, Amsterdam; American Elsevier, New York, 1970. MR **42** #7110; **44** #3538.

9. S. M. Nikol'skiĭ, *Properties of certain classes of functions of several variables on differentiable manifolds*, Mat. Sb. **33** (75) (1953), 261–326; English transl., Amer. Math. Soc. Transl. (2) **80** (1969), 39–118. MR **16**, 453.

Translated by K. MAKOWSKI

Trudy Mat. Inst. Steklov.
117 (1972)

Proc. Steklov Inst. Math.
117 (1972)

UNBOUNDED SOLUTIONS OF DEGENERATE ELLIPTIC EQUATIONS

UDC 518.333

G. N. JAKOVLEV

Abstract. In the halfspace $x_n > 0$ of the n-dimensional Euclidean space of points $x = (x', x_n) = (x_1, \cdots, x_{n-1}, x_n)$ we consider the equation

$$\sum_{i,j=1}^{n-1} \frac{\partial}{\partial x_i}\left(a_{ij}\frac{\partial u}{\partial x_j}\right) + \frac{\partial}{\partial x_n}\left(a\frac{\partial u}{\partial x_n}\right) = 0 \tag{1}$$

with measurable coefficients satisfying the conditions

$$a_{ij}(x) = a_{ij}(x), \quad a(x) = b(x')\,\varphi(x_n), \qquad \varphi(x_n) > 0$$

and

$$c_1 x_n^{2\alpha}|\xi|^2 \leqslant \sum_{i,j=1}^{n} a_{ij}\xi_i\xi_j + a\xi_n^2 \leqslant c_2 x_n^{2\alpha}|\xi|^2,$$

where $2\alpha > 1$, $|\xi|$ is the length of the vector $\xi = (\xi_1, \cdots, \xi_n)$ and c_1 and c_2 are positive constants.

Let

$$\Psi(x_n) = \int_{x_n}^{+\infty} \varphi(t)^{-1}\,dt.$$

It is proved that for any function $f_0(x') \in L_2^{1/2-\gamma}$, $2\gamma < 1$, there exists a generalized solution $u(x)$ of equation (1) such that $\Psi^{-1}(u) = f_0(x')$ for $x_n = 0$ and

$$\| x_n^{\gamma}\nabla(\Psi^{-1}u)\|_{L_2} \leqslant c\|f_0\|_{L_2^{\frac{1}{2}-\gamma}},$$

where the constant c does not depend on f_0. This solution is unique in the class of functions for which the expression $\|x_n^{\gamma}\nabla(\Psi^{-1}u)\|_{L_2}$ is finite.

Bibliography: 13 items.

Let E^n denote n-dimensional Euclidean space of points $x = (x_1, \cdots, x_n)$. On the half space $E_+^n = \{x: x_n > 0\}$ consider the equation

$$\sum_{i,j=1}^{n-1} \frac{\partial}{\partial x_i}\left(a_{ij}\frac{\partial u}{\partial x_j}\right) + \frac{\partial}{\partial x_n}\left(a\frac{\partial u}{\partial x_n}\right) = 0, \tag{1}$$

in which $a_{ij}(x) = a_{ji}(x)$ and for which the following ellipticity condition holds: there exist positive constants c_1 and c_2 such that for all $x \in E_+^n$ and all $\xi = (\xi_1, \cdots \xi_n)$

AMS (MOS) subject classifications (1970). Primary 35J70, 35D05.

$\in E^n$ the inequalities

$$c_1 x_n^{2\alpha} |\xi|^2 \leqslant \sum_{i,j=1}^{n-1} a_{ij}\xi_i\xi_j + a\xi_n^2 \leqslant c_2 x_n^{2\alpha} |\xi|^2 \tag{2}$$

hold, where $2\alpha > 1$ and $|\xi|$ is the length of ξ. As to the coefficient $a(x)$, we will assume that $a(x) = b(x')\varphi(x_n)$, $x' = (x_1, \cdots, x_{n-1})$ and $\varphi(x_n) > 0$. Note that we do not require any smoothness, except measurability, of the coefficients of (1).

Now let us introduce the definitions and notation required below.

A measurable function $u(x)$, $x \in E_+^n$, having generalized derivatives of first order in the sense of Sobolev [1, 2] which are locally summable on E_+^n is called a *generalized solution* of (1) if

$$\int \left\{ \sum_{i,j=1}^{n-1} a_{ij} \frac{\partial u}{\partial x_j}\frac{\partial \eta}{\partial x_i} + a\frac{\partial u}{\partial x_n}\frac{\partial \eta}{\partial x_n} \right\} dx = 0, \qquad \forall \eta \in C_0^1, \tag{3}$$

where the integration is performed over E_+^n and C_0^1 is the set of all finite continuously differentiable functions with support in E_+^n.

By $\mathfrak{L}_{2,\beta}^1$ we denote the set of $f(x)$, $x \in E_+^n$, belonging locally to $\mathfrak{L}_2$ on E_+^n and having generalized derivatives of first order such that

$$\|f\|_{1,\beta} \equiv \|x_n^\beta \nabla f\| \equiv \left(\int x_n^{2\beta} |\nabla f|^2\, dx\right)^{\frac{1}{2}} < +\infty, \tag{4}$$

where ∇f is the gradient of f.

It is well known [3] that for equation (1) with condition (2) the usual statement of the boundary value problem and of the Dirichlet problem in the class of generalized solutions in $\mathfrak{L}_{2,\alpha}^1$, $2\alpha > 1$, generally speaking, has no meaning. As will be shown, in this case the boundary value problem is correct in the class of unbounded (as $x_n \to 0$) solutions $u(x)$ satisfying

$$\left.\frac{u(x)}{\psi(x_n)}\right|_{x_n=0} = f_0(x'), \qquad x' \in E^{n-1}, \tag{5}$$

where $f_0(x')$ is a given function and ψ is defined by

$$\psi(x_n) = \int_{x_n}^{+\infty} \frac{dt}{\varphi(t)}, \tag{6}$$

and thus, as follows from (2),

$$c_3 x_n^{1-2\alpha} \leqslant \psi(x_n) \leqslant c_4 x_n^{1-2\alpha}. \tag{7}$$

The analogous problem on a bounded region, when the usual Dirichlet condition is defined on a part of the boundary, was posed for the first time by A. V. Bicadze

[4]. For certain equations other than (1) the problem of Bicadze was studied in [5], [6] and [7].

It is well known [3, 8] that for any function $f(x)$ in $\mathfrak{L}^1_{2,\gamma}$, $-1 < 2\gamma < 1$, the trace exists on the hyperplane $x_n = 0$; or, more precisely, there exists a function $f_0(x')$, $x' \in E^{n-1}$, such that for some function equivalent to $f(x)$ (which we will also denote by $f(x)$) there holds the equality

$$f(x', h) - f_0(x') = \int_0^h \frac{\partial f}{\partial x_n} dx_n,$$

and hence by the Hölder inequality

$$\lim_{h \to 0} \int_{E^{n-1}} |f(x', h) - f_0(x')|^2 dx' = 0.$$

In this case one writes

$$f(x)|_{x_n=0} = f_0(x') \quad \text{or} \quad f(x', 0) = f_0(x').$$

It is easy to see that $f_0(x') \in \mathfrak{L}_2(g')$ for any bounded region $g' \subset E^{n-1}$. Moreover, it is proved in [9, 10] that

$$\|f_0\|_{\frac{1}{2}-\gamma} \equiv \left\{ \sum_{i=1}^{n-1} \int_0^{+\infty} h^{2\gamma-2} dh \int_{E^{n-1}} |\Delta_i f_0(x')|^2 dx' \right\}^{1/2} \leqslant c_5 \|f\|_{1,\gamma},$$

where

$$\Delta_i f_0(x') = f_0(x_1, \ldots, x_i + h, \ldots, x_{n-1}) - f_0(x_1, \ldots, x_i, \ldots, x_{n-1}).$$

The set of $f_0(x')$ such that $\|f_0\|_{1/2-\gamma} < +\infty$, will be denoted by $\mathfrak{L}_2^{1/2-\gamma}$.

It is also well known [9, 10] that, conversely, for any function $f_0(x') \in \mathfrak{L}_2^{1/2-\gamma}$, $-1 < 2\gamma < 1$, one can find a function $f(x) \in \mathfrak{L}^1_{2,\gamma}$ such that $f_0(x')$ is the trace of $f(x)$ on the hyperplane $x_n = 0$ and

$$\|f\|_{1,\gamma} \leqslant c_6 \|f_0\|_{\frac{1}{2}-\gamma}.$$

Such a function, for example, is

$$f(x) = \frac{1}{x_n^{n-1}} \int_{x_1}^{x_1+x_n} d\xi_1 \ldots \int_{x_{n-1}}^{x_{n-1}+x_n} f_0(\xi_1, \ldots, \xi_{n-1}) d\xi_{n-1}. \tag{8}$$

Generally speaking, one can drop the condition $2\gamma > -1$, but, as follows from [11], in the case $2\gamma \leqslant -1$ the set of traces of functions in $\mathfrak{L}^1_{2,\alpha}$ consists only of constants. In the sequel the condition $2\gamma > -1$ will be relaxed.

Now we can precisely formulate the problem which is to be studied in this work.

PROBLEM 1. Let there be given a function $f_0(x') \in \mathfrak{L}_2^{1/2-\gamma}$, $2\gamma < 1$. It is required to find a generalized solution $u(x)$ of equation (1) such that $\psi^{-1}u \in \mathfrak{L}^1_{2,\gamma}$ and

$$\psi^{-1}u\,|_{x_n=0} = f_0(x'). \tag{9}$$

In place of this problem it is convenient to consider a somewhat different one which, as will be seen from our constructions, is equivalent to it. For its formulation we introduce some additional definitions and notation.

Let $\mathring{\mathfrak{L}}^1_{2,\gamma}$, $2\gamma < 1$, denote the set of $v(x)$ in $\mathfrak{L}^1_{2,\gamma}$ whose traces on $x_n = 0$ are identically zero. It is clear that $\|v\|_{1,\gamma}$ is a norm on $\mathring{\mathfrak{L}}^1_{2,\gamma}$ and that, by Hardy's inequality [12],

$$\|x_n^{\gamma-1}v\| = \left\| x_n^{\gamma-1}\int_0^{x_n} \frac{\partial}{\partial t} v\,(x',\ t)\,dt \right\| \leqslant \frac{2}{\varkappa}\left\| x_n^{\gamma}\frac{\partial v}{\partial x_n}\right\|,$$

where $\varkappa = \min\{1,\ 1-2\gamma\}$.

Hence the functionals

$$\|v\|_{1,\gamma} \text{ and } \|v\|_{1,\gamma} + \|x_n^{\gamma-1}v\| \tag{10}$$

are equivalent norms on $\mathring{\mathfrak{L}}^1_{2,\gamma}$.

PROBLEM 2. Let there be given a function $f(x) \in \mathfrak{L}^1_{2,\gamma}$, $2\gamma < 1$. It is required to find a generalized solution $u(x)$ of equation (1) such that

$$\psi^{-1}u - f \in \mathring{\mathscr{L}}^1_{2,\gamma}. \tag{11}$$

LEMMA. *The space* $\mathring{\mathfrak{L}}^1_{2,\gamma}$, $2\gamma < 1$, *is the completion of* C_0^1 *in the norm* (10).

PROOF. The fact that the completion of $\mathring{C}_0^1$ in the norm (10) lies in $\mathring{\mathfrak{L}}^1_{2,\gamma}$ follows immediately from the completeness of $\mathring{\mathfrak{L}}^1_{2,\gamma}$. Let us prove the converse that any function $v \in \mathring{\mathfrak{L}}^1_{2,\gamma}$ is the limit in the norm (10) of some sequence of functions from C_0^1. To this end it suffices to construct a sequence of finite functions in $\mathfrak{L}^1_{2,\gamma}$ with supports in E^n_+, since each such function, as is known [1], is the limit of Sobolev means.

Let $v(x)$ be any function in $\mathring{\mathfrak{L}}^1_{2,\gamma}$ and let $\xi_\delta(x_n)$, $\delta > 0$, be a continuous function equal to one if $2\delta \leqslant x_n \leqslant \delta^{-1}$, and to zero if $x_n \leqslant \delta$ or $x_n \geqslant 2\delta^{-1}$, and linear elsewhere. As is easy to see, each $v_\delta = v\xi_\delta$, $\delta > 0$, belongs to the Sobolev space W_2^1, i.e. it and its first derivatives belong to $\mathfrak{L}_2$ on E^n_+. Let us show that

$$\lim_{\delta\to+0}\|v_\delta - v\|_{1,\gamma} = 0. \tag{12}$$

Indeed,

$$\|v_\delta - v\|_{1,\gamma} \leqslant \|x_n^{\gamma}(1-\xi_\delta)\nabla v\| + \left\| x_n^{\gamma} v \frac{\partial \xi_\delta}{\partial x_n}\right\|,$$

and since the function $1-\xi_\delta$ is nonzero only off $2\delta < x_n < \delta^{-1}$ and $0 \leqslant \xi_\delta \leqslant 1$, it follows that

$$\lim_{\delta\to+0} \| x_n^\gamma (1-\xi_\delta) \nabla v \| = 0,$$

while for the second term we have

$$\left\| x_n^\gamma v \frac{\partial \xi_\delta}{\partial x_n} \right\|^2 = \frac{1}{\delta^2} \int_{E^{n-1}} dx' \int_\delta^{2\delta} x_n^{2\gamma} v^2 dx_n + \delta^2 \int_{E^{n-1}} dx' \int_{\frac{1}{\delta}}^{\frac{2}{\delta}} x_n^{2\gamma} v^2 dx_n$$

$$\leqslant 4 \int_{E^{n-1}} dx' \int_\delta^{2\delta} x_n^{2(\gamma-1)} v^2 dx_n + 4 \int_{E^{n-1}} dx' \int_{\frac{1}{\delta}}^{\frac{2}{\delta}} x_n^{2(\gamma-1)} v^2 dx_n \to 0.$$

Thus assertion (12) is proved.

Let $\zeta_A(t)$, $A>0$, be a continuous function equal to one if $-A \leqslant t \leqslant A$, and to zero off $[-2A, 2A]$, and linear elsewhere. Let us set $v_{\delta A} = v_\delta \zeta_A(|x|)$, where $|x|$ is the length of the vector x, and show that

$$\lim_{A\to+\infty} \| \nabla (v_{\delta A} - v_\delta) \| = 0. \tag{13}$$

Indeed,

$$\| \nabla (v_{\delta A} - v_\delta) \| \leqslant \| (1-\zeta_A) \nabla v_\delta \| + \| v_\delta \nabla \zeta_A \|.$$

Since the function $1-\zeta_A$ is nonzero only off $|x| > A$ and $0 \leqslant \zeta_A \leqslant 1$, we have

$$\lim_{A\to+\infty} \| (1-\zeta_A) \nabla v_\delta \| = 0.$$

For the second term we have

$$\| v_\delta \nabla \zeta_A \|^2 = \frac{1}{A^2} \int_{A<|x|<2A} v_\delta^2 | \nabla |x| |^2 dx \leqslant \frac{1}{A^2} \| v_\delta \|^2 \to 0$$

as $A \to +\infty$. Thus (13) is proved.

And now, using (12) and (13), by means of the usual diagonal process one constructs the desired sequence of finite functions in $\mathfrak{L}^1_{2,\gamma}$ with support in E^n_+. The lemma is proved.

THEOREM 1. *Let $u(x)$ be a solution of Problem 2. If $2(1-\alpha) \leqslant 2\gamma < 1$, then*

$$\| \psi^{-1} u \|_{1,\gamma} \leqslant \frac{c_2}{c_1} \frac{2\alpha-1}{1-2\gamma} \| f \|_{1,\gamma}. \tag{14}$$

PROOF. Let $v(x)$ be an arbitrary function in $\overset{\circ}{\mathfrak{L}}{}^1_{2,\delta}$ and let $\{\eta_k\}$ be a sequence of functions from C_0^1 converging to $v(x)$ in the norm (10). Let us set $u = w\psi$ and $\eta = \eta_k \psi^\epsilon$, where $\epsilon = (1 - 2\gamma)/(2\alpha - 1) \in (0, 1)$, and substitute into (3). Then after integration by parts we have

$$B_\varepsilon(w, \eta_k) = -(1 - \varepsilon) \int \psi^\varepsilon b \frac{\partial w}{\partial x_n} \eta_k dx, \tag{15}$$

where

$$B_\varepsilon(w, \eta_k) = \int \psi^{1+\varepsilon} \left\{ \sum_{i,j=1}^{n-1} a_{ij} \frac{\partial w}{\partial x_j} \frac{\partial \eta_k}{\partial x_i} + a \frac{\partial w}{\partial x_n} \frac{\partial \eta_k}{\partial x_n} \right\} dx.$$

Let us prove that in (15) one can pass to the limit under the integral as $k \to \infty$. Indeed, by inequalities (2) and (7) we have

$$|B_\varepsilon(w, \eta_k - v)| \leqslant c_2 c_4^{1+\varepsilon} \|w\|_{1,\gamma} \|\eta_k - v\|_{1,\gamma} \to 0$$

as $k \to \infty$. Further, note that from (2) there follows the existence of a constant c_7 such that $b(x') \leqslant c_7$. Hence

$$\left| \int \psi^\varepsilon b \frac{\partial w}{\partial x_n} (\eta_k - v)\, dx \right| \leqslant c_7 c_4^\varepsilon \|w\|_{1,\gamma} \|x_n^{\gamma-1}(\eta_k - v)\| \to 0$$

as $k \to 0$. Thus, passing to the limit in (15), we find that

$$B_\varepsilon(w, v) = (\varepsilon - 1) \int \psi^\varepsilon b \frac{\partial w}{\partial x_n} v dx, \quad \forall v \in \mathscr{L}^1_{2,\gamma}. \tag{16}$$

Here let us set $v = w - f$:

$$B_\varepsilon(w, w - f) = (\varepsilon - 1) \int \psi^\varepsilon b \frac{\partial w}{\partial x_n} (w - f)\, dx, \tag{17}$$

and transform the integral on the right:

$$\begin{aligned} \int \psi^\varepsilon b \frac{\partial w}{\partial x_n} (w - f)\, dx &= \frac{1}{2} \int \psi^\varepsilon b \frac{\partial}{\partial x_n} (w - f)^2\, dx \\ &+ \int \psi^\varepsilon b (w - f) \frac{\partial f}{\partial x_n} dx = \frac{1}{2} \int \frac{\partial}{\partial x_n} \{b\psi^\varepsilon (w - f)^2\}\, dx \\ &- \frac{1}{2} \int b (w - f)^2 \frac{\partial}{\partial x_n} \psi^\varepsilon dx + \int \psi^\varepsilon b (w - f) \frac{\partial f}{\partial x_n} dx. \end{aligned} \tag{18}$$

Since

$$\begin{aligned} &\psi^\varepsilon(\delta) \int_{E^{n-1}} b(x') \{w(x', \delta) - f(x', \delta)\}^2 dx' \\ &\leqslant c_4^\varepsilon c_7 \int_{E^{n-1}} \left\{ \int_0^\delta \frac{\partial}{\partial x_n} (w - f)\, dx_n \right\}^2 dx' \delta^{\varepsilon(1-2\alpha)} \\ &\leqslant c_7 c_4^\varepsilon \frac{1}{1 - 2\gamma} \int_0^\delta dx_n \int_{E^{n-1}} x_n^{2\gamma} \left(\frac{\partial w}{\partial x_n} - \frac{\partial f}{\partial x_n} \right)^2 dx' \to 0 \end{aligned}$$

as $\delta \to +0$, it follows that

$$\int \frac{\partial}{\partial x_n} \{b\psi^{\varepsilon}(w-f)^2\}\, dx \geqslant 0.$$

Noting this fact and the fact that $0 < \epsilon < 1$, we find after substitution of (18) into (17) that

$$B_{\varepsilon}(w, w-f) \leqslant \varepsilon(\varepsilon-1)\frac{1}{2}\int b\varphi^{-1}\psi^{\varepsilon-1}(w-f)^2\, dx + (\varepsilon-1)\int \psi^{\varepsilon} b\,(w-f)\frac{\partial f}{\partial x_n}\, dx.$$

Since

$$\left|\int \psi^{\varepsilon} b\,(w-f)\frac{\partial f}{\partial x_n}\, dx\right| \leqslant \frac{\varepsilon}{2}\int b\varphi^{-1}\psi^{\varepsilon-1}(w-f)^2\, dx + \frac{1}{2\varepsilon}\int b\varphi\psi^{1+\varepsilon}\left(\frac{\partial f}{\partial x_n}\right)^2 dx,$$

it follows that

$$B_{\varepsilon}(w, w-f) \leqslant \frac{1-\varepsilon}{2\varepsilon}\int a\psi^{1+\varepsilon}\left(\frac{\partial f}{\partial x_n}\right)^2 dx \leqslant \frac{1-\varepsilon}{2\varepsilon} B_{\varepsilon}(f, f)$$

and hence

$$B_{\varepsilon}(w, w) \leqslant B_{\varepsilon}(w, f) + \frac{1}{2}\left(\frac{1}{\varepsilon}-1\right) B_{\varepsilon}(f, f)$$
$$\leqslant \frac{1}{2} B_{\varepsilon}(w, w) + \frac{1}{2} B_{\varepsilon}(f, f) + \frac{1}{2}\left(\frac{1}{\varepsilon}-1\right) B_{\varepsilon}(f, f).$$

Thus

$$B_{\varepsilon}(w, w) \leqslant \frac{1}{\varepsilon} B_{\varepsilon}(f, f),$$

and so by inequalities (2) and the definition of ϵ we have

$$\| w \|_{1,\gamma} \leqslant \frac{c_2}{c_1}\frac{2\alpha-1}{1-2\gamma}\| f \|_{1,\gamma}. \tag{19}$$

Theorem 1 is proved.

COROLLARY. *Let $u(x)$ be a solution of Problem 1. If $2(1-\alpha) \leqslant 2\gamma < 1$, then there exists a constant c_8 such that*

$$\| \psi^{-1} u \|_{1,\gamma} \leqslant c_8 \| f_0 \|_{\frac{1}{2}-\gamma}.$$

THEOREM 2. *Let $2(1-\alpha) \leqslant 2\gamma < 1$. If $u_1(x)$ and $u_2(x)$ are two generalized solutions of equation* (1) *such that*

$$\psi^{-1}[u_2(x) - u_1(x)] \in \mathring{\mathscr{L}}^1_{2,\gamma},$$

then $u_1(x) \equiv u_2(x)$.

PROOF. Since $u_2(x) - u_1(x)$ is a generalized solution of equation (1), as seen in the proof of Theorem 1 we have

$$B_{\varepsilon}(w, v) = (\varepsilon-1)\int \psi^{\varepsilon} b \frac{\partial w}{\partial x_n} v\, dx, \quad \forall v \in \mathring{\mathscr{L}}^1_{2,\gamma}, \text{ where } w = (u_2 - u_1)\psi^{-1}.$$

Here, setting $v = w$, we find that

$$B_\varepsilon(w, w) = (\varepsilon - 1) \int \psi^\varepsilon b w \frac{\partial w}{\partial x_n} dx = \frac{1}{2}(\varepsilon - 1) \int \frac{\partial}{\partial x_n}(\psi^\varepsilon b w^2)\, dx - \frac{1}{2}(\varepsilon - 1) \int b w^2 \frac{\partial}{\partial x_n} \psi^\varepsilon dx.$$

We have shown that

$$\int \frac{\partial}{\partial x_n}(\psi^\varepsilon b w^2)\, dx \geqslant 0,$$

so that

$$B_\varepsilon(w, w) \leqslant -\frac{\varepsilon}{2}(1 - \varepsilon) \int b w^2 \varphi^{-1} \psi^{\varepsilon-1} dx,$$

i.e.

$$B_\varepsilon(w, w) + \frac{\varepsilon}{2}(1 - \varepsilon) \int b w^2 \varphi^{-1} \psi^{\varepsilon-1} dx \leqslant 0,$$

whence it follows that $w = (u_2 - u_1)\psi^{-1} \equiv 0$. Theorem 2 is proved.

THEOREM 3. *For any function* $f \in \mathcal{L}^1_{2,\gamma}$, $2(1 - \alpha) \leqslant 2\gamma < 1$, *there exists a solution of Problem* 2.

PROOF. Consider the equation (16):

$$B_\varepsilon(w, v) = (\varepsilon - 1) \int \psi^\varepsilon b \frac{\partial w}{\partial x_n} v\, dx, \qquad \forall v \in \mathring{\mathcal{L}}^1_{2,\gamma}.$$

Let $\{v_k\}$ be a linearly independent set of functions in C^1_0 such that its linear span is dense in $\mathring{\mathcal{L}}^1_{2,\gamma}$. An approximate solution of equation (16) will be sought in the form

$$w_N = \sum_{k=1}^{N} c_k^N v_k + f,$$

where the coefficients c_k^N are found from the system of linear equations

$$B_\varepsilon(w_N, v_i) = (\varepsilon - 1) \int \psi^\varepsilon b \frac{\partial w_N}{\partial x_n} v_i\, dx, \qquad i = 1, \ldots, N. \tag{20}$$

First of all let us show that the a priori estimate (19) holds for the approximate solution w_N. Indeed, let us multiply equations (20) by the c_i^N respectively and add. Then we obtain the equality

$$B_\varepsilon(w_N, w_N - f) = (\varepsilon - 1) \int \psi^\varepsilon b \frac{\partial w_N}{\partial x_n}(w_N - f)\, dx,$$

i.e. we obtain (17) where w_N appears instead of w. Hence the a priori estimate (19) holds with w replaced by w_N:

$$\|w_N\|_{1,\gamma} \leqslant \frac{c_2}{c_1} \frac{2\alpha - 1}{1 - 2\gamma} \|f\|_{1,\gamma}.$$

Thus we have the inequality

$$\left\| \sum_{k=1}^{N} c_k^N v_k \right\|_{1,\gamma} \leqslant c_9 \|f\|_{1,\gamma}, \tag{21}$$

from which, in particular, it follows that the system of linear algebraic equations (20) always has a solution, and a unique one.

From (21) and the equivalence of the norms (10) on $\mathring{\mathfrak{L}}^1_{2,\gamma}$ there follows the existence of a subsequence $\{w_{N_k}\}$ of approximate solutions which weakly converges in $W_2^1(g)$ for any bounded region g whose closure lies in the half space E^n_+, and strongly in $\mathfrak{L}_2(g_1)$ for any bounded region $g_1 \subset E^n_+$. Hence for the limit $w(x)$ there holds the inequality

$$\|w\|_{1,\gamma} \leqslant \frac{c_2}{c_1} \frac{2\alpha - 1}{1 - 2\gamma} \|f\|_{1,\gamma}. \tag{22}$$

After passing to the limit in (16) for the chosen subsequence we have

$$B_\varepsilon(w, v_i) = (\varepsilon - 1) \int \psi^\varepsilon b \frac{\partial w}{\partial x_n} v_i dx, \qquad \forall i = 1, 2, \ldots,$$

whence by the density of the linear span of the $\{v_k\}$ in $\mathring{\mathfrak{L}}^1_{2,\gamma}$ it follows that

$$B_\varepsilon(w, \eta) = (\varepsilon - 1) \int \psi^\varepsilon b \frac{\partial w}{\partial x_n} \eta dx, \qquad \forall \eta \in C_0^1.$$

From this condition we find, as was done in the proof of Theorem 1, that the function $w(x)$ satisfies equation (16).

Let us prove that $w - f \in \mathring{\mathfrak{L}}^1_{2,\gamma}$.

The existence of the trace $w(x', 0)$ for $w(x)$ follows from the fact that it belongs to $\mathfrak{L}^1_{2,\gamma}$. Hence, integrating the square of the obvious equality

$$w(x', 0) - f(x', 0) = w(x) - w_N(x) - \int_0^{x_n} \frac{\partial}{\partial t} w(x', t)\, dt + \int_0^{x_n} \frac{\partial}{\partial t} w_{N_k}(x', t)\, dt$$

over the cylinder $g_1 = \{x\colon x' \in g', 0 < x_n < \delta\}$, where g' is a bounded region in E^{n-1}, we obtain the inequality

$$\left\{ \int_{g_1} |w(x', 0) - f(x', 0)|^2 dx \right\}^{1/2} \leqslant \left\{ \int_{g_1} |w(x) - w_{N_k}(x)|^2 dx \right\}^{1/2}$$
$$+ \left\{ \int_{g_1} \left| \int_0^{x_n} \frac{\partial}{\partial t} w(x', t)\, dt \right|^2 dx \right\}^{1/2} + \left\{ \int_{g_1} \left| \int_0^{x_n} \frac{\partial}{\partial t} w_{N_k}(x', t)\, dt \right|^2 dx \right\}^{1/2},$$

from which it follows that

$$\left\{\int_{g_1} |w(x', 0) - f(x', 0)|^2 dx'\right\}^{1/2} \tag{23}$$
$$\leqslant \delta^{-1/2}\left\{\int_{g_1} |w(x) - w_{N_k}(x)|^2 dx\right\}^{1/2} + c\,\|f\|_{1,\gamma}\delta^{1/2-\gamma},$$

where c is a constant. Let $\epsilon > 0$ be given. Then for sufficiently small δ the second term on the right in (23) will be less than $\epsilon/2$ and for a fixed δ the first term converges to zero as $k \to \infty$. Hence

$$\int_{g'} |w(x', 0) - f(x', 0)|^2 dx = 0.$$

Our assertion is proved.

It remains to prove that $u = w\psi$ is a generalized solution of equation (1).

In equation (16) let v be any function in C_0^1. Set $u = w\psi$. Then after obvious transformations we obtain the equality

$$\int\left\{\sum_{i,j=1}^{n-1} a_{ij}\frac{\partial u}{\partial x_j}\frac{\partial}{\partial x_i}(v\psi^\epsilon) + a\frac{\partial u}{\partial x_n}\frac{\partial}{\partial x_n}(v\psi^\epsilon)\right\} dx = 0$$

for any $v \in C_0^1$. Since $\eta = v\psi^\epsilon$ will also be an arbitrary function in C_0^1, it also follows that $u = \psi w$ is a generalized solution of equation (1). Theorem 3 is proved.

REMARK 1. In exactly the same way as was done in this paper one can consider the boundary value problem for equation (1) on a bounded region $g \subset E_+^n$ a part of the boundary of which lies on the hyperplane $x_n = 0$. In this case for $x_n = 0$ condition (5) is defined, and on the part of the boundary lying in the halfspace $x_n > 0$ there is defined the usual Dirichlet condition.

REMARK 2. As follows from the author's works [11] and [13], the results of this work generalize to the case of equations of type (1) with a more general type of degeneracy.

BIBLIOGRAPHY

1. S. L. Sobolev, *Applications of functional analysis to mathematical physics*, Izdat. Leningrad. Gos. Univ., Leningrad, 1950; English transl., Transl. Math. Monographs, vol. 7, Amer. Math. Soc., Providence, R. I., 1963. MR 14, 565; 29 #2624.

2. S. M. Nikol'skiĭ, *Approximation of functions of several variables and imbedding theorems*, "Nauka", Moscow, 1969. (Russian)

3. L. D. Kudrjavcev, *Direct and inverse embedding theorems. Applications to solution of elliptic equations by variational methods*, Trudy Mat. Inst. Steklov. 55 (1959); English transl., Transl. Math. Monographs, Amer. Math. Soc., Providence, R. I. (to appear). MR 33 #7838.

4. A. V. Bicadze, *Equations of mixed type*, Izdat. Akad. Nauk SSSR, Moscow, 1959; English transl., Macmillan, New York, 1964. MR 29 #381.

5. S. A. Tersenov, *An elliptic type of equation degenerating at the domain boundary,* Dokl. Akad. Nauk SSSR **115** (1957), 670–673. (Russian) MR **19**, 965.

6. A. A. Vašarin and P. I. Lizorkin, *Certain boundary value problems for elliptic equations with strong degeneracy on the boundary,* Dokl. Akad. Nauk SSSR **137** (1961), 1015–1018 = Soviet Math. Dokl. **2** (1961), 376–379. MR **24** #A907.

7. G. N. Jakovlev, *A certain problem of A. V. Bicadze for degenerate elliptic equations,* Differencial′nye Uravnenija **4** (1968), 147–150 = Differential Equations **4** (1968), 76–78. MR **37** #614.

8. A. A. Vašarin, *Boundary properties of functions of class* $W_2^1(\alpha)$ *and their application to a boundary problem of mathematical physics,* Izv. Akad. Nauk SSSR Ser. Mat. **23** (1959), 421–454. (Russian) MR **22** #2808.

9. P. I. Lizorkin, *Boundary properties of a certain class of functions,* Dokl. Akad. Nauk SSSR **126** (1959), 703–706. (Russian) MR **21** #5700.

10. A. A. Vašarin, *The boundary properties of functions having a finite Dirichlet integral with weight,* Dokl. Akad. Nauk SSSR **117** (1957), 742–744. (Russian) MR **20** #1113.

11. G. N. Jakovlev, *On traces of functions whose derivatives are summable with some weight,* Imbedding Theorems and Their Applications, Proc. All-Union Sympos. Imbedding Theorems (Baku, 1966), "Nauka", Moscow, 1970, pp. 225–233. (Russian)

12. G. H. Hardy, J. E. Littlewood and G. Pólya, *Inequalities,* Cambridge Univ. Press, New York, 1934; Russian transl., IL, Moscow, 1948. MR 18, 722.

13. G. N. Jakovlev, *On the density of functions with compact support in weighted spaces,* Dokl. Akad. Nauk SSSR **170** (1966), 797–798 = Soviet Math. Dokl. **7** (1966), 1300–1302. MR **34** #6290.

Translated by J. MALEK

Trudy Mat. Inst. Steklov.
117 (1972)

Proc. Steklov Inst. Math.
117 (1972)

THE FIRST BOUNDARY VALUE PROBLEM FOR QUASILINEAR ELLIPTIC EQUATIONS OF SECOND ORDER

G. N. JAKOVLEV

UDC 518.333

Abstract. In a bounded subdomain G of n-dimensional space we consider the boundary value problem for the elliptic equation

$$-\sum_{i=1}^{n} \frac{\partial}{\partial x_i} F^i(u_x) = h, \qquad u_x = \left(\frac{\partial u}{\partial x_1}, \dots, \frac{\partial u}{\partial x_n}\right), \tag{1}$$

with the boundary condition

$$u = f \text{ on } \partial G. \tag{2}$$

The functions $F^i(y)$, $y = (y_1, \cdots, y_n)$, are supposed to be such that

$$|F^i(y)| \leqslant \mu (1 + |y|)^{p-1}, \qquad \sum_{i=1}^{n} F^i(y) y_i \geqslant \nu (1 + |y|)^p - \mu_1,$$

$$\frac{\partial F^i}{\partial y_j} = \frac{\partial F^j}{\partial y_i}, \qquad \left| \frac{\partial F^i(y)}{\partial y_j} \right| \leqslant \mu (1 + |y|)^{p-2},$$

where $p \geqslant 2$, $\mu > 0$, $\nu > 0$ and $\mu_1 > 0$. Equation (1) satisfies the following ellipticity condition:

$$\sum_{i,j=1}^{n} \frac{\partial F^i(y)}{\partial y_j} \xi_i \xi_j \geqslant \sum_{i=1}^{n} m_i^2(y_i) \xi_i^2,$$

where the functions $m_i(t) \geqslant 0$ are allowed to vanish but only so that the functions

$$M_i(t) = \int_0^t m_i(t)\, dt$$

remain strictly increasing.

It is shown that for $f \in W_p^{1-1/p}(\partial G)$ and $h \in W_{q,2}^1(G)$ $(1/p + 1/q = 1)$ there is a unique generalized solution $u \in W_p^1(G)$ of the problem (1), (2); moreover, u is such that $F^i(u_x) \in W_{q,1}^1(G)$ and $M_i(\partial u/\partial x_i) \in W_{2,1}^1(G)$.

Bibliography: 14 items.

Let E^n be the n-dimensional Euclidean space of points $x = (x_1, \cdots, x_n)$ and g a bounded domain in E^n with a sufficiently smooth boundary ∂g. For the sake of simplicity we shall assume that ∂g is an infinitely differentiable manifold of dimension $n - 1$. In the domain g we consider the elliptic equation

---

AMS (MOS) subject classifications (1970). Primary 35J60, 35D05.

$$-D_i F^i(u_x) = h(x), \tag{0.1}$$

where u_x is the gradient of the function $u(x)$, i.e. $u_x = (u_{x_1}, \cdots, u_{x_n})$, and $D_j F^i(u_x)$ is the derivative of the function $F^i(u_x) = F^i(u_{x_1}, \cdots, u_{x_n})$ with respect to the variable x_j; moreover, here and in all that follows the appearance of the same index twice indicates summation from 1 to n.

In the present article we study the first boundary value problem for equation (0.1) with the boundary condition

$$u(x) = f(x') \text{ on } \partial g, \tag{0.2}$$

where $f(x')$ $(x' \in \partial g)$ is a given function. Under the usual hypotheses on the functions $F^i(y)$ $(y \in E^n)$, $h(x)$ $(x \in g)$ and $f(x')$ $(x' \in \partial g)$, but without the assumption of the uniform ellipticity of the equation (0.1), we shall prove that the problem (0.1), (0.2) has a unique generalized solution $u(x)$ in the class $W_p^1(g)$; moreover, this solution is such that in g the $F^i(u_x)$ $(i = 1, \cdots, n)$, considered as functions of x, possess first generalized derivatives in the sense of S. L. Sobolev [1], and in any strictly interior subdomain of g these derivatives belong to the space L_q with $1/p + 1/q = 1$.

In its method the present work is close to M. I. Višik [2] (cf. also the monograph [3] of J. L. Lions); the main results obtained below are close to those of A. D. Džabrailov [4] and the author [5, 6]. More detailed references will be given in the process of exposition.

§1. Notation and the formulation of the problem

We assume that the functions $F^i(y)$ $(i = 1, \cdots, n)$ occurring in (0.1) are defined and continuous for all $y = (y_1, \cdots, y_n) \in E^n$, have first order derivatives in E^n and satisfy the following condition:

1) There exist real numbers $\mu > 0$ and $p \geqslant 2$ such that for all $y \in E^n$

$$\left\{\sum_{i=1}^n |F^i(y)|^2\right\}^{\frac{1}{2}} \leqslant \mu(1 + |y|)^{p-1}, \tag{1.1}$$

$$\left\{\sum_{i,\,j=1}^n |F^i_{,\,j}(y)|^2\right\}^{\frac{1}{2}} \leqslant \mu(1 + |y|)^{p-2}, \tag{1.2}$$

where $|y|$ denotes the length of the vector $y = (y_1, \cdots, y_n)$ and $F^i_{,j}(y) = F^i_{y_j}(y)$.

The equation (0.1) is supposed to be symmetric and elliptic. More exactly, we assume the following:

2) There exist nonnegative continuous functions $m_i(t)$ $(t \in E^1)$ such that for all $y \in E^n$ and $\xi \in E^n$

$$F^i_{,\,j}(y)\,\xi_i \xi_j \geqslant m_i^2(y_i)\,\xi_i^2, \tag{1.3}$$

where

$$F^i_{,j}(y) = F^j_{,i}(y). \tag{1.4}$$

The functions $m_i(t)$ can vanish at some points but only in such a way that the functions

$$M_i(t) = \int_0^t m_i(t)\,dt, \qquad i = 1, \ldots, n, \tag{1.5}$$

remain strictly increasing. Thus the functions $m_i(t)$ cannot be zero in a whole interval, and the inverse functions to (1.5) are defined.

Condition 2) shows that equation (0.1) in general is not uniformly elliptic, because the corresponding quadratic form can vanish at some points $y \in E^n$. However, we shall also assume the following:

3) There exist real numbers $\nu > 0$ and $\mu_1 \geqslant 0$ such that for every $y \in E^n$

$$F^i(y)\,y_i \geqslant \nu(1 + |y|)^p - \mu_1. \tag{1.6}$$

It is easy to see that this condition characterizes the behavior of $F^i(y)$ only for $|y| \to \infty$. Indeed, it always holds in any bounded region as long as condition (1.1) is satisfied.

Let us introduce the necessary definitions and notation.

$W^1_p(g)$ $(1 < p < \infty)$ denotes the usual Sobolev space [1]; its elements are functions $v(x) \in L_p(g)$ whose first order generalized derivatives $v_{,i} = v_{x_i}$ $(i = 1, \cdots, n)$ are in $L_p(g)$; in other words, those functions $v(x)$ for which the following norm is defined and finite:

$$\|v\|_{W^1_p(g)} = \|v\|_p + \|v_x\|_p;$$

where $\|v\|_p$ is the norm of $v(x)$ in $L_p(g)$, v_x is the gradient of the function $v(x)$ and $\|v_x\|_p$ is the norm of the vector v_x. It is well known that each function v from $W^1_p(g)$ takes on the boundary ∂g well-defined values called the trace of v on ∂g. $\mathring{W}^1_p(g)$ will denote the subspace of the space $W^1_p(g)$ consisting of functions whose trace on ∂g is zero. Obviously the expression $\|\eta_x\|_p$ defines a norm in the space $\mathring{W}^1_p(g)$. Moreover, if $\rho = \rho(x)$ denotes the distance of the point $x \in g$ to the boundary ∂g, then the following statement holds:

There exists a positive number c_1 such that

$$\|\rho^{-1}\eta\|_p \leqslant c_1 \|\eta_x\|_p, \quad \forall \eta \in \mathring{W}^1_p(g) \tag{1.7}$$

(for the proof, see, for example, [7]).

$W^r_p(\partial g)$, $1 < p < \infty$, $0 < r < 1$, will denote the space of functions $f(x') \in L_p(\partial g)$ for which the norm

$$\|f\|_{W_p^r(\partial g)} = \|f\|_{L_p(\partial g)} + \left(\int_{\partial g} dx' \int_{\partial g} \frac{|f(x') - f(y')|^p}{|x' - y'|^{n+pr}} dy' \right)^{\frac{1}{p}}$$

is finite; and $L_{q,\alpha}(g)$, $q = p/(p-1)$, will be the space of measurable functions $h(x)$ ($x \in g$) with norm $\|\rho^\alpha h\|_q$, where $\rho = \rho(x)$ denotes as above the distance of the point $x \in g$ to the boundary ∂g.

Given the functions $h(x) \in L_{q,1}(g)$ and $f(x') \in W_p^{1-1/p}(\partial g)$, a function $u(x) \in W_p^1(g)$ will be called a generalized solution of the boundary value problem (0.1), (0.2) if $f(x')$ is the trace of $u(x)$ on ∂g and

$$(F^i(u_x),\ v_{,i}) = (h,\ v), \qquad \forall v \in \overset{\circ}{W}{}_p^1(g), \tag{1.8}$$

where, for example, (h, v) means the integral of $h(x)v(x)$ over the domain g.

We claim that under the above hypotheses the last definition is correct. Indeed, by (1.1) the left-hand side in (1.8) is always finite, and for the right-hand side we get by (1.7)

$$|(h,\ v)| \leqslant \|\rho h\|_q \|\rho^{-1} v\|_p \leqslant c_1 \|\rho h\|_q \|v_x\|_p. \tag{1.9}$$

Furthermore, the uniqueness of generalized solutions to (0.1), (0.2) follows easily from conditions 1) and 2). Indeed, assume that in addition to $u(x)$ there is another solution $w(x)$, i.e. there exists a function $w(x)$ such that $w(x) - u(x) \in \overset{\circ}{W}{}_p^1(g)$ and

$$(F^i(w_x),\ v_{,i}) = (h,\ v), \qquad \forall v \in \overset{\circ}{W}{}_p^1(g). \tag{1.10}$$

Subtracting (1.10) from (1.8) and setting $v = u - w$, we obtain

$$(F^i(u_x) - F^i(w_x), \qquad u_{,i} - w_{,i}) = 0. \tag{1.11}$$

On the basis of condition (1.3) the integrand in (1.11) can be estimated from below as follows:

$$(u_{,i} - w_{,i})(F^i(u_x) - F^i(w_x))$$
$$\geqslant (u_{,i} - w_{,i})^2 \int_0^1 m_i^2(w_{,i} + t(u_{,i} - w_{,i}))\,dt = (u_{,i} - w_{,i}) \int_{w_{,i}}^{u_{,i}} m_i^2(t)\,dt.$$

This together with (1.11) yields

$$(u_{,i} - w_{,i}) \int_{w_{,i}}^{u_{,i}} m_i^2(t)\,dt = 0. \tag{1.12}$$

Since no function $m_i(t)$ is allowed to vanish on a whole interval, (1.12) implies that $u_{,i} = w_{,i}$ in g for $i = 1, \cdots, n$. Since $w - u \in \overset{\circ}{W}{}_p^1(g)$, this gives $u = w$ in g. The uniqueness of generalized solutions to (0.1), (0.2) is thus established.

In the formulation of the main theorem we shall need the following notation: $W^1_{q,\alpha}(g)$ will be the space of functions $h(x)$ $(x \in g)$ for which the norm

$$\|h\|_{W^1_{q,\alpha}(g)} = \|\rho^{\alpha-1}h\|_q + \|\rho^\alpha h_x\|_q$$

is well defined and finite.

THEOREM. *Let the functions* $F^i(y)$, $i = 1, \cdots, n$, *satisfy conditions* 1), 2) *and* 3); *and let the functions* $h(x)$ $(x \in g)$ *and* $f(x')$ $(x' \in \partial g)$ *be such that*

$$h(x) \in W^1_{q,2}(g),\quad f(x') \in W^{1-\frac{1}{p}}(\partial g),\qquad \frac{1}{p}+\frac{1}{q}=1. \tag{1.13}$$

Then the problem (0.1), (0.2) *has a unique generalized solution* $u(x) \in W^1_p(g)$. *Furthermore, this solution has the following properties: The functions* $F^i(u_x(x))$ *and* $M_i(u_{,i}(x))$ $(i = 1, \cdots, n)$ *as functions of* x *belong to the spaces* $W^1_{q,1}(g)$ *and* $W^1_{2,1}(g)$, *respectively. Moreover,* u *satisfies the inequality*

$$\|u\|^p_{W^1_p(g)} + \sum_{i=1}^n \{\|M_i(u_{,i})\|^2_{W^1_{2,1}(g)} + \|F^i(u_x)\|^q_{W^1_{q,1}(g)}\} \leqslant C\{1 + \|f\|^p_{W_p^{1-\frac{1}{p}}(\partial g)} + \|h\|^q_{W^1_{q,2}(g)}\}, \tag{1.14}$$

where the constant C *depends only on* n, p, μ, ν, μ_1 *and the domain* g.

Simple examples show that even for infinitely differentiable functions $h(x)$ $(x \in g)$ and $f(x')$ $(x' \in \partial g)$ the solution of an equation of type (0.1) need not have integrable derivatives of second order.

Consider, for instance, equation (0.1) with $F^i(y) = F(y_i)$, where $F(t) = t^3 \exp(-t^{-2})$ for $t \neq 0$ and $F(0) = 0$. In this case, obviously $p = 4$ and $|F^i(y)| \leqslant (1 + |y|)^3$; and

$$|F^i_{,j}(y)| \leqslant 3(1+|y|)^2,\qquad F^i_{,j}(y)\xi_i\xi_j \geqslant \xi_i^2 \exp(-y_i^{-2}),$$

i.e. conditions (1.1), (1.2) and (1.3) (with $m_i^2(t) = \exp(-t^{-2})$) are satisfied. It suffices to check (1.6) outside some cube. Let $y_i^2 > 1$, $i = 1, \cdots, n$; then

$$F^i(y)y_i = y_i^4 \exp(-y_i^{-2}) \geqslant \frac{1}{e\sqrt{n}}|y|^4.$$

Hence the equation satisfies conditions 1), 2) and 3). It is easy to see that the function

$$u(x) = \int_0^{|x|^2-1} t\left(2 + \cos\frac{1}{t}\right)dt + c,$$

where c is an arbitrary constant, represents a solution of this equation (with a C^∞ function $h(x)$), for instance, in the sphere $g = \{x: |x| < \sqrt{2}\}$. However, the second-order derivatives of u are not summable in g. Indeed, if all the second-order derivatives of u belonged to $L_1(g)$, then the function

$$\Delta u = \sum_{i=1}^{n} u,_{ii} = 2\{(n+2)|x|^2 - n\}\left(2 + \cos\frac{1}{|x|^2-1}\right) + \frac{4|x|^2}{|x|^2-1}\sin\frac{1}{|x|^2-1} \tag{1.15}$$

would also be an element of $L_1(g)$. Let us show that $\Delta u \notin L_1(g)$. The first summand in (1.15) is bounded in g and hence belongs to $L_1(g)$, but for the second summand we obtain

$$\int_g \frac{|x|^2}{||x|^2-1|}\left|\sin\frac{1}{|x|^2-1}\right| dx = n\sigma_n \int_0^{\sqrt{2}} \frac{r^{n+1}}{|r^2-1|}\left|\sin\frac{1}{r^2-1}\right| dr$$
$$\geqslant n\sigma_n \int_1^{\sqrt{2}} \frac{r}{r^2-1}\left|\sin\frac{1}{r^2-1}\right| dr = \frac{1}{2} n\sigma_n \int_1^{+\infty} \frac{|\sin t|}{t}\, dt = +\infty,$$

where σ_n is the volume of the n-dimensional unit sphere. Hence Δu cannot belong to $L_1(g)$.

The existence and uniqueness of generalized solutions in classes $W_p^1(g)$ to boundary value problems of type (0.1), (0.2) were studied in several articles, and by now the corresponding results are fairly complete (for a bibliography see [8]). For the proof of the existence of solutions one uses Galerkin's method. The main objective of the present work is to show that the solution of the problem (0.1), (0.2) is somewhat better than just being an element of $W_p^1(g)$; in particular, we shall show that it satisfies equation (0.1) almost everywhere. This will be established by a special choice of basis functions, which was suggested by M. I. Višik in [2]. In this work Višik considered a boundary value problem similar to (0.1), (0.2) in the case when the function $m^2(t) = \min\{m_i^2(t), i = 1, \cdots, n\}$ is such that $m^2(t) = c^2|t|^{p-2}$ for $|t| \geqslant R$ and $m^2(t) \geqslant c_0^2|t|^s$, $s \geqslant 0$, for $|t| < R$, with c, R, c_0 and s being constants. It was shown in [2] that a generalized solution $u(x) \in W_p^1(g)$ to this problem exists and is unique; and, moreover, that it is such that $M(u_{,i}) \in W_{2,1}^1(g)$, $i = 1, \cdots, n$, where $M(t)$ is related to $m(t)$ by formula (1.5). However the question of differentiability of the functions $F^i(u_x)$, $i = 1, \cdots, n$, was not considered in [2]. This problem was investigated by A. D. Džabrailov [4] and the author [6] for the case of elliptic but not uniformly elliptic equations of type (0.1). In these works the investigation was carried out by means of a different method and under more restrictive assumptions on the equation (0.1). These restrictions are roughly as follows: In condition 1) the sum $1 + |y|$ is to be replaced by $|y|$, and in condition 2) the functions $m_i^2(y_i)$ have to be replaced by the function $|y|^{p-2}$. Condition 3) is then automatically satisfied.

Finally let us mention that all statements in the sequel as well as the main theorem generalize easily to the more general elliptic equations and systems of the form

$$-D_i F^i(x, u, u_x) = h(x)$$

under the corresponding conditions on the functions F^i, which are analogous to those stated in [2].

§2. Auxiliary lemmas

LEMMA 2.1. *Let the quadratic form* $\Phi(\xi) = a_{ij}\xi_i\xi_j$, $\xi = (\xi_1, \cdots, \xi_n) \in E^n$, *be such that* $a_{ij} = a_{ji}$ *and* $\Phi(\xi) \geqslant 0$ *for all* $\xi \in E^n$. *Then*

$$|a_{ij}\xi_i| \leqslant \sqrt{a_{jj}}\sqrt{\Phi(\xi)}, \qquad \forall \xi \in E^n. \tag{2.1}$$

The proof is obvious.

Let us introduce the following notation: $v_{,i}$ denotes the first derivative of the function v with respect to the ith variable; similarly, $v_{,ij}$ is the second derivative of v with respect to the ith and jth variables. Furthermore, we set

$$|v_x| = \left(\sum_{i=1}^n v_{,i}^2\right)^{\frac{1}{2}}, \qquad |v_{xx}| = \left(\sum_{i,j=1}^n v_{,ij}^2\right)^{\frac{1}{2}}.$$

As usual, $\bar{g}$ means the closure of the domain g.

LEMMA 2.2. *If the function $f(x)$ is in $W^1_{q,\alpha}(g)$, $\alpha \geqslant 1$, and the function $\eta(x) \geqslant 0$ is continuously differentiable in the closed domain $\bar{g}$ and vanishes on ∂g, then the product $\eta^\alpha f(x)$ is in $\overset{\circ}{W}{}^1_q(g)$.*

PROOF. Since $\eta(x)$ is continuously differentiable in the closed bounded domain $\bar{g}$, its gradient η_x is bounded in g. Set $M = \max|\eta_x|$, $x \in \bar{g}$. Then

$$|\eta(x) - \eta(y)| \leqslant M|x - y|, \quad x \in \bar{g}, \quad y \in \bar{g}.$$

Substituting for y the closest point to x on the boundary ∂g, we get

$$\eta(x) \leqslant M\rho(x), \tag{2.2}$$

where $\rho(x)$ is the distance of x to the boundary ∂g.

Using inequality (2.2) it is easy to show that $\eta^\alpha f(x) \in W^1_q(g)$. Indeed,

$$\|\eta^\alpha f\|_q \leqslant M^\alpha \|\rho^\alpha f\|_q,$$
$$\|(\eta^\alpha f)_{,i}\|_q \leqslant \|\eta^\alpha f_{,i}\|_q + \alpha\|\eta^{\alpha-1}\eta_{,i} f\|_q \leqslant M^\alpha\|\rho^\alpha f_{,i}\|_q + \alpha M^\alpha \|\rho^{\alpha-1} f\|_q.$$

Hence the function $\eta^\alpha f(x)$ has a trace on ∂g. Thus it remains to show that this trace is zero.

The boundary ∂g of g being a sufficiently smooth $(n-1)$-dimensional manifold, we can find for each point $\mathcal{P} \in \partial g$ a neighborhood $O(\mathcal{P})$ such that the part of the boundary given by $\gamma_{\mathcal{P}} = \partial g \cap O(\mathcal{P})$ can be described (after renumbering the coordinates if necessary) by an equation of the form

$$x_n = \varphi(x'), \quad x' = (x_1, \ldots, x_{n-1}) \in g',$$

where the function $\varphi(x')$ is as smooth as we need. Moreover, we assume that there is a number $a_{\mathcal{P}} > 0$ such that for any $\delta \in (0, a_{\mathcal{P}})$ the set

$$g_\delta = \{x: \varphi(x') < x_n < \varphi(x') + \delta, \; x' \in g'\}$$

is a subset of g.

Let $\psi(x')$ be the trace of the function $\eta^\alpha f(x)$ on $\gamma_{\mathcal{P}}$. Then (cf. [9], p. 168) for almost all $x' \in g'$ and all $x_m \in (\varphi(x'), \varphi(x') + \delta)$ we have the equality

$$\psi(x') = \int_{\varphi(x')}^{x_n} (\eta^\alpha f)_{,n} \, dt - \eta^\alpha f(x),$$

which implies

$$\left(\int_{g_\delta} |\psi(x')|^q \, dx\right)^{\frac{1}{q}} \leqslant \left(\int_{g_\delta} \left| \int_{\varphi(x')}^{x_n} (\eta^\alpha f)_{,n} \, dt \right|^q dx\right)^{\frac{1}{q}} + \left(\int_{g_\delta} |\eta^\alpha f|^q \, dx\right)^{\frac{1}{q}}.$$

This inequality combined with Hölder's inequality and (2.2) gives

$$\delta^{\frac{1}{q}} \|\psi\|_{L_q(g')} \leqslant \delta \|(\eta^\alpha f)_{,n}\|_q + M^\alpha \delta \|\rho^{\alpha-1} f\|_q;$$

hence, letting $\delta \to 0$, we see that $\psi(x') \equiv 0$. The lemma is proved.

Let $W^2_{p,1}(g)$, $1 < p < \infty$, be the space of functions $F(x)$ $(x \in g)$ having generalized derivatives of second order in g and such that the norm

$$\|F\|_{W^2_{p,1}(g)} = \|F\|_p + \|F_x\|_p + \|\rho F_{xx}\|_p$$

is finite.

LEMMA 2.3. *For every function $f(x') \in W_p^{1-1/p}(\partial g)$ $(x' \in \partial g; 1 < p < \infty)$ one can construct a function $F(x)$ $(x \in g)$ from the space $W^2_{p,1}(g)$ such that the trace of F on ∂g is $f(x')$; and, moreover,*

$$\|F\|_{W^2_{p,1}(g)} \leqslant c_2 \|f\|_{W_p^{1-\frac{1}{p}}(\partial g)}. \tag{2.3}$$

The constant c_2 depends only on p and the domain g.

This lemma can be derived from Lemma 2.4 below by the usual method of partition of unity followed by pasting together the local results. However, before formulating Lemma 2.4 we need the following notation: Let Δ_{2a} be the cube with edge of length $2a$, i.e.

$$\Delta_{2a} = \{x\colon\ 0 < x_i < 2a,\quad i = 1, \ldots, n\},$$

then Δ'_{2a} will denote the face of Δ_{2a} which lies in the hyperplane $x_n = 0$. Similarly we define Δ_a, Δ'_a and Δ_1.

LEMMA 2.4. *For each function $f(x') \in W_p^{1-1/p}(\Delta'_{2a})$ there exists a function $F(x) \in W_p^1(\Delta_a)$ whose trace on Δ'_a is $f(x')$. Moreover, in Δ_a the function $F(x)$ has generalized derivatives of second order and satisfies the inequality*

$$\|F\|_{W_p^1(\Delta_a)} + \|x_n F_{xx}\|_{L_p(\Delta_a)} \leqslant c_3 \|f\|_{W_p^{1-\frac{1}{p}}(\Delta'_{2a})}, \tag{2.4}$$

where the constant c_3 depends only on n and p.

PROOF. As is well known (cf. [11]), for each function $f(x') \in W_p^{1-1/p}(\Delta'_{2a})$ one can find a function $\tilde{f}(x) \in W_p^1(\Delta_{2a})$ with trace on Δ'_{2a} being equal to $f(x')$ and satisfying the inequality

$$\|\tilde{f}\|_{W_p^1(\Delta_{2a})} \leqslant c_4 \|f\|_{W_p^{1-\frac{1}{p}}(\Delta'_{2a})}, \tag{2.5}$$

where c_4 depends only on n and p.

We claim that

$$F(x) = \int_{\Delta_1} \tilde{f}(x + x_n t)\, dt, \qquad t = (t_1, \ldots, t_n),$$

is the function we are looking for. (It should be mentioned that the idea of constructing $F(x)$ by the method of averaging with variable step is taken from L. D. Kudrjavcev [7].) The function $F(x)$ is defined in Δ_a. First let us show that its trace on Δ'_{2a} equals $f(x')$. Applying Minkowski's inequality we get

$$\begin{aligned}
&\left(\int_{\Delta'_a} |F(x) - f(x')|^p\, dx'\right)^{\frac{1}{p}} \\
&\quad \leqslant \left(\int_{\Delta'_a} \left| \int_{\Delta_1} [\tilde{f}(x + x_n t) - f(x' + x_n t')]\, dt \right|^p dx'\right)^{\frac{1}{p}} \\
&\quad + \left(\int_{\Delta'_a} \left| \int_{\Delta'_1} [f(x' + x_n t') - f(x')]\, dt' \right|^p dx'\right)^{\frac{1}{p}} \\
&\quad \leqslant \int_{\Delta'_1} \left(\int_{\Delta'_a} \left| \int_0^1 [\tilde{f}(x + x_n t) - f(x' + x_n t')]\, dt_n \right|^p dx'\right)^{\frac{1}{p}} dt' \\
&\quad + \int_{\Delta'_1} \left(\int_{\Delta'_a} | f(x' + x_n t') - f(x')|^p\, dx'\right)^{\frac{1}{p}} dt'.
\end{aligned}$$

By virtue of the absolute continuity of Lebesgue integrals the last integral tends to zero when $x_n \to 0$. For the first integral we get

$$\int_{\Delta_1'} \left(\int_{\Delta_a'} \left| \int_0^1 [\tilde{f}(x + x_n t) - f(x' + x_n t')]\, dt_n \right|^p dx' \right)^{\frac{1}{p}} dt'$$

$$\leqslant \left(\int_{\Delta_{2a}'} \left| \int_0^1 [\tilde{f}(x', x_n + x_n t_n) - f(x')]\, dt_n \right|^p dx' \right)^{\frac{1}{p}}$$

$$= \left(\int_{\Delta_{2a}'} \left| \int_0^1 dt_n \int_0^{x_n + x_n t_n} \tilde{f}_{,n}(x', \tau)\, d\tau \right|^p dx' \right)^{\frac{1}{p}}$$

$$\leqslant \left(\int_{\Delta_{2a}'} \left| \int_0^{2x_n} |\tilde{f}_{,n}(x', \tau)|\, d\tau \right|^p dx' \right)^{\frac{1}{p}} \leqslant (2x_n)^{\frac{1}{p}} \|\tilde{f}_{,n}\|_{L_p(\Delta_{2a})}.$$

Hence

$$\lim_{x_n \to 0} \int_{\Delta_a'} |F(x', x_n) - f(x')|^p\, dx' = 0,$$

i.e. $f(x')$ is the trace of $F(x)$ on Δ_a'.

Next we claim that $x_n |F_{xx}(x)| \in L_p(\Delta_a)$. Indeed, it suffices to estimate the norm of the second derivative with respect to x_n, for the estimates of the remaining second derivatives are similar if not easier. A direct differentiation yields

$$F_{,n}(x) = \frac{1}{x_n^n} \int_{x_1}^{x_1 + x_n} dt_1 \ldots \int_{x_n}^{2x_n} \tilde{f}_{,n}(t)\, dt_n + \frac{1}{x_n^{n+1}} \int_{x_1}^{x_1 + x_n} dt_1 \ldots \int_{x_n}^{2x_n} (t_i - x_i) \tilde{f}_{,i}(t)\, dt_n.$$

Having calculated the derivative $F_{,nn}$ we see that in order to estimate the norm of the function $x_n F_{,nn}(x)$ in the space $L_p(\Delta_a)$ it suffices to estimate the norms of the integrals

$$\frac{1}{x_n^n} \int_{x_1}^{x_1 + x_n} dt_1 \ldots \int_{x_n}^{2x_n} |\tilde{f}_{,i}(t)|\, dt_n = \int_{\Delta_1} |\tilde{f}_{,i}(x + t x_n)|\, dt,$$

$$\frac{1}{x_n^{n-1}} \int_{x_1}^{x_1 + x_n} dt_1 \ldots \int_{x_{n-1}}^{x_{n-1} + x_n} |\tilde{f}_{,i}(t', x_n)|\, dt_{n-1} = \int_{\Delta_1'} |\tilde{f}_{,i}(x' + t' x_n, x_n)|\, dt'.$$

Both integrals can be handled in the same fashion. Let us estimate, for instance, the first integral:

$$\left(\int_{\Delta_a}\left|\int_{\Delta_1}|\tilde{f}_{,i}(x+tx_n)|\,dt\right|^p dx\right)^{\frac{1}{p}}$$
$$\leqslant\int_{\Delta_1}\left(\int_{\Delta_a}|\tilde{f}_{,i}(x+tx_n)|^p\,dx\right)^{\frac{1}{p}}dt\leqslant\left(\int_{\Delta_{2a}}|\tilde{f}_{,i}(x)|^p\,dx\right)^{\frac{1}{p}}.$$

Thus we conclude that there is a constant $c_5 > 0$ such that

$$\|x_nF_{,nn}\|_{L_p(\Delta_a)}\leqslant c_5\|\tilde{f}_x\|_{L_p(\Delta_{2a})}.$$

Since the remaining derivatives can be estimated in a similar way, we obtain a positive constant c_6 such that

$$\|x_nF_{xx}\|_{L_p(\Delta_{2a})}\leqslant c_6\|\tilde{f}_x\|_{L_p(\Delta_{2a})}. \tag{2.6}$$

Inequality (2.4) then follows from (2.5), (2.6) and similar inequalities for $F(x)$ and $|F_x(x)|$. The lemma is proved.

Finally we shall derive one consequence (cf. [3], p. 53) of the theorem of Brouwer on fixed points of a continuous mapping (cf. [12], Russian p. 571).

Let $\xi = (\xi_1, \cdots, \xi_N)$ be an arbitrary point of the Euclidean space E^N and let $P(\xi) = (P_1(\xi), \cdots, P_N(\xi))$ be its image under the mapping $\xi \to P(\xi)$ which maps E^N into E^N. As usual $(P(\xi), \xi)$ will denote the scalar product of the vectors $P(\xi)$ and ξ, and $|P(\xi)|$ and $|\xi|$ will be the lengths of the corresponding vectors.

LEMMA 2.5. *Let $\xi \to P(\xi)$ be a continuous mapping of E^N into E^N. If there is a number $R > 0$ such that $(P(\xi), \xi) \geqslant 0$ for all ξ with $|\xi| = R$, then in the sphere $K = \{|\xi| \leqslant R\}$ there exists a solution of the equation $P(\xi) = 0$.*

PROOF. If $P(\xi) \neq 0$ in the sphere K, then

$$\xi \to -P(\xi)\frac{R}{|P(\xi)|}$$

is a continuous mapping of K into K. Hence by the theorem of Brouwer [12] there exists a fixed point ξ^0 of this mapping:

$$\xi^0 = -P(\xi^0)\frac{R}{|P(\xi^0)|}.$$

Taking the scalar product of each side with the vector ξ^0, we get

$$|\xi^0|^2 = -(P(\xi^0), \xi^0)R|P(\xi^0)|^{-1}.$$

Since $|\xi^0| = R$, the last equation implies the inequality $(P(\xi^0), \xi^0) < 0$, which is a contradiction. The lemma is proved.

§3. A priori estimates

LEMMA 3.1. *Let the functions* $F^i(y)$, $i = 1, \cdots, n$, *satisfy the conditions of* §1. *Furthermore, let* $\beta = \beta(x)$ *be a sufficiently smooth function satisfying the inequalities* $\tilde{c}_1\rho(x) \leqslant \beta(x) \leqslant \tilde{c}_2\rho(x)$, $\forall x \in g$, *where* $\tilde{c}_1$ *and* $\tilde{c}_2$ *are positive constants, and* $\rho(x)$ *is the distance of the point* x *to the boundary* ∂g. *Then for any function* $u(x) \in W^2_{p,1}(g)$ *the integral*

$$I(u) = \int_g \beta^2 F^i_{,k}(u_x)\, u_{,kj} u_{,ij}\, dx \tag{3.1}$$

is finite and the function $F^i(u_x)$ *belongs to the space* $W^2_{q,1}(g)$, *where* $q = p/(p-1)$. *Moreover, the following inequalities hold*:

$$\| \beta D_j F^i(u_x) \|_q \leqslant \sqrt{\mu}\, \sqrt{I(u)}\, \| 1 + |u_x| \|_p^{p\left(\frac{1}{q} - \frac{1}{2}\right)}, \tag{3.2}$$

where μ *is the constant from condition* 1) *of* §1.

(Let us recall that whenever the same index appears twice it means summation from 1 to n.)

PROOF. That the integral (3.1) is finite is almost obvious. Indeed, applying condition (1.2) and Hölder's inequality with exponents $p/2$ and $p/(p-2)$, we obtain the inequality

$$I(u) \leqslant \mu \int_g \beta^2 |u_{xx}|^2 (1 + |u_x|)^{p-2} dx \leqslant \mu \| \beta u_{xx} \|_p^2 \, \| 1 + |u_x| \|_p^{p-2},$$

which shows that the integral (3.1) converges.

By condition 2) of §1, the quadratic form $F^i_{,k}(y)\xi_i\xi_k$ satisfies all conditions of Lemma 1; hence

$$| D_{j_0} F^{i_0}(u_x) | = | F^{i_0}_{,k} u_{,kj_0} | \leqslant (F^i_{,k} u_{,kj_0} u_{,ij_0})^{\frac{1}{2}} | F^{i_0}_{,i_0} |^{1/2};$$

and since by (1.3)

$$| F^i_{,i}(u_x) | \leqslant \mu (1 + |u_x|)^{p-2},$$

the last two inequalities combined with Hölder's inequality (with appropriate exponents) imply the following chain of estimates:

$$\begin{aligned} \| \beta D_{j_0} F^{i_0}(u_x) \|_q^q &\leqslant \mu^{\frac{q}{2}} \int_g (\beta^2 F^i_{,k} u_{,kj} u_{,ij})^{\frac{q}{2}} (1 + |u_x|)^{\frac{p}{2}-1} dx \\ &\leqslant \mu^{\frac{q}{2}} \left(\int_g \beta^2 F^i_{,k} u_{,kj} u_{,ij} dx \right)^{\frac{q}{2}} \left(\int_g (1 + |u_x|)^p dx \right)^{1-\frac{q}{2}}. \end{aligned}$$

The lemma is proved.

Let $C_0^3(g)$ be the set of functions which are three times continuously differentiable in the closed domain $\bar{g}$ and vanish on ∂g. Let $B_\lambda u$ denote the linear operator

$$B_\lambda u = -\beta^2 \Delta u + \lambda u, \qquad \lambda > 0, \tag{3.3}$$

where $\beta = \beta(x)$ was defined in Lemma 3.1, λ is a number and Δu is the Laplace operator $\Delta u = D_j u_{,j}$. Moreover, set

$$Au = -D_i F^i(u_x), \tag{3.4}$$

where the functions $F^i(y)$, $i = 1, \cdots, n$, satisfy conditions 1), 2) and 3) of § 1.

LEMMA 3.2. *Let $f(x) \in W^2_{p,1}(g)$ and $v(x) \in C_0^3(g)$. Then for $u(x) = f(x) + v(x)$ we have*

$$(Au, v) \geqslant \frac{\nu}{q} \| 1 + |u_x| \|_p^p - b \| f_x \|_p^p - \mu_1 \text{ meas } g. \tag{3.5}$$

where $b = \mu^p \nu^{1-p} p^{p-2}$, and meas g *is the measure of the domain* g (μ *is the constant from condition* 1), *and* ν *and* μ_1 *are the same as in condition* 3) *of* § 1).

PROOF. First we claim that the functions $F^i(u_x)$, $i = 1, \cdots, n$, belong to the space $W^1_{q,1}(g)$. Indeed, we have $u(x) \in W^2_{p,1}(g)$, and by condition (1.1)

$$\| F^i(u_x) \|_q \leqslant \mu \| (1 + |u_x|)^{p-1} \|_q = \mu \| 1 + |u_x| \|_p^{p-1};$$

moreover, from (1.2) we get

$$\| \rho D_j F^i(u_x) \|_q = \| \rho F^i_{,k}(u_x) u_{,kj} \|_q \leqslant \mu \| \rho (1 + |u_x|)^{p-2} |u_{xx}| \|_q$$
$$\leqslant \mu \| \rho u_{xx} \|_p \| 1 + |u_x| \|_p^{p-2}.$$

Hence $F^i(u_x) \in W^1_{q,1}(g)$. Since $v(x) \in C_0^3(g)$, Lemma 2.2 shows that the functions $vF^i(u_x)$ are in the space $\overset{\circ}{W}{}^1_q(g)$ and satisfy the equation

$$\int_g D_i (vF^i(u_x))\, dx = 0,$$

which yields

$$(Au, v) = -(D_i F^i(u_x), v) = (F^i(u_x), v_{,i}).$$

In estimating the integrand of the last integral we apply inequalities (1.6) and (1.1):

$$F^i(u_x), v_{,i} = F^i(u_x) u_{,i} - F^i(u_x) f_{,i} \geqslant \nu (1 + |u_x|)^p - \mu_1 - \mu |f_x| (1 + |u_x|)^{p-1}.$$

The last summand can be estimated by Young's inequality:

$$\mu |f_x| (1 + |u_x|)^{p-1} \leqslant \frac{1}{p} \mu^p \delta^{-p} |f_x|^p + \frac{1}{q} \delta^q (1 + |u_x|)^p.$$

Choosing $\delta > 0$ so that $\nu/pq \leqslant \delta^q/q \leqslant \nu/p$, we then get

$$F^i(u_x)v_{,i} \geqslant \frac{\nu}{q}(1+|u_x|)^p - \mu_1 - b|f_x|^p,$$

from which inequality (3.5) follows after integrating over g. The lemma is proved.

LEMMA 3.3. *Let $f(x) \in W^2_{p,1}(g)$ and $v(x) \in C^3_0(g)$. Then for $u(x) = v(x) + f(x)$ we have*

$$\begin{aligned}-(Au, \beta^2\Delta v) \geqslant \frac{1}{2} I(u) - (44\varkappa + 3)\mu \| 1 + |u_x| \|^p_p \\ - 6\mu\varkappa \| f_x \|^p_p - 3\mu \| \beta f_{xx} \|^p_p,\end{aligned} \tag{3.6}$$

where $I(u)$ is defined by formula (3.1), *μ is the constant from condition* 1) *of* § 1, *and κ is such that*

$$|\beta_x|^2 \leqslant \varkappa, \qquad |(\beta^2)_{xx}| \leqslant \varkappa.$$

PROOF. Obviously

$$-(Au, \beta^2\Delta v) = (D_i(\beta F^i(u_x))', \beta\Delta v) - (\beta\beta_{,i}F^i(u_x), D_jv_{,j}). \tag{3.7}$$

As was shown in the proof of Lemma 3.2, the functions $\beta F^i(u_x)$ belong to the space $\overset{\circ}{W}{}^1_q(g)$, so that we can integrate by parts in the integrals on the right-hand side of (3.7). We obtain

$$-(Au, \beta^2\Delta v) = -(\beta F^i(u_x), D_i(\beta\Delta v)) + (D_j(\beta\beta_{,i}F^i(u_x)), v_{,j}).$$

Furthermore,

$$(\beta F^i(u_x), D_i(\beta\Delta v)) = (\beta\beta_{,i}F^i(u_x), D_jv_{,j}) + (\beta^2F^i(u_x), D_jv_{,ij}),$$

and by the same reason as above we can integrate by parts in the last integrals. Hence in the notation of (3.1) we finally get

$$\begin{aligned}-(Au, \beta^2\Delta v) = I(u) - (\beta^2 D_jF^i(u_x), f_{,ij}) \\ - 2(D_j(\beta\beta_{,j}F^i(u_x)), v_{,i}) + 2(D_j(\beta\beta_{,i}F^i(u_x)), v_{,j}).\end{aligned} \tag{3.8}$$

The last two integrals can be handled in the same fashion. Let us estimate, for instance, the second one.

Using hypotheses 1) and 2) of § 1, we can estimate the integrand of the last integral by the following chain of inequalities:

$$|v_{,j}D_j(\beta\beta_{,i}F^i(u_x))| \leqslant \frac{1}{2}|(\beta^2)_{,ij}F^i(u_x)v_{,j}| + |\beta\beta_{,i}v_{,j}F^i_{,k}(u_x)u_{,kj}| \leqslant$$

$$\leqslant \frac{1}{2}\mu\varkappa|v_x|(1+|u_x|)^{p-1}+|v_x|(\beta^2F^i_{,k}(u_x)u_{,kj}u_{,ij})^{\frac{1}{2}}(F^i_{,k}(u_x)\beta_{,i}\beta_{,k})^{\frac{1}{2}}$$

$$\leqslant \frac{1}{2}\mu\varkappa|v_x|(1+|u_x|)^{p-1}+\frac{1}{2}\varepsilon\beta^2F^i_{,k}(u_x)u_{,kj}u_{,ij}+\frac{1}{2\varepsilon}\mu\varkappa(1+|u_x|)^{p-2}|v_x|^2,$$

$$\varepsilon>0.$$

The first summand can be estimated by Young's inequality [13]; thus we get

$$\begin{aligned}|v_x|(1+|u_x|)^{p-1}&\leqslant(|u_x|+|f_x|)(1+|u_x|)^{p-1}\\&\leqslant(1+|u_x|)^p+\frac{1}{p}|f_x|^p+\left(1-\frac{1}{p}\right)(1+|u_x|)^p\\&=\left(2-\frac{1}{p}\right)(1+|u_x|)^p+\frac{1}{p}|f_x|^p.\end{aligned}$$

Similarly we can estimate the third summand:

$$\begin{aligned}|v_x|^2(1+|u_x|)^{p-2}&\leqslant 2(1+|u_x|)^p+2|f_x|^2(1+|u_x|)^{p-2}\\&\leqslant 2\left(2-\frac{2}{p}\right)(1+|u_x|)^p+2\frac{2}{p}|f_x|^p.\end{aligned}$$

Hence

$$\begin{aligned}|v_{,j}D_j(\beta\beta_{,i}F^i(u_x))|&\leqslant\frac{1}{2}\varepsilon\beta^2F^i_{,k}(u_x)u_{,kj}u_{,ij}\\&+\varkappa\mu\left(\frac{1}{2}+\frac{1}{2q}+\frac{2}{\varepsilon q}\right)(1+|u_x|)^p+\frac{1}{p}\mu\varkappa\left(\frac{1}{2}+\frac{2}{\varepsilon}\right)|f_x|^p.\end{aligned}$$

Setting $\epsilon = 1/5$, we obtain

$$|(D_j(\beta_{,i}\beta F^i(u_x)),\ v_{,j})|\leqslant\frac{1}{10}I(u)+11\mu\varkappa\|1+|u_x|\|_p^p+\frac{21}{2p}\mu\varkappa\|f_x\|_p^p. \tag{3.9}$$

It remains to estimate the second integral in (3.8). For the corresponding integrand we have

$$\begin{aligned}|\beta^2f_{,ij}D_jF^i(u_x)|&=|\beta^2F^i_{,k}(u_x)u_{,kj}f_{,ij}|\\&\leqslant\{\beta^2F^i_{,k}(u_x)u_{,kj}u_{,ij}\}^{\frac{1}{2}}\{\beta^2F^i_{,k}(u_x)f_{,kj}f_{,ij}\}^{\frac{1}{2}}\\&\leqslant\frac{\varepsilon}{2}\beta^2F^i_{,k}(u_x)u_{,kj}u_{,ij}+\frac{1}{2\varepsilon}\beta^2\mu|f_{xx}|^2(1+|u_x|)^{p-2}.\end{aligned}\tag{3.10}$$

The last summand can be estimated with the aid of Young's inequality:

$$\beta^2|f_{xx}|^2(1+|u_x|)^{p-2}\leqslant\beta^p|f_{xx}|^p+(1+|u_x|)^p. \tag{3.11}$$

Using (3.10) with $\epsilon = 1/5$ and (3.11), we obtain

$$|(\beta^2D_jF^i(u_x),\ f_{,ij})|\leqslant\frac{1}{10}I(u)+3\mu\{\|\beta f_{xx}\|_p^p+\|1+|u_x|\|_p^p\}. \tag{3.12}$$

It is easy to see that inequality (3.6) now follows from (3.8), (3.9) and (3.12). The lemma is proved.

LEMMA 3.4. *Let* $f(x) \in W^2_{p,1}(g)$ *and* $v(x) \in C^3_0(g)$. *If* $u(x) = f(x) + v(x)$ *and* λ *is such that*

$$\lambda \frac{\nu}{q} - 44\mu\varkappa - 3\mu \geqslant 1, \tag{3.13}$$

then

$$(Au,\ B_\lambda v) \geqslant \frac{1}{2} I(u) + \| 1 + |u_x| \|_p^p - b_\lambda \| f_x \|_p^p - 3\mu \| \beta f_{xx} \|_p^p - \lambda\mu_1 \operatorname{meas} g, \tag{3.14}$$

where

$$b_\lambda = 6\mu\varkappa + \lambda\mu^p \nu^{1-p} p^{p-2}. \tag{3.15}$$

This lemma is an immediate corollary of Lemmas 3.2 and 3.3.

LEMMA 3.5. *Let* $h(x) \in W^1_{q,2}(g)$ *and* $f(x) \in W^2_{p,1}(g)$. *Then for* $u(x) = f(x) + v(x)$ *with* $v(x) \in C^3_0(g)$ *we have*

$$(h,\ B_\lambda v) \leqslant \frac{1}{p} \| u_x \|_p^p + \frac{1}{p} \| f_x \|_p^p + \frac{2}{q} \{ \| \beta^2 h_x \|_q + (2\sqrt{\varkappa} + \lambda c_1^*) \| \beta h \|_q \}^q, \tag{3.16}$$

where $c_1^* = c_1 \tilde{c}_1^{-1}$.

PROOF. By the definition of $B_\lambda v$,

$$(h,\ B_\lambda v) = -(h\beta^2,\ D_j v_{,j}) + \lambda (h,\ v).$$

By Lemma 2.2, in the first integral we can integrate by parts:

$$(h,\ B_\lambda v) = (D_j (h\beta^2),\ v_{,j}) + \lambda (h,\ v).$$

The last summand has already been estimated (cf. (1.8)). The second summand can be estimated with the aid of Hölder's inequality:

$$(D_j (\beta^2 h),\ v_{,j}) = (\beta^2 h_{,j},\ v_{,j}) + 2 (\beta\beta_{,j} h,\ v_{,j}) \leqslant \{ \| \beta^2 h_x \|_q + 2\sqrt{\varkappa} \| \beta h \|_q \} \| v_x \|_p.$$

Hence

$$(h,\ B_\lambda v) \leqslant \{ \| \beta^2 h_x \|_q + (2\sqrt{\varkappa} + \lambda c_1^*) \| \beta h \|_q \} (\| u_x \|_p + \| f_x \|_p).$$

Applying Young's inequality, we obtain (3.16). The lemma is proved.

LEMMA 3.6. *Let the function* $u(x) = f(x) + v(x),\ x \in g$, *where*

$$f(x) \in W^2_{p,1}(g), \qquad v(x) \in C^3_0(g), \tag{3.17}$$

satisfy the condition

$$(Au,\ B_\lambda v) = (h,\ B_\lambda v), \tag{3.18}$$

where the operator $B_\lambda v$ *is defined by* (3.3), *the number* λ *satisfies* (3.13) *and the function* $h(x)$, $x \in g$, *is such that*

$$h(x) \in W^1_{q,\,2}(g), \qquad \frac{1}{p} + \frac{1}{q} = 1. \tag{3.19}$$

Then there exists a constant C_0 *such that*

$$\begin{aligned}\|u\|^p_{W^{(1)}_p(g)} + \sum_{i=1}^n \left\{\|M_i(u_{,i})\|^2_{W^1_{2,\,1}(g)} + \|F^i(u_x)\|^q_{W^1_{q,\,1}(g)}\right\} \\ \leqslant C_0\left\{1 + \|f\|^p_{W^2_{p,\,1}(g)} + \|h\|^q_{W^1_{q,\,2}(g)}\right\}\end{aligned} \tag{3.20}$$

for arbitrary functions $f(x)$ *and* $h(x)$ *satisfying conditions* (3.17) *and* (3.19). *The constant* C_0 *depends only on* n, p, μ, ν, μ_1 *and the domain* g. *The functions* $M_i(t)$ *are defined by* (1.5).

PROOF. From (3.14) and (3.16) we get

$$\begin{aligned} I(u) + \|1 + |u_x|\|^p_p \leqslant (2b_\lambda + 1)\|f_x\|^p_p + 6\mu\|\beta f_{xx}\|^p_p \\ + 4\{\|\beta^2 h_x\|_q + (2\sqrt{\varkappa} + \lambda c_1^*)\|\beta h\|_q\}^q + 2\lambda\mu_1 \operatorname{meas} g. \end{aligned} \tag{3.21}$$

Denoting the right-hand side of the last inequality by $C(f, h, g)$, we get from (3.2) and (3.21)

$$\|\beta D_j F^i(u_x)\|^q_q \leqslant n^q \mu^{\frac{q}{2}} C(f,\ h,\ g). \tag{3.22}$$

It follows from condition (1.3) that

$$\beta^2 F^i_{,k}(u_x)\, u_{,kj} u_{,ij} \geqslant \beta^2 \sum_{i,\,j=1}^n \{D_j M_i(u_{,i})\}^2,$$

and therefore

$$\sum_{i,\,j=1}^n \|\beta D_j M_i(u_{,i})\|^2_2 \leqslant I(u) \leqslant C(f,\ h,\ g). \tag{3.23}$$

Let us now estimate the norms of the functions $u(x)$, $F^i(u_x)$ and $M_i(u_{,i})$ in the spaces $L_p(g)$, $L_q(g)_n$ and $L_2(g)_n$ respectively.

Since $u(x) = v(x) + f(x)$ with $v(x) \in C^3_0(g)$, we have

$$\|u\|_p \leqslant \|v\|_p + \|f\|_p$$

so that by (1.7)

$$\|u\|_p \leqslant dc_1\|v_x\|_p + \|f\|_p \leqslant dc_1(\|u_x\|_p + \|f_x\|_p) + \|f\|_p, \tag{3.24}$$

where d is the diameter of g.

The estimate of the norm of $F^i(u_x)$ follows easily from (1.1):

$$\| F^i(u_x) \|_q^q \leqslant \mu \| 1 + |u_x| \|_p^p. \tag{3.25}$$

It remains to estimate the norms of the functions $M_i(u_{,i})$, $i = 1, \cdots, n$. From (1.3) and (1.2) it follows that

$$m_i^2(y_i) \leqslant \mu(1 + |y_i|)^{p-2};$$

hence

$$|M_i(y_i)| \leqslant \sqrt{\mu} \int_0^{|y_i|} (1+t)^{\frac{p}{2}-1} dt \leqslant \sqrt{\mu}(1 + |y_i|)^{\frac{p}{2}}.$$

Thus we get the inequality

$$\| M_i(u_{,i}) \|_2^2 \leqslant \mu \| 1 + |u_x| \|_p^p. \tag{3.26}$$

The inequality (3.20) is a consequence of (3.21)–(3.26). This proves the lemma.

§4. Proof of the main theorem

The proof will be based on Galerkin's method, where the approximate solutions to the equation (1.1) are sought in the form of linear combinations of functions from a family chosen in a particular way. The success of this approach depends basically on a successful choice of this family. In the present paper we use a method suggested by M. I. Višik in [2] (cf. also the monograph [3] of J. L. Lions).

It follows from a priori estimates (cf. §2) that we will be interested in solutions of the equation

$$B_\lambda \varphi = \psi, \tag{4.1}$$

where the operator B_λ was defined by the formula (2.3) and ψ is a sufficiently smooth function with support in the domain g. The next statement is an immediate corollary of the embedding theorems of S. L. Sobolev [1] combined with a lemma of M. I. Višik [2]:

LEMMA. *There exists a number $\lambda_0 > 0$ such that for each $\lambda \geqslant \lambda_0$ and any infinitely differentiable function $\psi(x)$ with support in g there exists a unique solution of equation* (4.1) *in the space $C_0^3(g)$.*

It should be stressed that λ_0 does not depend on $\psi(x)$.

Let $\{\psi^3(x)\}$ be a system of infinitely differentiable functions with supports in g such that the linear span of this system is dense in the space $\overset{\circ}{W}{}_p^1(g)$. Moreover, assume that every finite subset of this system is linearly independent. Let $\varphi^j(x)$ denote the solution in $C_0^3(g)$ of the equation

$$B_\lambda \varphi^j = \psi^j, \tag{4.2}$$

for which

$$\lambda \geqslant \max\{\lambda_0, \nu^{-1}q(1+44\mu\varkappa+3\mu)\}, \tag{4.3}$$

where λ_0 is the number from the lemma of Višik, μ is the constant from condition 1) of § 1, ν is the constant from (1.6) and κ is the constant from (3.6). By the above lemma, for the chosen value of λ the solution $\varphi^j(x)$ exists for each $j = 1, 2, \cdots$.

Let the function $f(x)$, $x \in g$, be an extension of the given function $f(x')$, $x' \in \partial g$, i.e. the function $f(x)$ $(x \in g)$ has a trace on ∂g equal to $f(x')$, and the norm of f in $W_{p,1}^2(g)$ can be estimated in terms of the norm of $f(x') \in W_p^{1-1/p}(\partial g)$ and a constant independent of $f(x')$, $x' \in g$. The existence of such a function follows from Lemma 2.3.

The approximate solution $u^N(x)$ of equation (0.1) will be sought in the form of a sum

$$u^N(x) = f(x) + v^N(x), \tag{4.4}$$

where $v^N(x)$ is a linear combination of the first N terms of the system $\{\varphi^j(x)\}$:

$$v^N(x) = \sum_{j=1}^N \xi_j \varphi^j(x). \tag{4.5}$$

The coefficients ξ_j can be found by solving the following system of nonlinear algebraic equations:

$$(Au^N, B_\lambda \varphi^j) = (h, B_\lambda \varphi^j), \qquad j = 1, \ldots, N. \tag{4.6}$$

We shall prove the solvability of system (4.4) by using Lemma 2.5:

Consider the mapping $\xi \to P(\xi)$, $P = (P_1, \cdots, P_N)$, of E^N into E^N, defined by the formulas

$$P_j(\xi) = (Au^N - h, B_\lambda \varphi^j), \qquad j = 1, \ldots, N.$$

This mapping is continuous; and, furthermore,

$$(P(\xi), \xi) = \sum_{j=1}^N P_j(\xi)\xi_j = (Au^N - h, B_\lambda v^N). \tag{4.7}$$

As was shown above (cf. Lemma 3.4), if λ satisfies condition (4.3), then

$$(Au^N, B_\lambda v^N) \geqslant \|u_x^N\|_p^p - b_\lambda \|f_x\|_p^p - 3\mu \|\beta f_{xx}\|_p^p - \lambda\mu_1 \operatorname{meas} g. \tag{4.8}$$

By Lemma 3.5,

$$(h,\ B_\lambda v^N) \leqslant \frac{1}{p}\|u_x^N\|_p^p + \frac{1}{p}\|f_x\|_p^p + \frac{2}{q}\{\|\beta^2 h_x\|_q + (2\sqrt{\varkappa} + \lambda c_1^*)\|\beta h\|_q\}^q. \tag{4.9}$$

Subtracting (4.9) from (4.8) and observing that

$$\|u_x^N\|_p^p \geqslant \frac{1}{2^{p-1}}\|v_x^N\|_p^p - \|f_x\|_p^p,$$

we obtain the inequality

$$(Au^N - h,\ B_\lambda v^N) \geqslant \frac{1}{2^p}\|v_x^N\|_p^p - c_7, \tag{4.10}$$

where the constant c_7 depends only on the boundary value problem under consideration, but not on $v^N(x)$.

Consider the function

$$F(\alpha) = \left\| \sqrt{\sum_{i=1}^{n} \left(\sum_{j=1}^{N} \alpha_j,\ \varphi^j_{,i} \right)^2} \right\|_p^p, \qquad \alpha = (\alpha_1,\ \ldots,\ \alpha_N).$$

This function is obviously continuous and nonnegative in E^N. We claim that F cannot vanish on the unit sphere $|\alpha| = 1$. Indeed, if $F(\alpha^0) = 0$ for some α^0, $|\alpha^0| = 1$, then $\Sigma_{j=1}^N \alpha_j^0 \varphi^j_{,i} \equiv 0$, $i = 1, \cdots, n$, and since the functions $\varphi^j(x)$ vanish on ∂g, this would imply $\Sigma_{j=1}^N \alpha_j^0 \varphi^j(x) \equiv 0$. But then for functions $\psi^j(x)$, $j = 1, \cdots, N$, we would get

$$\sum_{j=1}^N \alpha_j^0 \psi^j = \sum_{j=1}^N \alpha_j^0 B_\lambda \varphi^j = B_\lambda \left(\sum_{j=1}^N \alpha_j^0 \varphi^j \right) = 0,$$

which contradicts the linear independence of the system $\{\psi^j(x)\}$. Therefore there exists a number $\kappa_0 > 0$ such that

$$F(\alpha) \geqslant \varkappa_0, \qquad \forall \alpha,\ |\alpha| = 1.$$

Thus we have

$$\|v_x^N\|_p^p = |\xi|^p F\left(\frac{\xi}{|\xi|}\right) \geqslant \varkappa_0 |\xi|^p, \qquad \forall \xi \in E^N. \tag{4.11}$$

From (4.7), (4.10) and (4.11) we obtain

$$(P(\xi),\ \xi) \geqslant \frac{1}{2^p}\varkappa_0 |\xi|^p - c_7,$$

which implies $(P(\xi), \xi) \geqslant 0$ for $|\xi|^p = 2^p c_7 \kappa_0^{-1}$. Then by Lemma 2.5 there exists a point $\xi \in E^N$ such that $P(\xi) = 0$, i.e. the point ξ is a solution of the system (4.6). The existence of the approximate solution $u^N(x)$ given by the formulas (4.4) and (4.5) is thus established.

Multiplying the jth equation of system (4.6) by ξ_j and adding together the equations thus obtained, we arrive at the identity

$$(Au^N,\ B_\lambda v^N)=(h,\ B_\lambda v^N).$$

This shows that the approximate solutions $u^N(x)$ satisfy all conditions of Lemma 3.6; hence they also satisfy the inequalities

$$\|u^N\|^p_{W^1_p(g)}+\sum_{i=1}^n\left\{\|M_i(u^N_{,i})\|^2_{W^1_{2,1}(g)}+\|F^i(u^N_x)\|^q_{W^1_{q,1}(g)}\right\} \leqslant C_0\left\{1+\|f\|^p_{W^2_{p,1}(g)}+\|h\|^q_{W^1_{q,2}(g)}\right\}. \tag{4.12}$$

In view of well-known theorems on compactness, weak compactness and convergence almost everywhere we can choose from the sequence $\{u^N(x)\}$ a subsequence $\{u^k(x)\}$ with the following properties:

1) The sequence $\{u^k(x)\}$ converges strongly in $L_p(g)$ and weakly in $W^1_p(g)$ to some function $u(x)\in W^1_p(g)$ whose trace on ∂g is $f(x')$ $(x'\in\partial g)$.

2) The sequences $\{M_i(u^k_{,i})\}$, $i=1,\cdots,n$, converge strongly in $L_2(g_1)$ and weakly in $W^1_2(g_1)$ to certain functions $\mu^i(x)\in W^1_2(g_1)$, where g_1 is any region whose closure is contained in g.

3) The sequences $\{M_i(u^k_{,i})\}$, $i=1,\cdots,n$, converge in g almost everywhere.

4) The sequences $\{F^i(u^k_x)\}$, $i=1,\cdots,n$, converge strongly in $L_q(g_1)$ and weakly in $W^1_q(g)$ to certain functions $f^i(x)\in W^1_q(g_1)$ in any region g_1 whose closure is in g.

5) The sequences $\{F^i(u^k_x)\}$, $i=1,\cdots,n$, converge to $f^i(x)$ weakly in $L_q(g)$.

Furthermore, (4.12) implies that the limiting functions $u(x)$, $\mu^i(x)$ and $f^i(x)$ satisfy the inequality

$$\|u\|^p_{W^1_p(g)}+\sum_{i=1}^n\{\|\mu^i\|_{L_2(g_1)}+\|\rho\mu^i_x\|_{L_2(g_1)}\}^2 +\sum_{i=1}^n\{\|f^i\|_{L_q(g_1)}+\|\rho f^i_x\|_{L_q(g_1)}\}^q\leqslant D(f,\ h), \tag{4.13}$$

where $D(f,h)$ is the right-hand side of (4.12), $\rho=\rho(x)$ is the distance of $x\in g_1$ to ∂g, and g_1 is any region whose closure is in g. Since the right-hand side in (4.13) is independent of the choice of $g_1\subset\overline{g}_1\subset g$, we obtain, after taking the limits,

$$\|u\|^p_{W^1_p(g)}+\sum_{i=1}^n\left\{\|\mu^i\|^2_{W^1_{2,1}(g)}+\|f^i\|^q_{W^1_{q,1}(g)}\right\}\leqslant D(f,\ h). \tag{4.14}$$

Furthermore, in view of the definition of $\varphi^j(x)$, the equalities (4.6) can be rewritten as follows:

$$(F^i(u^N),\ \psi^j_{,i})=(h,\ \psi^j),\qquad j=1,\ \ldots,\ N.$$

Fix j and take the limit with respect to the chosen subsequence. Then for the functions $f^i(x)$, $i=1,\cdots,n$, we get the conditions

$$(f^i, \psi^j_{,i}) = (h, \psi^j), \qquad j = 1, 2, \dots,$$

Since the linear span of the system $\{\psi^j\}$ is dense in $\mathring{W}^1_p(g)$, the last condition is equivalent to the following:

$$(f^i, v_{,i}) = (h, v), \qquad \forall v \in \mathring{W}^1_p(g). \tag{4.15}$$

Let us show that $f^i(x) = F^i(u_x(x))$ and $\mu^i(x) = M_i(u_{,i}(x))$. By our hypothesis (cf. 1)) the inverse functions to $M_i(t)$ are defined and continuous; hence, by property 3) of sequences $\{M_i(u^k_{,i}(x))\}$, the sequences $\{u^k_{,i}(x)\}$ converge almost everywhere in g to some functions $u_i(x)$, $i = 1, \cdots, n$. Since the norms of functions $u^k_{,i}(x)$ in $L_p(g)$ are bounded, we can apply Fatou's Lemma (cf. [14]), which shows that the limiting functions $u_i(x)$ belong to $L_p(g)$ and their norms satisfy the same bounds. By Egorov's Theorem (cf. [14]), for each $\delta > 0$ there exists a measurable set $g_\delta \subset g$, meas $g_\delta < \delta$, such that on its complement $g - g_\delta$ the sequences $\{u^k_{,i}(x)\}$ converge uniformly to $u_i(x)$, $i = 1, \cdots, n$. Then

$$\|u^k_{,i} - u_i\|_{L_1(g)} \leqslant \|u^k_{,i} - u_i\|_{L_1(g-g_\delta)} + \delta^{\frac{1}{q}} (\|u^k_{,i}\|_p + \|u_i\|_p).$$

By a standard argument we then conclude that the sequences $\{u^k_{,i}(x)\}$ converge strongly in $L_1(g)$ to the functions $u_i(x)$, $i = 1, \cdots, n$, whence $u_i(x) = u_{,i}(x)$.

Since by the above argument the sequences $\{u^k_{,i}(x)\}$ converge to $u_{,i}(x)$, $i = 1, \cdots, n$, almost everywhere in g, and the functions $F^i(y)$ and $M_i(t)$ are continuous, the sequences $\{F^i(u^k_x(x))\}$ and $\{M_i(u^k_{,i}(x))\}$ $(i = 1, \cdots, n)$ must also converge almost everywhere in g to functions $F^i(u_x)$ and $M_i(u_{,i})$ respectively. From this, as well as from the strong convergence in $L_p(g_1)$ of the sequence $\{F^i(u^k_x(x))\}$ to the function $f^i(x)$, and the strong convergence in $L_2(g_1)$ of the sequence $\{M_i(u^k_{,i}(x))\}$ to $\mu^i(x)$, we conclude that $f^i(x) = F^i(u_x(x))$ and $\mu^i(x) = M_i(u_{,i}(x))$, for $i = 1, \cdots, n$.

Thus we have proved that, under the hypotheses: (i) $f(x') \in W^{1-1/p}(\partial g)$, (ii) $h(x) \in W^1_{q,2}(g)$, and (iii) the functions $F^i(y)$ $(y \in E^n)$ satisfy conditions 1), 2) and 3) of §1, there exists a solution $u(x) \in W^1_p(g)$ to the problem (0.1), (0.2), which moreover is such that the functions $F^i(u_x(x))$ and $M_i(u_{,i}(x))$, $i = 1, \cdots, n$ (as functions of $x \in g$), belong to the spaces $W^1_{q,1}(g)$ and $W^1_{2,1}(g)$ respectively.

It follows from the above construction of approximate solutions that the function $f(x)$ $(x \in g)$ extends the given function $f(x')$ $(x' \in \partial g)$ (cf. (4.4) and Lemma 2.3). The norm of the function $f(x) \in W^2_{p,1}(g)$ can be bounded by the norm of $f(x') \in W^{1-1/p}_p(\partial g)$; and the constant in this estimate does not depend on the choice of $f(x')$. Hence it follows from (4.14) that

$$\|u\|^p_{W^1_p(g)} + \sum_{i=1}^n \left\{\|M_i(u_{,i})\|^2_{W^1_{2,1}(g)} + \|F^i(u_x)\|^q_{W^1_{q,1}(g)}\right\} \leqslant$$

$$\leqslant C\left\{1+\|f\|^{p}_{W_p^{1-\frac{1}{p}}(\partial g)}+\|h\|^{g}_{W^1_{g,2}(g)}\right\},$$

where the constant C is independent of the choice of functions $f(x')$ $(x' \in \partial g)$ and $h(x)$ $(x \in g)$ in the corresponding spaces. The theorem is proved.

BIBLIOGRAPHY

1. S. L. Sobolev, *Applications of functional analysis in mathematical physics*, Izdat. Leningrad. Gos. Univ., Leningrad, 1950; English transl., Transl. Math. Monographs, vol. 7, Amer. Math. Soc., Providence, R. I., 1963. MR **14**, 565; **29** #2624.

2. M. I. Višik, *Quasilinear strongly elliptic systems of differential equations of divergent form*, Trudy Moskov. Mat. Obšč. **12** (1963), 125–184 = Trans. Moscow Math. Soc. **1963**, 140–208. MR **27** #6017.

3. J. -L. Lions, *Quelques méthodes de résolution des problèmes aux limites non linéaires*, Dunod; Gauthier-Villars, Paris, 1969. MR **41** #4326.

4. A. D. Džabrailov, *On investigation of certain classes of second order quasilinear elliptic equations*, Differencial'nye Uravnenija **5** (1969), 2245–2257 = Differential Equations **5** (1969), 1683–1693. MR **41** #7292.

5. G. N. Jakovlev, *A certain variational problem*, Differencial'nye Uravnenija **5** (1969), 1303–1312 = Differential Equations **5** (1969), 960–966. MR **41** #889.

6. ———, *Weak solutions of quasilinear elliptic systems of second order*, Differencial'nye Uravnenija **6** (1970), 157–163 = Differential Equations **6** (1970), 125–129. MR **40** #7629.

7. L. D. Kudrjavcev, *Direct and inverse embedding theorems. Applications to the solution of elliptic equations by variational methods*, Trudy Mat. Inst. Steklov. **55** (1959); English transl., Transl. Math. Monographs, Amer. Math. Soc., Providence, R. I. (to appear). MR **33** #7838.

8. O. A. Ladyženskaja and N. N. Ural'ceva, *Linear and quasilinear elliptic equations*, "Nauka", Moscow, 1964; English transl., Academic Press, New York, 1968. MR **35** #1955; **39** #5941.

9. S. M. Nikol'skiĭ, *Approximation of functions of several variables and embedding theorems*, "Nauka", Moscow, 1969. (Russian)

10. ———, *Properties of certain classes of functions of several variables on differentiable manifolds*, Mat. Sb. **33** (75) (1953), 261–326; English transl., Amer. Math. Soc. Transl. (2) **80** (1969), 39–118. MR **16**, 453.

11. E. Gagliardo, *Caratterizzazioni delle tracce sulla frontiera relative ad alcune classi di funzioni in n variabili*, Rend. Sem. Mat. Univ. Padova **27** (1957), 284–305. MR **21** #1525.

12. L. V. Kantorovič and G. P. Akilov, *Functional analysis in normed spaces*, Fizmatgiz, Moscow, 1959; English transl., Internat. Series of Monographs in Pure and Appl. Math., vol. 46, Macmillan, New York, 1964. MR **22** #9837; **35** #4699.

13. G. H. Hardy, J. E. Littlewood and G. Pólya, *Inequalities*, Cambridge Univ. Press, New York, 1934; Russian transl., IL, Moscow, 1948. MR **18**, 722.

14. I. P. Natanson, *Theory of functions of a real variable*, 2nd rev. ed., GITTL, Moscow, 1957; English transl., Ungar, New York, 1961. MR **26** #6309.

Translated by M. A. DOSTAL